# BIOLOGICAL ASPECTS OF FRESHWATER POLLUTION

Commission of the European Communities
Joint Research Centre Ispra (Varese), Italy

# ISPRA COURSES

Published for the Commission of the European Communities, Directorate General Scientific and Technical Information and Information Management, Luxembourg

# BIOLOGICAL ASPECTS OF FRESHWATER POLLUTION

Proceedings of the Course held at the Joint Research Centre of the Commission of the European Communities, Ispra, Italy, 5-9 June 1978

Edited by

O. RAVERA

Commission of the European Communities
Joint Research Centre, Ispra, Italy

Published for the
COMMISSION OF THE EUROPEAN COMMUNITIES
by
PERGAMON PRESS

OXFORD · NEW YORK · TORONTO · SYDNEY · PARIS · FRANKFURT

| | |
|---|---|
| U.K. | Pergamon Press Ltd., Headington Hill Hall, Oxford OX3 0BW, England |
| U.S.A. | Pergamon Press Inc., Maxwell House, Fairview Park, Elmsford, New York 10523, U.S.A. |
| CANADA | Pergamon of Canada, Suite 104, 150 Consumers Road, Willowdale, Ontario M2J 1P9, Canada |
| AUSTRALIA | Pergamon Press (Aust.) Pty. Ltd., P.O. Box 544, Potts Point, N.S.W. 2011, Australia |
| FRANCE | Pergamon Press SARL, 24 rue des Ecoles, 75240 Paris, Cedex 05, France |
| FEDERAL REPUBLIC OF GERMANY | Pergamon Press GmbH, 6242 Kronberg-Taunus, Pferdstrasse 1, Federal Republic of Germany |



First edition 1979

EUR 6392 en

**British Library Cataloguing in Publication Data**

Biological aspects of freshwater pollution.
1. Fresh-water ecology. 2. Water - Pollution
I. Ravera, O II. Commission of the European Communities
574.5'2632 TD423 79-40776

ISBN 0-08-023442-9

*In order to make this volume available as economically and as rapidly as possible the authors' typescripts have been reproduced in their original forms. This method has its typographical limitations but it is hoped that they in no way distract the reader.*



*Printed and bound in Great Britain by William Clowes (Beccles) Limited, Beccles and London*

# Contents

PREFACE vii

Studies on the effects of pollution on freshwater fish 1
J. F. de L. G. SOLBÉ

Zooplankton ecology and pollution studies 19
M. P. TEVLIN and MARY J. BURGIS

Systems of biological indicators for water quality assessment 39
G. PERSOONE and N. DE PAUW

The evolution of pollution evidenced by lake sediment pseudofossils 77
D. A. MURRAY

The impact of physical processes on the trophic state of a lake 93
D. M. IMBODEN and R. GÄCHTER

The study of eutrophication of algal benthos by essential metals in artificial rivers 111
E. EICHENBERGER

Heavy metal pollution in freshwater ecosystems 129
V. FÖRSTNER and F. PROSI

Radioactive contamination of the aquatic food-chain and consequent risk to man 163
A. A. CIGNA

Biological effects of ionizing radiations in aquatic environment 179
O. RAVERA

Methodology of waste water treatment 199
F. MALZ

# Preface

Since the second World War pollution and degradation of ecosystems have assumed a more general importance owing to the increase in human population and the growth and spread of industries and modern agricultural techniques. Water pollution means the discharge of noxious substances (e.g. biocides, detergents, heavy metals, oil, nutrients) and/or heat into a body of water in sufficiently large amounts as to produce deleterious effects on the aquatic environment.

The problem of pollution of fresh water is well known to be crucial, particularly for industrialized countries with a high population density. An increase in human population together with a rise in living standards causes an increase in the consumption of fresh water and a decrease in the quantity of clean water available. Because both these phenomena have the same basic cause, they combine to create a very dangerous situation in some areas. As a consequence, in recent years increased attention has been given to this problem by national governments and international organizations (e.g. CEC, FAO, UNESCO, OECD, WHO, MaB).

In spite of intensive studies on this subject, several fundamental aspects of the problem are still unclear. For example, the fate of most of the pollutants introduced into the natural environment is unknown and information about the relationship between the concentration of a given pollutant in the water and its biological effects is very scarse. Moreover, it is evident that the monitoring and control of polluted environments and any counteraction against pollution must be based on ecological principles. In the field of water pollution, the study of the biota is the only means of obtaining information on the hazard of pollutants. In fact, any judgement of the level of pollution in an ecosystem is based, not only on the concentration of toxic substance in the medium, but also on the knowledge (more or less complete) of the effects of a certain concentration of these substances on living organisms, animals and plants. More research is needed on the effects of pollution at population and community level, this being the basis for evaluating the actual damage produced by pollution on the biota.

With the abovementioned considerations in mind, it seemed opportune to organize the course "Biological aspects of fresh water pollution", on June 5th – 9th, 1978, at the Joint Research Centre (JRC), Ispra (Varese - Italy) in the framework of its Education and Training Program, realized by the so-called "ISPRA Courses". This advanced course was coordinated by the undersigned and given in the form of a series of lectures presented by 10 well-known experts in their specific field. The two most important aims of this course were:

1) to illustrate and discuss the fundamentals of the effects of pollution on fresh water organisms, populations and communities;
2) to supply an up-to-date picture of the research on this subject.

This book contains the lectures presented at this course reviewed by the authors. This book is, consequently, not a text on fresh water pollution: it tackles only a part of this extensive subject in the form of a series of problems.

The short and long term effects of toxic substances on fish is obviously one of the most important topics of fresh water pollution. The most significant results obtained in this field during the last 30 years by the Fish Section of the U.K. Water Research Centre are summarized very clearly. These studies, carried out in the laboratory, in the field and in artificial streams, made an important contribution to the elaboration of fresh water quality criteria for fish. This chapter covers the most important aspects of these problems. One chapter concerns lacustrine zooplankton and its importance in the study of fresh water pollution. Indeed, it may be used to detect even very low concentrations of some toxic substances and several pollutants may have a noxious influence on this central ring of the aquatic food network. The brief history of the zooplankton used as a pollution indicator is very interesting. The authors discuss the difficulties of discriminating between variations in the zooplankton community due to natural causes and those produced by pollution. This chapter concerns studies in the natural environment as well as laboratory experiments.

In another chapter, the most important biological methods developed at the international level for the assessment of the quality of surface waters are reviewed and discussed. The efforts made by the Commission of the European Communities to calibrate the methods commonly used and the results obtained, are outlined.

Because the community is strongly influenced by the physical and chemical characteristics of the environment in which it lives, it follows that variations in the environmental characteristics influence the structure of the community. On this basis it seems possible that the recent trophic history of a lake can be determined from analysis of the remains of organisms (pseudofossils) present in the sediments. To this end the remains of Cladocera and Chironomidae collected from the sediments of some Irish lakes have been studied. The results obtained are reported and discussed and are in good agreement with those obtained from analyses of some "biochemical pseudofossils" (pigments) present in the same sediments. In trophic evolution, physical processes are generally less studied than the chemical ones, but their importance is relevant. From the data from 62 European and North American lakes, the influence of the internal input of nutrients by vertical mixing on the annual primary production has been evaluated. If the mean annual charge of phosphorus is higher than a certain value, the influence of the internal input is abolished because the production rate is not limited by the nutrients but by the light intensity. The wide variations in primary production in lakes with similar nutrient charge is due to physical factors such as morphometrical characteristics, vertical mixing and redox potential of the interface water-sediment. This point of view may be the background for future research on the eutrophication problem.

One of the subjects also considered in this book is eutrophication caused by domestic sewage and pollution by heavy metals in running waters studied in artificial outdoor rivers. From a study of the effects on the benthic algae, several important points came to light. For example: heavy metals considered as nutrients as well as toxic substances, quantitative and qualitative changes in a benthic algae community, the influence of the season on metal toxicity, variations in the sensitivity of different algal species to the same metal, the recovery of the algal community from metal stress, the difficulty in employing and establishing lists of indicator organisms because the benthic community reflects the past situation. The most important point is the combined effects of eutrophication and heavy metals pollution.

The importance of heavy metals in fresh water pollution is treated in another chapter on a global scale for different water bodies (e.g. rivers, man-made lakes and natural lakes). Sediment analyses are considered to be very useful in assessing the sources, distribution and, in several cases, the potential hazard of this type of pollution. The water quality criteria (concerning heavy metals) proposed by different countries and organizations are compared and discussed. Among the most important points treated in this chapter were: metal speciation, seasonal cycle of metal transport, retention and release of heavy metals by and from the sediment and the transfer of heavy metals in aquatic food chains. The discussion of the different methods adopted for restoring lakes contaminated by mercury, is very relevant.

The need for clear and objective information on radioactive pollution and its biological consequences is well appreciated today. Two aspects of this problem are discussed in this book. One is the discharge of radionuclides to aquatic ecosystems and the protection of the environment (including man) from radiation, the other is the effects of ionizing radiations in aquatic biota. The description elaborating the "discharge formula" that fixes the quantity of radioactive substances which can be discharged into our environment is very accurate. I would like to mention that the "critical parameters" adopted in radioprotection and discussed in the book, can be also used for protecting the biota (and man) against non-radioactive pollution. The effects of ionizing radiations on aquatic populations are discussed. Genetic and somatic effects and the factors acting on them were evaluated from copious information obtained from laboratory experiments and the scant knowledge of the natural populations. One of the most important points in the chapter is the recovery-rate in connection with the demographic characteristics of the population.

The chapter on waste water treatment covers domestic sewage, industrial effluents and the influence of air pollution control on sewage treatment. The clear illustration of the subject with some well chosen examples covers a large proportion of the chemical and technological problems of waste water treatment. Of particular interest is the description of the new problems created in water treatment arising from air pollution control by wet techniques.

The text is complemented by numerous figures and tables. Many references (about 500) are reported at the end of each chapter. We expresse our gratitude to the authors who worked so hard in preparing their manuscripts, to Mr. B. Henry, Manager, Education and Training Program, who fully supported the course and book projects, and to his staff for the excellent organization of the course, and to Miss Lorenza Giannoni for her valuable help in organizing the preparation of the book.

*Oscar Ravera*

# Studies on the Effects of Pollution on Freshwater Fish

J. F. de L. G. Solbé

*Acting Head, Fish Section, Water Research Centre, Stevenage Laboratory, Elder Way, Stevenage, Herts, U.K.*

ABSTRACT

The formation of the United Kingdom's Water Research Centre and the role of the Centre's Fish Section (formerly a section of the Water Pollution Research Laboratory) are described. Work on the effects of pollution on freshwater fisheries has developed during the Section's 30 year existence from short-term studies of the lethal toxicity of single chemicals to long-term evaluation of the sub-lethal effects of mixtures of chemicals.

Field studies have been utilised to identify problems and to assess the relevance of the laboratory-derived data. Work in the laboratory, the field, and, as a compromise, artificial streams, is contributing to the development of water-quality criteria for freshwater fish. Work is also in progress on the use of fish in continuous, automatic monitors of water quality.

A bibliography of 90 of the Section's publications is included.

Keywords: freshwater fish, toxicity, acute studies, chronic studies, laboratory studies, field studies, artificial streams, water-quality criteria, environmental standards, bibliography.

INTRODUCTION

Fish have been valued for many years as excellent indicators of water quality. Their position in the aquatic ecosystem, generally at or close to the end of food 'chains', their sensitivity to a variety of pollutants, their size (making them easily visible when dead), and their devoted followers - the keenly-observant anglers - all contribute to their value in this respect.

Studies on the effects of pollution on fish have not only been made to 'calibrate' fish as indicators, but have also allowed the establishment of water-quality criteria to safeguard fish for their own sake. The critical assembly of data to this latter end has been particularly effectively co-ordinated by the European Inland Fisheries Advisory Commission. In the last 14 years EIFAC has published reports on the effects on fish of, for example, finely divided solids (EIFAC, 1964), pH value (EIFAC, 1968), ammonia (EIFAC, 1970), phenols (EIFAC, 1972), dissolved oxygen (EIFAC, 1973a), zinc (EIFAC, 1973b), copper (EIFAC, 1976), and cadmium

(EIFAC, 1977). Such reports have been taken into account by the ten Regional Water Authorities (RWAs) of England and Wales in their introduction of local measures for the control of pollution and, no doubt, by the equivalent authorities in other nations.

When the RWAs were established in the UK in April 1974 they took over control of the rivers (previously held by their predecessors, the River Authorities) and control of sewerage and sewage treatment which had been administered by local authorities. Although the RWAs inherited research facilities from the River Authorities and began to develop further facilities to serve their specific local requirements they required an establishment to cater for national research demands. This establishment, also created in April 1974, was called the Water Research Centre. It was formed by combining two existing laboratories, the Water Pollution Research Laboratory (WPRL), at Stevenage, and the Water Research Association (WRA), at Medmenham, with the Research Division of the Water Resources Board. The WPRL had been carrying out investigations for Central Government for 47 years, as far as 'dirty water' was concerned. The WRA was a rather younger 'clean water' research establishment whose subscribers included the water supply companies.

The WRC employs about 500 people in divisions specialising in water storage, treatment, and supply; river and coastal pollution; sewerage and sewage treatment; the disposal of sludge and refuse; water-quality surveillance and health; economics; and the effects of pollution on fish.

The following notes describe the development of the WPRL/WRC Fish Section and its evolving approaches to the problems of evaluating the impact of pollution on fisheries and, to return to my opening statement, to the investigation of ways in which fish can aid man in continuous automatic water-quality alarm systems.

## A BRIEF HISTORY OF THE WPRL/WRC FISH SECTION

Even before the creation of a 'Fish Section' in the late 1940s, *ad hoc* investigations of the effects of polluting substances on fish had been carried out by WPRL staff. For instance, data were published on the effects of potassium cyanide and *p*-cresol on trout (Southgate, Pentelow and Bassindale, 1933) and a paper was produced on the toxicity to fish of chlorinated sewage effluents (Allen, Blezard, and Wheatland, 1946). The first paper from the new Section appeared in 1952, and was concerned with the experimental methods, the interpretation of results, and variation of resistance to poison among a group of test fish (Herbert, 1952). At this early stage it was realised that 'fixed-volume' (static) tests suffered from severe limitations, that there was a need for oxygenation of the test waters without driving off volatile poisons, and that $CO_2$ production, pH value, and temperature should be controlled. Location of the median fish (e.g. the 6th to die in a batch of 11) was identified as representing a useful way of indicating the effects of a poison on a batch of fish in a statistically meaningful manner. The best estimate of the median period of survival of a batch of fish was considered to be the geometric mean of individual survival times because a log-normal distribution had been found when the number of fish overturning had been plotted against time. Not all variation noted was inherent; the possibility of acclimation had been recognised.

This methodology was employed in the first test programme, in which the toxicity of potassium cyanide to rainbow trout (*Salmo gardneri*) was investigated (Herbert and Merkens, 1952). Cyanide at that time was a constituent of effluents from gas works and of other industrial effluents. Acclimation of fish to low levels of the chemical increased their survival time in higher concentrations; larger fish survived longer than smaller fish of the same age. A linear relationship was found

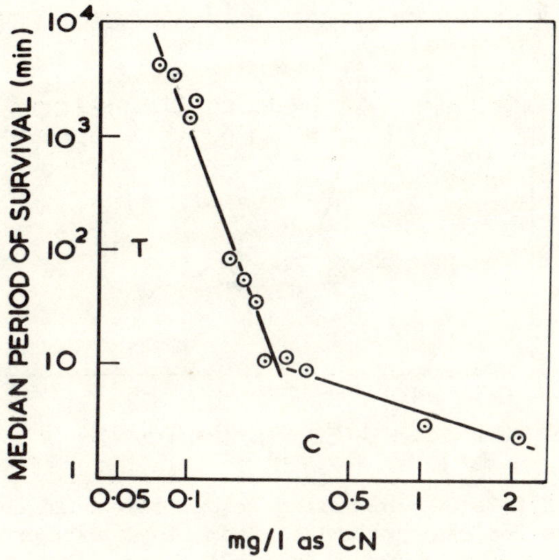


Fig. 1. Straight-line interpretation of toxicity test data (from Herbert and Merkens, 1952)

when the percentage of fish overturned (probability scale) was plotted against time in logarithms. An assumption was made that there was also a linear relationship between concentration of poison, C, and median period of survival, T, so that

$$n \log C + \log T = k$$

where n and k were constants (Fig. 1). This relationship only held for small lengths of the toxicity curve and of course did not allow for the existence of median lethal threshold concentrations (defined for the present purpose as concentrations of a chemical, under a given set of test conditions, below which half a batch of fish should survive for an extended period, illustrated in Fig. 2).

One of the first effects of an environmental variable on the toxicity of a poison studied by the Section also concerned cyanide. Downing (1954) established that survival times of rainbow trout in concentrations of potassium cyanide in the range 0.105–0.155 mg/l as CN increased as the concentration of dissolved oxygen (DO) in the water increased between 10 and 100 per cent of the air saturation value (ASV).

Ammonia is one of the most widespread of pollutants and was the next to be studied in detail by the Section. Work in the Laboratory showed the effects of $CO_2$ on the toxicity of ammonia to fish (Alabaster and Herbert, 1954; Allan, Alabaster and Herbert, 1954; Lloyd and Herbert, 1960; Pentelow, Allan, Herbert and Alabaster, 1958). This work marked an early stage in the long-term co-operation between the WPRL Fish Section and the Salmon and Freshwater Fisheries Laboratory of the Ministry of Agriculture, Fisheries and Food, and it involved many of the methods of approach utilised in later years.

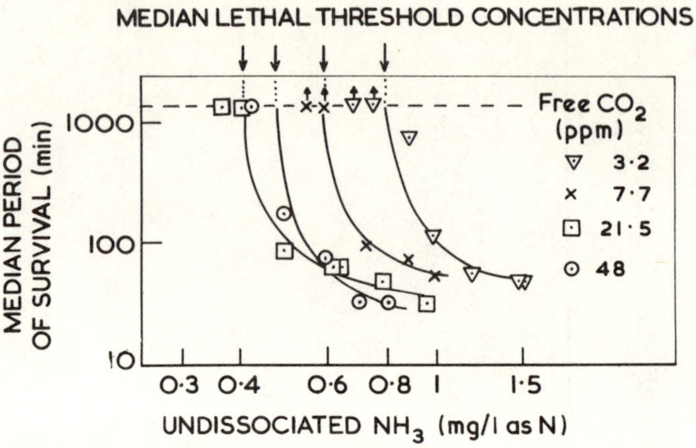


Fig. 2. Effect of free $CO_2$ on median lethal threshold concentration of ammonia for rainbow trout (from Lloyd and Herbert, Reprint No. 359)

Later studies with ammonia were concerned with the effects of DO concentration on toxicity to rainbow trout (Downing and Merkens, 1955) and to rainbow trout, perch (Perca fluviatilis), roach (Rutilus rutilus), and gudgeon (Gobio gobio). (Merkens and Downing, 1957.) Lloyd (1961a) assembled from his own and earlier work a graphical method for calculating the toxicity of ammonia to rainbow trout, which took into account the effects of pH value, temperature, DO, bicarbonate alkalinity and free $CO_2$. This method was incorporated later by Brown (1968) in the procedure for calculating the toxicity of a mixture of poisons to rainbow trout. Finally, concerning ammonia in the absence of other chemicals at poisonous concentrations, Ball (1967a) studied its effects on rainbow trout, perch, roach, bream (Abramis brama) and rudd (Scardinius erythrophthalmus). Ball found that the median lethal threshold concentrations for all these species were very similar, but they were reached quickly by rainbow trout and much more slowly by the other species (Fig. 3).

The effects of low concentrations of dissolved oxygen on fish, both in the absence of other stresses and influencing the toxicity of poisons, have been studied extensively by the Section. For trout and perch the inter-relationships of $CO_2$ and DO were examined by Alabaster, Herbert and Hemens (1957). Carbon dioxide reduces the affinity of the blood of many freshwater fish species for oxygen and the minimum non-lethal level of DO was indeed found to be higher as $CO_2$ concentrations were increased. In any potentially lethal situation fish live longer if the water temperature is lowered. Downing and Merkens (1957) confirmed this finding with rainbow trout, perch, roach, mirror carp (Cyprinus carpio), tench (Tinca tinca), dace (Leuciscus leuciscus), chub (Leuciscus cephalus) and bleak (Alburnus alburnus) and ranked the species in order of their tolerance, tench being the most tolerant and trout the most sensitive.

A further environmental variable, light, was considered by Alabaster and Robertson (1961) in their examination of the behaviour of roach, bream, and perch under the stresses of low DO and high temperatures. All the species were more active at dawn and dusk; perch and bream shoals tended to break up at night;

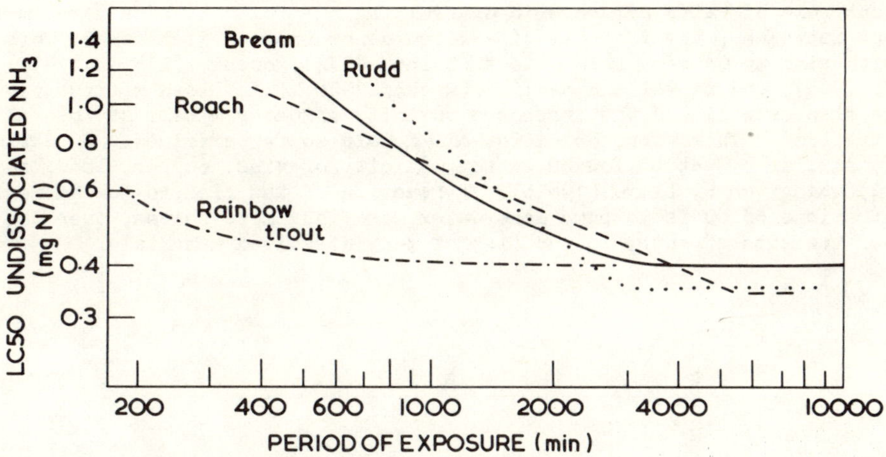


Fig. 3. Susceptibilities of different species to undissociated $NH_3$ (from Ball, 1967a)

non-lethal changes in temperature and DO caused the fish to become more active at any time in the 24 hours: such activity probably has survival value.

The Section has made a great many studies of the effects of detergents on fish over the years, but few have been published, being pieces of contract work for the manufacturer or for a Regional Water Authority. Herbert and others (1957) established the beneficial effect of sewage treatment in reducing the toxicity of methylene-blue active substances. Effects of chronic exposure (100 d) to zinc on the toxicity of a mixture of detergent and zinc, as seen by its action on the fine structure of rainbow trout gill tissue, were examined by Brown, Mitrovic and Stark (1968). The soft alkylbenzenesulphonate in mixture with zinc was more toxic to fish which had been previously exposed to zinc. The histopathological details resulting from exposure to these two materials were fully described.

Many large industrial plants in the UK are situated near major estuaries, and the effects of salinity both per se and on the toxicity of poisons have been studied occasionally. Generally speaking the toxicity of poisons is lowest when the salinity (i.e. osmotic pressure) of the medium is equal to that of the fishes' blood. For phenol this is not so, toxicity rising slightly with any rise in salinity (Brown, Shurben and Fawell, 1967). The oxygen requirements of migrating fish as they pass through estuaries were examined by Alabaster (1973).

A few more examples follow of the earlier work on single poisons before this summary deals with the toxicity of mixtures. Merkens (1958) elucidated the complex toxicity of chlorine, chloramines, and nitrogen trichloride to rainbow trout. Suspended solids (kaolin and diatomaceous earth) were studied by Herbert and Merkens (1961). There was evidence that abrasion of the gill epithelium by these solids could have encouraged microbial infections. China-clay wastes from the extensive workings in south-western England were examined in detail in the field. Although the invertebrate production was reduced considerably in the polluted streams, compared with the controls, electric fishing demonstrated that there were fish living in the polluted areas, but deriving quite a proportion of their food from allochthonous drift (Herbert and others, 1961). Lloyd (1960) published the first paper by the Section on heavy-metal toxicity, although studies of this type had been reported by Allan, Alabaster and Herbert (1954). Lloyd's work concerned the effects of zinc sulphate on rainbow trout and established the important role of water hardness in determining the toxicity of a heavy metal. As hardness increases, the toxicity of a given concentration decreases. This is the case with zinc up to very hard waters (Solbe, 1974), copper (Lloyd, 1961c), lead (Lloyd, 1965), and nickel and cadmium (Brown, 1968). In Lloyd's first paper on zinc he also established the increased toxicity of heavy metals at lowered DO concentrations. This, too, holds for all metals so far examined. An explanation for the similar effect of low DO on the toxicity of zinc, copper, lead, and phenol (Fig. 4) was given by Lloyd (1961b). A reaction of the fish to low ambient concentrations of DO is to pump more water, and thus more poison, over the gill surface, the site of entry to the fish of many dissolved materials.

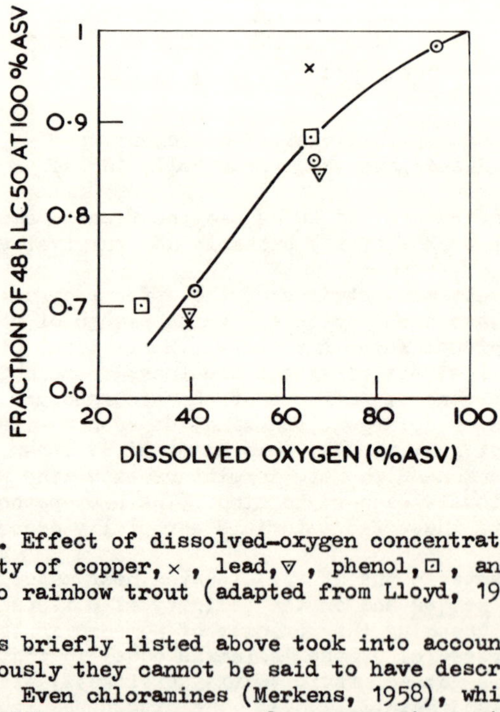


Fig. 4. Effect of dissolved-oxygen concentration on the toxicity of copper, ×, lead, ▽, phenol, ⊡, and zinc, ⊙, to rainbow trout (adapted from Lloyd, 1961b)

Although the studies briefly listed above took into account several environmental variables simultaneously they cannot be said to have described the toxicity of a mixture of poisons. Even chloramines (Merkens, 1958), which in solution may break down in complex equilibrium patterns, only represent one toxic species, hypochlorous acid (HOCl). It was not until Lloyd's work on zinc and copper sulphates that a start was made on the study of mixtures of poisons (Lloyd, 1961c). Four possible types of effect of a mixture of poisons may be distinguished: similar joint action ('addition'), in which the toxicity of the mixture could be predicted from a knowledge of the toxicity of its constituents; antagonism, where the mixture is less toxic than expected; synergism, where the mixture is more toxic; and independent joint action, in which the toxicity of the mixture resembles that of one or other of its constituents.

Expressing the toxicity of a mixture of copper and zinc as the sum of the fractions of the median lethal threshold concentrations of those two metals (as defined previously) it was found that in hard water their reaction in mixture was additive. That is, when $Zn_T$ and $Cu_T$ are the median lethal threshold concentrations and $Zn_S$ and $Cu_S$ are the concentrations used in the test mixture, any combination of $Zn_S/Zn_T + Cu_S/Cu_T = 1$ would have the same effect (that of just killing half a batch of fish in an extended period) whether values for $Zn_S/Zn_T$ and $Cu_S/Cu_T$ were, for instance, each 0.5 or were 0.6 and 0.4, 0.3 and 0.7, 0.8 and 0.2. In soft water, at higher concentrations of the metals there was a tendency towards synergistic action.

The general rule of additive action of poisons was found to apply most of the time for mixtures of ammonia and zinc (Herbert and Shurben, 1964), copper and ammonia, and zinc and phenol (Herbert and Vandyke, 1964) and even when 4 poisons (copper, phenol, zinc, and nickel) were mixed (Brown and Dalton, 1970).

The theory of summation of toxicity was put to the test on the extremely complex mixture of poisons found in spent still liquors from the distillation of coal (Herbert, 1962). The liquors contained ammonia, sulphides, cyanides, thiosulphates, thiocyanates, and phenols (phenol, cresols, xylenols, etc.), yet the summation theory gave reasonably good agreement with observations on fish on 32 out of 40 days in streams carrying the effluents. Tests of other effluents followed (Lloyd and Jordan, 1963, 1964), and any departures from the expected were as useful as any agreements with the theory: under-prediction indicated the presence of unsuspected poisons and over-prediction suggested previously unrecognised factors such as the complexing of metals, particularly copper. Generally these studies centred on the then grossly-polluted rivers which were fishless and whose waters were lethally toxic within the test period to fish held in cages in the river or, when low DO concentrations were likely to occur, in aquaria. If the sum of fractions (e.g. $Zn_S/Zn_T + Cu_S/Cu_T + \ldots$) equalled unity, half the batch of fish generally died in about 48 h. If the sum equalled a value greatly in excess of 1.0 there was no way of predicting when (under 48 h) the fish would die. If the sum approximately equalled 0.7 about 5 per cent of the fish had died within 48 h. Clearly, the method worked best when the fish died at around 48 h. In some of the rivers chosen the undiluted river water was lethal in 3 h. In these cases appropriate dilutions were made and the toxicity calculated on the resulting quality of the test water.

Edwards and Brown (1967) summarised the progress made in studying the effects of pollution on freshwater fisheries and decided that future studies must deal with the problems listed below.

(i) The chronic effects of long-term exposure of fish to low concentrations of poisons

(ii) Comparisons of the susceptibility to poisons of different species of fish and of different stages in their life cycles

(iii) The toxicity of mixtures of poisons

(iv) The effects of fluctuating concentrations of poisons

(v) Variations in sensitivity of different populations of fish resulting from their physiological condition, acclimation, and genetic adaptation

(vi) The validity of laboratory studies in interpreting effects of pollutants on natural populations

It would appear useful at this point, 11 years later, to assess what progress has been made in each of these areas at the Stevenage Laboratory and to add new areas of interest identified since 1967.

(i) Chronic Toxicity

If 'chronic exposure' for tests with species such as trout, which reach sexual maturity at the age of 2 or 3 years, is defined arbitrarily as exposure for at least 3 months, the first such experiments at Stevenage can be said to have been reported in the annual report of WPRL for 1966 (Ministry of Technology, 1967). Rainbow trout were exposed to concentrations of zinc from 0.6 to 2.0 mg/l in hard water for up to 6 months. In this period 20 to 35 per cent of fish died. One of the hopes in testing the chronic toxicity of poisons is that sub-lethal responses may become apparent. Condition factors (K), where $K = 100\ W/L^3$ (W = live weight, grams, and L = fork length, cm), were calculated at the end of the experiment and liver and kidney weights were compared with body weight. None of these parameters seemed affected by zinc but appetite was, being smaller at the highest concentration of zinc. Accumulation of metal in a number of organs and tissues was examined and histological studies were made. Highest concentrations of zinc occurred in gut tissue and lowest in muscle; fat and glycogen deposition did not occur in the liver of fish exposed to zinc.

Mortality, and acclimation to ammonia (see (v) below) of rainbow trout were studied over a 3-month period (Ministry of Technology, 1968). After 3-months exposure to 80 per cent of the 48-h LC50 of ammonia (concentration of undissociated ammonia equal to 0.23 mg N/l), about 15 per cent of the fish had died.

Depression of growth rate was noted in rainbow trout exposed to sub-lethal concentrations of phenol (1 and 2 mg/l) for 18 weeks (Department of the Environment, 1971).

Rudd were exposed to zinc for 255 days (Department of the Environment, 1972), for which period the LC50 was 10.5 mg/l. Histologically the only zinc-related damage observed was epidermal and gill epithelial hyperplasia. Differences in final biomass (total weight of surviving fish) were used as indicators of the long-term effect of this poison, and of ammonia, to rudd.

Attention then switched to the long-term effects of copper on rainbow trout (Department of the Environment, 1973) and for the past 4 years has focused on the effects of cadmium on brown and rainbow trout, exposure periods being of up to 2 years duration. A variety of possible responses was looked for, the most interesting being inhibition of maturation. Chronic tests are often so well 'sighted' now that no mortality occurs even in the highest concentration after 24 months exposure. Simplicity is the keynote of the apparatus used, simplicity in this case going hand-in-hand with reliability and accuracy. The dosing system used (Shurben, 1978) is a development of that of Stark (1967) (see Fig. 5).

(ii) Interspecific Comparisons of Sensitivity; Sensitivity at Different Stages in the Life Cycle

The first of these subjects is receiving attention again at present, following the work of Ball (1967a) on ammonia, described above, and (1967b) on zinc. Alabaster (1969) summarised progress in this direction at one of the biennial British Coarse Fish Conferences (now, the Freshwater Fish Conferences), at which workers studying all aspects of freshwater, non-salmonid fish, come together to discuss progress. There is a need for the continuation of such studies. Many of the 3 million UK anglers fish in lowland rivers inhabited by non-salmonid fish.

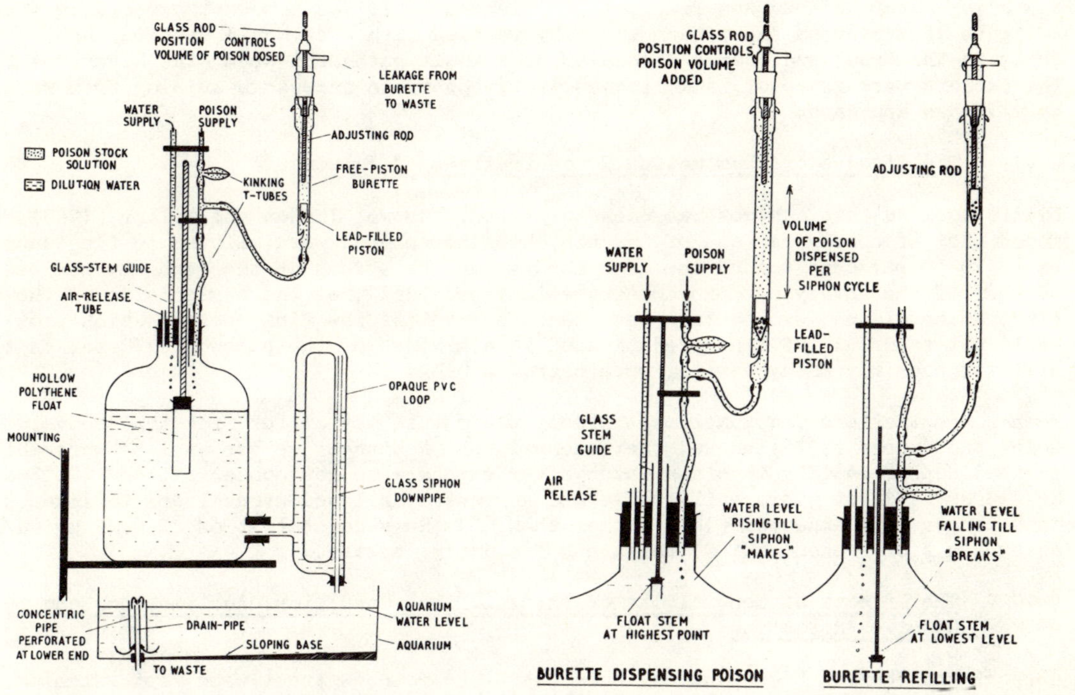


Fig. 5. Siphon doser (from Shurben, 1978)

Few toxicological studies on different stages in the life cycle of fish have yet been made. A comparison has been made of the toxicities of zinc, phenol, cadmium and nickel to rainbow and brown trout eggs and alevins (Ministry of Technology, 1967) in which it was shown that sensitivity to zinc poisoning decreased as embryonic development proceeded to hatching and then, after hatching, increased again throughout the alevin stage. Shaw and Brown (1971) examined the effects of copper and nickel on fertilisation of rainbow trout eggs: a concentration of 1 mg/l of either metal did not impair fertilisation, but the rate of development did increase, especially in the presence of copper.

## (iii) The Toxicity of Mixtures of Poisons

As indicated earlier, Brown (1968) gathered data on acute lethal toxicity of common pollutants to fish and presented a method by which the fisheries officer in the field could take the results of chemical and physical analysis of a water sample and predict (in simple terms) the toxicity to fish of that sample. The method used data on the poisons ammonia, phenol, cyanide, zinc, copper, cadmium, lead, and nickel and took into account the effects on their toxicity of temperature, pH value, total hardness, bicarbonate alkalinity, total dissolved solids, and concentrations of dissolved oxygen, as appropriate. The results were expressed as the calculated sum of the fractions of the predicted 48-h LC50 values of the poisons to rainbow trout.

Work preceded and has followed the publishing of that paper to establish the applicability of the method not just in describing the toxicity of the water but in trying to predict the status of a fishery from the calculated predicted toxicity (Brown, 1969). Alabaster, Garland, Hart and Solbe (1972) discovered empirically that where the sum of the fractions (see above) exceeded 0.28 for half the year at any given site the area was likely to be fishless. It was somewhat surprising that a figure derived from short-term toxicity testing with a salmonid species should apply to the trout <u>and</u> coarse fisheries of a whole catchment area (the River Trent). The authors were aware of, and discussed, the possible drawbacks arising from so cavalier an approach.

### (iv) The Effects of Fluctuating Concentrations of Poisons

Little work on this subject has been completed. Brown, Jordan and Tiller (1969) found that if concentrations of ammonia, zinc, or phenol were allowed to fluctuate by up to 50 per cent around the 48-h LC50 value the effect on the fish was similar to that of the unvarying mean concentration, provided that the periodicity of the fluctuation did not exceed the 'resistance period' of the fish, beyond which they would not recover. If zinc predominated in a mixture of the poisons (>70 per cent) effects, possibly antagonistic, were noted.

When concentrations fluctuate more widely the result may differ from that above. Solbe and Cooper (1976) showed that stone loach (*Noemacheilus barbatulus*) were able to shed copper rapidly in clean water after exposure to low concentrations (0.12 mg Cu/l) for 64 days. This ability, it was suggested, allowed stone loach to live 12 days longer than expected (18 d rather than 6 d) in a concentration of 0.49 mg Cu/l as a result of 2 short (<1 d) failures of a dosing system.

### (v) The Effects on Sensitivity of Physiological Condition, Acclimation, and Genetic Adaptation

Herbert and Shurben (1963) established that when rainbow trout were kept swimming at 85 per cent of their maximum sustainable velocity the toxicity of ammonia increased so that the 24-h LC50 was only 0.81 of the still-water value. Similarly, the 48-h of LC50 of zinc sulphate was reduced by a factor of 0.7. No increase in susceptibility was noted at swimming speeds lower than 55 per cent of maximum sustainable velocity. The differences were not considered large enough to warrant future consideration.

No mortalities associated with the stress of spawning have been observed yet in our long-term studies.

Acclimation here refers to physiological adaptation of a fish to low levels of a poison so that it can tolerate higher levels than it otherwise would. Reference has already been made to acclimation to KCN (Herbert and Merkens, 1952). When rainbow trout were exposed for 90 days to 5, 10, or 20 mg ammonia /l (as N) their resistance to acutely lethal concentrations of ammonia increased considerably. Fish acclimated to the two higher concentrations gave 48-h LC50 values of over 60 mg/l compared with 30 mg/l (as N) in unacclimated fish. However, this increased tolerance of ammonia was 'bought' at the expense of increased sensitivity to zinc, the 48-h LC50 for which was reduced to 75 per cent of its previous value (Ministry of Technology, 1968).

Acclimation, or physiological adaptation, must presumably involve the mobilisation of defence mechanisms appropriate to each poison, where such mechanisms have lain dormant. Recently, investigations have been carried out on the effects of pre-exposing fish to low levels of zinc and intermittent (but lethal if continued) concentrations of cadmium before prolonged continuous exposure to the same concentrations of the latter metal. The defence mechanism involved here may be the production of metal-binding proteins, stimulated by the pre-test routine. The effect was clear. In 7-week tests rainbow trout did not die in concentrations of cadmium which were twice the lethal threshold (Cooper and Solbe, 1978). Such tests may give us an insight into the toxic mechanisms of the poison, as well as the defensive reaction of the fish.

Genetic adaptation remains to be studied in detail, but a comparison has been made of the toxicity of zinc to brown trout from zinc-polluted and from unpolluted soft-water streams. No adaptation was found: the batches of fish from both streams were equally sensitive to zinc, but it was not certain that the fish in the polluted river had lived in the area for a long period.

### (vi) Assessing the Relevance of Laboratory Studies in Interpreting the Effects of Pollutants on Natural Populations

Many of the laboratory studies in the 1950s and early 1960s stemmed from the pioneering fieldwork on the River Colne (Allan, Alabaster and Herbert, 1954; Pentelow and others, 1958) and, in a sense, the heading (vi, above) also covers the work carried out when the summation theory was being developed in the laboratory and tested in grossly-polluted situations. In 1969 the first ecological studies of mildly-polluted waters began. The first site chosen for study was the Willow Brook, a hard-water stream polluted by effluents from domestic sewage works and steel works. During 5 years, water samples were taken every 2 or 3 weeks from seven sites on this stream (Fig. 6) and a picture of the physico-chemical quality of the system was built up. Toxicity values (sums of fractions of predicted 48-h LC50 values) were calculated and compared with twice-yearly observations on the fish community throughout the system. Some parts of the upper streams were fishless but water quality and the diversity of the fish (and invertebrate) communities improved as the water passed downstream. Observations on the fish were simple. They included species distribution, growth rate (including growth of tagged fish), survival of captive rainbow trout in cages, and accumulation of metals in fish tissues and organs. Fortunately, from the point of view of both the river and the project, the water quality improved markedly during the study period, and so did the 'status' of the fish communities. The work was briefly reported by Alabaster and others (1972) and more fully by Solbe (1973, 1977), and it gave rise to further studies in the laboratory. For example, it was necessary to know whether the equation relating log hardness and log 48-h LC50 held at the very high hardness values existing in the stream; Solbe (1974) showed that it did. It was also important to establish, as was the case, that the disappearance of stone loach for several years was caused by the observed concentrations of zinc (Solbe and Flook, 1975).

After beginning the Willow Brook studies another two rivers, the Churnet and the Tean in Staffordshire, received detailed attention. Neither produced quite so clear a picture of cause and effect as that for the Willow Brook, but observations on both have provided useful data for EIFAC publications (EIFAC, 1976, 1977) and the Tean study gave rise to a new approach for the Section - studies of artificial streams.

The artificial-stream experiment is a compromise between the controlled but unnatural laboratory study and the 'natural' but uncontrolled real situation of the polluted environment. In conjunction with the University of Aston in Birmingham, the WRC established three 300 m streams, 1.5 m wide, and divided into two riffle sections, 100 m long and 0.15 m deep, and two pool sections, 50 m long and 0.3 m deep. The streams were supplied with unpolluted water from the R. Tean and well-nitrified sewage effluent containing, among other things, cadmium, in the following proportions: Stream A, 100% river water; Stream B, 50% river water and 50% effluent; Stream C, 25% river water and 75% effluent. Each stream received $4.5 \times 10^6$ l/d.

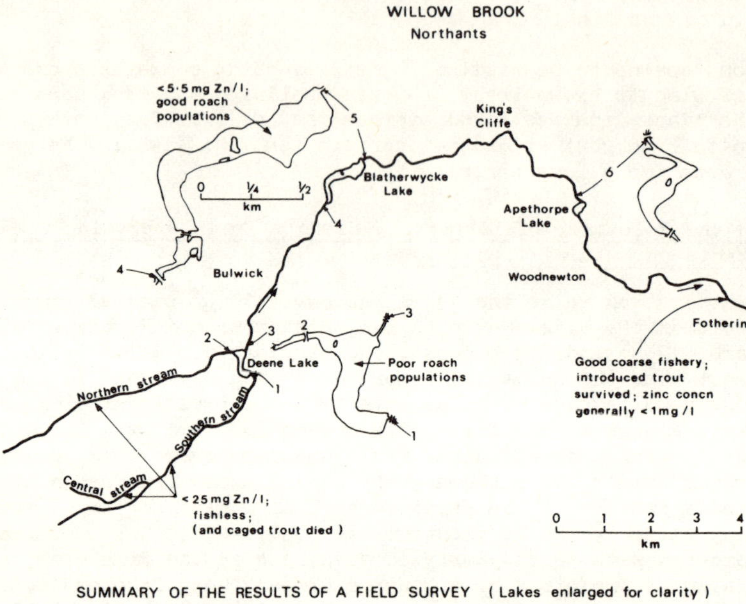


**Fig. 6. Summary of the results of a field survey (lakes enlarged for clarity)**

Algae and invertebrates were allowed to colonise the streams, and fish (rainbow or brown trout) were introduced to give the required biomass levels. The principal role of the study on the artificial streams is to establish the survival, growth rate, and metal accumulation of the fish. Detailed studies of the invertebrates have been included, partly to try to define the pathways by which metals reach fish tissues in a semi-natural environment.

In the first 3 years as well as making useful scientific observations a great deal was learnt about the management of such a large experimental facility. Problems which have been overcome include retaining fish in the channels while allowing large aggregations of filamentous algae to escape, and the handling of data arising from a programme of water-quality monitoring in which samples for some water-quality characteristics are collected every 15 min for 365 days each year! In the second 3-year period an additional experimental facility - large tanks holding 6 different qualities of water - is being established. The 75:25 stream is being changed to 25 per cent effluent, 75 per cent river water as a result of the findings to date.

Finally, we come to an area of research which makes use of the reactions of fish to a wide variety of potential pollutants in a continuous automatic alarm system. About 3 years ago the Section began to consider the use of fish in monitoring effluents. Shortly after, the Centre was asked to help a member Water Company to safeguard its potable-supply intake on a river from severe pollution incidents. A physico-chemical monitor was being installed but it was hoped that a biological monitor would offer an additional safety factor. Since a vertebrate, man, was to be protected it seemed sensible to use a vertebrate as a monitoring species. The rainbow trout was chosen, partly because of the large amount of information already existing on its sensitivity and partly because it is a 'good laboratory animal' – it can be handled and fed easily, and it can be confined in small volumes of water without becoming distressed.

Almost from the outset of the project we established that what we were looking for was an advanced ECG, gill-movement and whole-body activity monitor which would not subject the fish to any extra stress. (Some existing monitors use electric shocks to move a fish upstream in the monitor channel when the fish has fallen back downstream in response to the advance of a pollutant.) After about 12 months of work a system of channels, electrodes, amplifiers, frequency-selective electronic filters, and display apparatus had been developed to a satisfactory state so that biological evaluation could begin. In the design, each fish swims in its own 15-cm wide channel of about 60 cm length in a depth of about 7.5 cm of water. The electrodes are stainless steel plates placed above and below the swimming area of the fish. They pick up the μV signals from the contraction of the fish muscles and these signals are amplified and cleaned up by frequency filters before display or use in the 'alarm module'. At the moment the signals are integrated to give a single activity output for each fish and alarms can be sounded by death (lack of signal) or by two hyperactive states – the first when the fish is a little agitated and the second when it is frantic. A positive alarm would require 2 or even 3 of the three fish used to react within the same short period of time. To date, about 7 of these 'Mark I' Monitors have been produced for evaluation by the RWAs and the Water Companies.

A 'Mark II' version is being evolved simultaneously. This will automatically analyse the frequency and regularity of the heart-beats, gill movements, and 'coughing' of the fish. (Coughing is the expulsion of water forward from the gill chamber through the mouth, clearing debris from the gills. It appears to increase in frequency if the fish becomes stressed.) The electronics have been developed to a satisfactory stage and 2 types of monitor have been produced for organisations outside the WRC, particularly as a physiological tool in research laboratories. Meanwhile the use of a mini-computer to handle the data generated by the monitor during its evaluation has been arranged, which will greatly ease the analysis of the signals [which must at present (April, 1978) be read from recorder charts and then prepared for computation].

The small amount of evaluation completed to date suggests that we are on the right lines. Reaction times to pollutants vary between a few hours (30 μg Cd/l), 4 minutes (4 mg phenol/l), and instantaneous (for a harmless blue dye). Managements employing monitors are preparing procedures to be followed in the event of alarms and close liaison is being maintained via WRC.

## REFERENCES

Alabaster, J. S., and D. W. M. Herbert (1954). The influence of carbon dioxide on the toxicity of ammonia. Nature, 174, 404.

Alabaster, J. S., D. W. M. Herbert, and J. Hemens (1957). The survival of rainbow trout (*Salmo gairdnerii* Richardson) and perch (*Perca fluviatilis* L.) at various concentrations of dissolved oxygen and carbon dioxide. Ann. appl. Biol., 45, 177-188.

Alabaster, J. S., and K. G. Robertson (1961). The effect of diurnal changes in temperature, dissolved oxygen, and illumination on the behaviour of roach (*Rutilus rutilus* L.), bream (*Abramis brama* L.) and perch (*Perca fluviatilis* L.). Anim.Behav., 9, 187-192.

Alabaster, J. S. (1969). The comparative sensitivity of coarse fish and trout to pollution. Proc. 4th British Coarse Fish Conference.

Alabaster, J. S., J. H. N. Garland, I. C. Hart, and J. F. de L. G. Solbe (1972). An approach to the problem of pollution and fisheries. Symp. zool. Soc. Lond., No. 29, 87-114.
Alabaster, J. S. (1973). Oxygen in estuaries: requirements for fisheries. In Mathematical and Hydraulic Modelling of Estuarine Pollution. Water Pollution Research Technical Paper No. 13. H.M. Stationery Office, London. p.15.
Allan, I. R. H., J. S. Alabaster, and D. W. M. Herbert (1954). Recent studies on toxicity and stream pollution. Wat. sanit. Engr, 5, 109-112.
Allen, L. A., N. Blezard, and A. B. Wheatland (1946). Toxicity to fish of chlorinated sewage effluents. Surveyor, Lond., 105, 298.
Ball, I. R. (1967a). The relative susceptibilities of some species of freshwater fish to poisons. I. Ammonia. Wat. Res., 1, 767-775.
Ball, I. R. (1967b). The relative susceptibilities of some species of freshwater fish to poisons. II. Zinc. Wat. Res., 1, 777-783.
Brown, V. M., D. G. Shurben, and J. K. Fawell (1967). The acute toxicity of phenol to rainbow trout in saline waters. Wat. Res., 1, 683-685.
Brown, V. M., V. V. Mitrovic, and G. T. C. Stark (1968). Effects of chronic exposure to zinc on toxicity of a mixture of detergent and zinc. Wat. Res., 2, 255-263.
Brown, V. M. (1968). Calculation of the acute toxicity of mixtures of poisons to rainbow trout. Wat. Res., 2, 723-733.
Brown, V. M., D. H. M. Jordan, and B. A. Tiller (1969). The acute toxicity to rainbow trout of fluctuating concentrations and mixtures of ammonia, phenol and zinc. J. Fish Biol., 1, 1-19.
Brown, V. M. (1969). The prediction of acute toxicity of river waters to fish. Proc. 4th British Coarse Fish Conference.
Brown, V. M., and R. A. Dalton (1970). The acute lethal toxicity to rainbow trout of mixtures of copper, phenol, zinc and nickel. J. Fish Biol., 2, 211-216.
Cooper, V. A., and J. F. de L. G. Solbe (1978). Reducing the toxicity of cadmium sulphate to rainbow trout (Salmo gairdneri) by preliminary exposure of fish to zinc sulphate, with and without intermittent exposure to cadmium. A progress report. Water Research Centre Internal Laboratory Report ILR750.
Department of the Environment (1971). Water Pollution Research 1970. H.M. Stationery Office, London.
Department of the Environment (1972). Water Pollution Research 1971. H.M. Stationery Office, London.
Department of the Environment (1973). Water Pollution Research 1972. H.M. Stationery Office, London.
Downing, K. M. (1954). The influence of dissolved oxygen concentration on the toxicity of potassium cyanide to rainbow trout. J. exp. Biol., 31, 161-164.
Downing, K. M., and J. C. Merkens (1955). The influence of dissolved oxygen concentration on the toxicity of un-ionized ammonia to rainbow trout (Salmo gairdnerii Richardson). Ann. appl. Biol., 43, 243-246.
Downing, K. M., and J. C. Merkens (1957). The influence of temperature on the survival of several species of fish in low tensions of dissolved oxygen. Ann. appl. Biol., 45, 261-267.
Edwards, R. W., and V. M. Brown (1967). Pollution and fisheries: a progress report. Wat. Pollut. Control, 66, 63-78.
EIFAC (1964). Water quality criteria for European freshwater fish. Report on finely divided solids and inland fisheries. EIFAC Technical Paper No. 1. FAO, Rome.
EIFAC (1968). Water quality criteria for European freshwater fish. Report on extreme pH values and inland fisheries. EIFAC Technical Paper No. 4. FAO, Rome.

EIFAC (1970). Water quality criteria for European freshwater fish. Report on ammonia and inland fisheries. EIFAC Technical Paper No. 11. FAO, Rome.

EIFAC (1972). Water quality criteria for European freshwater fish. Report on monohydric phenols and inland fisheries. EIFAC Technical Paper No. 15. FAO, Rome.

EIFAC (1973a). Water quality criteria for European freshwater fish. Report on dissolved oxygen and inland fisheries. EIFAC Technical Paper No. 19. FAO, Rome.

EIFAC (1973b). Water quality criteria for European freshwater fish. Report on zinc and freshwater fish. EIFAC Technical Paper No. 21. FAO, Rome.

EIFAC (1976). Water quality criteria for European freshwater fish. Report on copper and freshwater fish. EIFAC Technical Paper No. 27. FAO, Rome.

EIFAC (1977). Water quality criteria for European freshwater fish. Report on cadmium and freshwater fish. EIFAC Technical Paper No. 30. FAO, Rome.

Herbert, D. W. M. (1952). Measurement of the toxicity of substances to fish. J. Proc. Inst. Sew. Purif., 60-68.

Herbert, D. W. M., and J. C. Merkens (1952). The toxicity of potassium cyanide to trout. J. exp. Biol., 29, 632-649.

Herbert, D. W. M., G. H. J. Elkins, H. T. Mann, and J. Hemens (1957). Toxicity of synthetic detergents to rainbow trout. Wat. Waste Treat. J., 6, 394-397.

Herbert, D. W. M., and J. C. Merkens (1961). The effect of suspended mineral solids on the survival of trout. Int. J. Air Wat. Pollut., 5, 46-55.

Herbert, D. W. M. (1962). The toxicity to rainbow trout of spent still liquors from the distillation of coal. Ann. appl. Biol., 50, 755-777.

Herbert, D. W. M., and D. G. Shurben (1963). A preliminary study of the effect of physical activity on the resistance of rainbow trout (Salmo gairdnerii Richardson) to two poisons. Ann. appl. Biol., 52, 321-326.

Herbert, D. W. M., and D. G. Shurben (1964). The toxicity to fish of mixtures of poisons. I. Salts of ammonia and zinc. Ann. appl. Biol., 53, 33-41.

Herbert, D. W. M., and J. M. Vandyke (1964). The toxicity to fish of mixtures of poisons. II. Copper-ammonia and zinc-phenol mixtures. Ann. appl. Biol., 53, 415-421.

Lloyd, R. (1960). The toxicity of zinc sulphate to rainbow trout. Ann. appl. Biol., 48, 84-94.

Lloyd, R. (1961a). The toxicity of ammonia to rainbow trout (Salmo gairdnerii Richardson). Wat. Waste Treat. J., 8, 278-279.

Lloyd, R. (1961b). Effect of dissolved-oxygen concentrations on the toxicity of several poisons to rainbow trout (Salmo gairdnerii Richardson). J. exp. Biol., 38, 447-455.

Lloyd, R. (1961c). The toxicity of mixtures of zinc and copper sulphates to rainbow trout (Salmo gairdnerii Richardson). Ann. appl. Biol., 49, 535-538.

Lloyd, R., and D. H. M. Jordan (1963). Predicted and observed toxicities of several sewage effluents to rainbow trout. J. Proc. Inst. Sew. Purif., 167-173.

Lloyd, R., and D. H. M. Jordan (1964). Predicted and observed toxicities of several sewage effluents to rainbow trout: a further study. J. Proc. Inst. Sew. Purif., 183-186.

Lloyd, R. (1965). Factors that affect the tolerance of fish to heavy metal poisoning. In Biological Problems in Water Pollution, 3rd Seminar, 1962. US Department of Health, Education and Welfare. pp.181-187.

Merkens, J. C., and K. M. Downing (1957). The effect of tension of dissolved oxygen on the toxicity of un-ionized ammonia to several species of fish. Ann. appl. Biol., 45, 521-527.

Merkens, J. C. (1958). Studies on the toxicity of chlorine and chloramines to the rainbow trout. Wat. Waste Treat. J., 7, 150-151.

Ministry of Technology (1967). Water Pollution Research 1966. H.M. Stationery Office, London.

Ministry of Technology (1968). Water Pollution Research 1967. H.M. Stationery Office, London.

Pentelow, F. T. K., I. R. H. Allan, D. W. M. Herbert, and J. S. Alabaster (1958). Sewage effluents and fish. J. Proc. Inst. Sew. Purif., 5-16.

Shaw, T. L., and V. M. Brown (1971). Heavy metals and the fertilization of rainbow trout eggs. Nature, 230, 251.

Shurben, D. G. (1978). A self-breaking concentric-siphon dosing apparatus. Lab. Pract., 27, 111-113.

Solbe, J. F. de L. G. (1973). The relation between water quality and the status of fish populations in Willow Brook. Wat. Treat. Exam., 22, 41-61.

Solbe, J. F. de L. G. (1974). The toxicity of zinc sulphate to rainbow trout in very hard water. Wat. Res., 8, 389-391.

Solbe, J. F. de L. G., and V. A. Flook (1975). Studies on the toxicity of zinc sulphate and of cadmium sulphate to stone loach Noemacheilus barbatulus (L.) in hard water. J. Fish Biol., 7, 631-637.

Solbe, J. F. de L. G., and V. A. Cooper (1976). Studies on the toxicity of copper sulphate to stone loach Noemacheilus barbatulus (L.) in hard water. Wat. Res., 10, 523-527.

Solbe, J. F. de L. G. (1977). Water quality, fish and invertebrates in a zinc-polluted stream. In J. S. Alabaster (Ed.), Biological Monitoring of Inland Fisheries. Applied Science Publishers, Barking, Essex. pp.97-105.

Southgate, B. A., F. T. K. Pentelow, and R. Bassindale (1933). The toxicity to trout of potassium cyanide and p-cresol in water containing different concentrations of dissolved oxygen. Biochem. J., 27, 983-985.

Stark, G. T. C. (1967). An automatic dosing apparatus made with standard laboratory ware. Lab. Pract., 16, 594-595.

OTHER PAPERS PRODUCED BY WPRL/WRC FISH SECTION
(Designated by Reprint Numbers)

568. Abram, F. S. H., and R. A. Dines (1968). A pneumatic thermostat. Lab. Pract., 17, 1355.

608. Abram, F. S. H. (1970). Automatic water mixing device. Lab. Pract., 19, 915-916.

623. Abram, F. S. H. (1971). An air-economizer apparatus for use in fish work. J. Fish Biol., 3, 203-205.

715. Abram, F. S. H. (1973). Apparatus for control of poison concentration in toxicity studies with fish. Wat. Res., 7, 1875-1879.

367. Alabaster, J. S., and A. L. Downing (1958). The behaviour of roach (Rutilis rutilis L.) in temperature gradients in a large outdoor tank. Proc. Indo-Pacific Fish. Coun., 3, 49-71.

576. Alabaster, J. S. (1969). Evaluating risks of pesticides to fish. Proc. 5th British Insecticide and Fungicide Conference, 370-377.

581. Alabaster, J. S. (1970). River flow and upstream movement and catch of migratory salmonids. J. Fish Biol., 2, 1-13.

595. Alabaster, J. S. (1970). Testing the toxicity of effluents to fish. Chemy Ind., 759-764.

633. Alabaster, J. S., J. H. N. Garland, and I. C. Hart (1971). Fisheries, cooling-water discharges, and sewage and industrial wastes. CEGB Symposium on Freshwater Biology and Power Generation.

664. Alabaster, J. S. (1972). Suspended solids and fisheries. Proc. R. Soc. B, 180, 395-406.

769. Alabaster, J. S. (1976). The water quality aspects of water transfers. Chemy Ind., 138-142.

533. Ball, I. R. (1967). The toxicity of cadmium to rainbow trout (Salmo gairdnerii Richardson). Wat. Res., 1, 805-806.

527. Brown, V. M., D. H. M. Jordan, and B. A. Tiller (1967). The effect of temperature on the acute toxicity of phenol to rainbow trout in hard water. Wat. Res., 1, 587-594.

606. Brown, V. M., D. G. Shurben, and D. Shaw (1970). Studies on water quality and the absence of fish from some polluted English rivers. Wat. Res., 4, 363-382.
714. Brown, V. M. (1973). Concepts and outlook in testing the toxicity of substances to fish. In Bioassay Techniques and Environmental Chemistry. Ann Arbor Science Publishers, Inc., Ann Arbor, Michigan. pp.73-95.
741. Brown, V. M., T. L. Shaw, and D. G. Shurben (1974). Aspects of water quality and the toxicity of copper to rainbow trout. Wat. Res., 8, 797-803.
1001. Brown, V. M. (1976). Advances in testing the toxicity of substances to fish. Chemy Ind., 143-149.
285. Herbert, D. W. M. (1955). Measuring the toxicity of effluents to fish. Analyst, Lond., 80, 896-898.
271. Herbert, D. W. M., and K. M. Downing (1955). A further study of the toxicity of potassium cyanide to rainbow trout (Salmo gairdnerii Richardson). Ann. appl. Biol., 43, 237-242.
323. Herbert, D. W. M., and H. T. Mann (1958). The tolerance of some freshwater fish for sea water. Salm. Trout Mag. No. 153, 99-101.
396. Herbert, D. W. M., J. S. Alabaster, M. C. Dart, and R. Lloyd (1961). The effect of china-clay wastes on trout streams. Int. J. Air Wat. Pollut., 5, 56-74.
397. Herbert, D. W. M., and A. C. Wakeford (1962). The effect of calcium sulphate on the survival of rainbow trout. Wat. Waste Treat., 8, 608-609.
398. Herbert, D. W. M. (1961). Freshwater fisheries and pollution control. Proc. Soc. Wat. Treat. Exam., 10, 135-161.
427. Herbert, D. W. M., and J. M. Richards (1963). The growth and survival of fish in some suspensions of solids of industrial origin. Int. J. Air Wat. Pollut., 7, 297-302.
441. Herbert, D. W. M., and A. C. Wakeford (1964). The susceptibility of salmonid fish to poisons under estuarine conditions. I. Zinc sulphate. Int. J. Air Wat. Pollut., 8, 251-256.
459. Herbert, D. W. M., and D. S. Shurben (1964). The toxicity of fluoride to rainbow trout. Wat. Waste Treat., 10, 141-142.
466. Herbert, D. W. M., and D. S. Shurben (1965). The susceptibility of salmonid fish to poisons under estuarine conditions. II. Ammonium chloride. Int. J. Air Wat. Pollut., 9, 89-91.
471. Herbert, D. W. M. (1965). Pollution and fisheries. In G. T. Goodman, R. W. Edwards, and J. M. Lambert (Ed.), Ecology and the Industrial Society. British Ecological Society Symposium No. 5. Blackwell Scientific Publications, Oxford. pp.173-195.
481. Herbert, D. W. M., D. H. M. Jordan, and R. Lloyd (1965). A study of some fishless rivers in the industrial midlands. J. Proc. Inst. Sew. Purif., 569-582.
571. Jackson, S., and V. M. Brown (1970). Effect of toxic wastes on treatment processes and watercourses. Wat. Pollut. Control, 69, 292-313.
449. Jordan, D. H. M., and R. Lloyd (1964). The resistance of rainbow trout (Salmo gairdnerii Richardson) and roach (Rutilus rutilus L.) to alkaline solutions. Int. J. Air Wat. Pollut., 8, 405-409.
359. Lloyd, R., and D. W. M. Herbert (1960). The influence of carbon dioxide on the toxicity of un-ionized ammonia to rainbow trout (Salmo gairdnerii Richardson). Ann. appl. Biol., 48, 399-404.
411. Lloyd, R., and D. W. M. Herbert (1962). The effect of the environment on the toxicity of poisons to fish. Instn. publ. Hlth. Engrs. J., 61, 132-145.
448. Lloyd, R., and D. H. M. Jordan (1964). Some factors affecting the resistance of rainbow trout (Salmo gairdnerii Richardson) to acid waters. Int. J. Air Wat. Pollut., 8, 393-403.
546. Mitrovic, V. V., V. M. Brown, D. G. Shurben, and M. H. Berryman (1968). Some pathological effects of sub-acute and acute poisoning of rainbow trout by phenol in hard water. Wat. Res., 2, 249-254.

727. Shaw, T. L., and V. M. Brown (1974). The toxicity of some forms of copper to rainbow trout. Wat. Res., 8, 377-382.

ACKNOWLEDGEMENT

This paper is published by permission of the Director, Water Research Centre.

# Zooplankton Ecology and Pollution Studies

**M. P. Tevlin* and Mary J. Burgis****

**Department of Applied Biology, Polytechnic of the South Bank, London S.E.1, England*
***Department of Biological Sciences, City of London Polytechnic, London E.1, England*

ABSTRACT

Zooplankton organisms occupy a central position in the food webs of aquatic ecosystems. This, plus their sensitivity to environmental factors, both natural and artificial, makes the Cladocera, particularly species of Daphnia, of considerable significance to pollution studies. They are especially useful in the detection of low levels of toxins and this is illustrated by a brief outline of the historical development of pollution studies using biological indicators.

Since the effects of pollutants on these organisms can only be assessed when the effects of natural environmental variation are taken into account, the complexities of natural population responses to environmental factors are reviewed. Pollution effects are discussed with reference to studies involving natural communities and laboratory assay procedures.

KEYWORDS

Zooplankton; Pollution; Daphnia;

## INTRODUCTION

The principal components of most zooplankton communities belong to four main taxonomic groups (the Rotifera, the Cladocera and the cyclopoid and calanoid Copepoda) some of whose most important characteristics, as far as their ecology is concerned, are listed in Table 1. Many species of all these groups are found in the littoral zones of lakes but relatively few are abundant in the true plankton. Among those which are abundant are Cladocera such as Daphnia hyalina and D. pulex, calanoids such as Diaptomus gracilis and cyclopoids such as Cyclops strenuus and Mesocyclops leuckarti. Here they occupy an intermediate position in the food web of the pelagic zone in that many of them feed largely on algae and bacteria, and in turn fall prey to numerous invertebrate and fish predators. This is particularly true of daphnids which are a very significant component of many aquatic ecosystems and on which a great deal of work has been done, both in the laboratory and in natural populations. Much of the following discussion will concern *Daphnia* spp. particularly but most of the general principles apply to all the filter feeders and to other components of the zooplankton also.

TABLE 1 Some Characteristics of Common Zooplankton Organisms

| | HERBIVORES | SIZE RANGE (approx.) | FEEDING METHOD | CARNIVORES (size) |
|---|---|---|---|---|
| | ROTIFERA eg. *Keratella*<br>*Brachionus* | 0.005-1.0mm | small particles<br>centrifugal precipitation | eg. *Asplanchna*<br>(*c* 2.0mm) |
| CRUSTACEA | CLADOCERA eg. *Daphnia*<br>*Bosmina* | 0.3-5mm | filtration with limb setae | eg. *Leptodora*<br>(10mm) |
| CRUSTACEA | COPEPODA Calanoida eg. *Diaptomus* | 0.5-2mm | filtration with mouth-part setae | |
| CRUSTACEA | COPEPODA Cyclopoida eg. *Cyclops* | 0.5-2.5mm | raptorial (grasping) feeders, able to tackle many types of food including those larger than themselves. Small forms tend to be herbivores and larger ones carnivores. | |

The limbs of Cladocera create a current of water through the carapace to serve for feeding and for respiration. This brings the body surface into intimate contact with large volumes of the surrounding medium which makes them particularly sensitive to changes in environmental conditions. They therefore serve as early indicators of pollution. Species of *Daphnia* have been extensively used for toxicity testing (eg. Cabejezek & Stasiak, 1960; Cabejezek *et al*, 1964; Malacea & Gruia, 1965; Crosby & Tucker, 1966; Frear & Boyd, 1967; Warnick & Bell, 1969; Biesinger & Christensen, 1972; Sandholm *et al*, 1973; Nebeker & Puglisi, 1974; Makrushin, 1976; Andrew *et al*, 1977; Terhaar *et al*, 1977; Tevlin, 1978) and they are particularly useful as target organisms in chronic sublethal bioassays.

It is because of this prime role in the study of chronic sublethal effects of toxins that relatively little work has been done on the direct effects of pollution on natural zooplankton populations. By the time toxic effects become manifest at population level the need for a sensitive early warning has passed, and the problem is then to remedy the effects of pollution rather than prevent them. Pollution studies have, therefore, become increasingly concerned with the early detection of low levels of toxins. To achieve this, indirect effects of natural variation in environmental factors, including both those characteristic of the water column (eg. Sprague, 1968; Ramamoothy & Kushner, 1975; Andrew et al, 1977; Zitko & Carson, 1976) and of the sediment (Pita & Hyne, 1975), on the toxin/test organism interaction need to be eliminated.

In order to separate the direct effects of pollution on the test organism from these indirect effects a large proportion of pollution investigations are necessarily laboratory studies involving groups of individual organisms, or microcosm studies designed to mimic natural ecosystems. This in no way invalidates the data produced, since there is no doubt that laboratory experiments form an indispensible part of ecological research (Salt, 1971). It does, however, make it more important than ever to ensure that the laboratory data can be readily related to natural populations.

To appreciate fully how important a contribution is made to such studies by work involving zooplankton ecology it is necessary to briefly outline the historical development of toxicity testing using biological indicators.

## TOXICITY TESTING AND BIOLOGICAL INDICATORS

The use of freshwater flora and fauna as indices of pollution has long been an invaluable complement to physico-chemical analyses since the detection of toxic agents in the water is of little ecological significance *per se*. The pertinence of such toxins lies in their effects on the biota, and bioassays have become increasingly important as more stringent control of water quality has become necessary (Little, 1977).

In the past bioassays were predominantly concerned with the acute effects of fairly high concentrations of toxins which resulted in heavy mortality of the test organisms. Such data yielded little information on the causes of death. More recently some authors have recommended that acute bioassays be extended to include a post-exposure recovery period for better evaluation of the impact of the toxin and perhaps to provide an insight into its mode of action (Hansen & Kawatski, 1976; Wright, 1976). The need for chronic sublethal assays became apparent as more and more toxic agents were shown to have important sublethal effects (Weber, 1973). The ability to not only detect toxicity at an early stage but also to investigate how the toxin exerts its effect, makes chronic sublethal data of greatest relevance to those concerned with the conservation of aquatic resources.

The choice of test organism is particularly important in chronic sublethal assays. It should be ecologically important, exhibit a prompt response to poisoning, lend itself readily to experimental manipulation and have a short life-span so that the effects of toxicity throughout its life-cycle can be detected by the investigator. Chronic assays using fish (Brungs, 1969; McKim & Benoit, 1971; Drummond et al, 1973; Eaton, 1974; Sellers et al, 1975) involve fairly elaborate procedures and are therefore expensive in terms of manpower, time and resources (see Weber, 1973). Their high trophic status also makes fish species remote from toxic effects transmitted from the base of the food web. Herbivorous invertebrate species, such as Cladocera, on the other hand are generally easier to work with and tend to be more sensitive to poisoning (Schweiger, 1957; Malacea & Gruia, 1965) due to their structural and physiological simplicity and their intimate contact with the surrounding medium (see p20). Variation in their response to toxins due to genetic variation and physiological condition is less than in macroinvertebrates and fish (Macek & Saunders, 1970) and, because of their strategic position in the food web, minor changes in their populations can have far-reaching consequences for the ecosystem as a whole. It is therefore important to detect any alteration in these organisms before those alterations lead to changes at the population level. Species of Daphnia are particularly suitable. Their short egg to egg development time (ca. 6-10 days at 20°C) permits investigation of the response to poisoning of different aged individuals within a reasonable experimental time. This is important since the age of crustaceans often strongly affects their response, (Little, 1977). As a laboratory test organism there is plenty of available literature, including culture and experimental techniques (Banta, 1921; Dewey & Parker, 1964; Taub & Dollar, 1964; D'Agostino & Provasoli, 1970; Murphy, 1970; Harvey, 1972); and the parthenogenetic ovoviviparity of Daphnia species ensures an abundant supply of genetically similar test organisms which differ only in size with age (Harvey, 1972).

The choice of parameters for investigation is also most important: the need to ultimately, predict chronic effects within a short period of time from limited data (Little, 1977) requires the use of those parameters which show a response by the test organism to the pollutant. It is thus necessary to combine the benefits of chronic sublethal data with the short experimental time normally associated with acute toxicity assays, and the problem is to select those criteria which give early but reliable indications that the test organism is adversely affected. Flerov (1974) has shown that the initial signs of toxin-induced metabolic disorder are behavioural, but these are difficult to quantify. More commonly, alterations in ultrastructure at light and electron microscope level are used to indicate poisoning (Baker, 1969; Martin, 1971; Schultz, 1975). An alternative approach is to use visible and readily quantifiable parameters such as growth and egg production (Biesinger & Christensen, 1972; Shcherban', 1972; McKim & Benoit, 1974) in conjunction with measurement of the physiological processes such as feeding and respiration upon which growth and fecundity depend (Arthur & Leonard, 1970; Scott & Major, 1972). These are precisely the kind of data that zooplankton ecologists have in abundance for Daphnia spp. under non-polluted conditions (Rigler, 1961; Schindler, 1968; Burns, 1969; Cowgill & Burns, 1975; Lampert, 1975; Kersting & Leeuw-Leegwater, 1976) and from these and a knowledge of zooplankton population dynamics we can deduce likely effects of alterations due to pollution.

## EFFECTS OF NATURAL ENVIRONMENTAL FACTORS

Before discussing the effects of pollution on zooplankton ecology it is first necessary to consider the natural environmental factors which may affect their ecology under non-polluted conditions. The results of toxicity studies done on natural communities may then be more readily interpreted, and laboratory studies considered in relation to natural situations.

One of the most significant environmental factors in natural waters is temperature, which varies both seasonally and with depth. Not only do basic metabolic functions such as respiration, of Cladocera such as Daphnia, vary with temperature but also all aspects of their growth, and this affects the population dynamics of the species and their ability to exploit their food supplies. For example, the duration of egg development is strongly related to temperature and it is now possible to derive, from a great deal of available data, some very useful general relationships for the prediction of egg development time in different groups of species (Bottrell et al, 1976) as illustrated in Fig.1.

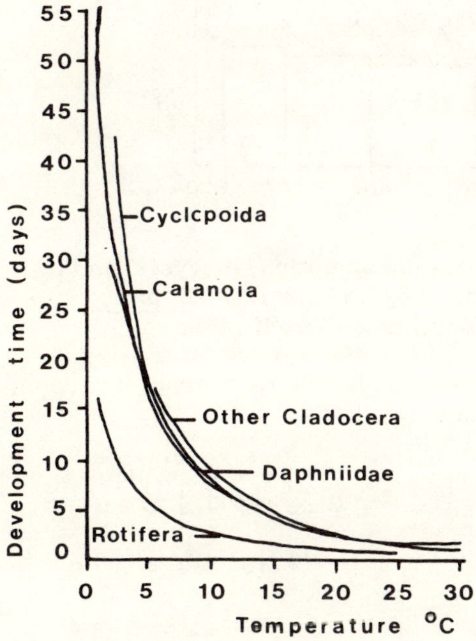


Fig.1. The duration of egg development for Daphindae, other Cladocera, Cyclopoida, Calanoida and Rotifera as predicted by regression equations derived from data accumulated from a variety of sources (from Bottrell et al 1976).

Once hatched the growth of Crustacea, including Cladocera, proceeds through a series of stepwise moults. A varying number of instars (which seems to be species specific - Bottrell, 1975) precedes maturity and, in Cladocera, at each moult thereafter a batch of eggs is produced. Copepoda cease to moult at maturity but lay successive batches of eggs. Figure 2 illustrates the first 600 hours of life of one Daphnia to show its growth in length, the increase in oxygen consumption, the number of moults and the number of young produced.

Thus, in Cladocera, the length of the juvenile instars and the interval between adult instars (both of which are strongly temperature dependant) can radically affect the number of eggs produced (which is also related to body length and food supply for any one batch) in a lifetime, and, therefore, the production and the age structure of the population.

Figure 3 shows the seasonal variation in size structure of a daphnid population in a reservoir. Three species of Daphnia were present in the year illustrated but temperature is only one of the factors determining this pattern.

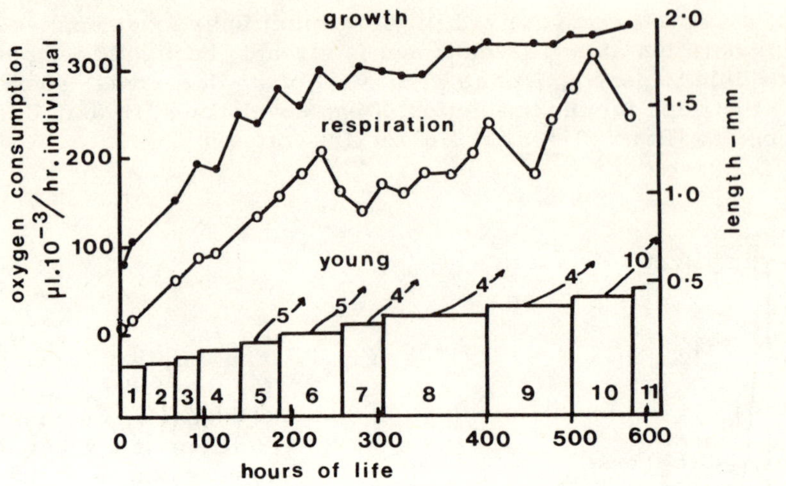


Fig.2. The body length, oxygen consumption, ecdysis and number of eggs produced by an individual *Daphnia hyalina* (number 11c) cultured at 20°C, fed on *Oocystis solitaria* and observed at daily intervals throughout its life span of 600h (redrawn from Duncan, Cremer & Andrew, 1970).

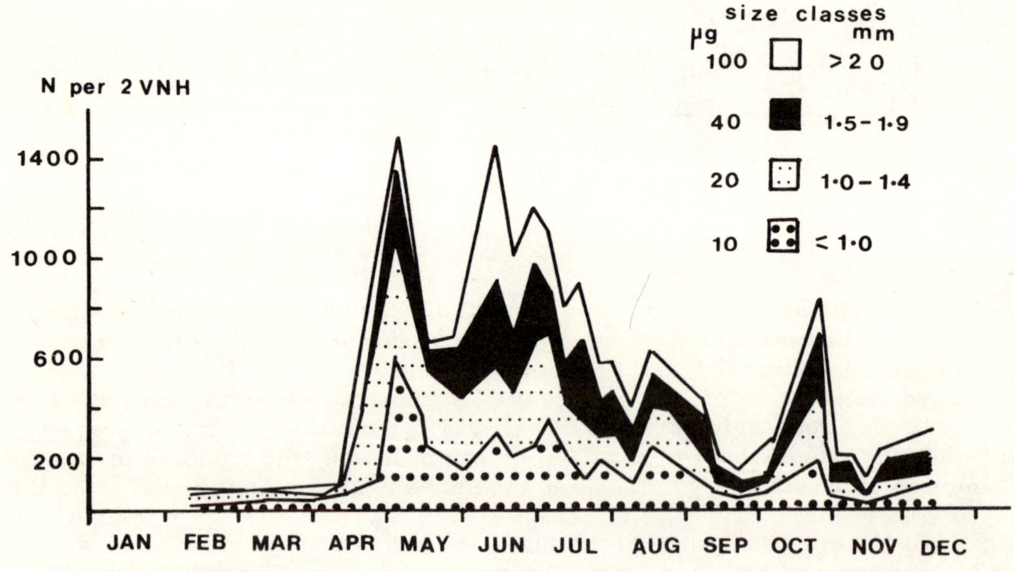


Fig.3. Seasonal variation in the size frequency distribution of the daphnid population in Queen Elizabeth II reservoir during 1972. VNH = vertical net haul. (from Duncan, 1975).

The size structure of a population may also be affected by food since the growth rate of individuals is, naturally, influenced by the amount of food available to them.

As shown by the experiments of Vijverberg (1976) (Fig.4) natural populations are not always growing at their maximum potential; presumably due to a scarcity of food in some natural lake waters.

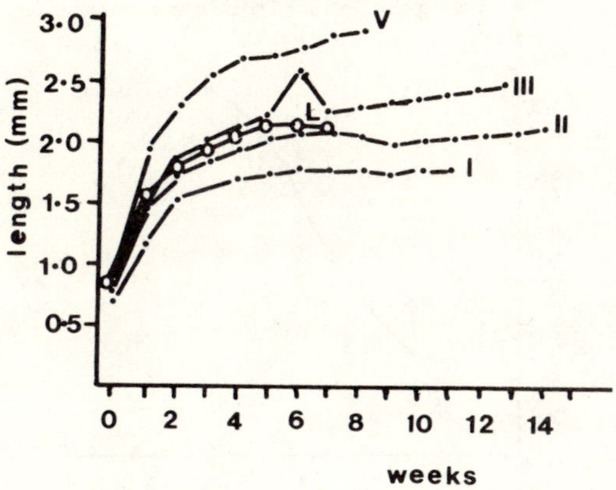


Fig.4. The effect of food regime on linear growth of Daphnia hyalina (redrawn from Vijverberg 1976). L is unfiltered lake water; I & II lake water filtered through 0.6 and 3.4 µm filters respectively; III & V are filtered lake water plus increasing concentrations of Scenedesmus added.

The size distribution and/or species composition of the zooplankton population may also be affected by predation. There is now ample evidence (eg Galbraith, 1967; Brooks, 1968) that planktivorous fish tend to prey selectively on the largest and most conspicuous components of the zooplankton, frequently the daphnids, and that when such fish are present the average size of the Daphnia is lower than when the fish are absent (eg. Duncan, 1975). Under very heavy predation the genus may disappear altogether and the balance of species in the whole zooplankton community changes. Invertebrate predators, such as Leptodora and some cyclopoid copepods tend to predate the smaller individuals of a population.

The resultant range of daphnid body sizes present in a population at any one time has a profound effect on the extent of their exploitation of the food supply. The filtration and feeding rate of individuals varies with body length as well as with temperature as demonstrated by the work of Burns and Rigler (1067) in Fig.5 and the maximum size of particles which can be ingested also varies with body size (Burns, 1968). Some examples of algal cell sizes ingested by Daphnia from a natural population are shown by the work of Nadin-Hurley and Duncan (1976) in Fig.7. In 68% of the Daphnia examined the maximum particle size observed was 20 µm in diametre although much larger algae were also present in the population.

Since even the largest Daphnia seem to be able to retain very small cells as well as a wide range of larger particles, which are unavailable to their smaller relatives, we are led from the preceding brief survey to the conclusion that in a natural habitat with plenty of edible particles (bacteria, algae and organic detritus) and no planktivorous fish, the individuals of the daphnid populations will grow large and exert a heavy grazing pressure on the algal populations. But this is only one extreme of a range of situations actually observed in natural planktonic communities (Zaret 1978). The complex interrelationships, which determine the wide variety of community structures observed, have been much discussed following the formulation of the "size efficiency hypothesis" of Brooks and

Dodson (1965).

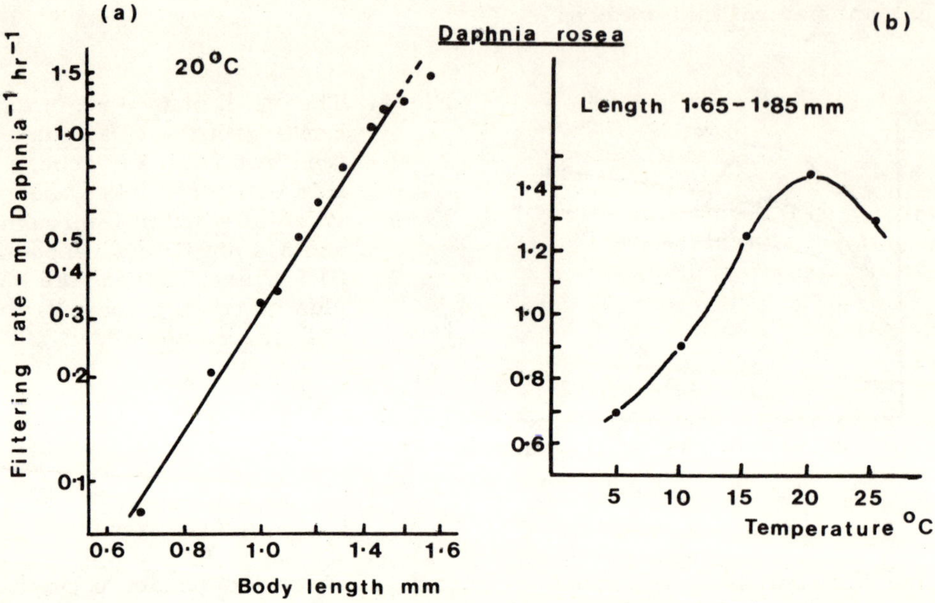


Fig.5. (a) The relationship between $\log_{10}$ filtering rate and $\log_{10}$ body length of Daphnia rosea at 20°C. (b) Relationship between filtering rate of Daphnia rosea adults and water temperature. In both cases feeding was carried out in membrane-filtered tap water containing 0.25 x $10^5$ cells/ml of Rhodotorula glutinis (from Burns and Rigler 1967).

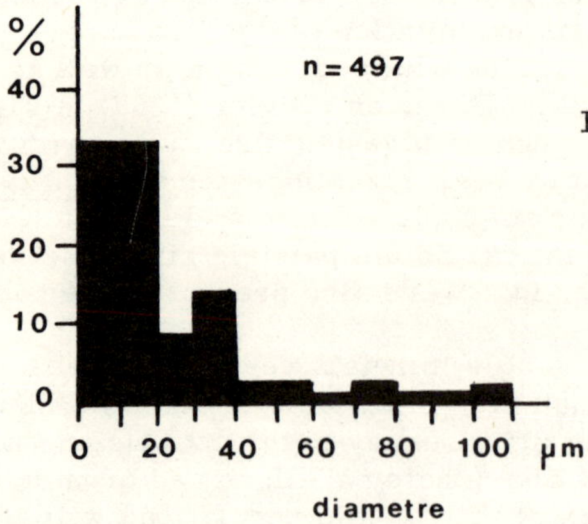


Fig.6. The percentage distribution of maximum particle sizes (width) in the guts of 497 Daphnia in the Queen Elizabeth II reservoir. (drawn from data given by Nadin-Hurley & Duncan 1976).

## POLLUTION EFFECTS

What then are the likely effects of pollution on these finely balanced interrelationships in the plankton?

Obviously any thermal pollution, such as that created below water-cooled industrial installations, would have a direct, and possibly lethal, effect. Temperature is an atypical pollutant insofar as it acts as a controlling factor without a threshold (Brett, 1969), and is of great significance in an environment where the fauna is almost entirely ectothermic. Under normal circumstances an increase in temperature generally increases the metabolic rate of such organisms, and it must be borne in mind that the rate of increase in temperature is as important as the extent of the increase, since most animals affected have limited powers of acclimation.

Temperature variation also affects the dissolved oxygen concentration, and the response of aquatic organisms to chemical toxins (Cairns, Heath and Parker,1975). Water at saturation holds ca. 13.2 mg. $l^{-1}$ oxygen at 4$^{\circ}$C but only 7.7 mg.$l^{-1}$ at 30$^{\circ}$C. Most aquatic animals are able to cope with the decreased availability of dissolved oxygen (DO) in the absence of additional stress but the concurrence of chemical pollution may have serious consequences. Shcherban' (1977) has shown that at temperatures of 10, 15, 25 and 30$^{\circ}$C the toxicity of Cu, Zn, Cd, Ni, and Mn sulphates to _Daphnia magna_ increased with increased temperature. Within the range 25-30$^{\circ}$C Cd toxicity increased ca. 4 fold, Cu and Zn ca. 2 fold, and Ni and Mn were a little more toxic at the higher temperature. Oseid & Smith (1975) showed that the combination of increased temperature with reduced DO increased toxicity of $H_2S$ to _Gammarus_, _Ephemera_ and _Hexagenia_. In fact the effects of dissolved oxygen and temperature are so manifold that they lend themselves more readily to modelling techniques than to biological monitoring as such (see Neal & Herwig, 1977).

The addition of nutrients to an unpolluted water body may affect not only the quantity of algae but also the species composition of the phytoplankton. The extent and timing of these changes may determine whether or not grazing by zooplankton exerts a controlling influence on the phytoplankton populations. If the process of eutrophication leads to prolonged dominance of the phytoplankton by large, colonial blue-greens these may well be unsuitable and unavailable as food for the Cladocera. The work of Schindler (1971) has shown that _Daphnia_ are able to digest only a small percentage of a blue-green species such as _Microcystis_ although they make better use of _Anabaena_. _Diaptomus_ is better able to assimilate both these species. These differences no doubt result from different types of feeding mechanisms and the differing structures of the algae. There is also evidence that algae such as _Microcystis_ are sometimes toxic to _Daphnia_ and other organisms (Maloney & Carnes, 1966; Stangenberg, 1968). Less direct effects of eutrophication may result from periods of reduced oxygen concentration in the water column, perhaps exacerbated by seasonally or artificially elevated temperatures.

With more clearly defined pollutants such as heavy metals, herbicides and insecticides the intimate contact of the organism with the surrounding medium is of vital importance. Where absorbtion of the toxin takes place through the body surface, uptake is potentially greatest during moulting. With heavy metals however there is evidence that a high proportion of the toxic ions are adsorbed directly onto the exoskeleton and do not necessarily enter the soft body tissue, unless they are first released into the medium and subsequently absorbed by the soft tissue (see Bryan, 1971). Tevlin (1978) showed that ca. 50% of the ion Cd taken up by *Daphnia magna* was taken up onto the carapace and Fowler and Small (1967) have shown that the largest fraction of $^{65}Zn$ taken up by euphausids was always associated with the exoskeleton. Furthermore, uptake by dead and live animals was statistically similar and therefore unlikely to be related to metabolism. In such situations the moulting process is important because deposition of moulted exoskeletal material promotes rapid transport of toxic ions to the sediment in a relatively harmless form (Martin, 1970).

The effects of particular pollutants on zooplankton organisms under laboratory assay conditions are adequately dealt with in several review papers (eg Bunting & Robertson, 1975; Herricks & Buikema, 1977; Leland *et al*, 1977; Little, 1977) and will not be individually discussed here. In summary of what emerges from such reviews, however, it is evident that most chronic toxicity studies have dealt largely with the distribution, accumulation and/or concentration of toxins within the test organisms, or with the translocation of toxins within the food chain by predation upon, or sedimentation of, all or part of those organisms (eg Nagy & Salanky, 1962; Brooks & Rumsby, 1965; Martin, 1970; Hiltibran, 1971; Bertine & Goldberg, 1972; Boothe & Knauer, 1972; Baudouin & Scoppa, 1974; Dean, 1974; Schultz, 1975; Schultz & Kennedy, 1976; Enk & Mathis, 1977). These studies were generally based upon physico-chemical detection and/or quantification of the toxin within the organism rather than the effects of poisoning on the organism.

In looking for a response by zooplankton organisms to poisoning the most reliable parameters seem to be growth and egg production (Canton *et al*, 1975; Hessov, 1975; Winner, 1976; Tevlin, 1978) because sublethal alterations in feeding, respiration etc. result in gross changes in growth and fecundity. For example, chronic sublethal Cd poisoning of *Daphnia magna* has been shown (Tevlin 1978a) to cause a significant decrease in filtration rate (P =<0.001) and a significant increase in respiration rate (P =<0.1), as a result of which there is insufficient energy to maintain growth and egg production at the level of unpoisoned animals (Fig. 7). Growth was therefore significantly reduced (P =<0.02), and both the percentage of females reaching sexual maturity and their mean egg numbers were reduced almost to zero (P =<0.001).

Most of the reported pollution studies on invertebrates in freshwater microcosms involve Cladocera, but deal simply with accumulation and biodegredation of the pollutants (Metcalf *et al*, 1973; Kanazawa, 1975; Yu & Sanborn, 1975; Yu *et al*, 1975). Where such studies have been used to evaluate environmental effects (Metcalf *et al*, 1973; Didia *et al*, 1975) the Cladocera and Diptera are shown to be the most sensitive test organisms, with other non-target organisms showing little or no mortality. *Daphnia* spp. in particular have been shown to be very sensitive eg. Baudouin & Scoppa (1974) investigated the toxicity of 12 metal ions to *Cyclops* sp., *Eudiaptomus* sp., and *Daphnia hyalina*, and for every metal ion *Daphnia* was the most sensitive.

The ecological impact of acute poisoning of the Cladocera in particular and zooplankton in general is not difficult to predict: the release of grazing pressure, at least until a more tolerant species of grazer takes over or the ecosystem adjusts in some other way; the disruption of the food web; and the collapse of fish populations at the top of the web, even if such collapse was only temporary.

The analogous impact of chronic sublethal poisoning on the zooplankton would be more insidious and possibly more far-reaching. Any reduction in body size of mature Cladocera and in recruitment of new-born to the population would result in a decrease in food availability for predator species. Even without accumulation and concentration of toxins in zooplankton organisms through prolonged exposure there seems little doubt that ingestion of large numbers of poisoned individuals could induce acute toxicity in predators. In this respect the kind of predator is important. Large predators such as fish, which ingest large numbers of whole zooplankters may rapidly show signs of poisoning; whereas smaller predators such as raptorial copepods and insect larvae, which ingest only small portions of the soft body may be slow to show such signs. In the latter case predation could promote rapid transport of the toxic agent within the prey to the sediment since small predators tend to kill the prey even though only part of it is ingested. Phytoplankton grazing would be reduced since poisoned organisms tend to feed less voraciously. The net result could well be to completely alter the species composition of the ecosystem without actually causing its collapse, and the gradual demise of the original, natural biota unless the cause of pollution was removed.

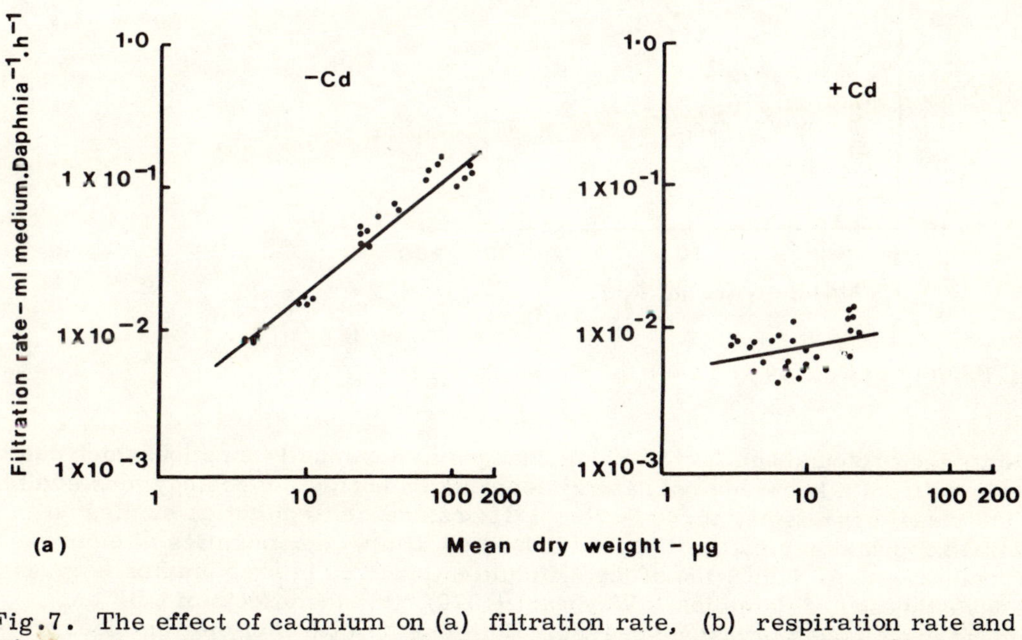


Fig.7. The effect of cadmium on (a) filtration rate, (b) respiration rate and (c) growth of Daphnia magna. Daphnia were reared with 0.1 ppm cadmium and in (a) fed on Chlorella grown with 0.1 ppm cadmium. Controls were reared without cadmium and fed on Chlorella grown without cadmium.

(Fig.7b and c - see over)

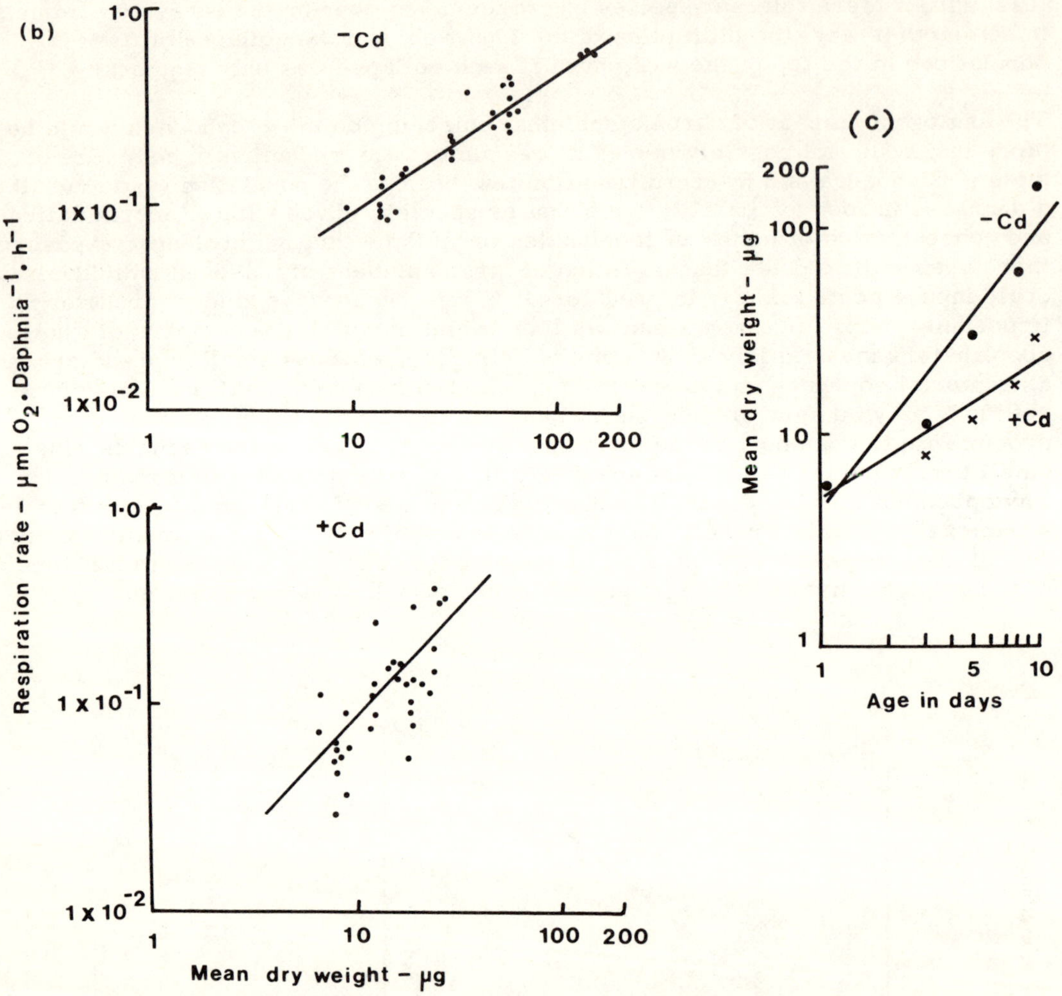


Fig.7b and c. (Details see over)

Despite the profound and far-reaching changes in community function which can be predicted from a knowledge of natural zooplankton population ecology plus studies on individual organisms, there is very little reference to pollution studies on natural zooplankton communities, or indeed on natural communities of other invertebrates. An indication of the difficulties involved in such studies is given by the investigation of Hamelink & Waybrant (1976), into the effects of DDE and Lindane in a flooded limestone quarry. Epilimnetic waters of the 15m deep, thermally stratified quarry were treated with these compounds and sedimented natural bottom mud, quarry walls, zooplankton and fish were sampled for estimation of toxin content: it was found that the behaviour of the toxins was apparently dictated by their physical properties and those of the system rather than persistence in the food chain. In other words test organisms cannot be isolated from the physico-chemical regime of their environment, and the results of such studies may often be misleading. In the particular case of a limestone quarry this is hardly surprising since lime is known to affect the toxicity of heavy metals by removing the metal ions from solution (Netzer et al, 1973); and the same is true of calcium carbonate which governs water hardness (Zitko & Carson, 1976; Hautala et al, 1977). Reactions such as chelation, adsorbtion and precipitation most frequently affect metal ion toxicity. Tevlin (1978b) has shown that the chelating effects of two amine buffers (HEPES and TES) and two soluble iron complexes (Fe-EDTA and Fe-gluconate) effectively reduce the toxicity of cadmium to Daphnia magna (Fig.8). However, chelating agents themselves (which, for example, frequently occur as components of modern detergents) may be toxic also (Chau & Shiomi, 1972; Childs, 1971). Nitrilotriacetic Acid (NTA, which was the most eligible replacement for the phosphate in detergents which were known to cause eutrophication problems) has been shown to have considerable complexing properties and to be toxic to freshwater invertebrates. Any chance of it reaching human populations should be avoided because of the risk of embryonic abnormalities (Thom 1971).

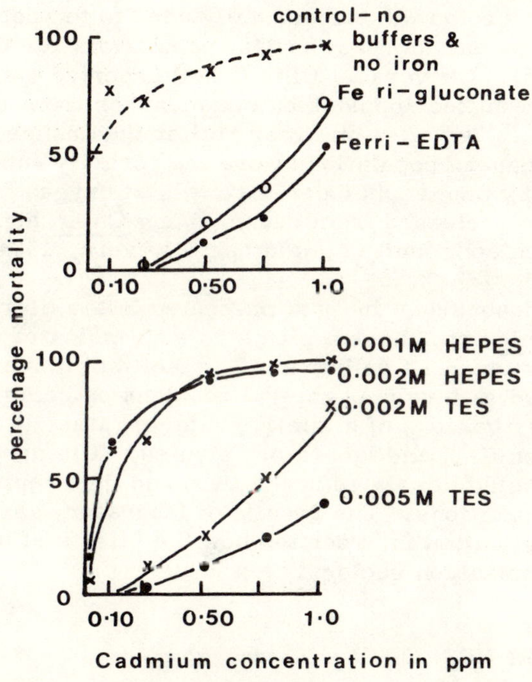


Fig.8. Forty-eight hour acute toxicity curves for Daphnia magna with Cadmium, in media containing different iron sources and different pH buffers: expressed as percentage mortality vs cadmium concentration (from Tevlin 1978b).

The known ability of natural waters to form less harmful chelates with chemicals added to them has led to specific studies to determine the apparent complexing capacity of lake waters (Chau, 1973; Gächter, Lum-Shue-Chan and Chau, 1974). Such chelating properties may work in two opposite ways: either they can remove toxic ions from free solution in the water column and form a less toxic chelate (Dwyer & Mellor, 1964); or they may mobilise toxic agents from the sediment where they had previously been safely 'bound' and thereby increase the level of free toxic ions in the water column. Even intrinsic components of the system such as moulted invertebrate exoskeletal material and bacteria and algae can affect the toxicity of pollutants by accelerating their deposition to the sediment (Martin, 1970; Winner et al, 1977).

Notwithstanding the problems involved in community studies of pollution Apostol (1975) evaluated the toxicity of ethanol-amine to several aquatic invertebrates, and concluded that Amoeba sp., Paramecium caudatum, Bodo sp. and Daphnia magna were the most sensitive organisms. During attempts to suppress mosquitos Mulla et al, (1975) showed that the synthetic pyrethroid FMC-33297 also suppressed non-target copepod populations.

The insecticide trichlorofon was successfully used to temporarily suppress crustacean populations and increase rotifer populations for the rearing of carp larvae (Tamas, 1975). Conversely Ellis (1976) reported simazine treatment for control of vegetation shifted zooplankton communities from rotifer dominance to crustacean dominance, but after one year rotifer dominance was re-established. Both rotifer and cladocera populations were temporarily suppressed in experimental ponds by 3mg copper per litre: whilst oxygen depletion, hydrogen production and copper released from decomposing Chara apparently suppressed the whole zooplankton community, (McIntosh & Kevern, 1974).

There seems little doubt from the data presented in the literature that Cladocera, and especially Daphnia spp., have a prime role as indicator organisms to provide early warning of low levels of pollution. The fact that there are severe limitations to pollution studies involving natural zooplankton communities should not preclude future investigations of a similar nature; rather it should promote closer liason between field and laboratory studies. The problems due to environmental variation should be isolated in the field and then reproduced under controlled laboratory conditions. The results of laboratory assays may then be related to the field situation in order to show the effects of particular toxins in terms of natural zooplankton ecology.

## REFERENCES

Andrew, R.W., Biesinger, K.E., and Glass, G.E. (1977). Effects of inorganic complexing on the toxicity of copper to Daphnia magna. Water Res., 11: 309-316.

Apostol, S. (1975). Ethanolamine toxicity to aquatic invertebrates. Stud. Cercet. Biol., 27: 345-351.

Arthur, J.W. and Leonard, E.N. (1970). Effects of copper on Gammarus pseudolimnaeus, Physa integra and Campeloma decisium in soft water. J. Fish. Res. Bd Can., 27: 1277-83.

Baker, J.T.P. (1969). Histological and electron microscopical observations on copper poisoning in the winter flounder, Pseudopleuronectes americanus. J. Fish. Res. Bd Can., 26: 2785-93.

Banta, A.M. (1921). A convenient culture medium for daphnids. Science, N.Y., 53: 557.

Baudouin, M.F. and Scoppa, P. (1974). Acute toxicity of various metals to freshwater zooplankton. Bull. environ. Contam. Toxicol., 12: 745-751.

Bertine, K.K. and Goldberg, E.D. (1972). Trace elements in clams, mussels, and shrimp. Limnol. Oceanogr., 17: 877-884.

Biesinger, K.E. and Christensen, G.M. (1972). Effects of various metals on survival, growth, reproduction and metabolism of Daphnia magna. J. Fish. Res. Bd Can., 29: 1691-1700.

Boothe, P.N. and Knauer, G.A. (1972). The possible importance of faecal material in the biological amplification of trace and heavy metals. Limnol. Oceanogr., 17: 270-274.

Bottrell, H.H. (1975). Generation time, length of life, Instar duration & Frequency of moulting, and their relationship to temperature in eight species of Cladocera from the River Thames, Reading. Oecologia (Berl.), 19: 129-140.

Bottrell, H.H., Duncan, A., Gliwicz, Z.M., Grygierek, E., Herzig, A., Hillbricht-Ilkowska, A., Kurasawa, H., Larsson, P. and Weglenska, T. (1976). A review of some problems in zooplankton production studies. Norw. J. Zool., 24: 419-456.

Brett, J.R. (1969). Résumé: Temperature and fish. Chesapeake Science, 10: 275.

Brooks, J.L. (1968). The effects of prey size selection by lake planktivores. Syst. Zool., 17: 273-291.

Brooks, J.L. and Dodson, S.I. (1965). Predation, body size & composition of plankton. Science, N.Y., 150: 28-35.

Brooks, R.R. and Rumsby, M.G. (1965). The biogeochemistry of trace element uptake by some New Zealand bivalves. Limnol. Oceanogr., 10: 521-527.

Brungs, W.A. (1969). Chronic toxicity of zinc to the fathead minnow, Pimephales promelas Rafinesque. Trans. Am. Fish. Soc., 98: 272-279.

Bryan, G.W. (1971). The effects of heavy metals (other than mercury) on marine and estuarine organisms. Proc. R. Soc. B., 177: 389-410.

Bunting, D.L. and Robertson, E.B. (1975). Lethal and sub-lethal effects of herbicides on zooplankton species. Water Resources Research Centre, University of Tennessee-Knoxville: Research Report 43.

Burns, C.W. (1968). The relationship between body size of filter feeding Cladocera and the maximum size of particle ingested. Limnol. Oceanogr., 13: 675-678.

Burns, C.W. (1969). Relationship between filtering rate, temperature and body size in four species of Daphnia. Limnol. Oceanogr., 14: 693-700.

Burns, C.W. and Rigler, F.H. (1967). Comparisons of filtering rates of Daphnia rosea in lake water & in suspensionsof yeast. Limnol. Oceanogr., 12: 492-502.

Cabejsek, I., Rybak, J.I. and Stanislawska, J. (1964). The effect of Aldrin on aquatic organisms. Roczn. panst. Zakl. Hig., 15: 495-501. (In: Wat. Poll. Absts. 39 (1966): 1405).

Cabejsek, I. and Stasiak, M. (1960). Investigations on the influence of some metals on aquatic life. Roczn Zakl. Hig. Warsz, 11: 303-312.

Cairns, J.Jr., Heath, A.G. and Parker, B.C. (1975). Temperature influence on chemical toxicity to aquatic organisms. J. Water Pollut. Control Fed., 47: 267-280.

Canton, J.H., Greve, P.A., Slooff, W. and Van Esch, G.J. (1975). Toxicity, accumulation and elimination studies of α-Hexachlorocyclohexane (αHCH) with freshwater organisms of different trophic levels. Water Res., 9: 1163-9.

Chau, Y.K. (1973). Complexing capacity of natural water - its significance and measurement. J. Chromat. Sci., 11: 579-584.

Chau, Y.K. and Shiomi, M.T. (1972). Complexing properties of nitrilotriacetic acid in the lake environment. Water Air Soil Pollut., 1: 149-164.

Childs, C.W. (1971). Chemical equilibrium models for lake water which contains NTA and 'normal' lake water. Proc. 14th Conf. on Great Lakes Research, pp.198-210.

Cowgill, U.M. and Burns, C.W. (1975). Differences in chemical composition between two species of *Daphnia* and some freshwater algae cultured in the laboratory. Limnol. Oceanogr., 20: 1005-1011

Crosby, D.G. and Tucker, R.K. (1966). Toxicity of aquatic herbicides to *Daphnia magna*. Science, N.Y., 154: 289-291.

D'Agostino, A.S. and Provasoli, L. (1970). Dixenic culture of *Daphnia magna* Strauss. Biol. Bull. (Woods Hole), 139: 485-494.

Dean, J.M. (1974). The accumulation of $^{65}Zn$ and other radio-nuclides by tubificid worms. Hydrobiologia, 45: 33-38.

Dewey, J.E. and Parker, B.L. (1964). Mass rearing of *Daphnia magna* for insecticide bioassay. J. Econ. Ent., 57: 821-825.

Didia, V., LaSalle, R. and Liem, K. (1975). The effects of Abate 2G mosquito larvicide on selected non-target organisms collected from forested temporary pools. Mosq. News, 35: 227-228.

Drummond, R.A., Spoor, W.A. and Olson, G.F. (1973). Some short-term indicators of sublethal effects of copper on brook trout, *Salvelinus fontinalis*. J. Fish. Res. Bd Can., 30 698-701.

Duncan, A. (1975). The importance of zooplankton in the ecology of reservoirs. In Proceedings of a Symposium, 'The effects of Storage on Water Quality'. Water Research Centre, Medmenham, England.

Duncan, A., Cremer, G.A. and Andrew, T. (1970). The measurement of respiratory rates under field and laboratory conditions during an ecological study on zooplankton. Polsk. Arch. Hydrobiol., 17: 149-160.

Dwyer, F.P. and Mellor, D.P. (Eds.). (1964). Chelating agents and metal chelates. Academic Press, N.Y.

Eaton, J.G. (1974). Chronic cadmium toxicity to the Bluegill (*Lepomia machrochirus* Rafinesque). Trans. Am. Fish. Soc., 103: 729-735.

Ellis, J., Higgins, E., Pruss, S. and Norton, J. (1976). Management of aquatic vegetation with Simazine. Proc. Northeast Weed Sci. Soc., 30: 86-91.

Enk, M.D. and Mathis, B.J. (1977). Distribution of cadmium and lead in a stream ecosystem. Hydrobiologia, 52: 153-158.

Flerov, B.A. (1974). Use of animal behaviour studies in water toxicology. Gidrobiol. Zh., 10: 114-120.

Fowler, S.W. and Small, L.F. (1967). Moulting of *Euphausia pacifica* as a possible mechanism for vertical transport of Zinc-65 in the sea. Int.J. Oceanol. Limnol., 1: 237-245.

Frear, D.E.H. and Boyd, J.E. (1967). Use of Daphnia magna for the microbioassay of pesticides I. Development of standardized techniques for rearing Daphnia and preparation of dosage-mortality curves for pesticides. J. Econ. Ent., 60: 1228-1236.

Gächter, R., Lum-Shue-Chan, K. and Chau, Y.K. (1974). Complexing capacity of the nutrient medium and its relation to inhibition of algal photosynthesis by copper. Schweiz. Z. Hydrol., 35: 252-261.

Galbraith, M.G. Jr. (1967). Size-selective predation on Daphnia by rainbow trout & yellow perch. Trans. Am. Fish. Soc., 96: 1-10.

Hamelink, J.L. and Waybrant, R.C. (1976). DDE and Lindane in a large scale model lentic ecosystem. Trans. Am. Fish. Soc., 105: 124-134.

Hansen, C.R. Jr. and Kawatski, J.A. (1976). Application of 24 hour post exposure observation to acute toxicity studies with invertebrates. J. Fish. Res. Bd Can., 33: 1198-1201.

Harvey, A.M. (1972). A semi-continuous culture technique for Daphnia pulex. J. Appl. Ecol. 9: 831-834.

Hautala, E., Randall, J., Goodvan, A. and Waiss, A. (1977). Calcium carbonate in the removal of iron and lead from dilute waste water. Water Res., 11: 243-5.

Herricks, E.E. and Buikema, A.S. (1977). Effects of pollution on freshwater invertebrates (review). J. Wat. Pollut. Control Fed., 49: 1493-1506.

Hessov, I. (1975). Toxicity of 5-hydroxymethyl-furfural and furfural to Daphnia magna. Acta pharmac. tox., 37: 94-96.

Hiltibran, R.C. (1971). Effects of cadmium, zinc, manganese and calcium on oxygen and phosphate metabolism of bluegill liver mitochondria. J. Wat. Pollut. Control Fed., 43: 818-823.

Kanazawa, J., Isensee, A.R. and Kearney, P.C. (1975). Distribution of carbaryl and 3,5 - xylylmethylcarbamate in an aquatic model ecosystem. J. agric. Fd Chem., 23: 760-763.

Kersting, K. and Leeuw-Leegwater, C. van der (1976). Effect of food concentration on the respiration of Daphnia magna. Hydrobiologia, 49: 137-142.

Lampert, W. (1975). A tracer study of carbon turnover in Daphnia. Verh. internat. Verein. Limnol., 19: 2913-2921.

Leland, H.V., Luoma, S.N. and Wilkes, D.J. (1977). Heavy metals and related trace elements (pollution review). J. Wat. Pollut. Control Fed., 49: 1340-1369.

Little, L.W. (1977). Bioassays - procedures and results (review). J. Wat. Pollut. Control Fed., 49: 1244-1254.

Macek, K.J. and Saunders, H.O. (1970). Biological variation in the susceptibility of fish and aquatic invertebrates to DDT. Trans. Am. Fish. Soc., 99: 89-90

Makrushin, A.V. (1976). Cladocera (Crustacea) as indicators of water pollution. Gidrobiol. Zh., 12: 101-104.

Malacea, I. and Gruia, E. (1965). Contributions to the study of the toxic effects of heavy metals on some aquatic organisms. In Wat. Poll. Absts., 38: 1567.

Maloney, T.E. and Carnes, R.A. (1966). Toxicity of a Microcystis water bloom from an Ohio pond. The Ohio Journal of Science, 66: 514-517.

Martin, J.H. (1970). The possible transport of trace metals via moulted copepod exoskeletons. Limnol. Oceanogr., 15: 756-761.

Martin, S. (1971). An analysis of the histopathological effects of copper sulfate on the asiatic freshwater clam, Corbicula fluminea (Müller). Ph D Thesis

Univ.Washington in Diss. Absts., 32: 1642-1643.
McIntosh, A.W. and Kevern, N.R. (1974). Toxicity of copper to zooplankton. J. Environ. Qual., 3: 166-170.
McKim, J.M. and Benoit, D.A. (1971). The effects of long-term exposure to copper on survival, growth and reproduction of the brook trout (Salvelinus fontinalis). J. Fish. Res. Bd Can., 28: 655-662.
McKim, J.M. and Benoit, D.A. (1974). Duration of toxicity tests for establishing 'no effect' concentrations for copper with brook trout (Salvelinus fontinalis). J. Fish. Res. Bd Can., 31: 449-452.
Metcalf, R.L., Kapoor, I.P., Lu, P-Y., Schuth, C.K. and Sherman, P. (1973). Model ecosystem pesticides. Environmental Health Perspectives; Experimental Issue No.4: 35-44.
Mulla, M.S., Darwazeh, H.A. and Majori, G. (1975). Field efficacy of some promising mosquito larvicides and their effects on non-target organisms. Mosq. News, 35: 179-185.
Murphy, J.S. (1970). A general method for the monoxenic cultivation of Daphnidae. Biol. Bull. (Woods Hole), 139: 321-332.
Nadin-Hurley, C.M. & Duncan, A. (1976). A comparison of daphnid gut particles with the sestonic particles present in two Thames Valley reservoirs throughout 1970 & 1971. Freshwat. Biol., 6: 109-123.
Nagy, Z. and Salanky, J. (1962). Distribution of cadmium in the organs of freshwater mussels when periodic activity is changed by cadmium chloride. Acta. biol. hung., 13, Suppl. 5: 72-73.
Neal, L. and Herwig, R. (1977). Water Pollution: oxygen sag and stream purification. J. Wat. Pollut. Control Fed., 49: 1299-1302.
Nebeker, A.V. and Puglisi, F.A. (1974). Effect of polychlorinated biphenyls (PCB's) on the survival and reproduction of Daphnia, Gammarus and Tanytarsus. Trans. Am. Fish. Soc., 103: 722-728.
Netzer, A., Bowers, A. and Norman, J.D. (1973). Removal of heavy metals from wastewaters by addition of lime. Proc. 16th Conf. Gt.Lakes Res. pp. 260-265.
Oseid, D.M. and Smith, L.L. (1975). Long-term effects of hydrogen sulphide on Hexagenia limbata (Ephemeroptera). Environ. Eutomol., 4: 15-18.
Pita, F.W. and Hyne, N.J. (1975). The depositional environment of zinc, lead and cadmium in reservoir sediments. Water Res., 9: 701-706.
Ramamoorthy, S. and Kushner, D.J. (1975). Heavy metal binding components of river water. J. Fish. Res. Bd Can., 32: 1755-1766.
Rigler, F.H. (1961). The uptake and release of inorganic phosphorus by Daphnia magna Straus. Limnol. Oceanogr., 6: 165-174.
Salt, G.W. (1971). The role of laboratory experimentation in ecological research. Res. Div. Monogr. Va. Polytech. Inst. St. Univ. No.3: 87-100.
Sandholm, M., Oksanen, H.E. and Pesonen, L. (1973). Selenium uptake by aquatic organisms. Limnol. Oceanogr., 18: 469-499.
Shcherban', E.P. (1972). Effect of atrazine and diuron on the productivity of Cladocera (Experimental studies). Gidrobiol. Zh., 8: 71-76.
Shcherban', E.P. (1977). Toxicity of some heavy metal ions for Daphnia magna Straus depending on temperature. Gidrobiol. Zh., (Kiev), 13: 86-91.
Schindler, D.W. (1968). Feeding, assimilation and respiration rates of Daphnia magna under various environmental conditions, and their relation to production estimates. J. Anim. Ecol., 37: 369-385.

Schindler, J.E. (1971). Food quality and zooplankton nutrition. J. Anim. Ecol., 40: 589-595.

Schultz, T.W. (1975). Some aspects of the fine structure of Daphnia pulex and the cytotoxic effects of 3-Amino-1, 2, 4, Triazole. Ph.D. Thesis, Univ. of Tennessee, Knoxville.

Schultz, T.W. and Kennedy, J.R. (1976). Cytotoxic effects of the herbicide 3-amino - 1, 2, 4 triazole on Daphnia pulex (Crustacea : Cladocera). Biol. Bull. (Woods Hole), 151: 370-385.

Schweiger, G. (1957). The toxic action of heavy metal salts on fish and organisms on which fish feed. Arch. Fischer., 8: 54-78.

Scott, D.M. and Major, C.W. (1972). The effect of copper (II) on survival, respiration and heart rate in the common blue mussel, Mytilus edulis. Biol. Bull. (Woods Hole), 143: 679-688.

Sellers, C.M., Heath, A.G. and Bass, M.L. (1975). Effect of sublethal concentrations of copper and zinc on ventilatory activity, blood oxygen, and pH in Rainbow Trout (Salmo gairdneri). Water Res., 9: 401-408.

Sprague, J.B. (1968). Promising anti-pollutant: chelating agent NTA protects fish from copper and zinc. Nature, 220: 1345-1346.

Stangenberg, M. (1968). Toxic effects of Microcystis aeruginosa Kg. extracts on Daphnia longispina O.F.Müller and Eucypris virens Jurine. Hydrobiologia, 32: 81-87.

Tamas, G. (1975). Directed formation of Rotifer zooplankton in fry ponds. Gidrobiol. Zh., 11: 70-72.

Taub, F.B. and Dollar, A.M. (1964). A Chlorella-Daphnia food chain study: the design of a compatible chemically defined culture medium. Limnol. Oceanogr., 9: 61-74.

Terhaar, C.J., Ewell, W.S., Dziuba, S.P., White, W.W. and Murphy, P.J. (1977). A laboratory model for evaluating the behaviour of heavy metals in an aquatic environment. Water Res., 11: 101-110.

Tevlin, M.P. (1978a). The physiological effects of chronic sublethal cadmium poisoning on Daphnia magna and some ecological implications. Unpublished Ph D. Thesis, Council for National Academic Awards, U.K.

Tevlin, M.P. (1978b). An improved experimental medium for freshwater toxicity studies using Daphnia magna. Water Res., 12: 1017-1024.

Thom, N.S. (1971). Nitrilotriacetic acid: a literature survey. Water Res., 5: 391-399.

Vijverberg, J. (1976). The effect of food quantity & quality on the growth, birth-rate & longevity of Daphnia hyalina Leydig. Hydrobiologia, 51: 99-108.

Warnick, S.L. and Bell, H.L. (1969). The acute toxicity of some heavy metals to different species of aquatic insects. J. Wat. Pollut. Control Fed., 41: 280-284.

Weber, C.I. (1973). Biological monitoring of the aquatic environment. Spec. tech. Publs. Am. Soc. Test. Mater., 528: 46-60.

Winner, R.W. (1976). Toxicity of copper to daphnids in reconstituted and natural waters. Ecol. Res. Service U.S. Environmental Protection Agency, 600/3 - 76 - 051, Washington, D.C.

Winner, R.W., Keeling, T., Yeager, R. and Farrel, M.P. (1977). Effect of food type on the acute and chronic toxicity of copper to Daphnia magna. Freshwat. Biol., 7: 343-349.

Wright, A. (1976). The use of recovery as a criterion for toxicity. Bull. Environ. Contam. Toxicol., 15: 747-749.

Yu, C-C. and Sanborn, J.R. (1975). The fate of Parathion in a model ecosystem. Bull. Environ. Contam. Toxicol., 13: 543-550.

Yu, C-C., Hansen, D.J. and Booth, G.M. (1975). The fate of Dicamba in a model ecosystem. Bull. Environ. Contam. Toxicol., 13: 280-283.

Zaret, T. (1978). A predation model of zooplankton community structure. Verh. Internat. Verein. Limnol., 20:

Zitko, V. and Carson, W.G. (1976). A mechanism of the effects of water hardness on the lethality of heavy metals to fish. Chemosphere, 5: 299-303.

# Systems of Biological Indicators for Water Quality Assessment[1]

G. Persoone and N. De Pauw

*Laboratory for Biological Research in Aquatic Pollution, J. Plateaustraat 22, B-9000 Ghent/Belgium*

ABSTRACT

The paper reviews and comments on the major methods which have been worked out at the international level for the assessment of the quality of surface waters with the aid of biological-ecological criteria.
The endeavours of the Commission of the European Economic Communities aiming at the intercalibration of methods used in different countries, and the trends which emerged from recent exercises carried out in different hydrographic basins are outlined briefly.
Keywords : Biological indicators, water quality assessment methods.

## INTRODUCTION

In the first pages of every classic textbook on Ecology, one can read that the living part of ecosystems is composed of "organisms" which form "populations" and that these populations are grouped in distinct "biological communities".
From the functional point of view every organism belongs basically to one of the three categories : producers, consumers or reducers.
Looking at the relationships between free living aquatic organisms (Fig. 1) it is clear that there is a variety of categories of organisms which are potentially present in each aquatic ecosystem, and that these groups differ very much by their increasing levels of organization and morphological complexity.
Bearing in mind the complexiety of the biological part of the ecosystem, the two basic questions which constitute the backbone of this review can be formulated :
- Are there communities, populations or species of aquatic organisms which can serve as indices of pollution?
- Is it possible to evaluate the degree of pollution in a body of water by the characteristics of the aquatic communities living there?

To answer these questions, we should first of all, consider how biological communities originate and how they change.
WARREN (1971) wrote the following in this regard : "At a location having not too extreme physical and chemical characteristics, where plant and animal species have an opportunity to colonize, an assemblage of species will occur after a period of time. Complex interrelationships will develop among these species and between them and the conditions and resources of the location.

[1] The present review has been restricted deliberately to methods based on the morphological-taxonomical analysis of aquatic communities. Methods based on physiological or biochemical criteria are not considered.

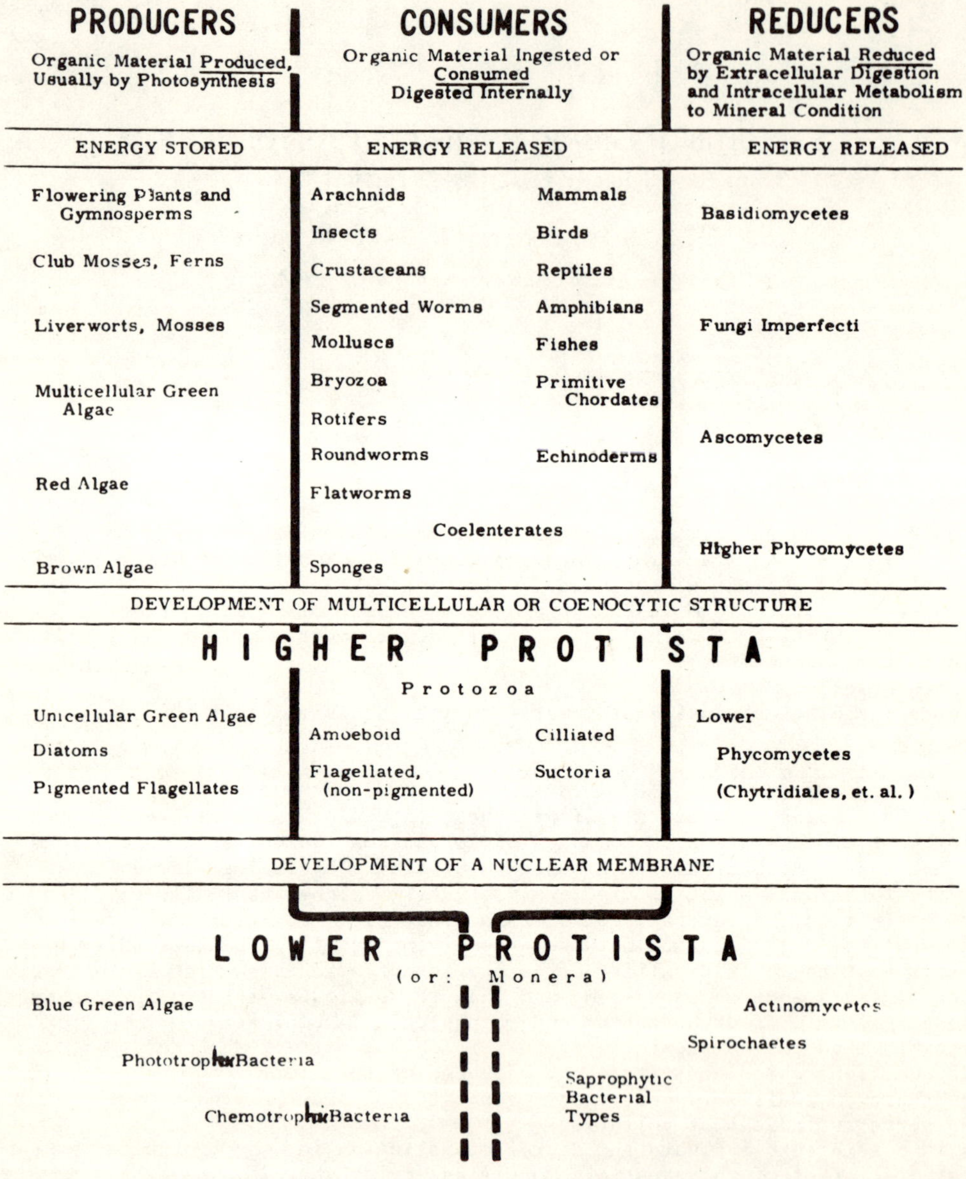


Fig. 1. Relationships between free living aquatic organisms Energy flows from left to right, general evolutionary sequence is upward.
(From : Training Manual on Water Pollution, 1970, Environmental Protection Agency, U.S.A.)

After a period of time, if conditions are not too variable, this assemblage will take on rather definite characteristics. It can then be considered a biological community ... A biological community, being dependent on the conditions and resources of its location, may change if these change. Such a community change is a reflection of changes in the plant and animal populations that compose the community. Because of complex interrelationships, a change in conditions or resources that causes one or more of these populations to change, may cause many of the populations to change and be replaced by others. This will result in another assemblage of species distinguishable as a different community."
From WARREN's considerations we remember that any change in the environment (natural or man-made) will bring about changes in individual organisms, in their populations and in their communities.
WARREN concludes : "Since the environment at a location in large part determines what organisms are able to inhabit that location, the inhabitants should be biological indices of environmental changes."

## SAPROBIC SYSTEMS

### Qualitative

Freshwater biologists have been concerned since the middle of the 19th century, with the changes which pollutional conditions bring to natural communities of aquatic organisms.
KOLENATI (1848) and COHN(1853) (both quoted by LIEBMANN,1962) noticed that certain organisms show a certain relation to the purity or respectively pollution of water, in other words they discovered that the organisms present in polluted water differed from those found in clear water.
It should be emphasized that at that time there was no treatment of sewage at all and that surface waters were gradually exposed to an ever increasing pollution.
MEZ, a student of COHN, published in 1898 the first book dealing with microscopical analyses of water, in which he describes four categories of organisms :

1. those found in very pure water
2. those present in slightly polluted water
3. those of fairly polluted water
4. those of very strongly polluted water

It were, however, KOLKWITZ and MARSSON (1902) who made the definite break-through with the concept of "biological indicators of pollution" in their so-called "saprobic system".
This system is based on different zones of organic enrichment, each of which is characterized by specific plant and animal species.
Their concept is based on the finding that, when sewage or other putrescible organic wastes are discharged into a river, this leeds to sequences of events in time and distance of flow which create different environmental conditions and thus result in different aquatic communities in successive reaches of the river.
As these sequences show a gradual purification of the water, KOLKWITZ and MARSSON introduced the concept of biological selfpurification, recognizing 3 levels (or degrees, or zones) in this process :

a. a zone with predominance of reduction processes which they called polysaprobic level
b. a zone where the reduction gradually ceases and is taken over by oxidative processes = the mesosaprobic level
c. a zone with only oxidative processes = the oligosaprobic level.

The term "saprobia" was introduced by KOLKWITZ and MARSSON to express the dependence of the organisms on decomposing organic substances as sole source of food.
In 1908 and 1909 the same authors subdivided the mesosaprobic level into an $\alpha$ and a $\beta$ part, the $\alpha$ part being the closest to the polysaprobic level.

They also added a category of extremely pure water, the katharobic level. In their 2 papers : "Ecology of plant saprobia" (1908) and "Ecology of animal saprobia" (1909) KOLKWITZ and MARSSON publish long lists of species of plants and animals they believed to be associated with each of these zones and they classify the organisms accordingly as : polysaprobic, mesosaprobic and oligosaprobic. It is from these papers that the concept of species, being biological indicators or indices of pollution has originated.
Several other systems of zonation of rivers receiving organic wastes have been proposed in the 2 decades following KOLKWITZ and MARSSON's publications. WARREN (1971) using data of WHIPPLE et al. (1927) gives the following comparative scheme of the different zonations, proposed at that time (Fig. 2).

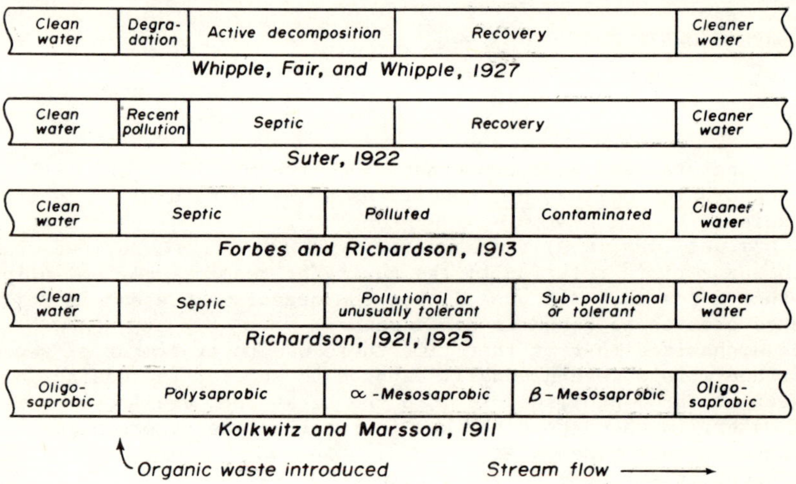


Fig. 2. Systems suggested by various investigators for classifying the zones of rivers receiving sewage.

It is obvious that all the systems are similar in principle, but differ in nomenclature and delimitation of zones.
We should now for the purpose of this review theoretically comment on the extensive lists of organisms proposed by KOLKWITZ and MARSSON and relate them to the physical and chemical characteristics of the saprobic zones which these authors have proposed.
Instead we shall restrict ourselves to a general but modern graphical presentation of the ecological characteristics and biological components of the successive zones which are normally encountered, in time or in space, during the process of biological selfpurification of a river polluted with organic (biodegradable) wastes.
The following scheme (Fig. 3) and text originates from a document worked out in 1970 by Dr. H.W. JACKSON of the USA Environmental Protection Agency, and is based

on unpublished data from C.M. TARZWELL (Robert A. Taft Engineering Center, Cincinnati, Ohio, U.S.A.) and an excellent review paper of BARTSCH and INGRAM (1959). The subdivision follows the zonation proposed by SUTER and MOORE (1922).

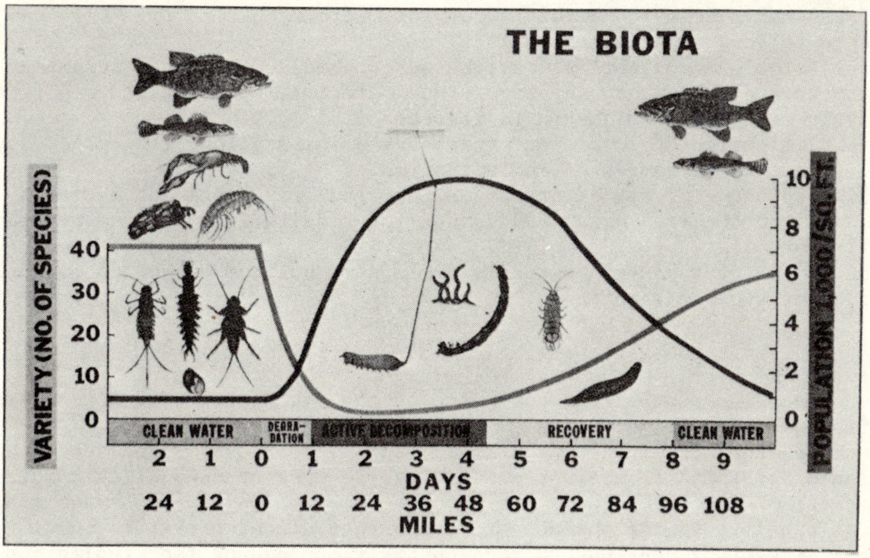


Fig. 3. Relations between variety and abundance (production) of aquatic life, as organic pollution (discharged at mile 0) is carried down a stream. Time and distance scales are only relative and will be found to differ in nearly every case. After Bartsch and Ingram (1959).

1. <u>The starting point</u>

A. A normal unpolluted stream is assumed as a starting point (Fig. 3).
B. The cycle of life is in reasonably stable balance.
C. A great variety of life is present, but no one species or type predominates.
D. The organisms present are adjusted to the normal ranges of physical and chemical factors characteristic of the region.

2. Pollution

A. With the introduction of organic pollution (Fig. 3, day 0), a succession of fairly well organized events is initiated.
B. Numbers of species present tend to decrease with pollution.
C. Numbers of individuals of each species tend to increase with pollution.
D. Ratio's between types of organisms are disturbed by pollution.

3. The zone of recent pollution

A. The zone of recent pollution begins with the act of pollution, the introduction of excessive organic matter : food for microorganisms (Fig. 3, Day 0).
B. There follows a period of physical mixing.
C. Many animals and plants are smothered or shaded out by the suspended material.
D. With this enormous new supply of food material, bacteria and other saprophytic microorganisms begin to increase rapidly.
E. The elimination of intolerant predatory animals allows the larger scavengers to take full advantage of the situation.
F. This explosive growth of organisms, particularly fungi and bacteria, draws heavily on the free dissolved oxygen for respiration, and may eventually eliminate it.
G. The number of types of organisms diminishes but numbers of individuals of tolerant types may increase.

4. The septic zone

A. The exact location of the beginning of the septic zone, if one occurs, varies with season and other circumstances (Fig. 3, day 1).
B. Lack of free dissolved oxygen kills many microorganisms and nearly all larger plants and animals, again replenishing the mass of dead organic material.
C. Varieties of both macro- and microorganisms and adjustable types (facultative) that can live in the absence of free oxygen (anaerobic) take over.
D. These organisms continue to feed on their bonanza of food (pollution) until it is depleted.
E. The numbers of types of organisms is now at a minimum, numbers of individuals may or may not be at maximum.

5. The recovery zone

A. The septic zone gradually merges into the recovery zone (Fig. 3, day 4)
B. As the excessive food reserves diminish so do the numbers of anaerobic organisms and other pollution tolerant forms.
C. As the excessive demand for oxygen diminishes, free dissolved oxygen begins to appear and likewise oxygen requiring (aerobic) organisms.
D. As the suspended material is reduced and available mineral materials increase due to microbial action, algae begin to increase, often in great abundance.
E. Photosynthesis by the algae releases more oxygen, thus hastening recovery.
F. Since algae require oxygen at all times for respiration (like animals), heavy concentrations of algae will deplete free dissolved oxygen during the night when it is not being replenished by photosynthesis.
G. Consequently this zone is characterized by extreme diurnal fluctuations in dissolved oxygen.
H. With oxygen for respiration and algae, etc. for food, general animal growth is resumed.
I. The stream may now enter a period of excessive productivity which lasts until the accumulated energy (food) reserves have been dissipated.

6. Clean water zone

A. Clean water conditions again obtain when productivity has returned to a normal, relatively poor level, and a well balanced varied flora and fauna are present. (Fig. 3, day "10"). Conditions may usually be characterized as follows :

B. General features : similar to upstream clean water except that it is now a larger stream.

C. Characteristic biota : similar to upstream clean water fauna and flora except that species include those indigenous to a larger stream.

An excellent, and very often used graphical representation of the succession of physical, chemical and biological events occuring during the process of biological self-purification has been worked out by HYNES (1960) (Fig. 4).

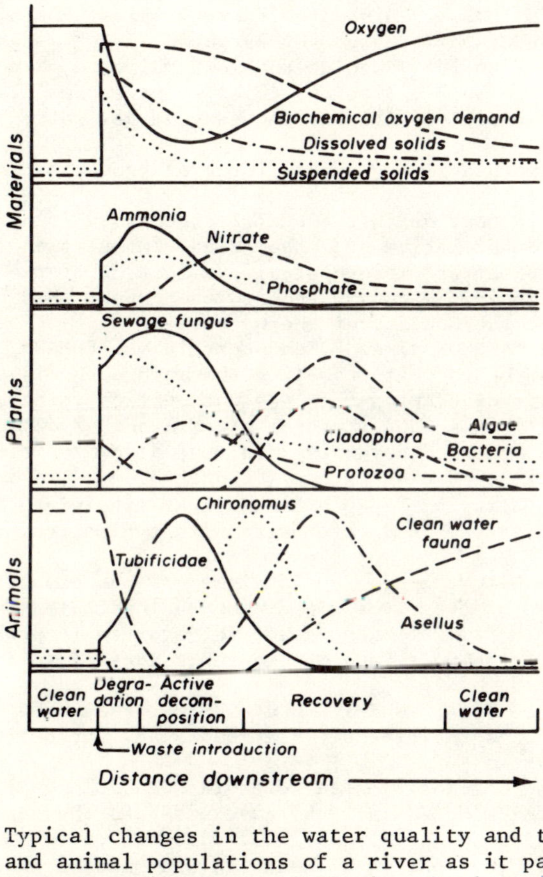


Fig. 4. Typical changes in the water quality and the plant and animal populations of a river as it passes through various zones following the introduction of sewage

Summarizing the previous descriptions, it can be said (as KOLKWITZ and MARSSON had already found out) that bacteria and protozoa predominate in the zone where oxygen becomes severely limiting; more downstream, these groups decline in numbers, and the algae, benefiting from the increase in inorganic nutrients (resulting from the bacterial mineralization of biodegradable wastes) become first abundant and then decline in their turn, after exhaustion of the supply of food. Where the oxygen concentration is very low, the number of species of larger invertebrates is greatly reduced, but those which are able to tolerate low oxygen concentrations (by morphological or physiological adaptations) become numerically very abundant since there is plenty of food for them and little interspecific competition.

Finally, when the original natural conditions are restored the number of species increases and the strong numerical preponderance of particular species tends to disappear.

Digging back into the history of the concept of biological indicators of pollution, after the death of MARSSON the system of saprobic organisms was revised by KOLKWITZ in a paper published in 1935, and in 1950 in his book entitled : "Ökologie der Saprobien".

Another German scientist, the late Professor LIEBMANN, well-known for his famous handbook on the biology of clean and polluted waters ("Handbuch der Frischwasser und Abwasserbiologie", 1962) retained the four classical zones proposed by KOLKWITZ and MARSSON, but called them "Wasserguteklassen" = classes of water quality, ranking from I to IV (IV being the most polluted, I the cleanest) and allocated them typical colors : blue for class I, green for class II, yellow for class III and red for class IV, for the easy visualisation of water quality surveys.

LIEBMANN's system leans heavily on the ciliates as "good indicators" of the state of pollution.

It should be emphasized that neither KOLKWITZ and MARSSON's nor LIEBMANN's systems consider any quantitative or mathematical formulation for determining the degree of pollution after microscopical examination of the biota. Fundamentally each organism is considered and the expert uses his personal experience to evaluate the quality of the water.

However, KOLKWITZ and MARSSON already, and certainly LIEBMANN, have emphasized that in the evaluation of the water quality the main emphasis should not be laid on the individual organisms but on the communities (biocoenoses).

LIEBMANN, for example in his 1962 Treatise gives a number of charts with drawings of the major indicator organisms in typical communities of more or less polluted standing or running waters (Figures 5, 6, 7, 8).

In 1964 the danish phycologist FJERDINSTAD further subdivided the 4 zones of the original saprobic system in 9 categories with typical communities (see Table 1).

The major differences with the previous systems are the creation of a new category for that part of the river containing the undiluted faecal matter where oxidation and mineralization have not yet started (coprozoic zone) and further subdivision of the poly- and mesosaprobic zones into an $\alpha$ , $\beta$ and $\gamma$ part.

Contrary to his predecessors, KOLKWITZ and MARSSON, FJERDINGSTAD only uses a few indicator organisms (mainly bacteria and algae) for delimitation of the 9 zones in his system.

Last but not least the Czechoslovakian scientist SLADECEK in 1965 (and before him his compatriot SRAMEK HUSEK in 1956) have extended the saprobic system by including new categories for highly polluted waters.

The original nomenclature proposed by KOLKWITZ and MARSSON is conserved in the so-called "limnosaprobity" which groups all the waters polluted by biodegradable wastes and showing some degree of selfpurification.

The katharobity comprises the very pure waters, i.e. those waters without any trace of pollution (underground waters, springs, etc...).

The eusaprobity refers to waters highly polluted by biodegradable domestic and industrial wastes.

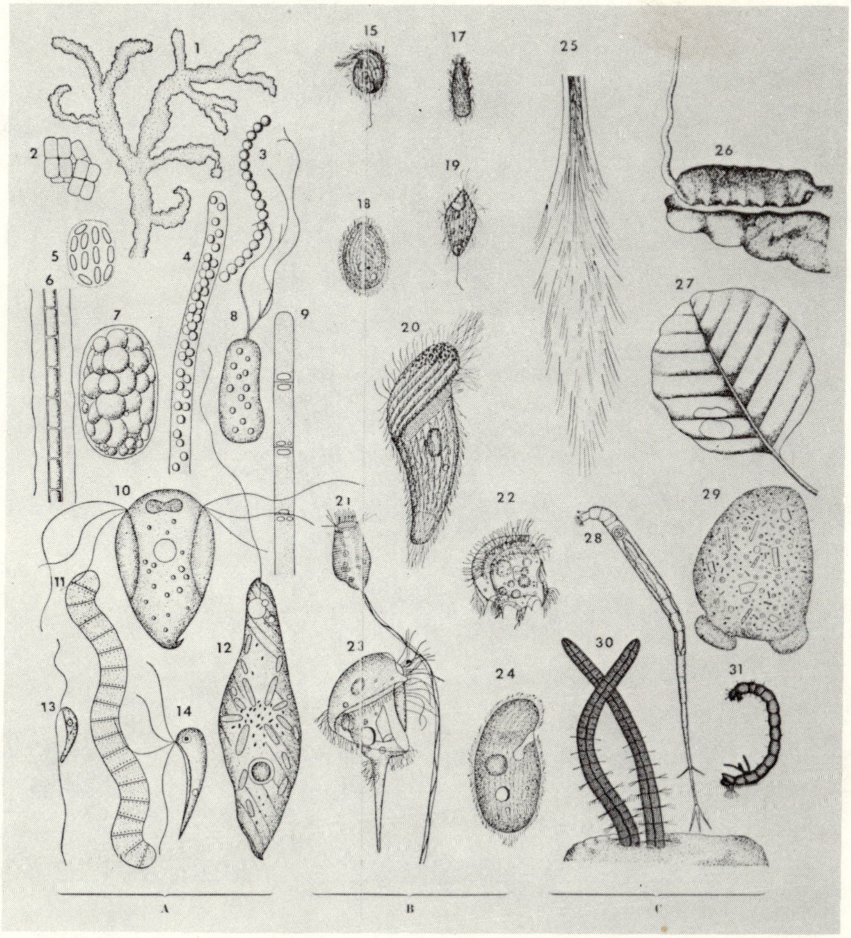


Fig. 5. *POLYSAPROBIC ORGANISMS*

(Indicators for extremely polluted waters)

A

1. Zoogloea ramigera
2. Sarcina paludosa
3. Streptococcus margaritaceus
4. Beggiatoa alba
5. Chlorobacterium aggregatum
6. Sphaerotilus natans
7. Achromatium oxaliferum
8. Chromatium okenii
9. Oscillatoria putrida
10. Trigonomonas compressa
11. Spirulina jenneri
12. Euglena viridis
13. Bodo putrinus
14. Tetramitus pyriformis

B

15. Hexotricha caudata
17. Enchelys vermicularis
18. Glaucoma scintillans
19. Trimyema compressa
20. Metopus es
21. Vorticella microstoma
22. Saprodinium dentatum
23. Caenomorpha medusula
24. Colpidium colpoda

C

25. Sphaerotilus natans
26. Larva of Eristalis tenax
27. Lamprocystis roseo-persicina
28. Rotaria neptunia
29. Pelomyxa palustris
30. Tubifex rivulorum
31. Chironomus thummi

- The organisms are not drawn in proportion to each other

(After LIEBMANN, 1962)

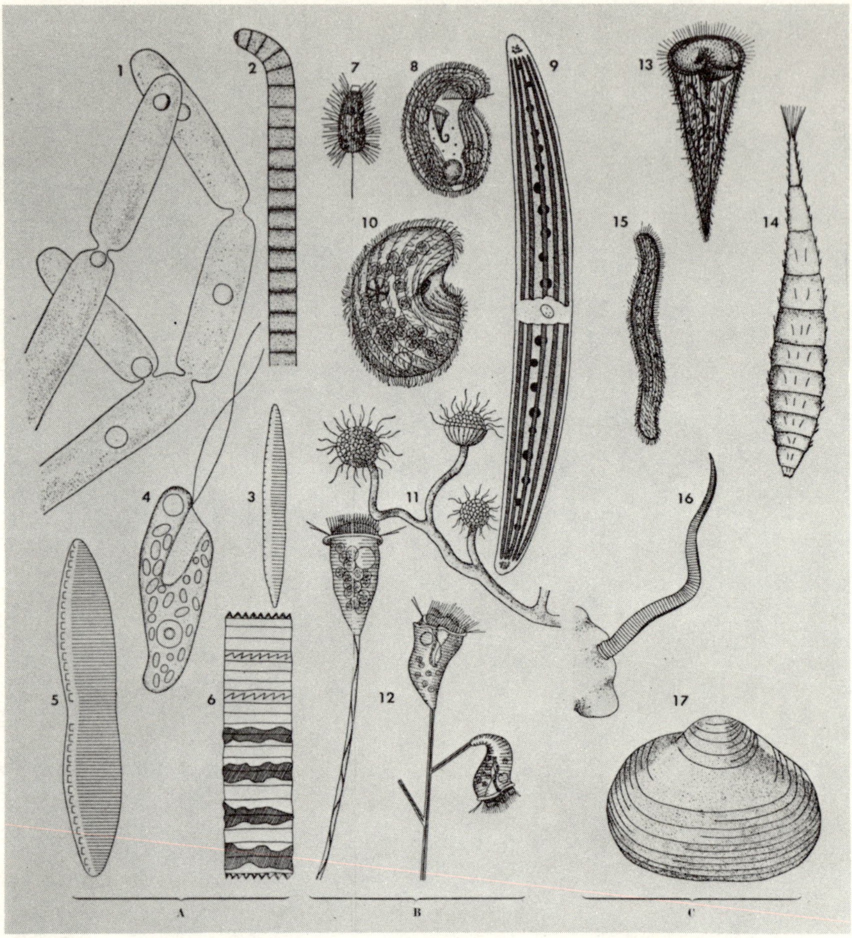


Fig. 6. *α-MESOSAPROBIC ORGANISMS*

(Indicators for highly polluted waters)

A

1. Leptomitus lacteus
2. Oscillatoria formosa
3. Nitzschia palea
4. Chilomonas paramaecium
5. Hantzschia amphioxys
6. Stephanodiscus hantzschii

B

7. Uronema marinum
8. Chilodonella uncinata
9. Closterium acerosum
10. Colpoda cucullus
11. Anthophysa vegetans
12. Vorticella convallaria

C

13. Stentor coeruleus
14. Stratiomys chamaeleon larva
15. Spirostomum ambiguum
16. Herpobdella atomaria
17. Sphaerium corneum

- The organisms are not drawn in proportions to each other

(After LIEBMANN, 1962)

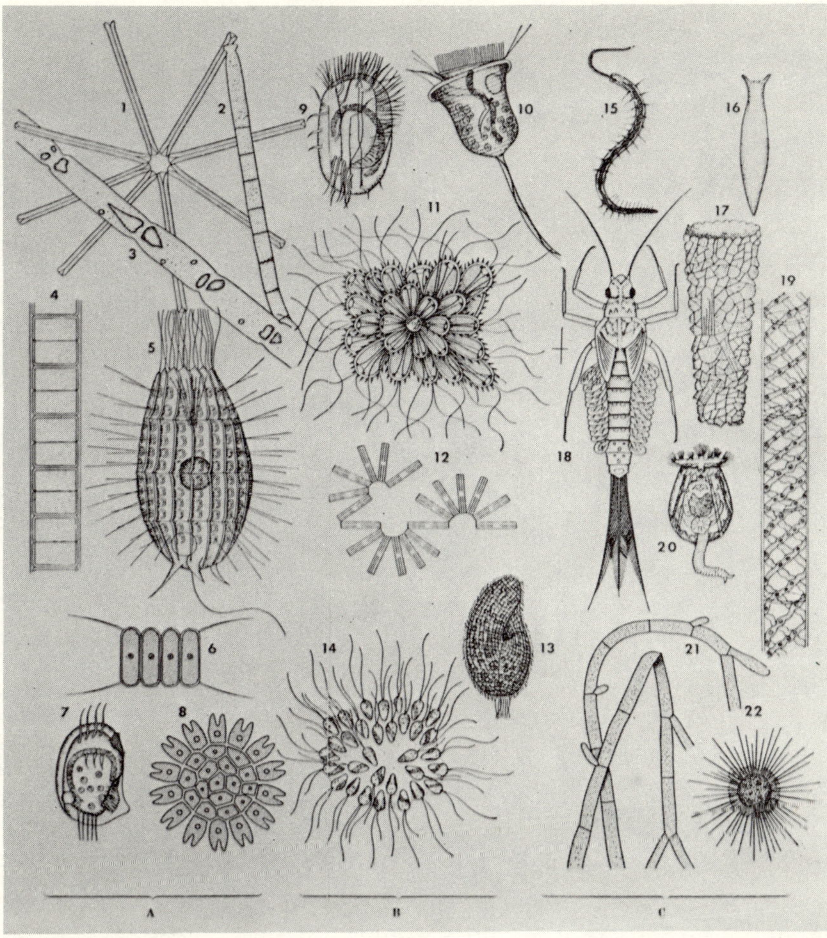


Fig. 7. <u>β-MESOSAPROBIC ORGANISMS</u>

(Indicators for moderately polluted water)

<u>A</u>

1. Asterionella formosa
2. Oscillatoria rubescens
3. Oscillatoria redekei
4. Melosira varians
5. Coleps hirtus
6. Scenedesmus quadricauda
7. Aspidisca lynceus
8. Pediastrum boryanum

<u>B</u>

9. Euplotes charon
10. Vorticella campanula
11. Synura uvella
12. Tabellaria fenestrata
13. Paramaecium bursaria
14. Uroglena volvox

<u>C</u>

15. Stylaria lacustris
16. Polycelis cornuta
17. Hydropsyche lepida (case of pupa)
18. Cloëon dipterum larva
19. Spirogyra crassa
20. Brachionus urceus
21. Cladophora crispata
22. Actinosphaerium eichhorni

- The organisms are not drawn in proportions to each other.

(After LIEBMANN, 1962)

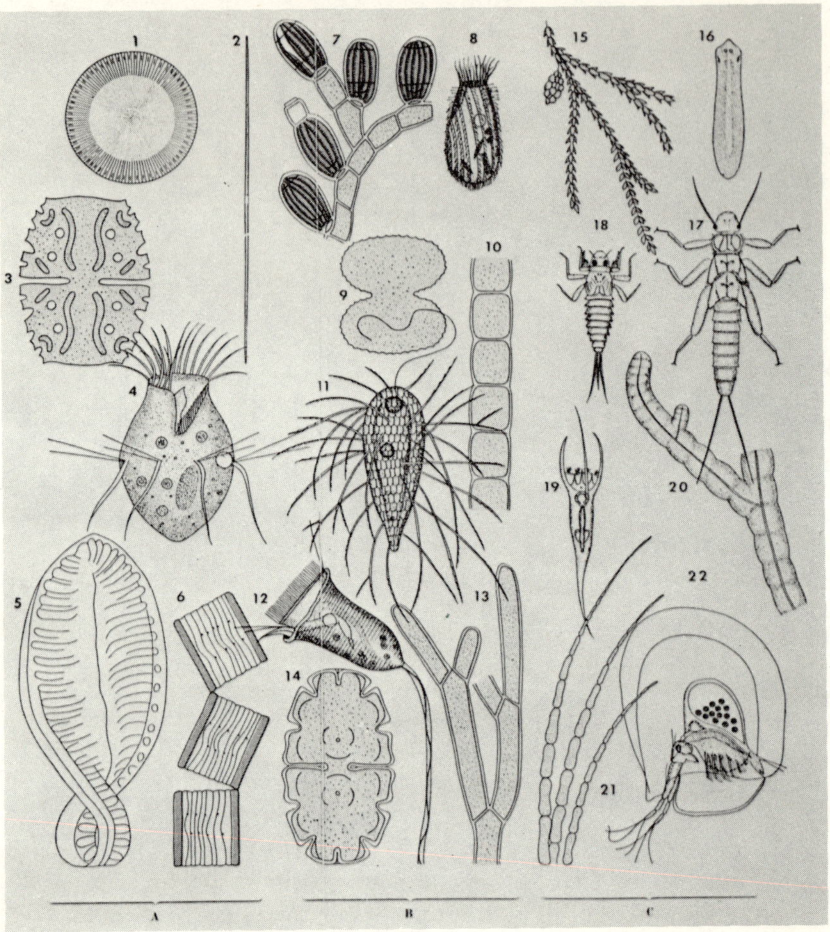


Fig. 8. *OLIGOSAPROBIC ORGANISMS*

(Indicators for scarcely polluted water)

A

1. Cyclotella bodanica
2. Synedra acus var. angustissima
3. Micrasterias truncata
4. Halteria cirrifera
5. Surirella spiralis
6. Tabellaria flocculosa

B

7. Bulbochaete mirabilis
8. Strombidinopsis gyrans
9. Staurastrum punctulatum
10. Ulothrix zonata
11. Mallomonas caudata
12. Vorticella nebulifera var. similis
13. Cladophora glomerata
14. Euastrum oblongum

C

15. Fontinalis antipyretica
16. Planaria gonocephala
18. Larva of Oligoneuria rhenana
17. Larva of Perla bipunctata
19. Notholca longispina
20. Batrachospermum vagum
21. Lemanea annulata
22. Holopedium gibberum

- The organisms are not drawn in proportions to each other

(After LIEBMANN, 1962)

TABLE 1. Outline of the saprobic zones of Fjerdingstad (1964) with their characteristic communities (quoted according to Schwoerbel 1970)

| ZONE | COMMUNITIES [x] |
|---|---|
| I. Coprozoic zone | Community of bacteria or of flagellate *Bodoibal* or both |
| II. α -polysaprobic zone | 1. *Euglena* community<br>2. Community of rhodo- and thiobacteria |
| III. β -polysaprobic zone | 1. *Beggiatoa* community<br>2. Community of *Thiothrix nivea* *Euglena* community |
| IV. γ -polysaprobic zone | 1. Community of *Oscillatoria chlorina*<br>2. Community of *Sphaerotilus natans* |
| V. α -mesosaprobic zone | Community of *Ulothix zonata* (high trophy) or *Oscillatoria benthonicum* or *Stigeoclonium tenue* |
| VI. β -mesosaprobic | Community of *Cladophora fracta* or *Phormidium* |
| VII. γ -mesosaprobic zone | Community of Rhodophyceae (*Batrachospermum monoliforme* or *Lemanea fluviatilis)* or community of Chlorophyceae *(Cladophora glomerata* or *Ulothrix zonata)* |
| VIII. Oligosaprobic | Community of Chlorophyceae (*Draparnaldia glomerata)* or a pure community of *Meridion circulare* or community of Rhodophyceae *(Lemanea annulata, Batrachospermum vagum,* or *Hildenbrandia rivularis)* or community of *Vaucheria sessilis* or *Phormidium inundatum* |
| IX. Katharobic zone | Community of Chlorophyceae *(Chlorotylium cataractum* and *Draparnaldia plumosa* or community of Rhodophyceae *(Chantransia chalybdea* and *Hildenbrandia rivularis* or community of encrusting algae *(Chamaesiphon polonicus* and various species of *Cladothrix)* |

[x] 1, 2 and 3 are various degrees of impurity within a zone

SLADECEK divides the latter zone into 4 parts :

a. the isosaprobity characterized by ciliate populations as dominants, accompanied by colorless flagellates and bacteria.
b. the metasaprobity is the zone of colorless flagellates accompanied by huge developments of bacteria.
c. the hypersaprobity with bacteria and fungi as only living organisms.
d. the ultrasaprobity is the abiotic (=azoic) zone which preceeds the processes of degradation and selfpurification.

As can be seen in Table 2 SLADECEK also introduces a category for waste waters where there is no biochemical decomposition (i.e. waters containing e.g. toxic substances, inorganic materials, radioactive wastes, etc...) and calls it transsaprobic. He distinguishes an antisaprobic zone where organisms are killed by the poisons present and the radiosaprobic zone with radioactive contamination.

TABLE 2. Synthesis of the extended saprobity system of Sladecek (1965)

| Katharobic conditions (very clean water) | Katharobic conditions | |
|---|---|---|
| Limnosaprobic conditions (contaminated surface and ground water) | Oligosaprobic<br>β-mesosaprobic<br>α-mesosaprobic<br>Polysaprobic | Corresponding to the system of KOLKWITZ and MARSSON |
| Eusaprobic conditions (domestic and industrial waste water, the basis of bacterial destruction) | Isosaprobic<br>Metasaprobic<br>Hypersaprobic<br>Ultrasaprobic | Zone of ciliates<br>Zone of colourless flagellates<br>Zone of bacteria<br>Azoic zone but not toxic yet |
| Transsaprobic conditions (without bacterial breakdown) | Antisaprobic<br>Radiosaprobic | Toxic zone<br>Radioactive waste water |

As SLADECEK himself admits, toxicity as well as radioactivity do not fit very well in the system of saprobity and even interfere with it. We shall discuss this very important and even critical point in detail further on.
An important fact in the work of SLADECEK is that besides the indicator organisms, he also allocates physico-chemical characteristics to each of the zones which he defines and gives as well some rough quantitative estimations of the number of biological indicators (including the bacteria) present.
Several other European scientists, have proposed modifications to these systems on which, however, we cannot enter here in detail.
For more information we refer to BICK (1963), SCHWOERBEL (1970) and SLADECEK (1973).

Let us quote, however, the work of CASPERS and co-authors (1960, 1962, 1966a,b, 1967) for their particular, new interpretation of saprobity, based on physiological principles (the classic concept is based on ecological ones). CASPERS departs from the definition of trophism proposed by ELSTER (1958) which describes trophism as "the intensity of organic primary production", to deduce that saprobity is the intensity of the decomposition of dead organic matter.
In an international Symposium in Prague in 1966 the following definition of saprobity was adopted as a compromise of several other definitions which had been proposed formerly :
"Within the bioactivity of a body of water, Saprobity is the sum total of all those metabolic processes which are the antithesis of primary production. It is therefore the sum total of all those processes which are accompanied by a loss of potential energy."
In a paper of 1966a and 1967 CASPERS and KARBE give a summarizing table of demarcating of classic "saprobic zones" according to the trophic structure of the communities of organisms (Table 3).

TABLE 3. Trophic structure of different saprobity levels (from Caspers and Karbe 1966a, 1967).

| Saprobity level | Structure of the communities of organisms |
|---|---|
| I. β-oligosaprobic | Balanced relationship between producers, consumers and decomposers; communities of organisms are poor in individuals but there is a moderate variety of species, small biomass and low bioactivity. |
| II. α-oligosaprobic | Balanced relationship between producers, consumers and decomposers; communities of organisms are rich in individuals and species, with a large biomass and high bioactivity. |
| III. β-mesosaprobic | Substantially balanced relationship between producers, consumers and decomposers; a relative increase in the abundance of decomposers and, accordingly, of the consumers living of them; communities of organisms are rich in individuals and species, with a large biomass and high activity. |
| IV. α-mesosaprobic | Producers decline as compared with an increase in consumers and decomposers; mixotrophic and amphitrophic forms predominate among the producers; communities of organisms rich in individuals but poor in species, with a large biomass and extremely high bioactivity; still only few species of macroorganisms; mass development of bacteria and bacteria-eating ciliates. |
| V. β-polysaprobic | Producers drastically decline; communities of organisms are extremely rich in individuals but poor in species, with a large biomass and high bioactivity; macrofauna represented only by a few species of tubificids and chironomids; as in IV these are in great abundance; mass development of bacteria and bacteria-eating ciliates. |
| VI. α-polysaprobic | Producers are absent; the total biomass is formed practically solely by anaerobic bacteria and fungi; macroorganisms are absent; flagellates outnumber ciliates amongst the protozoa. |

Although this saprobic-trophic subdivision worked out by CASPERS is excellent from the theoretical point of view, the author admits that it is not possible to give presently exact data to distinguish the individual levels. As a consequence SLADECEK (1973) pertinently concludes that "... CASPERS' scheme is more a programme than a functioning system ...".

## Quantitative

During the establishment of saprobic systems, several methods have been proposed to quantify the indicator organisms found, and to reduce the large number of data to simple figures with the aid of mathematical formulas.
In the method proposed by KNÖPP (1954, 1955), the species found at each sampling point are listed, and their estimated frequency is valued, using a scale of 7 degrees :

1 = one single finding
2 = few
3 = few to average numbers
4 = average numbers
5 = average to many
6 = many
7 = abundant

The indicator value of each species is looked up in the saprobity system of KOLKWITZ, using the 4 symbols $o$, $\beta$, $\alpha$ and $p$ for oligosaprobic, $\beta$- and $\alpha$- saprobic and polysaprobic respectively.
The frequency numbers of all the species are added up for each saprobic zone : $\Sigma o$, $\Sigma\beta$, $\Sigma\alpha$ and $\Sigma p$.
For a visual representation of the state of pollution of a particular biotope KNÖPP plots the so-called "positive" values ($\Sigma o$ and $\Sigma\beta$ representative for no pollution or minute pollution) above the axis, and the negative values ($\Sigma\alpha$ and $\Sigma p$) below the axis.
The points of the same saprobic classes are joined, showing the geographical evolution of the pollution (Fig. 9A).
KNÖPP also introduced the concept of relative purity ("relative Güte") and relative pollution ("relative Belasting") which can be calculated from the summations of abundance of the indicators in the 4 saprobic classes.

$$\text{Relative purity (in \%)} = \frac{\Sigma\ (o + \beta)}{\Sigma\ (o + \beta + \alpha + p)} \times 100$$

$$\text{Relative loading (in \%)} = \frac{\Sigma\ (o + \alpha)}{\Sigma\ (o + \beta + \alpha + p)} \times 100$$

This relative loading is also called and used now as Biologically Effective Organic Load (B.E.O.L.) (KNÖPP, 1954).
A graphical representation of the evolution of the "relative purity" in the River Main in Germany is given in Fig. 9B.

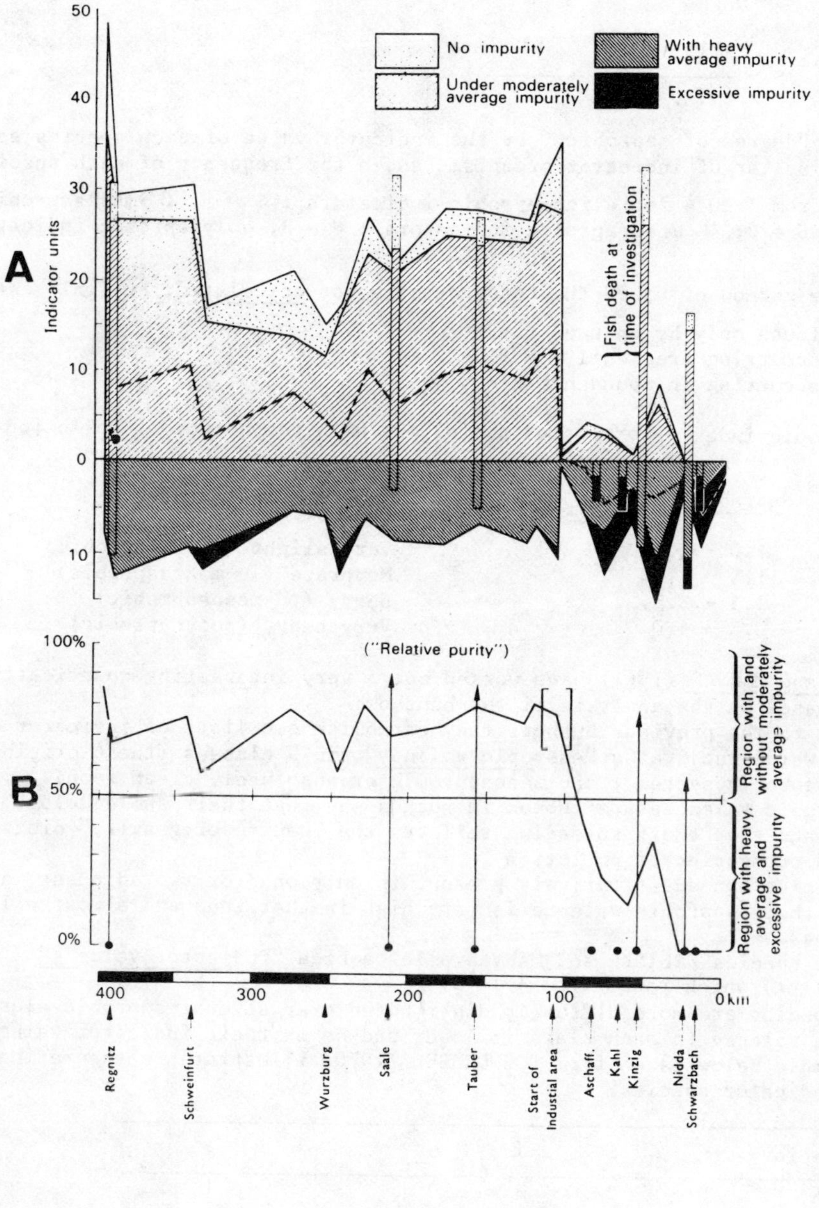


Fig. 9A. Longitudinal section of the quality of the water of the River Main according to KNÖPP's graphic method.

Fig. 9B. Representation of the relative purity of the same stretch of river. (modified from KNÖPP, 1955).

In 1955 PANTLE and BUCK proposed the saprobic index which can be calculated with the formula :

$$S = \frac{\Sigma . s. h.}{\Sigma . h.}$$

s is the "degree of saprobity" or the indicator value of each species according to LIEBMANN's list of indicator organisms and h the frequency of each species found.

s ranks from 1 to 4 for oligosaprobic indicators : S = 1; β -mesosaprobic indicators : S = 2; α -mesosaprobic indicators : S = 3; polysaprobic indicators : S = 4.

As in the method of KNÖPP the fequences are not calculated, but only estimated.

species found only by chance h = 1
species occurring frequently h = 3
species occurring in abundance h = 5

The saprobic index S correlates as follows with the degree of pollution :

| Saprobic index | Degree of pollution |
|---|---|
| 1.0 - 1.5 | Very slight (oligosaprobic) |
| 1.5 - 2.5 | Moderate (β -mesosaprobic) |
| 2.5 - 3.5 | Heavy (α -mesosaprobic) |
| 3.5 - 4.0 | Very heavy (polysaprobic) |

ZELINKA and MARVAN (1961) have worked out a very interesting quantitative saprobic method based on the analysis of the benthos.
Contrary to the previous authors they establish own lists of indicator organisms each of which receives a "saprobic valency" in 5 classes (the 4 original classes of the saprobic system + one xenosaprobic or absolutely clean zone).
The saprobic valences are chosen in such a way that their sum equals 10 for each species and that their spreading reflects the "centre of gravity" distribution in the 5 categories of pollution.
Some species are quantitatively present in only one (or two adjacent) zones.
As such their saprobic valence is very high in that zone and almost nil in the other ones.
To those species ZELINKA and MARVAN allot a high "indicator value g" (or indicative weight) which ranks from 1 to 5.
Other species are more diffusely distributed over several saprobic classes. Their saprobic valency in each class is lower and so is their indicator value.
The example below taken from SCHWOERBEL (1970) illustrates a theoretical case for 5 indicator species.

| Species | x | o | β | α | p | g |
|---|---|---|---|---|---|---|
| $A_1$ | 3 | 3 | 3 | 1 | | 1 |
| $A_2$ | 5 | 5 | | | | 3 |
| $A_3$ | 4 | 6 | | | | 3 |
| $A_4$ | 2 | 7 | 1 | | | 3 |
| $A_5$ | x | 8 | 2 | | | 4 |

x indicates an isolated finding

$A_5$ is the species with the highest indicator value (g=4) since it has a high saprobic valence (8) in the oligosaprobic zone.
The frequence h, i.e. the exact number of each indicator species found, is determined in each sample and multiplied by the saprobic valence of the species for each of the 5 pollution zones.
If, in the theoretical example given above, the abundance for indicator species $A_1$ is 69, 31 for $A_2$, 30 for $A_3$, 42 for $A_4$ and 8 for $A_5$, the following results are obtained (after SCHWOERBEL, 1970).

| Species | Abundance | Multiplication results for | | | | |
|---|---|---|---|---|---|---|
| | (h) | x | o | β | α | p |
| $A_1$ | 69 | 207 | 207 | 207 | 69 | |
| $A_2$ | 31 | 465 | 465 | | | |
| $A_3$ | 30 | 360 | 540 | | | |
| $A_4$ | 42 | 252 | 882 | 126 | | |
| $A_5$ | 8 | + | 256 | 64 | | |
| Total | | 1284 | 2350 | 397 | 69 | 0 |
| Average | | 3.13 | 5.73 | 0.97 | 0.17 | 0 |

The totals, for each category, are finally divided by hg, i.e. the sum of the abundances of the different indicators encountered multiplied by their respective indicator values.

In our example hg = (69 x 1) + (31 x 3) + (30 x 3) + (42 x 3) + (8 x 4) = 410.

We thus obtain for each of the 5 pollution zones the value :

$$X = \frac{\Sigma \; x\,hg}{\Sigma \; hg} = 3.13$$

$$O = \frac{\Sigma \; o\,hg}{\Sigma \; hg} = 5.73$$

etc...

The highest value of the 5 numbers calculated characterizes the category (and as such the degree) of pollution of the site under investigation (in our particular example : oligosaprobic).
The method of ZELINKA and MARVAN is more time consuming than that of PANTLE and BUCK, but on the other hand seems to be more exact since it is based on precise quantitative calculations of the frequency of occurrence of the species, instead of rough estimations.
It should be mentioned here that DITTMAR, in 1959, almost simultaneously with ZELINKA and MARVAN, proposed another method very similar to that of the latter authors.

Last but not least the extensive work of SLADECEK and co-authors (see review in SLADECEK, 1973), should certainly be given thorough attention to.
The existing list of indicator organisms is revised, detailed and extended by these scientists for each organism. The saprobic valence, in 10 points (in the sense of ZELINKA and MARVAN) is given, as well as the indicator value (Indikationsgewicht G). The saprobic index S is extended to cover all the subdivisions and zones of SLADECEK's saprobic classification.
According to SLADECK (1973), the saprobic indices have the following values :

| | |
|---|---|
| Xenosaprobic organisms | 0 up to 0.5 |
| Oligosaprobic organisms | 0.51 up to 1.5 |
| β-mesosaprobic organisms | 1.51 up to 2.5 |
| α-mesosaprobic organisms | 2.51 up to 3.5 |
| Polysaprobic organisms | 3.51 up to 4.5 |
| Isosaprobic organisms | 4.51 up to 5.5 |
| Metasaprobic organisms | 5.51 up to 6.5 |
| Hypersaprobic organisms | 6.51 up to 7.5 |

In his remarkable work of 1973 : "System of Water Quality from the Biological Point of View", SLADECEK lists approximately 2000 indicator organisms from bacteria to fishes.

## Criticisms to and sources of error in the saprobic systems

Although widely adopted and utilized by many hydrobiologists in continental Europe, the saprobic system has always met criticisms and objections of many scientists, especially in Great Britain and the U.S.A.
SLADECEK, in his 1973 review patiently lists 33 objections, on which he comments one by one.
As it is impossible to go into detail on each of these points we shall briefly discuss some of the major drawbacks of saprobic systems, quoting the 5 reasons mentioned by BARTSCH and INGRAM in 1966, as the major objections formulated by U.S.A. and Canadian hydrobiologists against the application of saprobic systems.

### 1. Necessity to identify organisms as to species with only poor taxonomic tools available

There can be no doubt that this is the major weakness of any system of the "saprobic" type.
Even without considering the time involved to determine single organisms to the species level, it is clear that especially with lower organisms (protozoans and algae) there are still endless discussions about the nomenclature and the taxonomic position of many species.
Moreover, there are numerous disagreements among scientists about the "ecological valency", i.e. the place of the organisms into one or another of the categories (or zones) of the saprobic systems.
Last but not least it suffices to have a brief glance at the long lists of organisms listed by SLADECEK to realize that a team of trained taxonomists is necessary to analyse the biological community of relatively pure water or sediment samples.
All these criticisms have been very well summarized by a biologist in the following statement : "Too much time, too many people, too much money".

2. Paucity of basic knowledge on the ecological characteristics and requirements of individual species and communities

This criticism is based on the fact that the saprobic system implies more knowledge than presently exists, on the ecological characteristics, the environmental requirements, the tolerance limits and the physiological responses of organisms to pollution.
The difficulties are inherent to the famous ecological "law of interaction" which says that biotic responses of organisms to one factor are also influenced by changes of other factors. Indeed, considering only the physical factors stream, velocity and temperature, it has been demonstrated by AMBUHL (1959) and ZIMMERMANN (1961) that, on the basis of indicator organisms, the same pollution is more favourably estimated in rapidly flowing water and at lower temperature than it is in slowly running water at a higher temperature.

3. Failure to properly quantitate species relationships

It is well known to any hydrobiologist that aquatic biotopes, not the least the running waters, are a mosaic of microhabitats. To achieve a reasonable accuracy in the qualitative and quantitative biological assessments considerable numbers of samples need to be taken.
HYNES (1964) emphasizes that biological sampling is a highly skilled job, much more complicated than the simple filling of bottles for chemical analysis or monitoring of measuring equipment.

4. Failure to accommodate the uniqueness of each stream, each pollutant, each problem

Many North American biologists have expressed the conviction that each aquatic system is unique. Nowhere in the world can we find 2 streams, or lakes, or ponds with exactly the same physico-chemical characteristics, thus with the same aquatic biocoenoses. The effect of exactly the same quantity of a pollutant must thus be different. As a result, saprobic systems are theoretically only applicable for the aquatic biota from which the method was described originally.

5. Restriction of the use of saprobic systems to pollution caused by municipal sewage or similar acting organic wastes

As mentioned in the beginning of this chapter on saprobic systems, the early students of water pollution dealt mainly with pollution by degradable organic wastes and more particularly domestic sewage.
Today, there are numerous other types of pollutants : inorganic and organic, degradable and non-degradable, toxic and non-toxic. In the majority of cases the pollution is of a mixed nature : domestic sewage + industrial wastes.
As admitted by SLADECEK, the original concept of biological indicators of pollution (expressed in saprobic systems) is unfortunately not applicable to industrial wastes or to domestic wastes mixed with toxic pollutants.
Notwithstanding these and other criticisms which SLADECEK (1973) partly agrees with, this author thinks that "each system must be applied reasonably, neither rigidly nor stiffly ...
At present it cannot be replaced by any other method ... In the saprobic system we see one interesting phenomenon : the more it is criticized and rejected, the more it flourishes and gains in importance ...".

## OTHER SYSTEMS OF INDICATOR ORGANISMS

Independently of the saprobic systems, several other methods were elaborated mainly in North America, but also in Europe.
Once again it is out of the scope of this review to try to mention and discuss all of them. In the following pages we shall comment briefly on some of the most known.
A first important statement is that, contrary to the saprobic systems, which are focusing mainly micro-organisms (protozoans and algae) as indicator organisms, most of the other systems are utilizing benthic macro-invertebrates as major indicators.
In 1928 already the American scientist RICHARDSON classified the benthic invertebrates of the Illinois River according to their tolerance with regard to the introduction of sewage into flowing waters. He subdivided the organisms in a pollutional group, a sub-pollutional group and cleaner water species.
It is interesting to note that despite his remarkable scientific work, RICHARDSON himself had but few illusions about the value of the indicator organisms concept : "... As a matter of fact the number of small bottom dwelling species of the fresh waters of our distribution area that can be safely regarded as having even a fairly dependable individual index value in the present connection is surprisingly small; and even those few have been found in Illinois to be reliable as index species only when used with the greatest caution, and when checking with other indicators".
According to SLADECEK (1973), most of the methods proposed by PATRICK (1949), BECK (1954, who worked out a "Biotic Index"), WURTZ (1955) and BEAK (1964), are in fact very compatible with the modern versions of the saprobic system.
MACKENTHUN, in an excellent book "The Practice of Water Pollution Biology" (1969) gives a nice visual representation of some typical "sensitive, intermediate and tolerant bottom dwelling macro-invertebrates" (Fig. 10).
The subdivisions are based on the concept, known for a long time already, that the usual order of disappearance of macro-invertebrates subsequent to pollution is as follows :

Stoneflies (Plecoptera)
Mayflies (Ephemeroptera)
Caddiflies (Trichoptera)
Amphipods
Isopods
Midges (Diptera)
Oligochaetes

It is the latter concept which inspired the British scientist WOODIWISS in 1964 to develop the Trent Biotic Index, originally worked out specifically for use in the Trent River Authority area in England, but presently adopted in many other countries.
Samples are taken with a handnet (780 μm). Kick and handsamples are taken in such a way as to include material from all the macrohabitats and to give a cross-sectional representation of the fauna.
With this method, streams can be classified according to the presence or absence of key groups of the following categories of bottom invertebrates (some of which are to be identified to the species level, others to the genus, family or even higher taxa) :

- Each known species of Plathyhelminthes
- Annelida (excluding genus Nais)
- Genus Nais
- Each known species of Hirudinea
- Each known species of Mollusca
- Each known species of Crustacea
- Each known species of Plecoptera
- Each known species of Ephemeroptera (excluding Baetis rhodani)

- Baetis rhodani
- Each family of Trichoptera
- Each species of Neuroptera larvae (alder-fly)
- Family Chironomidae (midge larvea) except Chironomus Ch. thummi
- Chironomus Ch. thummi (blood worms)
- Family Simuliidae (black-fly larvae)
- Each known species of other fly larvae
- Each known species of Coleoptera
- Each known species of Hydracarina (water-mites)

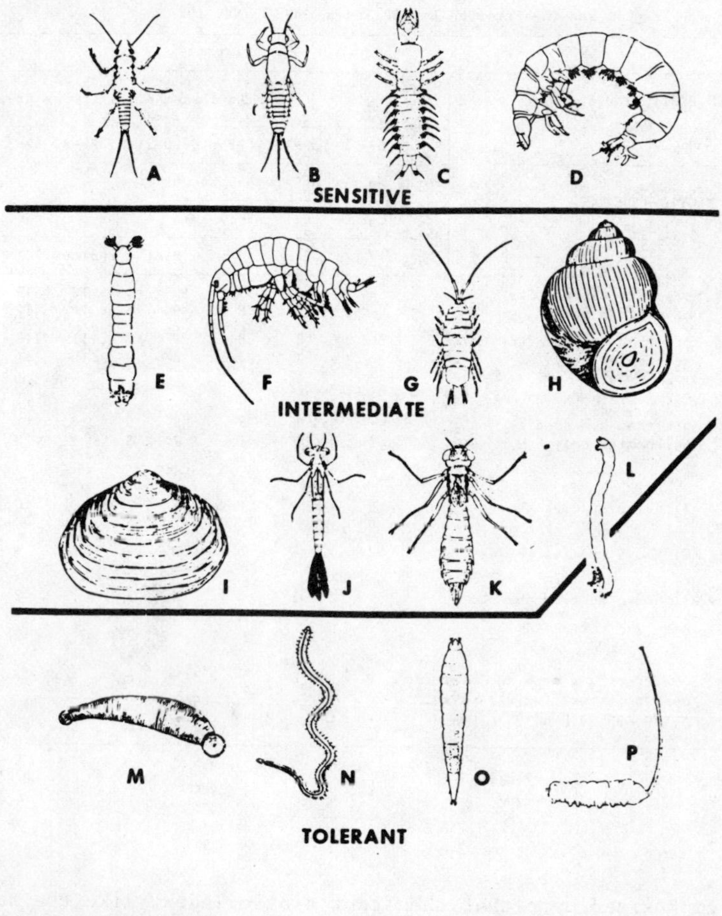


Fig. 10. Sensitive, intermediate and tolerant bottom dwelling macro-invertebrates (from MACKENTHUN, 1969)

A. Stonefly nymph (Plecoptera)
B. Mayfly naiad (Ephemeroptera)
C. Hellgrammite or Dobsonfly larvae (Megaloptera)
D. Caddisfly larvae (Trichoptera)
E. Black fly larvae (Simuliidae)
F. Scud (Amphipoda)
G. Aquatic sowbug (Isopoda)
H. Snail (Gastropoda)
I. Fingernail clam (Sphaeriidae)
J. Damselfly nymph (Zygoptera)
K. Dragonfly nymph (Anisoptera)
L. Bloodworm or midge fly larvae (Chironomidae)
M. Leech (Hirundinea)
N. Sludgeworm (Tubificidae)
O. Sewage fly larvae (Psychodidae)
P. Rat-tailed maggot (Tubifera-Eristalis)

From the specific groups and the total number of groups present, the Biotic Index, which ranges from 0 to 10, can be extrapolated with the aid of Table 4. Clean streams score an index close to 10; this index drops to lower values with increasing pollution and the heavy polluted streams average an index of 1-2. Recently, as a result of criticism from European colleagues, on the grounds of compression at the top of the scale, WOODIWISS (1978) has extended his Trent Biotic Index to cover a range of water qualities from 0 to 15 (Table 4).

TABLE 4. Species and groups of organisms considered in the Trent Biotic Index (WOODIWISS, 1964) and the Extended Biotic Index (WOODIWISS, 1977)

| | | | | | | | | | | | | | |
|---|---|---|---|---|---|---|---|---|---|---|---|---|---|
| EXTENDED BIOTIC INDEX | | | Total number of groups present | | | | | | | | | | |
| | | | 0-1 | 2-5 | 6-10 | 11-15 | 16-20 | 21-25 | 26-30 | 31-35 | 36-40 | 41-45 | .... |
| TRENT BIOTIC INDEX | | | Total number of groups present | | | | | | | | | | |
| | | | 0-1 | 2-5 | 6-10 | 11-15 | 16+ | | | | | | |
| | | | Biotic indices | | | | | | | | | | |
| Clean | Plecoptera nymphs present | More than one species | – | 7 | 8 | 9 | 10 | 11 | 12 | 13 | 14 | 15 | |
| | | One species only | – | 6 | 7 | 8 | 9 | 10 | 11 | 12 | 13 | 14 | |
| Organisms in order of tendency to to disappear as degree of pollution increases | Ephemeroptera nymphs present (excluding *Baetis rhodani*) | More than one species | – | 6 | 7 | 8 | 9 | 10 | 11 | 12 | 13 | 14 | |
| | | One species only | – | 5 | 6 | 7 | 8 | 9 | 10 | 11 | 12 | 13 | |
| | Trichoptera larvae or *Baetis rhodani* present | More than one species | – | 5 | 6 | 7 | 8 | 9 | 10 | 11 | 12 | 13 | |
| | | One species only | 4 | 4 | 5 | 6 | 7 | 8 | 9 | 10 | 11 | 12 | |
| | *Gammarus* present | All above species absent | 3 | 4 | 5 | 6 | 7 | 8 | 9 | 10 | 11 | 12 | |
| | *Asellus* present | All above species absent | 2 | 3 | 4 | 5 | 6 | 7 | 8 | 9 | 10 | 11 | |
| | Tubificid worms and/or red Chironomid larvae present | All above species absent | 1 | 2 | 3 | 4 | 5 | 6 | 7 | 8 | 9 | 10 | |
| Polluted | All above species absent | Some organisms such as *Eristalis tenax* not requiring dissolved oxygen may be present | 0 | 1 | 2 | – | – | – | – | – | – | – | |

It should be emphasized here that the Trent Biotic Index, like the saprobic system does not take into account the relative abundance of the organisms present.
In the "Score system" proposed in 1970 by the Scottish scientist CHANDLER, 5 levels of abundance are recognized based on the numbers of organisms taken during 5-minute sampling with a hand net (Table 5).
After determination of the species or groups found and their abundance, scores can be looked up in Table 6, where the animals are listed according to increasing tolerance to organic pollution.
By adding up the individual scores for the different groups present one finally obtains the index for the particular station.

Table 5. Levels of abundance considered in the score system of CHANDLER (1970)

| Aquatic macro-invertebrates | | |
|---|---|---|
| Level | Number of individuals per 5-min sample | Remarks |
| Present | 1 to 2 | May be drift fauna from upstream |
| Few | 3 to 10 | Probably indigenous, but rare |
| Common | 11 to 50 | |
| Abundant | 51 to 100 | |
| Very abundant | more than 100 | |

The CHANDLER score increases with increasing abundance for "clear water" species whereas it decreases with increasing abundance for pollution tolerant species. Once again, SLADECEK (1973) points out that the system of CHANDLER is only a variety of a saprobic system, more precisely of the 10 points procedure of ZELINKA and MARVAN, and repeats that : "... each effort to avoid the saprobic system leads only to a modification of it".

Last but not least the Frenchman TUFFERY and VERNEAUX developed in 1968 an own method for the "Determination of River Water Biological Quality".

The method is in fact derived from the Trent Biotic Index and relies on sampling of the benthic and periphytic macrofauna in the 2 "facies" of a particular site in the river : the flowing (lotic) facies and the more quiet (lentic) facies. Depending on the depth and the current speed of the river 3 samples are taken in each facies with a Surber sampler (1 $ft^2$, 300 μm), or a grab (2 $dm^3$).

The necessity of considering the 2 facies of a biotope is claimed by the authors to be of great importance since the benthic biocoenoses in the currents are much different from those in the "still" portions of the same locale.

As the method was developed for field work, with limited possibilities for the analysis, it relies on benthic invertebrates larger than 0,5 mm, which are determined "up to the limit of easy identification".

According to the category of organisms, the identification will be as precise as the species level or as rough as the family.

TUFFERY and VERNEAUX consider that the essential point is not to arrive at the exact taxonomic determination, but to identify the largest number of "systematic units" in each group of organisms.

TABLE 6. Species and groups considered in the score system of CHANDLER (1970)

| Groups present in the sample | Increasing abundance | | | | |
|---|---|---|---|---|---|
| | Present | Few | Common | Abundant | Very Abundant |
| | Points scored | | | | |
| Each species of *Planaria alpina,* Taenopterygidae, Perlidae, Isoperlidae, Chloroperlidae | 90 | 94 | 98 | 99 | 100 |
| Each species of Leuctridae, Capniidae and Nemouridae (exd. *Amphinemura)* | 84 | 89 | 94 | 97 | 98 |
| Each species of Ephemeroptera (exd. *Baetis)* | 79 | 84 | 90 | 94 | 97 |
| Each species of cased Caddis, Megaloptera | 75 | 80 | 86 | 91 | 94 |
| Each species of *Ancylus* | 70 | 75 | 82 | 87 | 91 |
| Each species of *Rhyacophila* (Trichoptera) | 65 | 70 | 77 | 83 | 88 |
| Genera of *Dicranota, Limnophera* | 60 | 65 | 72 | 78 | 84 |
| Genera of *Simulium* | 56 | 61 | 67 | 73 | 75 |
| Genera of Coleoptera, Nematoda | 51 | 55 | 61 | 66 | 72 |
| Genera of *Amphinemura* (Plecoptera) | 47 | 50 | 54 | 58 | 63 |
| Genera of *Baetis* (Ephemeroptera) | 44 | 46 | 48 | 50 | 52 |
| Genera of *Gammarus* | 40 | 40 | 40 | 40 | 40 |
| Each species of uncased Caddis (exd. *Rhyacophila)* | 38 | 36 | 35 | 33 | 31 |
| Each species of Tricladida (exd. *P alpina)* | 35 | 33 | 31 | 29 | 25 |
| Genera of Hydracarina | 32 | 30 | 28 | 25 | 21 |
| Each species of Mollusca *(exd. Ancylus)* | 30 | 28 | 25 | 22 | 18 |
| Each species of Chironomids (exd. *C riparius)* | 28 | 25 | 21 | 18 | 15 |
| Each species of *Glossiphonia* | 26 | 23 | 20 | 16 | 13 |
| Each species of *Asellus* | 25 | 22 | 18 | 14 | 10 |
| Each species of leech, exd. *Glossiphonia, Haemopsis* | 24 | 20 | 16 | 12 | 8 |
| Each species of *Haemopsis* | 23 | 19 | 15 | 10 | 7 |
| Each species of *Tubifex sp.* | 22 | 18 | 13 | 12 | 9 |
| Each species of *Chironomus riparius* | 21 | 17 | 12 | 7 | 4 |
| Each species of *Nais* | 20 | 16 | 10 | 6 | 2 |
| Each species of air breathing species | 19 | 15 | 9 | 5 | 1 |
| No animal life | 0 | 0 | 0 | 0 | 0 |

According to these authors the limits of practical determination in the field are as follows (Table 7) :

TABLE 7. Limits of practical determinations of organisms in the field (according to TUFFERY and VERNEAUX, 1968)

| ORDERS | LIMITS |
|---|---|
| PLECOPTERA | genus |
| TRICHOPTERA | family or genus depending on the cases |
| EPHEMEROPTERA | genus |
| ODONATA | genus |
| COLEOPTERA | family |
| MOLLUSCA | genus or species depending on the cases |
| CRUSTACEA | family |
| MEGALOPTERA | genus |
| HEMIPTERA | genus |
| DIPTERA | family, sub-family or tribe depending on the cases |
| PLANARIDAE | genus or species depending on the cases |
| HIRUDINEA | genus or species depending on the cases |
| OLIGOCHAETA | family |
| NEMATODA | presence |
| HYDRACARI | presence |

Following the example of the Trent Biotic Index TUFFERY and VERNEAUX have drawn a tableau from which a Biotic Index can be calculated according to the faunistic groups found and the number of systematic units present in each particular group (Table 8).
The french rivers are considered to be polluted when the index is lower than 5 (part to the left of the black line). When the index is equal to 5, there is pollution, but at that moment it is interesting to consider the 2 "facies" (lotic and lentic) each of which with its own index : $I_{lo}$ and $I_{le}$, especially if there happens to be a marked difference between the biocenoses in both facies.
If the difference between both indices is superior to 2, generally there is pollution if at least one of both indices is lower than 5.
TUFFERY and VERNEAUX have pointed out that the applicability of their method is limited to rather shallow rivers.

TABLE 8. Standard table of determination of biotic indices according to TUFFERY and VERNEAUX (1968)

| I | | II | III | | | | |
|---|---|---|---|---|---|---|---|
| Faunistic groups | | | Total number of systematic units present | | | | |
| | | | 0-1 | 2-5 | 6-10 | 11-15 | 16 et + |
| | | | Biotic Index | | | | |
| 1. Plecoptera or Ecdyonuridae | 1 | 1 S.U.* | - | 7 | 8 | 9 | 10 |
| | 2 | 1 S.U. | 5 | 6 | 7 | 8 | 9 |
| 2. Trichoptera with cases | 1 | 1 S.U. | - | 6 | 7 | 8 | 9 |
| | 2 | 1 S.U. | 5 | 5 | 6 | 7 | 8 |
| 3. Ancylidae Ephemeroptera | 1 | 2 S.U. | - | 5 | 6 | 7 | 8 |
| (Ecdyonuridae excepted) | 2 | 2 or 2 S.U. | 3 | 4 | 5 | 6 | 7 |
| 4. Aphelocheirus Odonata or Gammaridae or Mollusca (Sphaeridae excepted) | 0 | All S.U. above absent | 3 | 4 | 5 | 6 | 7 |
| 5. Asellus or Hirudinae or Sphaeridae or (Aphelocheirus excepted) | 0 | All S.U. above absent | 2 | 3 | 4 | 5 | - |
| 6. Tubificidae or Chironominae of the Thumni plumosus group | 0 | All S.U. above absent | 1 | 2 | 3 | - | - |
| 7. Eristalinae | 0 | All S.U. above absent | 0 | 1 | 1 | - | - |

* S.U. : Systematic Unit chosen.

## DIVERSITY INDICES

As pointed out in the Introduction, biological communities change with the changes in the environment in which they occur. From all methods discussed or quoted above one can extrapolate the same ecological rule : reduction of the number of species, and proliferation of some of the remaining ones, as a result of the introduction of biodegradable wastes into rivers or lakes. The concept of "pollutional stress" with subsequent decrease in the variety of species led the American scientist PATRICK and co-authors in 1954 to use the diatom community, growing on slides suspended in water for fixed periods, as a biological measure for pollution.

By plotting the number of species per interval, against the number of specimens per species, PATRICK found out that in an ordinary diverse habitat the mode of the resulting curve is high, and the curve itself short; in other words many species occur in small numbers and none is very abundant.

In a severe habitat on the contrary the mode is low and the curve long : there are few rare species and a few are present in large numbers (Fig. 11).

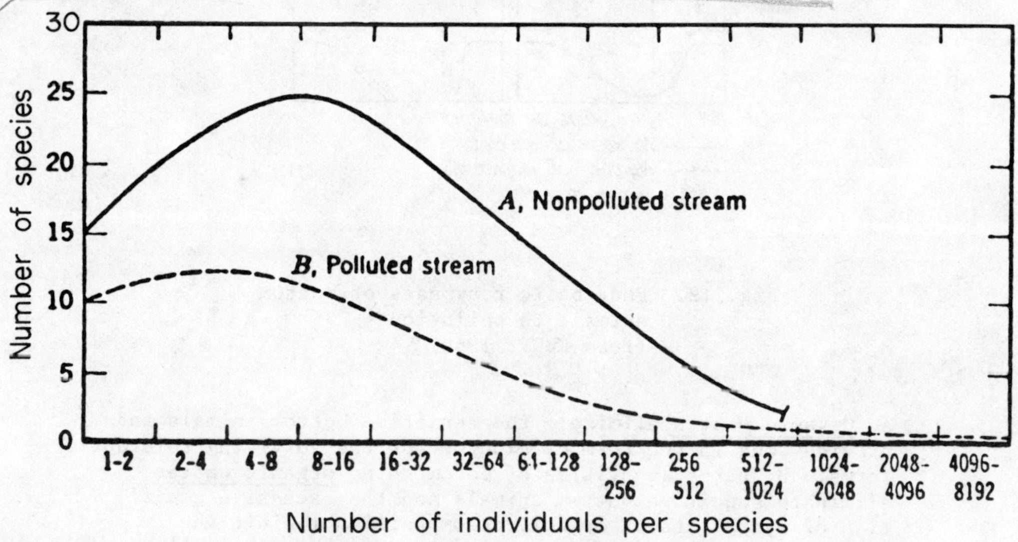


Fig. 11. A graphic comparison of diatom communities from two different environments. (After Patrick and co-workers, 1954).

When toxic materials are discharged, the community change is, however, quite different and results in a decrease of the number of species as well as the number of individuals.

Figure 12 (from KEUP, 1966) illustrates both phenomena.

The decrease in number of species inspired the Germain scientist KOTHÈ in 1962 to elaborate a "species - deficit - method" ("Artenfehlbetrag") as a biological procedure to quantify pollution.

The difference in the number of species occuring at a sampling site upstream of a point of discharge of wastes ($A_1$) and other ones (downstream ($A_X$) is expressed as a percentage of $A_1$, and expresses the "species deficit"

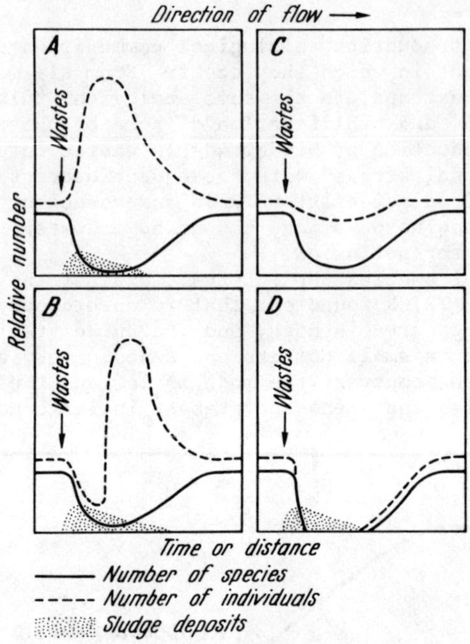


Fig. 12. Four basic responses of bottom animals to pollution (from KEUP, 1966)

A. Organic wastes eliminate the sensitive bottom animals and provide food in the form of sludges for the surviving tolerant forms. B. Large quantities of decomposing organic wastes eliminate sensitive bottom animals and the excessive quantities of byproducts of organic decomposition inhibit the tolerant forms ; in time, with natural stream purification, water quality improves so that the tolerant forms can flourish, utilizing the sludges as food. C. Toxic materials eliminate the sensitive bottom animals ; sludge is absent and food is restricted to that naturally occuring in the stream, which limits the number of tolerant surviving forms. Very toxic materials may eliminate all organims below a waste source. D. Organic sludges with toxic materials reduce the number of kinds by eliminating sensitive forms. Tolerant survivors do not utilize the organic sludges because the toxicity restricts their growth.

$$F = \frac{A_1 - A_X}{A_1} \times 100$$

F = 100 means total suppression of the biocoenosis ;
0 = no change of the biological communities

The results can be plotted graphically, with the length of the river stretch considered on the abscissa.
The weakness of the species deficit method lies in the fact that it is necessary to establish a species standard (from samplings at point no. 1, upstream the pollution outfall) which must be representative for the river stretch examined ; if the hydrographic character of the river changes, it is normal that even without pollution, the composition of the biological communities will change accordingly.
The changes in diversity of biological communities, as a result of the introduction of wastes in an aquatic ecosystem, have inspired several authors to utilize diversity indices as a type of biological indicator of pollution.
All of these indices are mathematical expressions of the relationship between the number of species (S) and the abundance (N).
WILHM and DORRIS (1968) calculated diversity indices (based on information teory) for biological communities above and below points of discharge of wastes in fresh as well as marine waters.
From table 9 it is clear that the community diversity decreases below the waste outfall ; with time and distance, the index, gradually comes back to values close to the original ones.

TABLE 9. Changes in diversity indices, $\bar{d} = -\Sigma(n_i/n)\log_2(n_i/n)$, based on information theory, with distance of flow below points of waste discharge into natural and other waters.
After WILHM and DORRIS (1968).

| Water | Waste | Diversity index ($\bar{d}$) | | | |
|---|---|---|---|---|---|
| | | Above outfall | Near outfall | Downstream | |
| Skeleton Creek | Domestic and oil refinery | | 0.84 | 1.59 | 3.44 |
| Skeleton Creek | Domestic and oil refinery | 3.75 | 0.94 | 2,43 | 3.80 |
| Otter Creek | Oil field brines | 3.36 | 1.58 | | 3.84 |
| Refinery Ponds | Oil refinery | | 0.98 | 2.79 | 3.17 |
| Keystone Reservoir | Dissolved solids | | 0.55 | | 3.01 |
| Alamitos Bay | Oil field brines | | 1.49 | 2.50 | |
| Alamitos Bay | Oil field brines | | 1.44 | 2.70 | |
| Alamitos Bay | Storm sewer | | 1.45 | 2.81 | |

Values of less than one are typical for heavily polluted zones, values of 1 to 3 for zones of moderate pollution, and values of more than 3 for clean waters.
To conclude we shall quote CAIRNS and co-authors (1968, 1971) who developed an elegant and simple method for the estimation of relative differences in biological diversity : the Sequential Comparison Index (S.C.I.).
According to the authors this index fills the need for a rapid numerical method of assessing the biological consequences of pollution. The S.C.I. method has the advantage that it requires no taxonomic expertise from the investigator and as such saves considerable times in analyzing the aquatic community.
The method is based on a modification of the "sign test" and the "theory of runs". The only criterion is the capability of the student to recognize differences in shape, color and size of organisms.
For teaching purposes the authors utilize a tableau with imaginary organisms (which they call "critterkids") drawn at random (Fig. 13).
A worksheet has to be filled out with only 2 symbols X and O, which have no other significance but to symbolize a change of kind of the organism analyzed with the previous one. If it is similar, it is part of the same run, and the symbol does not change ; if not, it is part of a new run and one utilizes the other symbol. The greater the number of runs per number of specimens examined, the greater the biological diversity, and so the healthier the community.
The SCI diversity index is equal to the number of runs divided by the number of specimens, times the number of kinds (taxa or species) of organisms present.

$$SCI = \frac{N_{runs}}{N_{specimens}} \times N_{taxa}$$

The number of kinds of organisms (taxa) is determined by the investigator after he has finished determining the number of runs.
Obviously the greater the taxonomic training of the student the greater his accuracy in determining the number of taxa present in a sample.
CAINRS claims that for reasonably healty stations approximaterly 250 specimens are necessary to calculate a moderately reliable index.
The S.C.I. index can be calculated for micro- as well as for macro invertebrate communities.
From our own experience we can mention that one of the major drawbacks of this simple and interesting method is the inherent difficulty of random sampling of individuals in invertebrate communities composed of species of widely diverging sizes.

## CONCLUSIONS AND PERSPECTIVES

The only conclusions which we would like to forward at the end of this short and incomplete review are 2 quotations of scientists which devoted the best of their efforts to the very difficult and extremely complex problem of trying to reduce a lot of biological data into a single system or a mathematical expression.
BARTSCH and INGRAM wrote the following in 1966 : "It appears that there are about as many procedures as the number of biologists working in the field. Many of these procedures are continually being modified so that this obviously is not an appropriate time to consider a single scheme for a standard method. In fact, one can well question whether a standard method is possible or even desirable."
SLADECEK (1973) is even more pessimist : "A great individualism among the hydrobiologists inhibits any standardization of procedures, classifications or schemes. Everybody tries to erect his own system, method, apparatus, or at least to modify

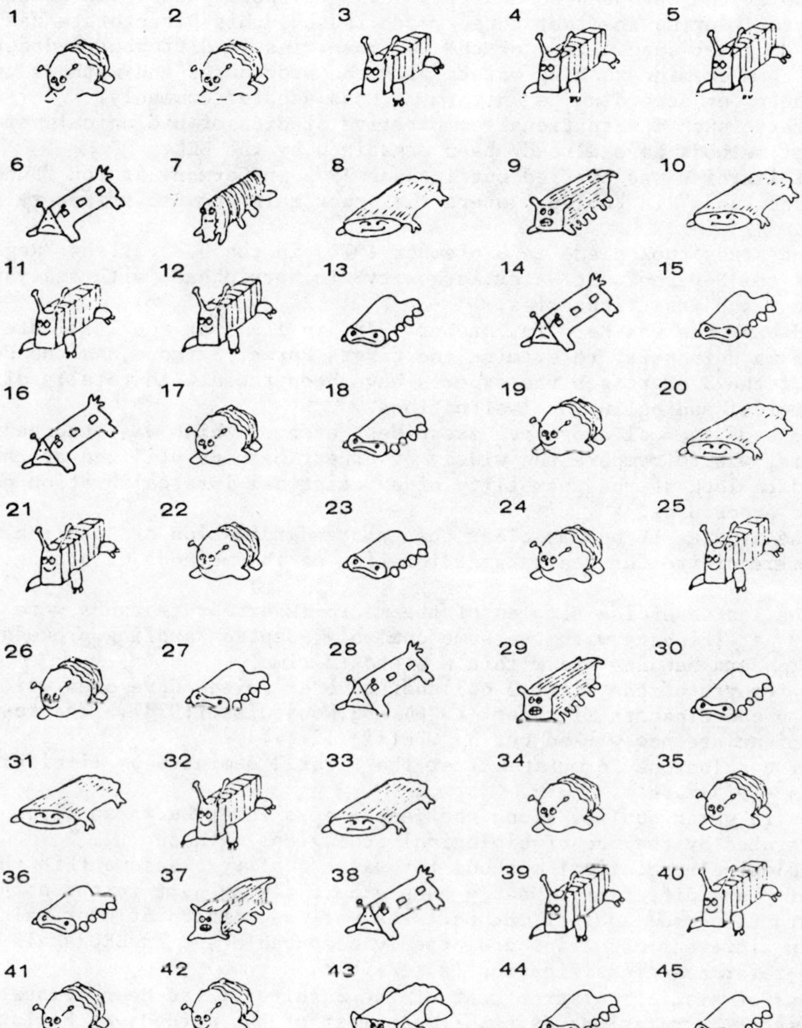


Fig. 13. "CRITTERKIDS" for teaching the SEQUENTIAL COMPARISON INDEX (from CAIRNS and co-workers, 1968)

something, because he needs something new, individual, to acquire a scientific degree or reputation...".
Considering this confused situation, one can but applaud the initiative of the Commission of the European Communities and more specifically the Health Protection Directorate. During the last three years indeed, this Directorate has endavoured to bring together specialists of the EEC countries in different hydrographic basins and made them examine running waters with the procedures and methods customary in their country or according to uniform methods adopted commonly.
To date three such international comparative studies of biological water quality assessment methods have already been organized by the EEC.
The first exercise was carried out in June 1975 in Germany at the "Bundesanstalt für Gewässerkunde" in Koblenz, where the lower Main and its tributary the Kahl were examined.
The second study took place in September 1976, in the U.K. at the "Regional Laboratory of the Severn-Trent Water Authority" in Nottingham, with analysis of the River Trent and its tributaries.
The third exercise was held in October 1978, in Italy at the Institute of Ecology of the Parma University to examine the rivers Parma, Stirone and the Po.
In each of the 3 exercises the experts have been faced with totally different biotypological and pollutional situations.
The purpose of the collaborative exercises, each of which was attended by more than 30 experts, was to compare the widely divergent methods utilized at the national level and to look at the possbility of a reciprocal intercalibration of the different procedures.
During the studies it became clear that a standardization of sampling methods was afirst prerequisite for the intercalibration of the methods of biological water quality.
During the last exercise samples of the macro-invertebrate fauna were taken by the different participants with the same commonly adopted sampling procedure i.e. with a standard handnet and within a standard time.
Detailed reports of the first 2 collaborative exercises have been written by the respective coordinators Tittizer (1976) and Woodiwiss (1978). The results of the third seminar are now worked out by Ghetti (1978).
The major conclusions formulated after the first 2 seminars by Tittizer (1976) and Woodiwiss (1978) are :

- Changes in water quality along the lenght of a large European river can be demonstrated by the use of biological-ecological methods.
- The biological-ecological methods for water quality assessment in the European Community are different. Nearly each expert taking part in the comparative investigation used another method. The various methods of expressing results used in different countries are broadly comparable and an EEC scale of comparability-reference classification is feasible.
- By means of statistical treatment of the results it has been demonstrated that significant correlations exist between most of the methods of biological assessment but that many are not linear.
- For the evaluation of water quality macrozoobenthic assessments were employed by most of the experts ; only a few used microphyton and microzoobenthic assessments (plankton or periphyton) for this purpose.
- The most commonly used assessments are the biotic index group of methods.
- The level of identification attempted for the different categories of organisms varies from country to country and reflects in part the quality and local availability of satisfactory taxonomic keys.
- Biological assessment values can be influenced by the level of identification attempted or achieved by different biologists.
- The Working group should consider the need for a list of the major groups of organisms giving sensible (i.e. reasonably attainable) levels of identification which should be attempted for routine surveillance and particularly as a basis for reporting at the international level.

- Different methods of sampling employed by biologists can give rise to significantly different biological analysis and consequently biological assessments.
- The working group should therefore consider ways of defining standard sampling procedures in order to reduce the effect of this variable to a minimum.

Let us conclude this review with the pertinent phrase of KNÖPP (1976) after the first comparative study in Koblenz, which in our opinion reflects at best the possibilities and limitations of biological water quality assessment methods : "biological-ecological analysis cannot replace chemical, physical or biochemical analysis, but can indeed significantly complement them. Only the complete spectrum of information provides an image of the quality condition with respect to water management uses".

## REFERENCES

Ambühl, H. (1959). Die Bedeutung der Ströming als Ökologischer Faktor. Schweiz. Z. Hydrol., 21, 133-264.

Bartsch, A.F., and W.M. Ingram (1959). Stream life and the pollution environment. Public Works, 90, 104-110.

Bartsch, A.F., and W.M. Ingram (1966). Biological analysis of water pollution in North America. Verh. Internat. Verein. Limnol., 16, 786-800.

Beak, T.W. (1964). Biological measurement of water pollution. Chem. Eng. Progress., 60, 39-43.

Beck, W.M. (1954). Studies in stream pollution biology. I. A simplified ecological classification of organisms. Quart. J. Fla. Acad. Sci., 17, 211-227.

Cairns, J., D.W. Albaugh, F. Busey, and M.D. Chaney (1968). The sequential Comparison Index. A simplified method to estimate relative differences in biological diversity in stream pollution studies. J. Wat. Poll. Contr. Fed., 40, (9), 1607-1613.

Cairns, J., and K.L. Dickson (1971). A simple method for the biological assessment of the effects of waste discharges on aquatic bottom dwelling organisms. J. Wat. Poll. Contr. Fed., 43, (5), 755-772.

Caspers, H., and L. Karbe (1966a). Proposals for a saprobiological classification of waters. WHO-EBL-66.80 (Geneva), 1-28.

Caspers, H., and L. Karbe (1966b). Trophie und Saprobität als stoffwechseldynamischer Komplex. Gesichtspunkte für die Definition der Saprobitätsstufen. Arch. Hydrobiol., 61, 453-470.

Caspers, H., and L. Karbe (1967). Vorschläge für eine saprobiologische Typisierung der Gewässer. Int. Revue ges. Hydrobiol., 52, 145-162.

Caspers, H., and H. Schulz (1960). Studien zur Wertung der Saprobiensysteme. Erfahrungen an einem Stadtkanal Hamburgs. Int. Revue ges. Hydrobiol., 45, 535-565.

Caspers, H., and H. Schulz (1962). Weitere Unterlagen zur Prüfung der Saprobiensysteme. Int. Revue ges. Hydrobiol., 47, 100-117.

Chandler, J.R. (1970). A biological approach to water quality management. Water Poll. Control., 69, 415-422.

Dittmar, H. (1959). Reicht das bisherige Saprobiensystem für die Gütebeurteilung eines Gewässers aus ? Forschung und Beratung, A(8), 263-265.

Elster, H.J. (1958). Das limnologische Seetypensystem, Rückblick und Ausblick. Verh. Internat. Verein. Limnol., 13, 101-120.

Fjerdingstad, E. (1964). Pollution of streams estimated by benthal phytomicroorganisms. I. A saprobic system based on communities of organisms and ecological factors. Internat. Rev. Hydrobiol., 49, 63-131.

Ghetti, P.F. (1978). 3th Technical Seminar - Biological Water assessment methods (Parma, 8-13 October, 1978). Working Paper. Commission of the European Communities, Environment and Consumer Protection Service.

Hynes, H.B.N. (1960). The biology of polluted waters. Liverpool Univ. Press., 202 pp.

Hynes, H.B.N. (1964). The interpretation of biological data with reference to water quality. In Symposium Environmental Measurements - Valid Data and Logical Interpretation. Public.Health Service - Publi. 999-WP-15 : 289-298.

Keup, L.E. (1966). Stream Biology for assessing sewage treatment plant efficiency. Water Sewage Works, 1-7.

Knöpp, H. (1954). Ein neuer Weg zur Darstellung biologischer Vorfluteruntersuchungen, erläutert an einem Gütelängschnitt des Maines. Die Wasserwirtschaft, 45, 9-15.

Knöpp, H. (1955). Grundsätzliches zur Frage biologischer Vorfluteruntersuchungen, erläutert an einem Gütelängschnitt des Maines. Arch. Hydrobiol. Suppl., 22, 383-368.

Knöpp, H. (1976). Comparative study of biological Water Assessment Methods. Demonstration on the river Main. In R. Amavis and J. Smeets (Eds.): Principles and Methods for determining ecological criteria on hydrobiocoenoses, Commission of the European Communities, Pergamon Press, Oxford, New York. pp. 263-272.

Kolkwitz, R. (1935). Pflanzenphysiologie 3rd ed., Fisher Verlag, Jena.

Kolkwitz, R. (1950). Okologie der Saprobien. Schr.-R. Ver. Wasser-, Bodem- und Lufthygiene, 4, 1-64.

Kolkwitz, R., and M. Marsson (1902). Grundsätze für die biologische Beürteilung des Wassers nach seiner Flora and Fauna. Mitt. Prüfungsanst. Wasserversorg. Abwasserreinig., 1, 33-72.

Kolkwitz, R., and M. Marsson (1908). Ökologie der pflanzlichen Saprobien. Ber. dt. Bot. Ges., 26A, 505-519.

Kolkwitz, R., and M. Marsson (1909). Ökologie der tierischen Saprobien. Internat. Rev. Hydrobiol., 2, 126-152.

Kothé, P. (1962). Der "Artenfehlbetrag", ein einfaches Gütekriterium und seine Anwendung bei biologischen Vorfluteruntersuchungen. Dt. Gewässerkundl. Mittl., 6, 60-65.

Liebmann, H. (1962). Handbuch der Frischwasser- und Abwasserbiologie, Vol. I, 2nd edit., R. Oldenbourg, München, 588 pp.

Mackenthun, K.M. (1969). The practice of water pollution biology. FWPCA, Washington, 281 pp.

Margalef, R. (1958). Temporal succession and spatial heterogenity in natural phytoplankton. In Perspectives in marine biology. Univ. of California Press. pp. 323-49.

Mez, C. (1898). Mikroskopische Wasseranalyse. Springer Verlag, Berlin.

Pantle, R., and H. Buck (1955). Die biologische Uberwachung der Gewässer und die Darstellung der Ergebnisse. Gas and Wasserfach, 96, 604.

Patrick, R. (1949). A proposed biological measure of stream conditions, based on a survey of the Conestaga basin, Lancaster Country, Pennsylvania. Proc. Acad. Nat. Sci. Philadelphia, 101, 277-341.

Patrick, R., M.H. Hohn, and J.K. Wallace (1954). A new method for determining the pattern of the diatom flora. Nat. Philad. Acad. Sci., 259, 1-12.

Richardson, R.E. (1928). The bottom fauna of the middle Illinois River 1913-1925 : its distribution, abundance, valuation and index value in the study of stream pollution. Bull. Illinois State Nat. Hist. Survey, 17, 387-475.

Sladecek, V. (1965). The future of the Saprobity System, Hydrobiologia, 25, 518-533.

Sladecek, V. (1973). System of Water Quality from the biological point of view. Arch. Hydrobiol. Beih, 7, I-IV, 1-218.

Schwoerbel, J. (1970). Methods of Hydrobiology - Freshwater Biology. Pergamon Press, London, 200 pp.

Sramek-Husek, R. (1956). Zur biologischen Charakteristik der höheren Saprobitätsstufen. Arch. Hydrobiol., 51, 376-390.

Suter, R., and R. Moore (1922). Stream Pollution Studies. State of N.Y. Conservation Commission. Albany N.Y., p. 3-27.

Tittizer, T. (1975). Comparative study of biological - ecological water assessment methods - Practical demonstration on the River Main (2-6 June, 1975). General Report. Commission of the European Communities Health and Safety Directorate, 55 pp.

Tuffery, G., and J. Verneaux (1968). Methode de détermination de la qualité biologique des eaux courantes. Exploitation codifiée des inventaires de la faune de fond. Ministère de l'agriculture (France), Centre National d'Etudes techniques et de recherches technologiques pour l'agriculture, les forêts et l'équipement rural "C.E.R.A.F.E.R.", Section Pêche et Pisciculture, 23 pp.

Warren, C.E. (1971). Biology and Water Pollution Control. W.B. Saunders Cy., Philadelphia, London, 434 pp.

Whipple, G.G., G.M. Fair and M.C. Whipple (1927). Microscopy of drinking water. J. Wiley, N.Y., 586 pp.

Wilhm, J.L., and T.C. Dorris (1968). Biological parameters for water quality criteria. Bioscience, 18, 477-481.

Woodiwiss, F.S. (1964). The biological system of stream classification used by the Trent River Board. Chemistry Industry, 14, 443-447.

Woodiwiss, F.S. (1976). Comparability study of Biological - Ecological Assessment Methods - Technical Seminar Nottingham, 26th Sept. - 1st. Oct., 1976 - Background information, Commission of the European Communities, Environment and Consumer Protection Service, 89 p.

Woodiwiss, F.S. (1978). Comparative study of Biological - ecological water quality assessment methods. Second practical demonstration, Nottingham (20 sept. to 1 Okt., 1976). Summary Report. Commission of the European Communities, Environment and Consumer Protection Service.

Wurtz, C.B. (1955). Stream biota and stream pollution. Sewage Industr. Wastes, 27, 1270-1278.

Zelinka, M., and P. Marvan (1961). Zur Präzisierung der biologischen klassifikation der Reinheid fliessender Gewässer. Arch. Hydrobiolog., 57, 389-407.

Zimmerman, P. (1961). Der Einfluss der Strömung auf die Zusammensetzung der Lebensgemeinschaften im Experiment. Schweiz. Z. Hydrol., 23, 1-81.

# The Evolution of Pollution Evidenced by Lake Sediment Pseudofossils

**D. A. Murray**

***Dept. of Zoology, University College Dublin, Dublin 4, Ireland***

## ABSTRACT

The micro-organisms that live within a lake are, in many cases, sensitive to changing environmental conditions and thus may be used as "indicator" organisms because there is a general equilibrium between a biotic community and the environment. Any change in the environment will alter the structure and perhaps composition of the community. Sediments that accumulate in a lake basin are derived from many sources and are the result of many physical, chemical and biological processes and are deposited, for the most part, chronologically. The remains of many organisms which live within a lake are preserved within the accumulating sediment as pseudofossils. The majority of pseudofossils seen in sediments belong to the Cladocera (Crustacea) and Chironomidae (Diptera). The use of pseudofossils of these groups, together with other sediment parameters especially phaeopigments, is outlined in this paper to demonstrate how such a study can provide evidence of changing conditions within a lake ecosystem.

Keywords: eutrophication, pseudo-fossils, Planktonic pigments, Sediment, Cladocera, Chironomids.

## INTRODUCTION

In their natural state lakes exhibit different degrees of trophic status ranging from the unproductive oligotrophic type through mesotrophic to the highly productive eutrophic type. During the last two or three decades many lakes in developed countries of the world have shown signs of an increasing rate of enrichment or eutrophication. Some lakes, which, under purely natural conditions would be categorized as oligotrophic, have reached a stage of high productivity. These changes, brought about by man's influence, are mainly due to an excessive input of plant nutrients. This paper presents some of the evidence of the evolution of eutrophication which may be derived from sediment analysis.

The term eutrophication refers to a change in status of a lake. Lakes may be oligotrophic, mesotrophic or eutrophic at the present time. Biological methods to assess the present state of a lake tell us little about its previous conditions. A lake which is eutrophic at the present time may be so as a result of natural evolution or as a result of man's influence (culturally eutrophic). This important distinction has practical implications in water management. In the naturally eutrophic lake reduction of nutrient input would have little effect on the trophic state, while in the artificially enriched lake a reduction could cause the lake to become less rich. It becomes apparent that if some historical perspective is available a more accurate assessment of the trophic status of a particular water body may be made. Occasionally, data on previous conditions within a lake are available and it is possible to see if changes in trophic status have occurred. In many cases, however, information on earlier conditions is not available and use must be made of the information contained in the lake sediments.

## SEDIMENT COMPOSITION AND EUTROPHICATION

The sediments that accumulate in a lake basin are derived from two main sources, but a variety of processes exist to modify the contribution from either source. Sediments are derived from sources external to the lake - allochthonous - and sources within the lake itself - autochthonous. The former include materials such as sand, silt, clay, minerals etc., from leaching and erosion, plant and animal remains or by-products and, more recently, human wastes. The autochthonous contribution to sediment comprises the remains of organisms which live in the lake and materials which are elaborated by these organisms. The nature of the material being deposited is controlled by the geology of the catchment area, climate, and a range of abiotic and biotic factors operating both in the lake itself and in it's watershed. Somewhat similar factors are involved in determining lake productivity or trophic status. Three basic factors, edaphic, morphometric and climatic, in a complex range of parameters influence lake productivity (Fig. 1). The consequences of changes in the main factors or modifications in derivatives of these may be interdependant because of the nature of inter-relationships involved. The sediments, however, result from the sum total of factors involved and thus reflect these processes. With the continued addition of sediments to the floor of a lake it follows that all lakes, to a greater or lesser extent, are gradually filling in. This gradual accumulation of sediments is part of the natural process of lake evolution.

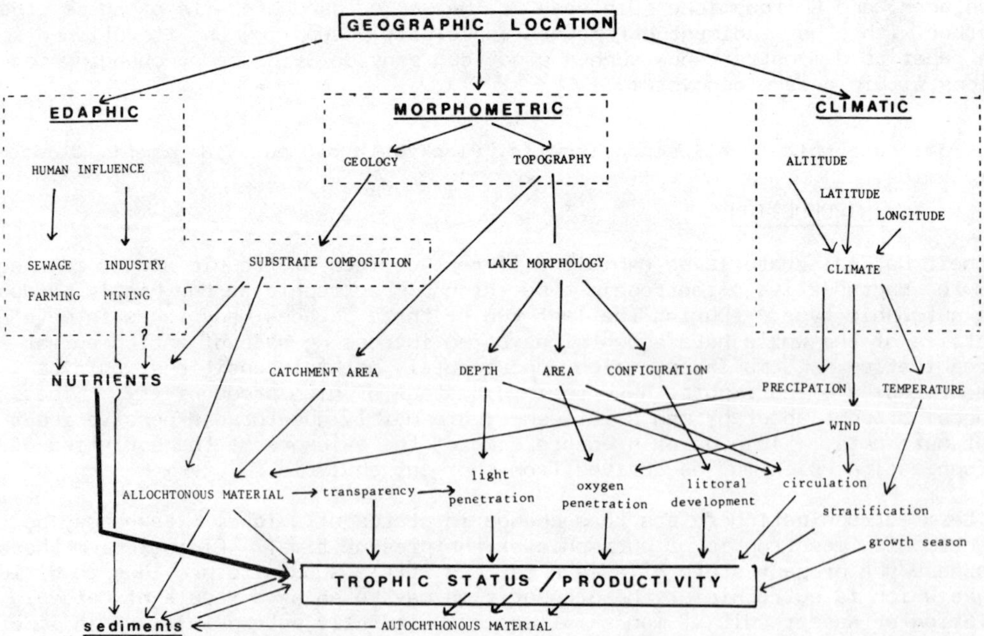


Fig. 1. Some factors influencing trophic status and lake sediment composition.

There are many schemes for delimiting lake zones, but basically two strata of water exist in all lakes (Fig. 2). In the trophogenic zone plant life can flourish. This is the euphotic area occupying the zone which receives sufficient light to allow photosynthesis. Beneath this lies the tropholytic zone where decomposition of sinking organic matter, produced in the trophogenic zone, occurs. The amount of oxygen used in this oxidative decomposition is related to the biomass of material falling from the trophogenic zone. As a lake gradually fills in the tropholytic zone decreases in volume, while the volume of the trophogenic may remain static. Thus, in a shallow lake, and especially one which stratifies in summer, there is less oxygen available for biological decay than in a deeper lake with similar trophogenic volume. Consequently periods of anoxia may occur in enriched or eutrophic lakes.

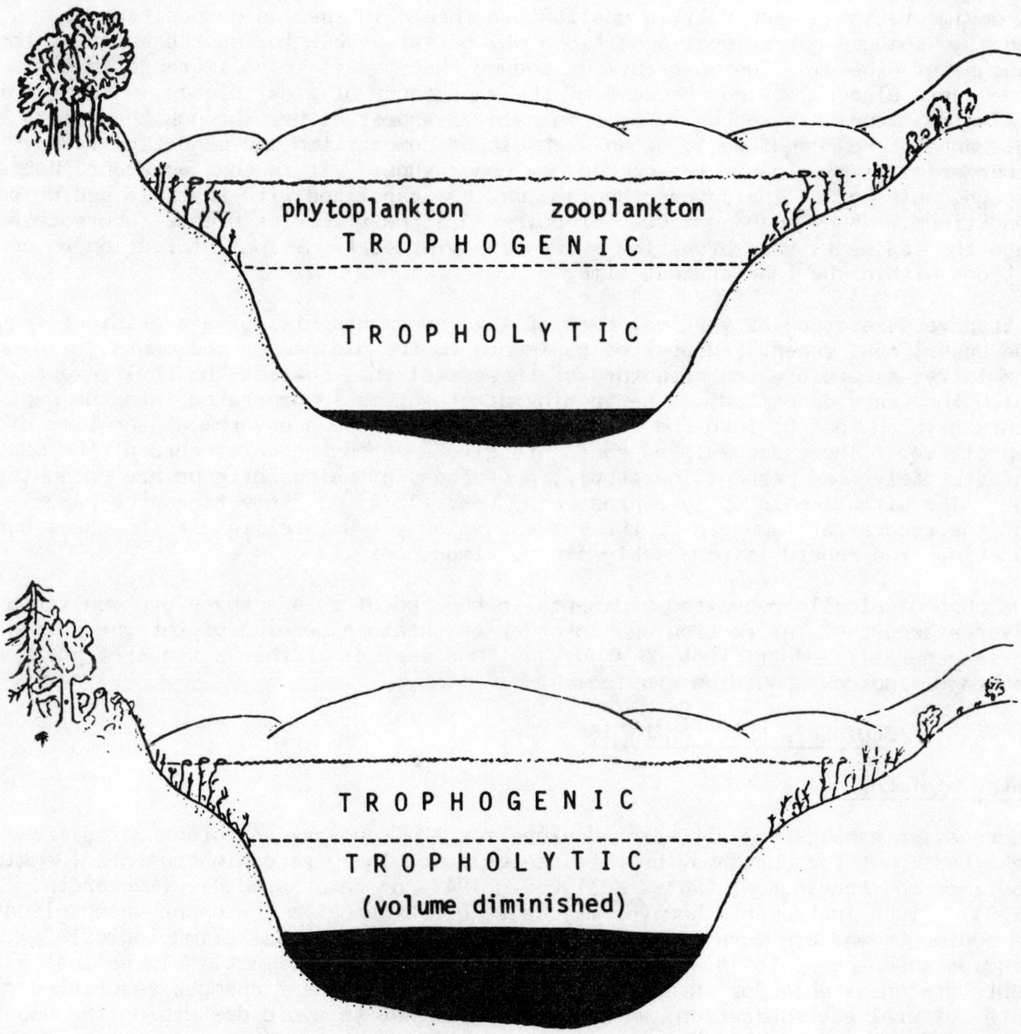


Fig. 2. Lake zonation showing decrease in relative volume of the tropholytic with age as sediments accumulate in the basin.

Allied to this concept of morphological eutrophication is the effect of nutrients on the system. Shallow lakes with a mean depth of less than 18m tend to be eutrophic, deeper lakes oligotrophic. However, if the drainage area provides an insufficient amount of nutrients a shallow lake may not become eutrophic. Conversely, a deep lake may become artifically eutrophic if supplied with unusual amounts of nutrients. It is apparent that lakes differ in their individual capability to absorb increases in nutrient level but shallow lakes are at a greater risk to such increases. During cultural eutrophication the condition of a lake changes and these changes result in two basic trends, which are also reflected in the sediments. Firstly, higher concentrations of nutrients become apparent, resulting in a quantitative increase in biomass with higher productivity rates and increased population densities of some species. As a direct consequence of increased productivity the rate of incorporation of the remains of these organisms into the sediment will also increase, thus contributing to an increase in sedimentation. Secondly, a qualitative change in species composition can occur. Changed nutritional conditions may be favourable for some species at the expense of others. The most obvious change that occurs is an increase in the blue green algae - obvious because of the appearance of algal blooms. Changes in the animal fauna also occur. Some animals disappear either through their intolerance of new conditions, or as a result of competition and predation from other animals which can tolerate the new conditions. It is thus apparent that a recognisable set of indicator organisms may be associated with specific ecological conditions. When such indicator organisms die their remains become incorporated into the sediment, thus providing a measurable property giving information on conditions within the lake at that time.

A transverse section or vertical core of lake sediments will give a gradient from the upper, new, recently deposited sediments to the old deeper sediments. The chemistry, minerology and structure of these sediments reflect the environment in which they were deposited. The remains of organisms incorporated into the sediments give an insight into the biological community and ecosystem at the time of deposition. These accumulated sediments give a chronological record of the past. Unfortunately, and perhaps inevitably, occasional omissions or gaps are encountered. The situation is aptly expressed by Frey (1974) as "some times the pages (of the record) are smudged or the words form as yet unintelligible languages but sometimes the record is remarkably clear, almost epic."

The chronologically deposited sediments in the bed of a lake therefore contain a diverse amount of information and in order to obtain a meaningful interpretation it is generally insufficient to study just one aspect; rather a range of parameters or indices should be studied.

## SEDIMENT CORE ANALYSIS

### Methods of Coring

There exist a number of differing devices for the recovery of intact cores from lake beds. Large cores may be obtained using cable operated instruments (Jenkin, Mortimer and Pennington, 1941; Kullenberg 1947) or compressed air (Mackereth, 1958). Such instruments however may cause disturbance in the upper unconsolidated sediments and are more suited to classical Quarternary paleolimnological studies spanning up to 15,000 years and perhaps 10m of sediment. These instruments are unsuitable for investigations dealing with recent changes associated with cultural eutrophication, where it is important to avoid disturbing the upper sediments.

Corers designed to obtain upper levels of sub-aqueous sediment are of varying types. Some use extension rods as in the modified Livingstone corer (Rowley and Dahl, 1956). Cable operated corers are less complicated. The Jenkin sampler

(Mortimer, 1942) obtains intact cores, but only up to 20 to 30cm in depth. The Shapiro corer (Edmondson, 1971) uses weights to drive a core tube past a stationary piston into the sediment. Similarly the Davis/Doyle corer (Davis and Doyle, 1969) uses lead weights and incorporates a signal device to indicate when the instrument has reached the sediment surface. The corer described by Murray (1976) uses elastic rubber to propel a core tube past a stationary piston into the sediment. Most of these corers have advantages as well as disadvantages in mode of operation, but all have the common feature of obtaining virtually undisturbed upper sediment profiles.

## Sediment Dating

For all sediment core analyses it is important that accurate dating be carried out, so that the results obtained from chemical or biological analyses may have a meaningful interpretation with a time perspective. The events associated with cultural eutrophication are relatively recent, so that the older classical methods of paleoecological dating are of comparatively little use. Analysis of sediments for $^{14}C$ is of little value for events of the last 150 years. Similarly, the construction of pollen diagrams may in many cases yield poor results, unless some major vegetational change has occurred in the vicinity of a lake, for example, forest felling or plantation, or introduction of exotic species. In the U.S.A. the abrupt appearance and increase of *Ambrosia* pollen (ragweed) in sediment profiles is associated with the arrival of european settlers and land disturbance for agriculture. This marker has been used in many American investigations (Bradbury and Megard, 1972).

Occasionally special events other than vegetational changes provide time marks. Hutchinson (1970) demonstrated that the construction of the via Cassia created a major disturbance in the Lago di Monterosi, a closed basin of volcanic origin. Following the construction of this road in 171 B.C. productivity increased markedly leading to a maximum sediment organic content. However, this hardly comes within the sphere of recent times. Documented records of natural events such as landslides or artificial raising or lowering of water level can give valuable information. In 1916 the level of Lake Washington was lowered, resulting in the redeposition of littoral sand in the profundal sediments; this horizon is apparent when organic content as loss in ignition is determined (Edmondson, 1969).

If the annual addition to sediment remained as a distinct layer it would be possible to count these layers to establish an absolute chronology. Such layers do occasionally occur as varves in some lakes, and are the result of recurring seasonal events. Even where such layers are not visible to the naked eye the use of x-ray densitometry and scanning electron microscopy may reveal cryptic laminations (Edmondson and Allison, 1968; Edmondson, 1975).

In the absence of obvious time markers it is possible to obtain accurate dates for sediment cores by analysing the radionuclides of $^{137}Cs$ and $^{210}Pb$. Sediments up to 120 years old have been dated by a lead isotope technique (Robbins and Eddington, 1975; Pennington and co-workers, 1976; Oldfield, Appleby and Battarbee, 1978). Time horizons for the more recent sediments deposited within the last 24 years have been established by determining the distribution of fall-out $^{137}Cs$ derived from nuclear weapons testing. Significant levels of this radionuclide were first detected in 1954 and reached a peak in 1963. According to Armstrong and Gloyra (1969) caesium contained in rain falling on a lake's surface is absorbed on suspended material and becomes part of the process of sediment deposition. Ritchie and McHenry (1973) suggested that caesium becomes attached to fine particles on the soil surface in the lake's catchment and is subsequently transported to the lake. This method of dating has proved successful in U.S.A. (Ritchie and McHenry, England (Pennington, Cambray and Fisher, 1973), Italy (Ravera and Premazzi, 1971) and Ireland (Fig. 3) and probably has world wide application.

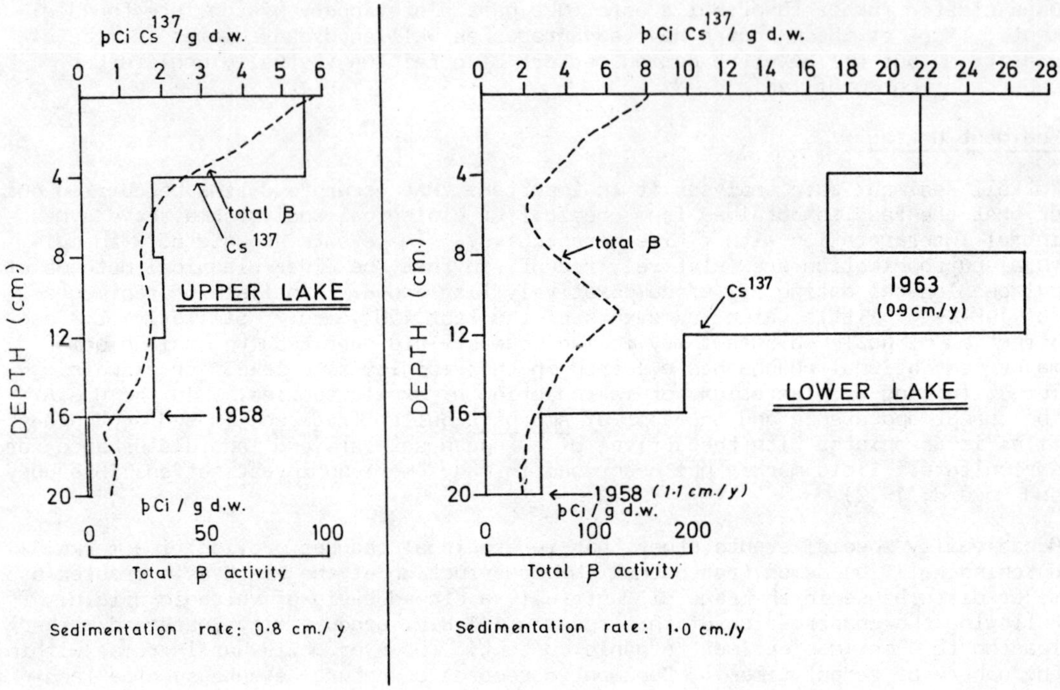


Fig. 3. Distribution of $^{137}$Cs in sediments from two Irish lakes to estimate sedimentation rate.

Lake Sediment Fossils or Pseudofossils

The morphological and biochemical remains of organisms in sediments constitute a fossil record. Practically all groups of organisms living in inland waters leave at least some morphological remains and or organic biochemical breakdown products, all of which are potentially useful as pseudofossil indicators of previous lake productivity.

Morphological pseudofossils. Long term changes in climate and changes brought about by eutrophication induce changes in the biota of a lake. The animal remains of this biota which are most frequently encountered in lake sediments belong to the Cladocera and Chironomidae. A review of the assemblage of microfossils likely to be encountered has been given by Frey (1964).

Cladoceran pseudofossils. The Cladocera, in common with other Crustacea, grow by periodic ecdysis of the chitinous exoskeleton. The cast exuviae, and ultimately the organism when it dies, are rapidly disarticulated into their component parts.

These cladoceran pseudofossil remains, generally present in numbers up to several thousand per $cm^3$ of fresh sediment, are sufficiently abundant to allow the construction of close interval stratigraphs. Many of the cladocera are sensitive to changing environmental conditions and are very useful as indicator organisms. Changes in extant plankton associated with eutrophication of Lake Lugano have been reported by Ravera (1977). It is apparent, therefore, that this group of aquatic organisms is important in attempts to demonstrate previous conditions of a lake.

The Cladocera may be divided into two distinct habitat groups - planktonic forms in the open water and substrate forms which are found mostly in the littoral zone. The more important planktonic cladocera belong to the families Bosminidae and Daphniidae. The substrate cladocera chiefly belong to the Chydoridae, but occasionally Sida (Family Sididae) is an important component. The exoskeletal remains of cladocera preserved in lake sediments include the shell, ephippia, head shield, post abdomen, mandible and a few minor fragments, such as antennules and legs. Identification to species of the fragments is not always possible. All skeletal components of Bosminidae may be preserved in the sediment, but only the head shield may be identified to species. Because of this, most identifications proceeds only as far as coregoni group or the longirostris group, which can be separated easily by the location of the lateral pores on the head shield. The daphniid remains are also difficult to identify to species level and mostly such remains may be identified to genus only. A much more satisfactory situation exists with the Chydoridae, where, as a result of morphological differentiation of exoskeletal component parts, most species can be identified with certainty. Frey (1976) pointed out that some cladoceran species have more value as indicators of changes due to cultural eutrophication than others, but that for general paleoecological studies interpretation of results should preferably be on a community basis. The work of Whiteside (1970) demonstrates that definable cladoceran communities may be associated with lakes of different degrees of productivity. Of particular relevance to the present discussion is the representation of two species, Chydorus sphaericus and Alona rectangula, which have a dominant representation in polluted waters. These two species are cited by Frey as being closely associated with enriched conditions. Both are substrate dwellers and are normally part of the littoral community. However, when algal blooms develop as a result of eutrophication, these species may become planktonic, using algae as a substrate and increase rapidly in numbers. This situation is reported at the present time in Lake Lugano (Ravera, 1977). The applications of these observations may be seen in sediment cores. In a study of Lake Shagawa, U.S.A. Bradbury and Megard (1973) and Bradbury and Waddington (1973) record a marked increase in abundance of C. sphaericus coinciding with an increase in nutrient input. Recent investigations in L. Ennell in Ireland reflect a similar situation, where C. sphaericus increases in number in association with eutrophication (Fig. 4). In L. Leane another Irish lake the distribution of A. rectangula provides evidence of an earlier phase of ecosystem disturbance (Fig. 5).

As indicated above a community approach to paleoecological studies may in some cases be more meaningful than a study of the distribution of single or a few indicator species. Multiple discriminant analysis can be a useful exercise by using community structure of sediment cladoceran microfossils to gain detailed insight into past conditions. Goulden (1969) has shown that in cladoceran populations both the number of species (S) and the Shannon-Wiener diversity function (H) tend to increase in log-linear fashion. Goulden's results indicate that the cladoceran taxocene strives toward an equilibrium over time and that this trend may be related to the MacArthur broken stick statistical model. (See also Frey 1974, 76). This observation has an important bearing when studying "disturbed" ecosystems such as culturally eutrophicated lakes. For M individuals distributed among N species the theoretical distribution of relative abundance, according to the MacArthur broken stick model, can be calculated and compared with the actual composition. In an undisturbed situation the agreement is very close Departures from this agreement may be the result of a number of

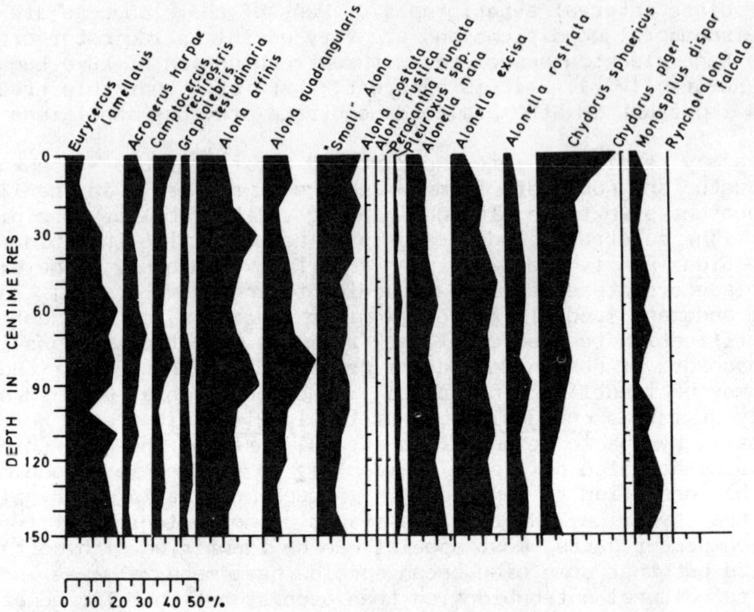


Fig. 4. Cladoceran stratigraphy in core from L. Ennell, Ireland.

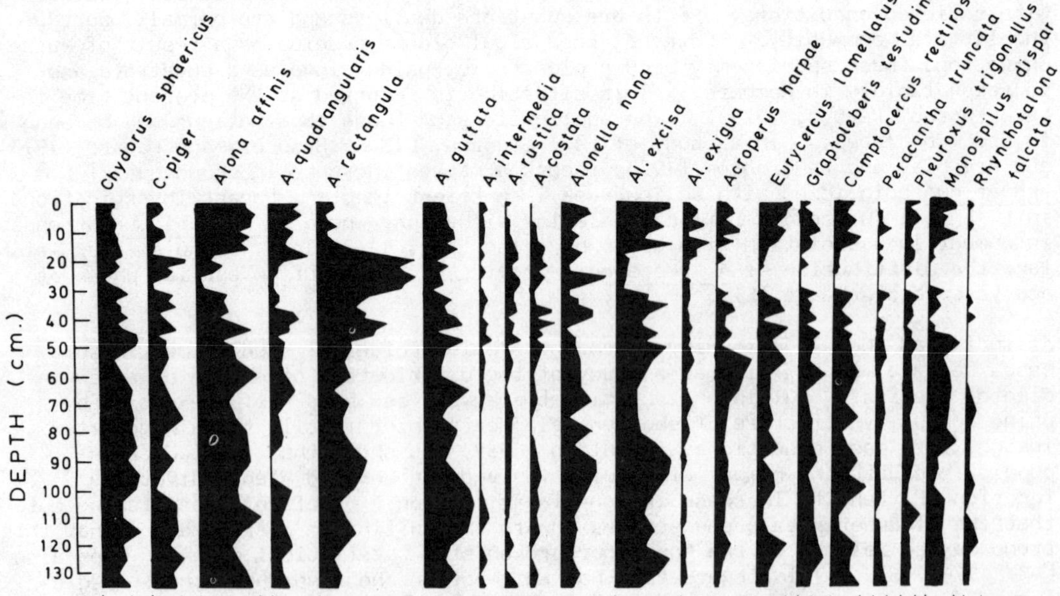


Fig. 5. Cladoceran stratigraphy in core from L. Leane, Ireland.

influences including disturbances caused by agriculture (tillage, tree felling etc.) or eutrophication.

Chironomid (Diptera) pseudofossil remains. The value of larval remains of this aquatic insect group lies in their use as indicator organisms. Of particular importance are the deep water profundal species whose distribution and abundance is related to the seasonal minimum oxygen content of the hypolimnion.

In the successional changes of a lake undergoing eutrophication oxygen in the hypolimnion may become periodically depleted and thus its availability is a limiting factor. As the oxygen content declines species requiring higher levels of oxygen are replaced by those able to tolerate progressively lower levels. The classical typology of lakes, based on chironomid species present in the profundal sediments, as defined by Thienemann, places lakes in a series. Oligotrophic lakes have a profundal fauna dominated by Tanytarsus species, mesotrophic lakes have Stictochironomus and Sergentia, while eutrophic lakes have Chironomus larvae. The typology has been refined by Brundin (1956) and Saether (1974). Although it is apparent that any rapid change in trophic status will be reflected in the chironomid fauna certain difficulties remain with this group. The taxonomic diversity of these insects is very much greater than the cladocera but much of this diversity is apparent in the adult stage and thus of no use in sediment studies. The larval head capsules which are preserved as pseudofossils, exhibit a smaller morphological differentiation than the cladocera. Very often the study of chironomid remains must be carried out in association with some specialist in the group. Allied to these difficulties the low numbers of remains encountered (usually less than 100/cm$^3$ of sediment) make this a difficult group to work with. Nevertheless much useful substantiative evidence may be obtained from analysis of their pseudofossils in sediment cores. A review of the paleolimnology of the group has been given by Frey (1964), Stahl (1969) and Hofmann (1971). One of the more recent studies of chironomid pseudofossils to interpret the typological succession of a lake is the work of Warwick (1975). In the Bay of Quinte, Lake Ontario, five phases of development are recognised in connection with human influence in the area. Following tree felling in the immediate vicinity the chironomid community first became more oligotrophic as a result of increased inorganic input to the lake. Subsequently an imbalanced oligotrophic phase is recognised, a brief mesotrophic phase and finally the eutrophic stage of the upper sediments. The time span involved in the succession is only 200 years. Clearly the Chironomidae have a use as indicators of changing trophic status.

Some cautionary words are required for the interpretation of the pseudofossils of Cladocera and Chironomidae, particularly in association with eutrophication or pollution. Many factors may be involved in changing the biotic community of a lake. Changes brought about by natural processes such as climate or landslides may be such as to mask the effects of nutrient enrichment. The response of certain invertebrate species to predation by fish must also be considered, especially if fish stocks change radically as a result of man's activities (overfishing, introduction of new species etc.). The effect of fish predation on populations of Bosmina (Cladocera) has been shown by Brooks (1969). Thus in any biological core stratigraphy the interpretation of results must consider whether changes in the distribution with time of indicator species is caused directly by the input of nutrients per se or by the indirect result of eutrophication on some other predator, competitor or prey species?

Biochemical "pseudofossils." Organic matter in lake sediments can have its origin from within or outside the lake. Photosynthetic pigments, or their derivatives, from bacteria, algae and higher plants are found in sediments long after the disintegration of the organisms that produced them. There are thus two potential sources of pigments in lake sediments - terrestrial and aquatic. There has been much discussion on the relative importance of both contributions in different lakes. Studies on sedimentary pigments by two dimensional chromatography to establish pigment diversity have yielded results which support the idea that most of the sedimentary organic matter in productive lakes is autochthonous rather than allochthonous (Gorham 1960, Sanger and Gorham 1972, Gorham and co-workers, 1974).

The concentration of sediment chlorophyll degradation products in relation to the total organic content has proved to be an indication of trophic conditions (Fogg and Belcher 1961; Gorham 1960, 1961; Wetzel 1970; Sanger and Gorham 1972). A good correlation was found between algal standing crop, epilimnetic chlorophyll and sedimentary pigments in some of the lakes of the English lake district (Gorham and co-workers, 1974). The results obtained by these authors give validity to the use of sedimentary pigments to elucidate past productivity. In general, pigments may be easily and rapidly extracted by 90% acetone and concentrations measured by spectrophotometer at wavelength of 480nm, 665nm and 750nm for carotenoids, chlorophyll derivatives and bacterio chlorophylls (Vallentyne 1955, Gorham and Sanger 1964, 1967; Sanger and Gorham 1972).

The use of sediment pigments as trophic indicators is dependant on a reasonable preservation in the sediments of the lake bed. The cool dark and anoxic conditions prevailing in the profundal of a lake are ideal for pigment preservation. Results of sediment pigment determinations in short cores from some Irish lakes are shown (Fig. 6). Sediment pigment studies are not restricted to recent sediments, Wetzel (1970) recovered pigment from sediments deposited 14,000 years before present, while such pigments have also been recovered from sediments at all levels in the 200m core from Lake Biwa in Japan.

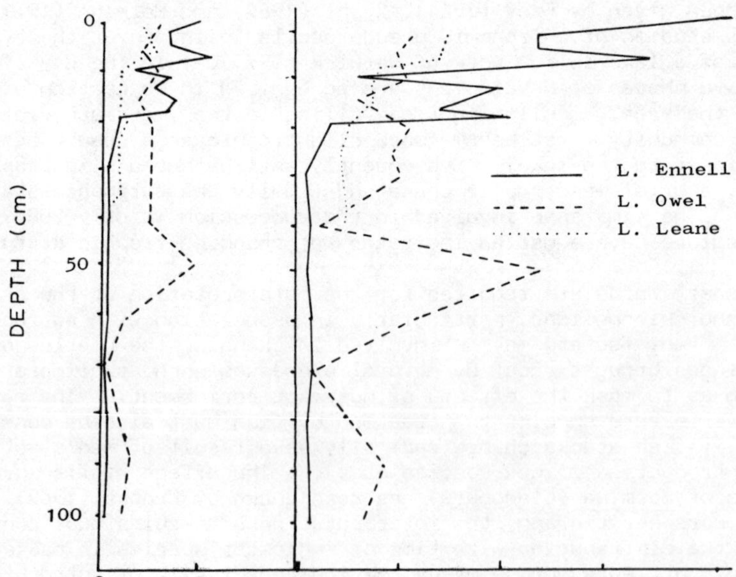


Fig. 6. Concentrations of sediment chlorophyll degredation products (left side) and sediment carotenoids (right side) from three Irish lakes. Results expressed as units per gm. organic weight.

Specific pigments are of use in certain circumstances. Oscillaxanthine is a pigment associated with certain blue green algae, for example, it occurs in Oscillatoria rubescens and O. agardhii, both of which are frequently nuisance species associated with algal blooms. Oscillaxanthine has been extracted from Lake Washington cores (Griffiths et al 1975) and the results clearly reflect the period when Oscillatoria was dominant in the lake. Cores from L. Ennell in Ireland, analysed for this pigment, likewise document the comparatively recent increase in the population of Oscillatoria (Fig. 7).

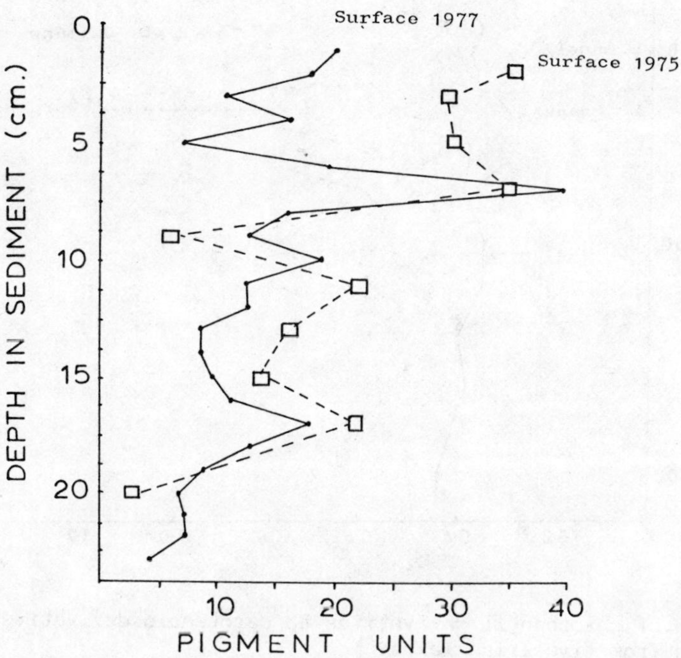


Fig. 7. Vertical distribution of Oscillaxanthine (a, 1977 core units/5gm and Carotenoids (b, 1975 core units $gm^{-1}$) in cores taken at the same location in L. Ennell two years apart. The vertical displacement of the two lines is arranged to compensate for addition of sediment to the lake bed based on the known recent sedimentation rate obtained from $^{137}Cs$ analysis.

Finally, in relation to pigments, mention should be made of the ratio of chlorophyll derivatives to carotenoids as an index of productivity. Fogg and Belcher (1961) and Moss (1968) have shown that the above ratio of these pigments decreases with increasing lake productivity and decreasing hypolimnetic oxygen. Gorham and co-workers (1974) quote values of 0.61 to 1.37, (mean .74) for six rocky unproductive lakes and values of 0.33 to 0.79, (mean .53) for seven productive lakes of the English lake district.

Current studies in Ireland give values in the surface sediment (reflecting the recent productivity) ranging from 0.17 for the highly enriched L. Ennell to 0.80 to the oligotrophic upper lake at Killarney. Similar analyses of sediments from L. Comabbio, Varese and Monate (Italy) give values of 0.25, 0.33 and 0.56 respectively to substantiate the present classification, i.e. eutrophic to oligotrophic.

Analysis of this ratio at close intervals in short cores may provide a useful index to changing productivity and, if time horizons are established, to the rate of change. Results of such analysis from five Irish lakes are given (Fig. 8). It is obvious that, of the five profiles shown, the rate of change has been most marked in L. Ennell. This trend is consistent with results obtained from other analyses - supporting the use of sediment pigments as productivity indices.

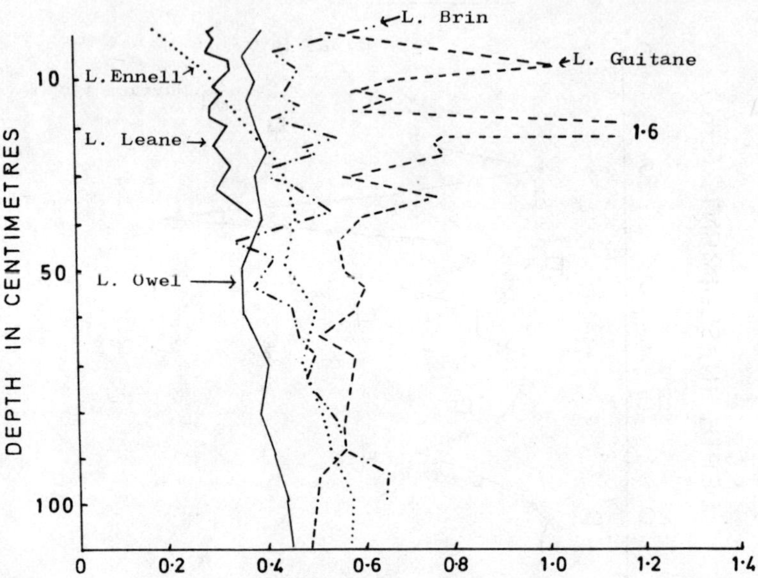


Fig. 8. Ratio of chlorophyll derivatives to carotenoid derivatives in sediment cores from five Irish lakes.

The results obtained by Gorham and co-workers (1974) show highly significant correlation coefficients between sediment pigments and mean standing crops of algae and strongly supports the use of pigments as indices of lake productivity. Frey (1974) concurs with this view and, furthermore, draws attention to the applicability of such studies in determining past levels of productivity.

Normally, the results of an ecological survey concerned with pollution or eutrophication can demonstrate that a problem has arisen, due to man's influence, and clearly show the nature of the problem and suggest remedies. However, a study of the chronologically deposited sediments can detect signs of an impending eutrophication problem. The present short survey demonstrates that much information may be obtained about the past trophic status of a lake by analysis of its sediments. From a careful analysis of a few selected parameters it is possible to say in advance, that a problem would be likely to occur if the present state of nutrient input were to be maintained. The establishment of a lake's evolutionary history will show if the lake has a latent eutrophication tendency. If the sedimentary record contains evidence of an accelerating productivity this may be considered a warning sign.

The present paper has been limited to biological pseudofossils and no mention has been made of the worth of preparing chemostratigraphies for sediment cores. Chemical analysis of sediments may give important additional evidence to changing environmental conditions. Ideally core analysis should encompass a range of biological indices associated with chemistry of the sediments. The main limitation is the time and effort input needed to obtain meaningful results. Nevertheless the sediments accumulating on the floor of a lake contain a vast amount of information and a record of how man has polluted, and is continuing to pollute, his aquatic environment.

ACKNOWLEDGEMENTS

The following sources of funds for the Irish paleolimnological studies referred to in this paper are gratefully acknowledged; National Science Council of Ireland, URG 31/74 and 51/76; part of the work has been carried out under contracts 095-74 -11 ENV. EIR and 217-77-1 ENV. EIR of the Environmental Research Programme of the European Economic Community.

REFERENCES

Bradbury, J.P. and Megard, (1972). Stratigraphic Record of Pollution in Shagawa Lake, Northeastern Minnesota. Bull. Geol. Soc. Am. 83 2639-2648.

Bradbury, J.P. and Waddington, J., (1973). The impact of European settlement on Shagawa Lake, Northeastern Minnesota, U.S.A. In Quarternary Plant Ecology. Birks & West (Ed.). Blackwell, London.

Brooks, J.L., (1969). Eutrophication and changes in the composition of the Zooplankton, in Eutrophication, causes, consequences, correctives. Proc. Internat. Symp. Nat. Acad. Sci. Wash. DC.

Brundin, (1956). Die bodenfaunistischen Seetypen und ihre Anwendbarkiet Auf die Südhalbkugel. Rep. Inst. Freshw. Res. Drottingholm 37.

Davis, R.B. and Doyle, R.W., (1969). A Piston Corer for Upper Sediment in Lakes. Limnol & Oceanogr., 14, 643-647.

Edmondson, W.T., (1969). Cultural Eutrophication. Mitt. int. Ver. Limnol 17, 19-32.

Edmondson, W.T., (1975). Microstratification of Lake Washington sediments. Ver. int. Ver. Limnol 19, 770-775.

Edmondson, W.T. and Winberg, G.G., (1971). IBP Handbook 17, Blackwell, London.

Edmondson, W.T. and Allison, D.E., (1970). Recording densitometry of X-radiographs for the study of cryptic laminations in the sediment of Lake Washington. Limnol. Oceanogr. 15 138-144.

Fogg, G.E. and Belcher, J.H., (1961). Pigments from the bottom deposits of an English lake. New Phytol 60, 129-142.

Frey, D.G., (1964). Remains of animals in Quarternary Lake deposits and their interpretation. Ergebn. Limnol 2, 1-114.

Frey, D.G., (1969). The rationale of Paleolimnology. Mitt int. Ver. Limnol. 17, 7-18.

Frey, D.G., (1974). Paleolimnology. Mitt. int Ver. Limnol. 20, 95-123.

Frey, D.G., (1976). Interpretation of Quarternary Paleocology from Cladocera and midges and prognosis regarding usability of other organisms. Can. J. Zool. 54, 2208-2226.

Gorham, E., (1960). Chlorophyll derivatives in surface muds from the English Lakes. Limnol. Oceanogr. 5, 29-33.

Gorham, E., (1961). Chlorophyll derivatives, sulphur and carbon in sediment cores from two English Lakes. Canad. J. Bot. 39, 333-338.

Gorham, E., Lund, J., Sanger, J. and Dean, W., (1974). Some relationships between algal standing crop, water chemistry and sediment chemistry in the English Lakes. Limnol. Oceanogr. 19, 601-617.

Gorham, E. and Sanger, J., (1964). Chlorophyll derivatives in Woodland Swamp. In Y. Miyake and T. Koyama, (Ed.), "Recent researches in the fields of hydrosphere, atmosphere and nuclear geochemistry" - Maruzen.

Gorham, E. and Sanger, J., (1967). Plant pigments in woodland soils. Ecology 48, 306-308.

Goulden, C.E., (1969a). Interpretative studies of Cladoceran microfossils in lake sediments. Mitt. int. Ver. Limnol 17, 43-55.

Goulden, C.E., (1969b). Temporal changes in diversity, and stability in ecological systems. Woodwell and Smith (Eds.) Brookhaven Symposium No. 22, VII.

Griffiths, M., Perrot, P., and Edmondson, W.T., (1975). Oscillaxanthin in the sediment of Lake Washington. Limnol. Oceanogr. 14.

Hofmann, W., (1971). Die postglaziale Entwicklung der Chironomiden und Chaoborus fauna des Schöhsees. Arch. Hydrobiol. Suppl. (40).

Hutchinson, G.E., (1970). Janula: an account of the history and development of the Lago di Monterosi, Latium, Italy. Trans. Amer. Phil. Soc. NS. 60 (4), 1-178.

Jenkin, B., Mortimer, C., and Pennington, W., (1941). The study of lake deposits. Nature, 147, 496-500.

Kullenberg, B., (1947). The Piston Core Sampler Svenska Hydrograf. Biol. Komm. Skrifter Ser. 3 Hydrograf 1.

Mackereth, F.J., (1958). A portable core sampler for lake deposits. Limnol. Oceanogr. 3, 142-145.

Mortimer, C.H., (1942). The exchange of dissolved substances between mud and water in lakes. J. Ecol. 30. 147-201.

Moss, B., (1968). Studies on the degredation of chlorophyll a and carotenoids in fresh water. New Phytol. 67, 49-59.

Murray, D.A., (1976). A lightweight corer for sampling soft subaqueous deposits. Limnol. Oceanogr. 21, 341-344.

Oldfield, F., Abbleby, P. and Battarbee, R., (1978). Alternative 210pb dating: results from the New Guinea Highlands and Lough Erne. Nature, 271, 339-342.

Pennington, W., Cambray, R.S., Eakens, J. and Harkness, D., (1976). Rationuclide dating of the recent sediments of Blelham Tarn. Freshwater Biology 6, 317-331.

Ravera, O., (1977). Effects of eutrophication on the Zooplankton of a sub-Alpine lake: Lake Lugano - In L. Downey and G. NiUid (Eds.) Lake Pollution pre vention through eutrophication control. National Science Council of Ireland, Stationary Office, Dublin.

Ravera, O., and Premazzi, G., (1971). A method to study the history of any persistant pollution in a lake by the concentration of $^{137}Cs$ from fall out. Publication No. 679. Biology Directorate EEC 1971.

Rowley, J.R., and Dahl, A., (1956). Modifications in design and use of the Livingstone piston sampler. Ecology 37, 849-851.

Robbins, J.A. and Eddington, D.N., (1975). Determination of recent sedimentation rates in Lake Michigan, using $^{210}pb$ and $^{137}Cs$. Geochemica et. Cosmochima Acta, 39, 285-300.

Ritchie, J.C., and McHenry, J.R., (1976). Rapid method for determining recent deposition rates of freshwater sediments. In H. Golterman (Ed.), Interactions between Sediments and Water. Junk. The Hague. pp. 203-207.

Saether, O.A., (1974). Nearctic chironomids as indicators of lake typology. Verh. int. Ver. Limnol. 19, 3127-3133.

Sanger, J. and Gorham, E., (1972). Stratigraphy of fossil pigments as a guide to the post glacial history of Kirchner Marsh, Minnesota. Limnol. Oceanogr. 17, 840-854.

Stahl, J.B. (1969). The uses of chironomids and other midges in interpreting lake histories. Mitt. int. Ver. Limnol. 17, 111-125.

Vallentyne, J.R., (1955). Sedimentary chlorophyll determination as a paleobotanical method. Canad. J. Bot. 33, 304-313.

Warwick, W.F., (1974). The impact of man on the Bay of Quinte. Lake Ontario, as shown by subfossil chironomid succession. Verh. int. Ver. Limnol. 19, 3134-3141.

Wetzel, R.G., (1970). Recent and postglacial production rates of a marl lake. Limnol. Oceanogr. 15, 491-503.

## APPENDIX

### a) Preparation of samples for cladoceran pseudofossils.

1 ml sample of wet sediment, heat with 10 - 15 ml 10% KOH for 30 min. Wash through a 50um sieve. Add 10 ml 10% HCl. Heat for 15 - 30 mins. Wash through 50um sieve and store in a known volume. From this volume, which is equivalent to a known amount of sediment 0.25 ml aliquots are transferred to slides and mounted in glycerine jelly.

The use of phase contrast microscospy greatly facilitates identification of remains.

### b) Preparation of samples for chironomid pseudofossils.

Sediment samples for chironomid pseudofossils are treated chemically in much the same manner as for cladoceran remains. One ml portions of wet sediment are heated with 10 - 15 ml 10% KOH to deflocculate the sediment and clear the chironomid larval head capsules. Sample is then sieved through 80um, the residue is washed, concentrated by centrifuge (4000 - 4000 r.p.m.). Residue treated with 10% HCl. Remains are hand picked with fine entomological needles and mounted with a suitable mountant.

Identification based on various keys. One of the problems in dealing with this group is the diversity of keys.

### c) Oscillaxanthine extraction.

The method used is that of Griffiths and Edmondson, 1975. The pigment is extracted from one cm core slices and the wet sediment is centrifuged before use to remove excess water. The carotenoids are extracted from the mud by stirring and grinding in ethanol, using a mortor and pestle. The mixture is then centrifuged and the supernatant poured off. Extraction is repeated using batches of fresh ethanol until the supernatant is colourless. The extracts are then combined and evaporated to dryness in a rotary evaporator (Büchi Rotavapor - R) keeping the temperature below 40°C. The apparatus is then filled with nitrogen and the pigment redissolved in a minimum amount of ethanol; (0.2 ml) Carotenoids tend to be unstable in the presence of oxygen and light and so all manipulations are carried out in dim light. The extracts are shielded from light using silver foil and when not used immediately the extracts are stored at -20°C under nitrogen.

Oscillaxanthin is separated from other carotenoids using thin layer chromatography. Under a stream of nitrogen the concentrated crude ethanol extract is streaked onto a 10 x 20 cm cellulose thin layer plate. Depending on the extract used aliquots of between 10 - 30u litres are applied to the plate. The oscillaxanthin is then separated from the other carotenoids by developing the chromatogram in 6% ethanol in benzene. However, more complete separation may be obtained if the chromatogram is first developed in 0.5% ethanol in benzene, dried rapidly in a stream of nitrogen and redeveloped in 6% ethanol in benzene. After chromatography, the red band of oscillaxanthin is scraped off the plate and the pigment eluted with 1.0 ml ethanol. The pigment may be identified by its spectral properties in different solvents and also through comparison with authentic oscillaxanthin obtained from unicultures of _Oscillatoria agardii_.

The evaluation of pollution

# The Impact of Physical Processes on the Trophic State of a Lake

D. M. Imboden and R. Gächter

*Swiss Federal Institute for Water Resources and Water Pollution Control, EAWAG, 8600 Dübendorf, Switzerland*

ABSTRACT

Physical processes such as the intensity of vertical mixing, influence the annual primary production ($\Sigma P$) which is chosen as trophic state index. Measured values of $\Sigma P$ can be split into two components, a minimum productivity ($\Sigma P^{o}$) calculated with a simple epilimnic one-box model, and an "internal productivity" ($\Sigma P^{1}$) attributed to internal nutrient supply by vertical mixing. A statistical analysis for $\Sigma P$ versus $L_P$, the annual P-input per unit lake surface, based on 62 lakes from North America and Europe, reveals a tendency for $\Sigma P$ to increase with decreasing mean depth. For $L_P > 3$ g P $m^{-2}yr^{-1}$, $\Sigma P$ reaches a saturation value of about 400 gC $m^{-2}yr^{-1}$, while the influence of $\Sigma P^{1}$ disappears due to light limitation. For an intermediate $L_P$, the large variation of $\Sigma P$ among individual lakes with the same loading shows the importance of other factors such as morphometry, intensity of vertical mixing (reflecting exposure to wind), and the redox potential at the sediment-water interface. The later controls the redissolution of mineralized nutrients at the sediment surface.

Keywords:

Physical processes, Eutrophication, Primary productivity, Sediment-water interaction, Phosphorus, Trophic state index, Redissolution of nutrients, vertical mixing (in lakes).

INTRODUCTION

In many areas of the world, lake eutrophication represents one of the central problems of water pollution control. In most cases, the increasing input of nutrients has been identified as the direct cause of eutrophication. Under the influence of environmental stress, lakes show great differences in their response. Since the optimal planning of water pollution control bases on the understanding and prediction of lake response, a main task of limnologists and engineers consists in explaining the individual behavior of lakes in terms of typical characteristics such as the morphology of the lake and its drainage basin, or of meteorological and geochemical factors.

This chapter deals with physical phenomena which may be of importance for chemical and biological processes (Fig. 1). In connection with ecological investigations, the physical properties of the lake are often assumed as being unaffected by anthropogenic influences.

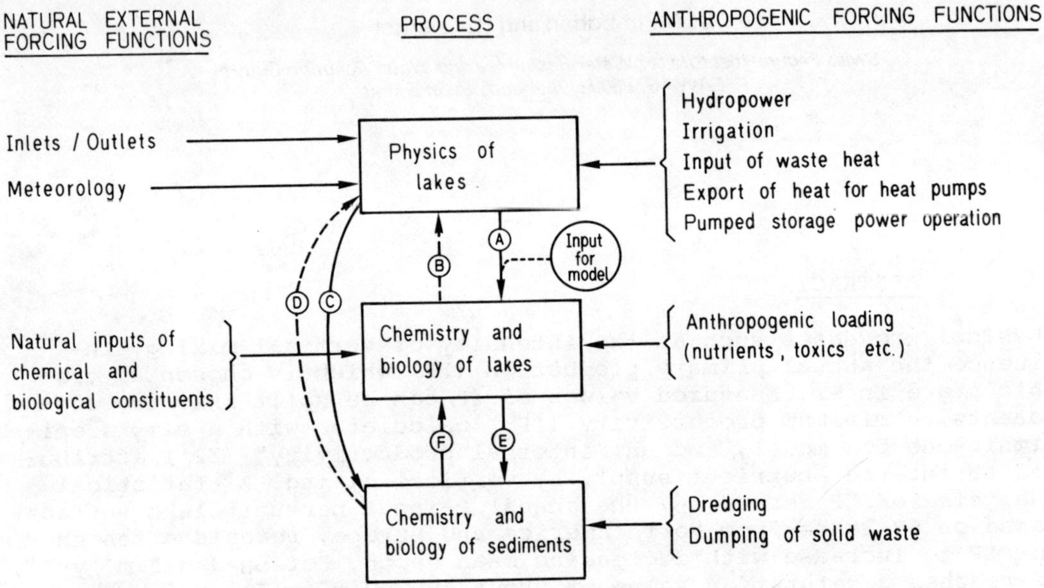


Fig. 1. Physical processes in lakes influence chemical and biological processes (interaction A,C) whereas feedback mechanisms (B,D) are generally weak. From Imboden, 1979.

Although there are many physical models of lakes, expressing various degrees of complexity, direct coupling of physical and chemical-biological models is still problematic. There exists, however, a need for such coupling due to the growing number of inadvertent anthropogenic influences on the mixing regime in a lake such as:

- Input of waste heat from thermal power plants or other industrial processes
- Withdrawal of heat for heat pump operation
- Change of hydraulic kinetic energy input by channeling rivers or changing inlet/outlet position
- Operation of a pumped storage power station on the lake (Imboden, 1979).

Based on the relationships shown in Fig. 1, one would expect to find various synergetic effects between physical, chemical and biological perturbations. At least for one physical parameter, temperature, this has already been recognized and found its way into legal regulations.

For example, in 1968 the upper limit for the relative temperature increase in rivers produced by cooling of thermal power plants was proposed by a committee of experts appointed by the Swiss government to be set at three degrees Celsius. This margin, however, should be applied only to those rivers with low chemical and biological pollution. For higher levels of pollution, due to the synergetic effects between physical and other parameters, the limit should be reduced.

In this chapter, two aspects of physical perturbations will be analyzed in more detail. The direct effect of temperature on biology will however not be further discussed. One aspect concerns hydrodynamics (lake circulation), especially the problem of thermal stratification and vertical transport. The other subject concerns the interaction between sediments and water. However, before talking about physics, a definition is needed for the term "trophic state". This complex matter is the subject of the next section.

## TROPHIC STATE INDEX AND PRIMARY PRODUCTIVITY

Concerning the historical development of the term "trophic state" I quote from the introductory remarks of Imboden and Gächter (1978):

"There is no generally accepted unique definition of a lake's trophic state, nor is there a quantitative measure. The eutrophication concept has been confused since 1921 when Thienemann and Naumann amalgamated their different definitions of a lake's trophic state. Their concept failed most probably because it was based on instantaneous parameters such as oxygen tension, nutrient concentration and phytoplankton density. As an empirical and therefore static concept it does not support any functional relationship between the different state variables, nor the prediction of the dynamics of the system's response to internal or external perturbations. Birge and Juday (1927), Strøm (1931) and Hutchinson (1957) introduced the concept of rates, i.e. the velocity of exchange between various mass and energy compartments. By doing so, it became possible to relate concentrations to supply and consumption as well as phytoplankton density to primary production."

The classical trophic state indices are connected to external parameters by the rate (process) of primary production (Fig. 2).

In accordance with Rodhe's (1969) qualitative definition of eutrophication as an increase in a lake's primary productivity, Imboden and Gächter (1975, 1978) have used primary production per unit lake surface and time, $\Sigma P$, as a quantitative measure of a lake's trophic state.

Since in most cases phosphorus has been found to be the controlling nutrient, at least during most periods of the year, the natural approach would be to look for a connection between $\Sigma P$ and the annual phosphorus input per unit surface ($L_p$) or per unit volume ($l_p$), respectively. Imboden and Gächter (1975, 1978) using 27 lakes from Europe and North America found a medium correlation between log $L_p$

and log $\Sigma P$ (correlation coefficient 0.85) and a somewhat stronger correlation between log $l_P$ and log $\Sigma P$ (0.91).

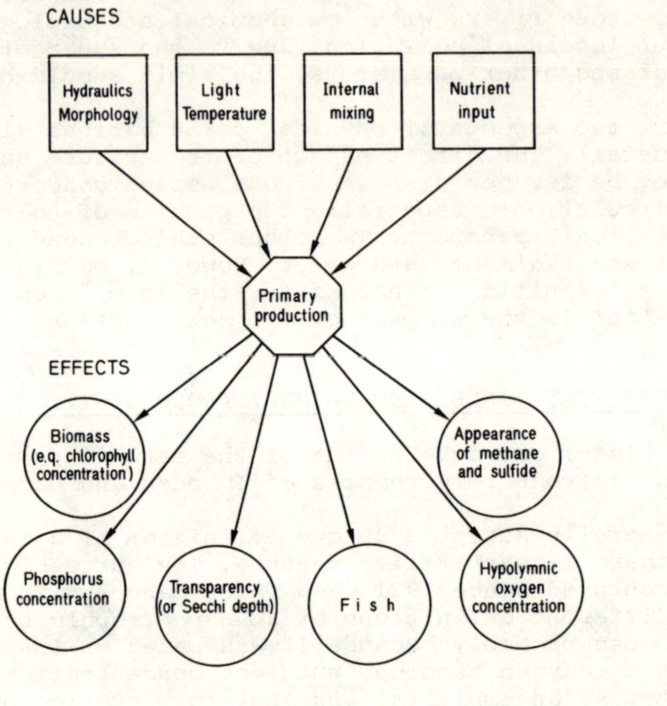


Fig. 2. Relationship among various trophic state indicators.

In the mean time, more productivity data have become available. In Fig. 3 is shown the correlation between log $L_P$ and log $\Sigma P$ for 62 lakes (North America and Europe). In comparison to the earlier statistical analysis the correlation coefficient slightly dropped to 0.78 and the exponent became larger (0.70 instead of 0.35) indicating a stronger increase of $\Sigma P$ with rising loading $L_P$. The correlation between log $l_P$ and log $\Sigma P$ shows a similar trend, i.e. a shift to a lower correlation and a larger exponent.

Schindler and Fee (1974) have derived a fit between log $L_P$ and log $\Sigma P$ which is nearly identical with the results of Fig. 3. However, Straškraba (1976) with good reason critisized this kind of correlations since it is mainly the logarithmic operation which produces fairly acceptable results whereas a direct correlation between $L_P$ and $\Sigma P$ exhibits enormous scattering of the data. Evidently, primary productivity is not simply a function of phosphorus loading, although regression analysis between $\Sigma P$ and various characteristic lake parameters always exhibit the highest correlations between $\Sigma P$ and $L_P$. Thus, in the context of the subject of this chapter the following questions have to be answered:

1. On which lake characteristics (hydraulic loading, mean depth,

vertical mixing, morphology, climate etc.) does the relation between $L_P$ and $\Sigma P$ depend?

2. Are some of the observed relations possibly originating from the influence of vertical mixing on productivity?
3. What kind of anthropogenic activities may affect the mixing regime of a lake and what are the consequences for the trophic state?

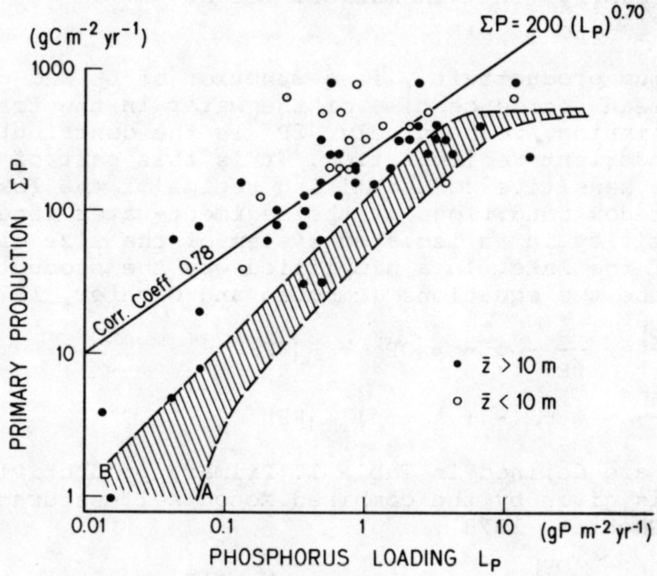


Fig. 3. Correlation between annual P-loading per surface area ($L_P$) and primary productivity ($\Sigma P$) for various European and North American lakes. Shaded area: Results from steady-state productivity model (eq. 5) for epilimnic residence times $\tau_E$ of water between 0.06 yr (curve A) and 15 yr (curve B). Shallow lakes (mean depth $\bar{z}$<10 m) show a slight tendency toward higher productivities.

Brylinsky and Man (1973) have identified the solar input as being the strongest factor regulating productivity, but this would be less the case with our data, which in contrast to those used by the authors mentioned, originate from a relatively homogeneous climate zone. Another factor has been brought forth by Vollenweider (1968), who found an increase of "nutrient tolerance" with increasing lake depth. Indeed, the data in Fig. 3 show a slightly higher productivity for shallow lakes, a tendency mainly associated with eutrophic lakes. A multidimensional regression analysis results in the equation ($\bar{z}$: mean depth)

$$\log \Sigma P = 2.6 - 0.24 \log \bar{z} + 0.66 \log L_P \qquad (1)$$

(correlation coefficient: 0.76)

but again the scattering is far too large in order to be acceptable for a "unique" theory of primary productivity.

However, the existence of a lower limit for $\Sigma P$ as a function of $L_P$ (Fig. 3) is a salient feature which suggests the following statement (valid for lakes in the moderate climate zone):

Statement: Given $L_P$, there exists a minimum productivity $\Sigma P^o$. Internal factors, especially the recirculation of nutrients from the hypolimnion in shallow and small lakes, may increase the minimum productivity significantly. In mathematical terms:

$$\Sigma P = \Sigma P^o + \Sigma P^1 \tag{2}$$

$\Sigma P^o$, the minimum productivity, is a function of $L_P$ and probably of $\tau$ or $\tau_E$, the mean residence time of the water in the total lake or in the epilimnion, respectively. $\Sigma P^1$ is the contribution from the internal nutrient recirculation. It is this part of $\Sigma P$ which is expected to be sensitive to the mixing regime of the lake, its morphometry and redox conditions at the sediment-water interface. $\Sigma P^o$ is the productivity in an isolated system of the size of the trophogenic layer of the lake. In a simplified way the productivity can be expressed by the two equations (Imboden and Gächter, 1975)

$$\frac{d\,[OP]}{dt} = \frac{L_P}{z_E} - \phi - \lambda\,[OP] + R\,[PP] \tag{3a}$$

$$\frac{d\,[PP]}{dt} = \phi - (R + \lambda + \sigma)\quad [PP] \tag{3b}$$

The variables are defined in Table 1. Primary productivity per unit surface area is given by the combined Monod/self-saturation equation (Imboden and Gächter, 1978):

$$\phi = \frac{\mu}{z_E}\ \frac{[OP]}{K_M + [OP]} \cdot \frac{[PP]}{\varepsilon_o + \beta\,[PP]} = \frac{1}{z_E}\ \Sigma P_P \tag{4}$$

The steady-state concentrations of eq. (3), given by quadratic equations for [OP] and [PP], are used to calculate from eq. (4) the steady-state productivity per unit area and time (expressed as phosphorus uptake $\Sigma P_P$). An average Redfield factor of 40 is chosen to convert $\Sigma P_P$ into carbon uptake $(\Sigma P)$[1].

Mixing depth $z_E$ and production rate $\phi$ show significant seasonal variations with the result that most of the yearly productivity occurs during the stagnation period, i.e. during about 8 months. Average parameter values from Imboden and Gächter (1978) are listed in Table 1. In a simplified way, the annual primary productivity is calculated from the steady-state concentrations $[OP]_s$ and $[PP]_s$ :

$$\Sigma P_s = 40 \cdot T_{St} \cdot \mu_{St}\ \frac{[OP]_s}{K_M + [OP]_s} \cdot \frac{[PP]_s}{\varepsilon_o + \beta\,[PP]_s} \tag{5}$$

where $T_{St}$ = 240 days/yr is the average duration of stagnation and $\mu_{St}$ the mean uptake rate during this period. The range of $\Sigma P_s$ as a

---

[1]Note that in the phosphorus model $\Sigma P_P$ does not describe the much faster epilimnic P-cycle but the net P-uptake which acts more or less in parallel with the carbon uptake.

function of $L_P$ and $\tau_E$ is amazingly similar to the lower boundary of measured values (Fig. 3). Both show a nearly linear relationship (exponent near 1) between $L_P$ and $\Sigma P$ and an upper saturation around 400 gC $m^{-2}yr^{-1}$. No doubt that this simple steady-state model gives a fairly reliable picture of $\Sigma P^o$. The product of [OP] and [PP] in eq. (4) is partly responsible for productivity being less sensitive than the individual parameters, [OP] or [PP], to the actual evolution of the system from the beginning of stagnation ([OP] large, [PP] small) to its end.

The role of internal phosphorus supply can be made more transparent by the so-called P-efficiency $n_P$, defined as the average frequency each phosphorus atom is used in the photosynthetic process before leaving the lake through an outlet or by irreversible sedimentation:

$$n_P = \Sigma P_P / L_P = \frac{\Sigma P}{40 \cdot L_P} \qquad (6)$$

In Fig. 4 are plotted values of measured $n_P$ (the same set of lakes is used as in Fig. 3) and $n_P$ values computed from the steady-state model (eq. 5). For large $L_P$, measured values agree wiht the steady-state model indicating that $\Sigma P$ is controlled by light (self-shadowing) whereas internal nutrient recirculation does not influence a lake which is already too richly nourished.

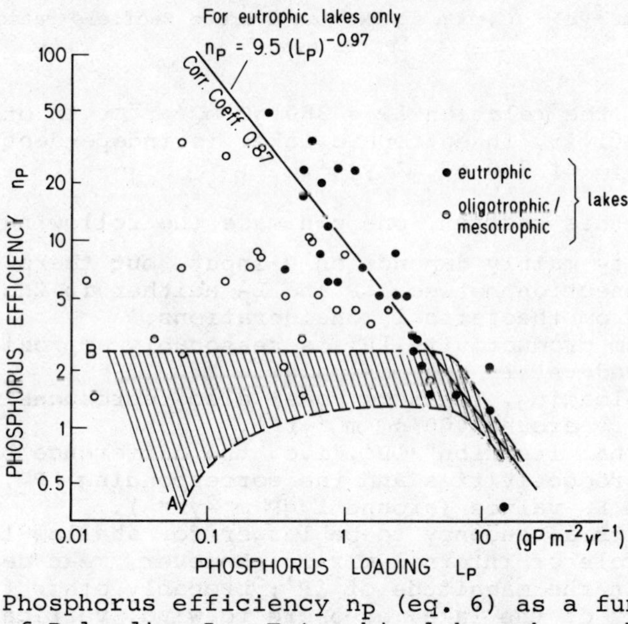


Fig. 4. Phosphorus efficiency $n_P$ (eq. 6) as a function of P-loading $L_P$. Eutrophic lakes show a high correlation between $L_P$ and $n_P$. Shaded area (curve A and B) are the results from the steady-state model as in Fig. 3.

The scattering of the $n_P$ values as a function of $L_P$ is significantly reduced if only eutrophic lakes are analyzed. The linear regression found between log $L_P$ and log $n_P$ for eutrophic lakes is nearly

TABLE 1 Definitions and Values of Parameters for simple Phosphorus Lake Model (Imboden and Gächter, 1975, 1978)

| | | | |
|---|---|---|---|
| [OP] | | $mg\ m^{-3}$ | concentration of dissolved phosphorus |
| [PP] | | $mg\ m^{-3}$ | concentration of particulate phosphorus (as measure for biomass) |
| $L_P$ | | $mgP\ m^{-2}yr^{-1}$ | Annual P-input in lake per unit surface area |
| $z_E$ | 6 | m | depth of epilimnion |
| $\lambda = \frac{1}{\tau_E}$ | | $yr^{-1}$ | exchange of epilimnic water by river through flow |
| R | 0.07 | $d^{-1}$ | rate constant of mineralization of organic material |
| $\sigma$ | 0.025 | $d^{-1}$ | loss of biomass from epilimnion by sedimentation |
| $\phi$ | | $mgP\ m^{-3}d^{-1}$ | P-uptake taken as measure for primary production |
| $\mu$ | 1 | $d^{-1}$ | rate constant for P-uptake |
| $K_M$ | 1.5 | $mg\ P\ m^{-3}$ | Michaelis constant |
| $\varepsilon_o$ | 0.3 | $m^{-1}$ | extinction coefficient of water without biomass |
| $\beta$ | 0.015 | $m^2(mgP)^{-1}$ | specific influence of biomass on extinction |
| $\Sigma P_P$ | | $mgP\ m^{-2}yr^{-1}$ | Uptake of P per unit time and water column |
| $\Sigma P$ | | $mgC\ m^{-2}yr^{-1}$ | Carbon uptake (an average Redfield ratio $\Sigma P = 40\ \Sigma P_P$ is used) |

identical with the relation $\Sigma P = 380\ gC\ m^{-2}yr^{-1}$. In other words, primary productivity in eutrophic lakes is independent of $L_P$ and has a mean value of $380\ gC\ m^{-2}yr^{-1}$.

In concluding this section, one can make the following statements:

1. Productivity mainly depends on P-input, but there exists no unique connection between $\Sigma P$ and $L_P$ neither in the measured data nor from theoretical considerations.
2. The minimum productivity $\Sigma P^o$ is reasonably approximated by a simply steady-state one-box model.
3. At high P-loading, the same model also reproduces the saturation effect on $\Sigma P$ around $400\ gC\ m^{-2}yr^{-1}$.
4. The "internal fraction" $\Sigma P^1$, i.e. the difference between the measured productivities and the corresponding $\Sigma P^o$, is largest for medium $L_P$ values (around $1\ gP\ m^{-2}yr^{-1}$).
5. $\Sigma P^1$ exhibits a tendency to be larger for shallow lakes, indicating the role of internal mixing. However, mean depth alone cannot explain the magnitude of $\Sigma P^1$; probably other factors such as morphometry of the lake, exposure to wind, vertical mixing intensity and the redox conditions at the sediment-water interface are of equal importance.
6. As exemplified by the curves A and B in Fig. 3, hydraulic loading is only of limited influence on $\Sigma P$, although extreme values for $\tau_E$ have been used for the model.

## VERTICAL TRANSPORT

The statistical analysis of measured productivity data from various lakes provided some basis for speculation on the possible influence of vertical mixing on the trophic state. In this section a case study together with theoretical consideration on factors responsible for the intensity of vertical mixing are presented with the goal to further substantiate the supposition.

Two mixing regimes will be considered:

(1) Vertical mixing during stagnation
(2) Thermocline erosion and winter circulation (complete or partial turnover)

### Vertical mixing during stagnation

The ratio of euphotic to mixing depth, $\alpha$, is a measure for the most direct influence of vertical mixing on productivity. If the euphotic depth is defined as the 5 percent level of the incoming radiation, it follows[2]

$$\alpha = \frac{3}{\varepsilon z_E} \quad \text{with} \quad \varepsilon = \varepsilon_o + \beta[PP] \qquad (7)$$

For $\alpha \ll 1$, a situation typical for winter conditions, the produced biomass is permanently mixed into great depth preventing the build-up of a dense algal population (Riley, 1963; Stefan, Skoglund, and Megord, 1976). In most mesotrophic and eutrophic lakes, $\alpha$ is about 1 during stagnation. Ultra-oligotrophic lakes with low $\varepsilon_o$ as e.g. Lake Tahoe, California-Nevada may have their productivity maximum well below the thermocline (Imboden and co-workers, 1977) which creates a rather different situation for internal nutrient supply.

The position of the thermocline is the result of a dynamic equilibrium between two kinds of forces which drive the motion of fluid in a lake (Fig. 5). First, the exchange of thermal energy at the surface produces vertical density gradients equivalent to a change of the potential energy of the water column. Second, the mechanical energy flux (turbulent kinetic energy from wind force and inlets) tends to destroy the density gradients.

In most lakes the production of turbulent kinetic energy is dominated by the effect of wind-induced shear stress. Thus, the larger wind fetch of large lakes generally results in greater mixing depth and higher vertical mixing intensity below $z_E$. However, several authors have used mean depth $\bar{z}$ as the characteristic quantity correlated to various mixing parameters such as mean vertical eddy diffusion coefficient in the hypolimnion (Mortimer, 1942), vertical entrainment of hypolimnic water into the epilimnion (Blanton, 1973), or vertical exchange coefficient in the thermocline (Snodgrass

---

[2] The symbol $z_E$ is also used as mixing depth during the circulation period when possibly $z_E = z_{max}$

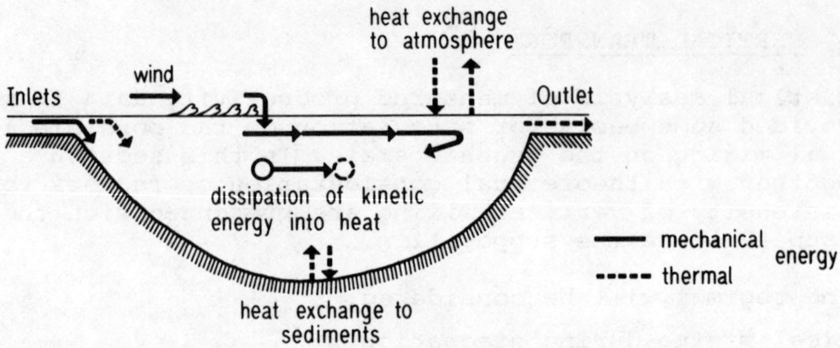


Fig. 5. Thermal and mechanical energy flux in a lake. From Imboden, 1979.

and O'Melia, 1975). The hidden positive correlation between surface area and mean depth suggests that the size of the lake is at least partially responsible for the behavior of vertical mixing intensity with mean depth.

One-dimensional thermocline models were first derived by Munk and Anderson (1948) and Kraus and Turner (1967). Several refinements of this model exist, but they all have in common the failure to adequatly describe vertical mixing in the hypolimnion. Octavio Hurley, Jirka and Harleman (1977) from MIT hoped to overcome this difficulty by introducing into the model vertical advection resulting from the different depths of input and output flows. While this remedy works for the two examples analyzed by the MIT group (both artificial reservoirs, 20 and 40 meters deep, with large subsurface withdrawal) it does certainly not help for deep natural lakes. For instances, in Lake Lucerne (maximum depth of 210 meters), using the MIT-model the vertical diffusion of heat below the thermocline turns out to be larger by two to three orders of magnitude than molecular diffusion (Imboden, 1979).

Only the complete understanding of the three-dimensional hydrodynamic behavior of a lake would give the adequate instrument to replace empirical parameters (coefficient of eddy diffusion) by basic physical principles. At present, the state of the art of physical modelling is far from reaching this general goal, and it may never reach it. Thus we still depend on measured values of eddy diffusivity and on empirical correlations to other quantities.

From the standpoint of turbulence theory, the gradient Richardson number (Ri) would be the adequate parameter for such a correlation (Kato and Phillips, 1969). It is defined as (u:horizontal advection velocity)

$$Ri = \frac{N^2}{(\partial u/\partial z)} \tag{8}$$

where $N^2$ is the (Brunt-Väisälä) stability frequency

$$N^2 = \frac{g}{\rho} \frac{\partial \rho}{\partial z} \tag{9}$$

However, since the measurement of shear ($\partial u/\partial z$) is very complicated and only rarely undertaken, a correlation between eddy diffusivity and the static parameter $N^2$ is often attempted, instead (Jassby and Powell, 1975). An example is given in Fig. 6.

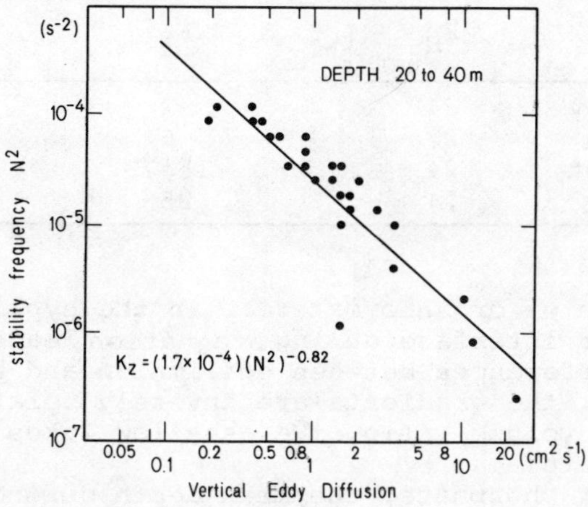


Fig. 6. Example of an empirical correlation between the Brunt-Väisälä stability frequency $N^2$ and the vertical eddy diffusion coefficient. From Imboden, 1979.

The coefficient of vertical eddy diffusivity ($K_Z$) is usually calculated from the temporal evolution of vertical temperature distribution (Hutchinson, 1957). In addition, natural and artificial chemical tracers can be used.

In the small and highly eutrophic Greifensee (Switzerland), Imboden and Emerson (1978) have measured the vertical distribution of natural radon-222, a decay product of radium-226. Since radon is a noble gas, it is less reactive than radium and thus diffuses from the sediments (where it is produced) into the overlying water column.

Vertical transport in Greifensee is an ideal example to demonstrate the influence of vertical mixing on productivity. The results of the mentioned investigation summarized in Table 2, demonstrate that in the second half of the stagnation period, internal phosphorus loading, i.e. the flux of phosphorus from the hypolimnion into the trophogenic layer ($F_I$), reaches a size comparable to the external P-input (rivers, sewage).

Since the vertical material flux in the hypolimnion and through the thermocline is the product of the local $K_Z$ and the local vertical concentration gradient ($\partial C/\partial z$),

$$F = -K \frac{\partial C}{\partial z} \quad \text{(Flux per unit area and time),} \qquad (10)$$

the following general statements can be derived from the case study of Greifensee:

TABLE 2 Internal Loading of Phosphate $F_I$ (Flux through the Thermocline) compared to the external Nutrient Loading $F_E$ (taken from measurements in 1968) in Greifensee for the Summer 1975. From Imboden and Emerson (1978).

| | $F_E$ (mgP $m^{-2}$ $day^{-1}$) | $F_I$ | $F_I:F_E$ |
|---|---|---|---|
| April/May | 24 | 2 | 0.1 |
| May/June | 24 | 5 | 0.2 |
| June/Sept | 24 | 15±7 | 0.6 |
| Oct./Nov | 24 | 25 | 1 |

(1) Mineralization of organic material in the hyplimnion and at the sediment-water interface during stagnation leads to vertical concentration differences between epilimnion and hypolimnion. Qualitatively, the gradients are inversly related to the hypolimnic/epilimnic volume ratio, i.e. shallow lakes generally have larger gradients.

(2) In the case of phosphate, the mean depth dependency of vertical gradients is further enhanced by the factor of hypolimnic $O_2$ depletion. Under anaerobic conditions released phosphate at the sediment surface is not bound by surface absorption mechanisms. The occurence of anaerobic conditions is more probable in shallow lakes since, given a certain production of biomass at the surface, it is the size of the hypolimnic oxygen reserves (i.e. of the hypolimnic volume) which determines at what time of the summer anaerobic conditions occur.

(3) As a consequence of (1) and (2), the trophic level of shallow lakes should show a larger sensitivity with respect to external nutrient input than deep lakes, which is in accordance with the early findings by Vollenweider (1968).

(4) One effect partially counteracting the above statements consists in the hidden correlation between lake mean depth and surface area. As mentioned before, large lakes (which are often the deeper ones) generally have a higher vertical mixing intensity which would favor transport from the sediments through the thermocline into the trophogenic layer. However, this effect seems to be less pronounced than the other ones.

## Winter circulation

Another shortcoming of the existing one-dimensional thermocline models lies in the adequate description of hypolimnion erosion and complete turnover. During the cooling period, where convective turbulence may become more important than wind-induced turbulence, the vertical propogation of turbulent kinetic energy apparently is very complex (Imboden, 1979). It is not yet possible to exactly predict those meteorological conditions under which total turnover occurs in certain lakes. Of course, there exist many cases where total turnover occurs yearly and very regularly but there are many meromictic and facultatively holomictic lakes, where this is not the case. It is of great importance to understand natural and possible man-made causes for the absence of total mixing, since the accumulation of hypolimnic oxygen deficits during several years may trigger a positive feedback mechanism which can stabilize the lake in an unfavorable way (Fig. 7).

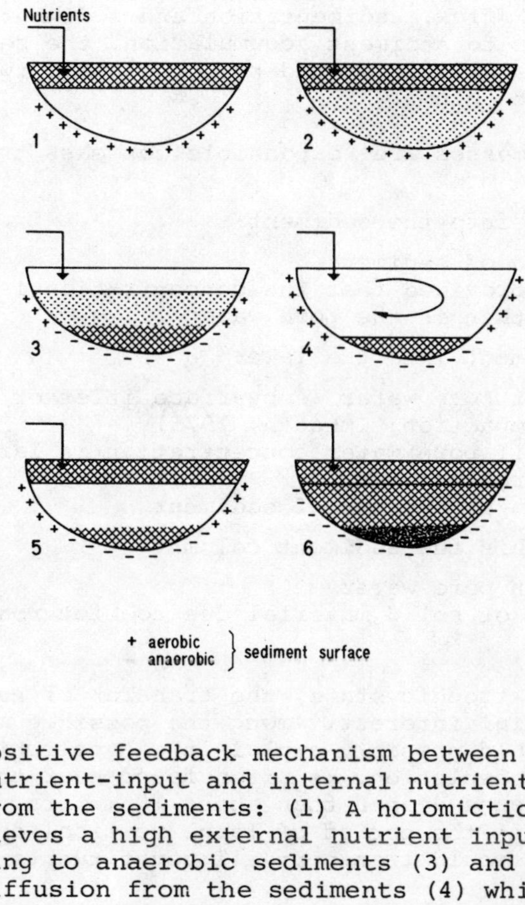


Fig. 7. Positive feedback mechanism between external nutrient-input and internal nutrient-diffusion from the sediments: (1) A holomictic lake recieves a high external nutrient input (2) leading to anaerobic sediments (3) and nutrient diffusion from the sediments (4) which prevents total turnover and (5) keeps the sediments anaerobic (6). In the following year, the stabilazing chemocline becomes even stronger.

## SEDIMENT-WATER INTERACTION

As mentioned in the preceeding section, the material flux through the thermocline into the trophogenic layer depends on both the diffusion coefficient and the concentration gradient (eq. 10). The evolution of vertical gradients in a stratified system is the result of depth-depending source and sink functions as exemplified by the vertical distribution of primary productivity. Boundary fluxes at the sediment surface represent another important cause for the appearance of vertical concentration gradients. Indeed, measurements in different lakes often show significant gradients above the lake bottom; either the concentrations increases toward the interface indicating that the sediment acts as a source (phosphate, methane, sulfide) or it decreases if the sediment acts as a sink (dissolved oxygen).

Beside diffusional flux, sedimentation and sediment resuspension are of importance. Due to sediment accumulation, the sediment-water interface represents a moving boundary between the two systems, lake water and sediment column.

The following processes are responsible for mass transfer between both systems:

1. From the lake into the sediment:
   - accumulation of sediment
   - diffusion, provided that the concentration is larger in the free water than in the pore water
2. From the sediment into the lake:
   - advection of pore water (subsurface inlet or/and effect of sediment compaction; Imboden, 1975)
   - diffusion, if pore water concentration is larger than hypolimnic concentration
   - resuspension or erosion of sediment
3. Transport within the sediment column:
   - diffusion in pore water
   - "diffusion" of solid material due to bioturbation
   - advection.

In the context of trophic state, the transfer of nutrients and oxygen is of special interest. Among the possible mechanisms listed, only diffusion could possibly explain the hypolimnic phosphate increase and oxygen depletion, as often found near the bottom during the summer. For lakes deeper than about 10 meters, resuspension does hardly occur during the period of large vertical density gradients, whereas advection would transport all constituents into the same direction.

The conclusion concerning the transport mechanisms seems often to be contradicted by measurements of nutrient concentration profiles in the pore water of lake sediments. Following eq. 10, transport should occur against the concentration gradient. However, even in lakes which are influenced by large sediment diffusion inputs, phosphate in sediments often decreases with depth.

For instance in Greifensee, the test lake discussed in the preceeding section, profiles of dissolved phosphate in the interstitial water of the sediment (Emerson, 1976) show decreasing concentrations with increasing depth (Fig. 8). Only at the surface layer does the gradient permit diffusional flux into the open lake water. Since concentration undergoes a significant decrease from the sediment pore water ($\sim$3.5 mg P/l) to the overlying water ($\sim$1 mg P/l). The conclusion for this and similar lakes is that the phosphate flux

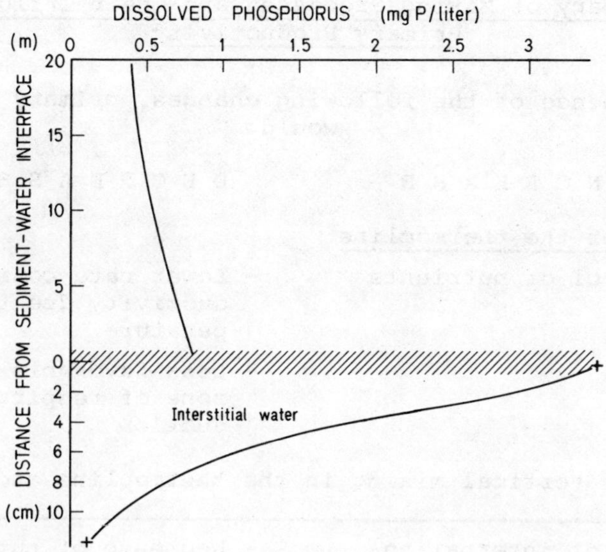


Fig. 8. Typical vertical profile of dissolved phosphorus in Greifensee in autumn (Imboden and Emerson, 1978) and in interstitial water of sediments (Emerson, 1976). The actual size of the gradient at the interface cannot be calculated because it strongly depends on the thickness of the boundary layer. Note the different spatial scale for lake and sediments.

from the sediments must have its origin in the uppermost layer whereas older parts of the sediment column are shielded off by a downward diffusional flux. In fact, the flux may just be produced by the decomposition of organic material at the very surface of the sediments, a supposition which recieves some credibility by the synchronism between productivity maxima (leading to high sedimentation of organic debris) and phosphate flux from the sediments found in Greifensee (Imboden and Gächter, 1978).

During the circulation period or in lakes relatively shallow compared to the range of downward moving turbulence, sediment resuspension accompanied by an input of phosphorus into the hypolimnion may be important as, for instance, shown for Lake Erie by Lam and Jaquet (1976). Evidently, in such lakes intensive vertical mixing produced by storms may lead to an immediate and large nutrient input into the epilimnion and trigger an algal bloom.

## CONCLUSIONS

Is it possible to quantify the combined effects of the described mechanisms on primary productivity and trophic state? Or is it, at least, possible to predict whether productivity would increase or

TABLE 3 A Summary of Mixing Processes and their Influence on Primary Productivity

Under the influence of the following changes, primary productivity would

| INCREASE | DECREASE |
|---|---|
| **A Deepening of the thermocline** | |
| - Higher pool of nutrients | - Lower rate constant for productivity due to lower temperature<br>- Less favorable ratio between zone of respiration and production |
| **B Increase of vertical mixing in the thermocline and the hypolimnion** | |
| - Increase of internal nutrient loading | - Decrease of nutrient flux from the sediments since the sediment surface remains aerobic during a longer period<br>- Decrease of biomass density by dilution due to mixing |
| **C Increase of the probability that the lake undergoes total turnover during winter (of importance only for meromictic lakes)** | |
| - Recycling of hypolimnic nutrient pool leading to larger initial concentrations in spring<br>- Turbulence decreases nutrient retention (lower sedimentation velocity and/or resuspension of sediments) | - Decrease of sediment boundary flux due to higher hypolimnic oxygen concentrations |

decrease as the result of a certain change in the vertical mixing regime of a lake? It may be possible under certain unique conditions, but in most cases a careful weighing between opposite effects is necessary.

In Table 3, a summary of the relevant mechanisms and their sensitivity with respect to vertical mixing is given. Only a refined lake model would be able to predict the synergetic behavior of the system as a whole. Such an extensive treatment would, of course, lie beyond the intention of this contribution.

## Acknowledgement

This work has been supported by the Swiss National Science Foundation within the framework of the National program on Lake Mixing.

## REFERENCES

Birge, E. A., and C. Juday (1927). The organic content of the water of small lakes. Proc. Am. Philos. Soc., 66, 357-372.

Blanton, J. O. (1973). Vertical entrainment into the epilimnia of stratified lakes. Limnol. Oceanogr., 18, 697-704.

Brylinsky, M., and K. H. Mann (1973). An analysis of factors governing productivity in lakes and reservoirs. Limnol. Oceanogr., 18, 1-14.

Emerson, S. (1976). Early diagenesis in anaerobic lake sediments: chemical equilibria in interstitial waters. Geochim. Cosmochim. Acta, 40, 925-934.

Hutchinson, G.E. (1957). A treatise on Limnology. Vol. 1. Wiley, New York.

Imboden, D. M. (1975). Interstitial transport of solutes in non-steady state accumulating and compacting sediments. Earth Planet. Sci. Lett., 27, 221-228.

Imboden, D. M. (1979). Modelling of vertical temperature distribution and its implication on biological processes in lakes. In S. E. Jørgensen (Ed.), State-of-the-Art of Ecological Modelling. Pergamon Press, Oxford.

Imboden, D. M., and S. Emerson (1978). Natural radon and phosphorus as limnologic tracers: horizontal and vertical eddy diffusion in Greifensee. Limnol. Oceanogr., 23, 77-90.

Imboden, D. M., and R. Gächter (1975). Modelling and control of lake eutrophication. Int. Fed. Automatic Control (IFAC), Proc. 6th Triennal World Congress, August 1975, Boston/Cambridge, Part 3, 61.1.

Imboden, D.M., and R. Gächter (1978). A dynamic lake model for trophic state prediction. Ecol. Modelling, 4, 77-98.

Imboden, D.M., R. F. Weiss, H. Craig, R. L. Michel, and C. R. Goldman (1977). Lake Tahoe geochemical study. 1. Lake chemistry and tritium mixing study. Limnol. Oceanogr., 22, 1039-1051.

Jassby, A., and T. Powell (1975). Vertical patterns of eddy diffusion during stratification in Castle Lake, California. Limnol. Oceanogr., 20, 530-543.

Kato, H., and O. M. Phillips (1969). On the penetration of a turbulent layer into stratified fluid. J. Fluid Mech., 37, 643-655.

Kraus, E. B., and J. S. Turner (1967). A one-dimensional model of the seasonal thermocline. II. The general theory and its consequences. Tellus 19, 98-106.

Lam, D. C. L., and J. M. Jaquet (1976). Computations of physical transport and regeneration of phosphorus in Lake Erie, fall 1970. J. Fish. Res. Board Can., 33, 550-563.

Mortimer, C. H. (1942). The exchange of dissolved substances between mud and water in lakes. J. Ecol., 30, 147-201.

Munk, W. H., and E. R. Anderson (1948). Notes of a theory of the thermocline. J. Mar. Res., 7, 276-295.

Octavio Hurley, K. A., G. H. Jirka, D. R. F. Harleman (1977). Vertical heat transport mechanisms in lakes and reservoirs. MIT Dep. of Civil Engineering, Report No. 227.
Riley, G. A. (1963). Theory of food-chain relations in the ocean. In N. M. Hill (Ed.), The sea, Vol. 2, Interscience, New York. pp. 438-463.
Rodhe, W. (1969). Crystallization of eutrophication concepts in Northern Europe. In Eutrophication: causes, consequences, correctives. Proceedings of a symposium, National Academy of Sciences, Washington D.C. pp. 50-64.
Schindler, D. W., and E. J. Fee (1974). Primary production in freshwater. Proc. First Int. Congress Ecology, The Hague. pp. 155-158.
Snodgrass, W. J., and C. R. O'Melia (1975). Predictive Model for phosphorus in lakes. Environ. Sci. Technol., 9, 937-944.
Stefan, H., T. Skoglund, and R. O. Megard (1976). Wind control of algae growth in eutrophic lakes. J. Environ. Engineering Div., ASCE, 102, 1201-1213.
Straškraba, M. (1976). Empirische und analytische Eutrophierungsmodelle. EUTROSYM, : Proc. Int. Symp. Eutrophication and Rehabilitation of Surface Water, Karl-Marx-Stadt, Vol. 3, pp. 365-385.
Strøm, K. M. (1931). Feforvatn. A physiographical and biological study of a mountain lake. Arch. Hydrobiol., 22, 491-536.
Vollenweider, R. A. (1968). Water management research. OECD Rep. 68.27., Paris.

# The Study of Eutrophication of Algal Benthos by Essential Metals in Artificial Rivers

**Elie Eichenberger**

*EAWAG, Federal Institute for Water Resources and Water Pollution Control, Experimental Station Tüffenwies, Zürich, Switzerland*

ABSTRACT

The causes of eutrophication of rivers are being studied in artificial outdoor rivers. Small amounts of domestic sewage added to groundwater stimulates the growth of algae. This eutrophying effect cannot be duplicated by the supply of the same quantity of $PO_4$-P and/or $NO_3$-N as is found in the active sewage. The supply of a mixture of the 6 essential heavy metals Fe, Mn, Co, Cu, Mo and Zn on the other hand does affect the growth activity of the algae occurring naturally in the channels, whereby each metal most likely has a different effect. This results, depending on the season and the amount of mixture added, in changes of the sociological composition of the algal benthos and can considerably enhance the biomass accumulation. Due to the characteristic dose response of single algal species to individual metals the role of essential heavy metals in eutrophication is likely to be rather complex.

KEYWORDS

Eutrophication, artificial rivers, algal benthos, sewage, phosphate, essential heavy metals, growth, species distribution, toxicity, stress recovery.

## 1. EUTROPHICATION IN RIVERS

Changes in the pattern of vegetation have recently been observed in many rivers all over the world. Whereas some plants have lost ground, others have expanded in area and density to become occasionally troublesome weeds (Mitchell, 1974; Eichenberger, 1976). This increase in plant biomass - generally called eutrophication - can have many causes, such as the introduction of a foreign plant species, a modification of the hydraulic conditions, shifts in light-climate or some alteration of the water chemistry as a consequence of changes in land use or waste input.

Very frequently eutrophication is understood as a consequence of the enrichment of the water with nutrients; this correlation has been so stressed, that in German the 2 separate concepts, a) the increased production of plant biomass and b) the increased supply of nutrients are expressed by one and the same expression "Eutrophierung".

When man interferes with a river, by damming or polluting for example, his activities affect an array of parameters, reflecting physical, chemical and biotic modifications. The problem which confronts us is the detection of those factors which contribute most to the observed reactions of the life community.

## 2. THE USE OF ARTIFICIAL RIVERS IN EXPERIMENTAL RIVER ECOLOGY

### 2.1. The design of artificial river systems

We can distinguish 4 main groups of factors which determine the structure of algal communities in running waters (Table 1), chemical, climatic, hydraulic and biotic factors.

TABLE 1 Factors Affecting the Development of Algal Communities in Running Water

| main factor | important variables |
|---|---|
| chemical factors | basic nutrients for the synthesis of structural elements and energy carriers |
| | growth substances which modify the use of the basic nutrients |
| | inhibiting substances ranging from specific antogonists of biological origin to manmade toxic compounds |
| climatic factors | light as energy source and as ecological factor influencing the distribution of heterotrophic microphytes and animals |
| | temperature |
| hydraulic factors | kinetic energy of the waterflow which determines the structure of the bottom substrate and affects the stability of the underground and of the vegetation |
| | turbulence which controls the mass exchange between the organisms and the surrounding water thus modifying the effect of the chemical factors |
| biotic factors | competition between plants |
| | grazing by herbivores |

The recognition of the factors responsible for eutrophication in running waters present methodological difficulties because so many variables influence the behaviour of any particular river. In addition there occurs much random variation or periodic fluctuation of essential parameters. For this reason artificial rivers have been conceived and built to meet specific ecological requirements.

Our artificial river system at the Research Station Tüffenwies in Zürich consists of 9 parallel channels made of asbestos-cement with a square profil of 20 x 20 cm and a length of 65 m. Groundwater from a nearby well with only minor yearly fluctuation in chemical composition except for $NO_3$ and oxigen (Wuhrmann and Eichenberger, 1975) is discharged at a constant rate of 1.5 l/sec and can be enriched by dosage of chemicals. All the artificial rivers have the same inclination of 1.5 ‰.

The bottoms are covered with pebbles of 2-3 cm diameter. Flow velocities without vegetation are about 15 cm/sec and drop depending on the amount and the structure of the algal cover and reach a minimum of about 7 cm/sec when the channels start to overflow.

Once water quality is under control the most important variable out of door is the climate. The use of a multichannel system offers the possibility to study different chemical conditions simultaneously under a uniform set of climatic variables.

A further variable of importance is the density of grazing insects which vary in time and along the channels. As the aim of our experiments was to study the direct effect of essential metals on algal benthos we excluded insects from our channels by a regular application of insecticides (once a week at night for 1 hours 9 mg rotenone/l).

### 2.2. The measure of biomass production

The organic biomass derived from photosynthetic production in algal benthos is subjected on the whole to 4 different fates. If not metabolized as energy donors a part will be worked over by herbivores; some will thus be eaten, some will accidentally drift away and join other spontanously detached material to constitute the microphytic drift or export. The remaining accumulated biomass forms the standing crop.

At the time the experiments to be discussed were carried out (1969-1974) no reliable technique to determine microphytic drift was available. The standing crop was estimated by harvesting all material in 1/2 m long segments (1/10 $m^2$ surface) placed at regular intervals along the section investigated. The biomass is expressed as organic dry weight (odw), that is ash free dry weight determined after ignition at 550°C; its C content is very nearly 50 %.

Biomass accumulation, expressed in g odw/$m^2$.day, is determined by the change of the standing crop in a given time interval.

The loss by export can be considerable and depends on sociology, age and insect density of the algal cover (Eichenberger and Schlatter, 1978). The biomass accumulation is therefore not identical with what is usually understood as productivity.

Each time before starting a new experiment the vegetation was completely removed by thorough mechanical abraison of channels and pebbles.

### 2.3. The description of sociology

The algal benthos of our channels is composed of a relative large number of species; however only a few of them contribute more than 5 % of the biomass. Even so, the quantification of the most important species of a benthic community on natural substrate presents a very tricky problem mainly for the following reasons:
1. on the centimeter scale the algal community is inhomogeneous and presents a variable mosaic of species even when the overall aspect gives the impression of great homogeneity.
2. along a channel of a length of 65 m there occur considerable changes in the composition of the vegetation.
3. significant measures of predominance such as determination of volume are very demanding and labor intensive.

In view of these difficulties the sociology will be described mainly in terms of the distribution of the green filamentous algae and the diatoms which form more or less defined floating streamers or settle on filamentous algae.
The density of the vegetation is estimated on a subjective scale ranging from 1 to 10 (Table 2).

TABLE 2 Scale for the Classification of the Density of the Vegetation

| density estimate | phenomenology of vegetation | approximate fraction of highest possible vegetation density |
|---|---|---|
| 1 | isolated occurrence | < 1/20 |
| 2 | few small patches | 1/20 |
| 3 | patchy, very thin growth | 1/10 |
| 4 | patchy, thin growth | 1/5 |
| 5 | even, thin growth | 1/3 |
| 6 | patchy, medium growth | 2/5 |
| 7 | even, medium growth | 1/2 |
| 8 | dense growth | 2/3 |
| 9 | very dense growth | 4/5 |
| 10 | complete colonization of the available room | 1 |

In communities dominated by one species there is a good linear correlation between the density estimate and the standing crop. Although the procedure gives a good idea of the distribution of the forms which determine the aspect of the algal cover, the values cannot be converted to standing crop data.
The species distribution is in general based on single microscopic observations of samples considered representative.

## 3. THE IMPLICATION OF ESSENTIAL METALS IN EUTROPHICATION

In artificial rivers the addition of small amounts of domestic sewage to ground water enhances the development of algae (Wuhrmann and Eichenberger, 1975). The most effective sewage concentration for promoting algal growth depends on several factors. The seasons for example affect the presence of the algae (Eichenberger, 1972); thus the stimulation of algal growth by a given sewage increment can vary within wide limits (Fig. 1).

On the basis of physiological knowledge different possible modes of action of domestic sewage on algae can be postulated and then experimentally verified. We can distinguish between the direct supply of some compound essential to growth and the indirect effects which may influence speciation of elements already present and will thus render the elements more or less available to the living cells. Some of the more likely effects would comprise:
a) direct growth promotion by
- essential macronutrients like carbon, phosphorous, nitrogen or iron; in particular instances the supply of the more readily metabolized ionic species may be the limiting factor (for example $NH_4$ as preferred nitrogen source)
- essential micronutrients like boron, cobalt, copper, manganese, molybdenum and zinc

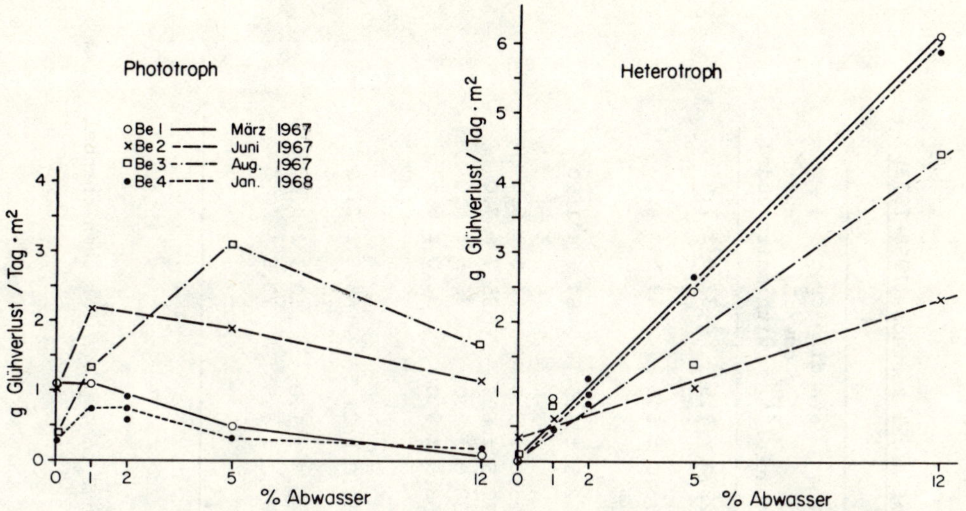


Fig. 1 Development of the phototrophic and the heterotrophic biomass as a function of the sewage load and the season growth rate : g organic substance/day and $m^2$ overgrown surface. Duration of colonisation : 20 days. (from Eichenberger, 1972)

- organic growth substances of vitamin-like character, which have been shown essential in the culture of some algae (Schwarz, 1969) and may be limiting in some natural conditions.

b) indirect effects like

- solubilization of metals by naturally occuring or man made complexing agents thus increasing the supply of essential micronutrients
- precipitation and coprecipitation of inhibitory metals and organic substances by different compounds, possibly iron hydroxides, phosphates or carbonates (Stumm, 1972); this could result in partial or complete detoxification by removal of substances either naturally present or introduced by man.

In a number of experiments at different seasons the ground water of the artificial river containing about 15 µg $PO_4$-P and .6 mg $NO_3$-N/l has been supplied with additional phosphates and nitrates either alone or in combination up to the concentrations found in the most effective sewage load (Table 3) (200 µg $PO_4$-P and 3 mg $NO_3$-N/l). This addition of phosphorus and of nitrogen had no effect on the sociological structures of the community nor on the biomass accumulation, although considerable P accumulation was observed changing the C/P ratio within the algae from around 180 to 80. Thus in the lotic system studied 15 µg $PO_4$-P/l and .6 mg $NO_3$-N/l are not limiting algal growth. So we concluded that there must exist some other compounds than these nutrients $PO_4$ and $NO_3$ which are responsible for the observed growth stimulation of algae by sewage. In an earlier publication it was reported that mixtures of essential heavy metal salts can enhance algal growth (Wuhrmann and Eichenberger, 1975) in our artificial rivers fed with ground water. This lecture is a presentation of some of our experiments in the years 1972 to 74 which suggested that the essential heavy metals play some role in the eutrophication of rivers.

TABLE 3 The Effect of Phosphate-P and Nitrate-N on the Growth of Algal Benthos (1968-1972)

| exp | time month | age days | predominant algae | addition µg/l P | addition µg/l N | number plots[3] co[2] | number plots[3] salt | standing crop g odw/m² control[2] | standing crop g odw/m² with salts added |
|---|---|---|---|---|---|---|---|---|---|
| Eu 3 | Sep-Jan | 106 | Melosira, Tribonema, Stigeoclonium | 200 | - | 12 | 12 | 29.0 ± 7.5 | 32.0 ± 7.0 |
| Eu 4 | Mar-Apr | 32 | Diatoma hiemale, D.vulgare, Melosira, Hydrurus | 100 | 3000 | 8 | 8 | 55.0 ± 6.0 | 49.5 ± 4.5 |
| Eu 5 | Mai-Jun | 28 | Melosira, D.hiemale, D.vulgare | 100 | 3000 | 8 | 8 | 63.5 ± 6.5 | 64.5 ± 15.0 |
| Eu 6 | Sep-Oct | 64 | Tribonema | 100 | 3000 | 8 | 8 | 23.0 ± 9.0 | 24.0 ± 6.0 |
| Eu 7 | Dec-Feb | 61 | Tribonema, Melosira, | 200 | 3000 | 4 | 8 | 37.5 ± 6.5 | 35.7 ± 7.0 |
| Eu 8 | May-Jun | 34 | D.hiemale, D.vulgare | 200 | 3000 | 4 | 8 | 38.0 ± 3.0 | 36.5 ± 3.0 |
| Eu 9 | Jan-Mar | 59 | Ulothrix, Hormidium, Synedra, then Oscillatoria limosa | 100 | 2000 | 12 | 12 | 53.5 ± 9.0 | 52.0 ± 6.5 |
| Eu 10 | Aug-Sep | 20 | Tribonema, Melosira, Oscillatoria | 100 | 1000[1] | 6 | 6 | 60.0 ± 21.5 | 65.5 ± 21.5 |
| Eu 11 | Dec-Jan | 40 | Melosira | 200 | 2000 | 4 | 8 | 6.5 ± 1.0 | 10.5 ± 4.0 |
| Eu 12 | Jun-Jul | 39 | D.hiemale, Tribonema Hormidium, Ulothrix | 200 | 2000 | 4 | 8 | 67.0 ± 7.5 | 70.5 ± 10.5 |

1) N supplied as $NH_4$

2) control : ground water without any additions

3) for each experimental condition 2 to 3 channels were run with 2 to 4 plots for each channel

± standard deviation

There are 6 heavy metals which are considered essential to plants (Fe, Mn, Co, Cu, Mo and Zn) (Eyster, 1964; Bowen, 1966). They are added in various proportions to algal culture solutions (Table 4) in general in the following order of decreasing concentration Fe> Mn> Mo, Co, Cu and Zn. In the experiments to be discussed we have maintained an atomic ratio of 10 Fe:5 Mn:1 Co:1 Cu : 1 Mo : 1 Zn.

TABLE 4 The Concentrations of Essential Heavy Metals in some Culture Solutions and in our Channels

| | culture solution µg/l | | | channel water Tüffenwies | | |
|---|---|---|---|---|---|---|
| | 1 | 2 | 3 | supply by solution FeMe* µg/l | µMol/l | groundwater** µg/l |
| Fe | 1100 | 10 | 800 | 5.6 | .1 | 2-5 |
| Mn | 200 | 2 | 1200 | 2.8 | .05 | $<$ 1 |
| Mo | 200 | 10 | - | .96 } | | $\sim$ 1 |
| Co | 20 | 5 | 3 | .59 } | .01 | $<$ 1 |
| Cu | 20 | 3 | 1.2 | .64 } | | $\leqslant$ 1 |
| Zn | 20 | 10 | 150 | .65 } | | 1-3 |

1 Monodus standard medium, Miller and Fogg 1957
2 seawater Sverdrup et al 1942
3 seawater Provasoli et al 1957
(columns 1-3 from Eyster, 1964)
* The metal solutions were made up with the following salts: Fe-EDTA, $MnSO_4$, $ZnSO_4$, $CoNO_3$, Ammonium Molybdate; iron was the only element given in complexed form.
** order of magnitude

## 4. THE EFFECT OF A STANDARD MIXTURE OF ESSENTIAL METALS ON ALGAL BENTHOS

### 4.1. The effect of various concentrations on composition and production of the algal biocenosis

In 2 separate experiments the standard solution FeMe containing the 6 essential heavy metals (Table 4) was added to channels in concentrations ranging from 1/2 to 3x FeMe (i.e. 0.05 to 0.3 µmolar in respect to iron).

The supply of the metals elicited two kinds of responses,
a) a very marked change in the specific composition of the biocenosis and
b) some alteration of the productivity of the biocenosis (Fig. 2).

In early summer when light intensities are high (3000 $kcal/m^2.day$) and the water temperature between $10^o$ and $15^oC$ the highly productive algal community fed with pure ground water is dominated by Diatoma hiemale, which makes up more than 90 % of the biomass. As additional species we find the diatoms Synedra ulna and several species of Nitzschia and some filamentous algae such as Ulothrix zonata, Hormidium rivulare and Tribonema vulgare.

In the channels with added essential metals the initial growth consists of Diatoma hiemale as in the ground water control; but after a few days there occurs a very conspicuous spread of green filamentous algae. At first Hormidium rivulare is predominant, later Tribonema vulgare is the prevaling filamentous species.

The metal induced shift in the ratio of diatoms to filamentous algae was observed in two consecutive experiments. Temperature seems to modify the effect. In the first experiment in June at water temperatures from $10^{\circ}$ to $12^{\circ}C$ the filamentous algae were stimulated in all channels with added metals; however they became dominant only at higher metal concentrations ($\geqslant$ FeMe 2x). A month later, in the second experiment, when the temperature was about $4^{\circ}C$ higher, even low concentrations (FeMe 1/2x) gave the full shift in the composition of the community (Fig. 2).

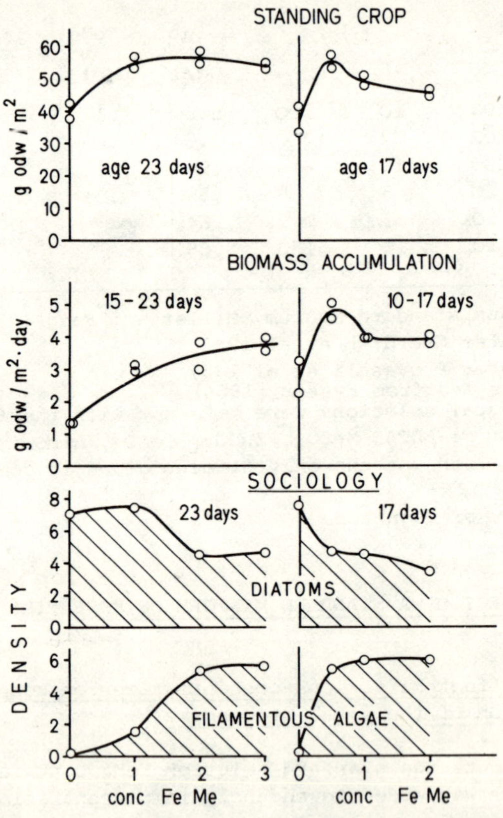


Fig. 2 The effect of different concentrations of the solution FeMe (see Table 4) on the development of the algal benthos in June (left) and July (right) 1973. Biomass: each condition 2 channels each 4 plots; Sociology: average of 2 channels with each 3 plots.

The addition of metals results in all channels in a greater biomass accumulation than in ground water alone. The observed changes in production are the result of two antithetic reactions, i.e. an inhibition of Diatoma hiemale and a stimulation of Tribonema vulgare. Whether the spread of a species results alone from actual stimulation of its growth or is at least in part due to a changed competitive situation needs further clarification, based an acurate determination of the quantitative distribution of the important species in the course of the formation of the algal communities.

## 4.2. The modification of the effect of essential metals by climatic conditions

The climatic conditions in our outdoor rivers varies throughout the year between wide limits. The water temperature has a minimum of 5°C in April and a maximum of 18-20°C in September. The light intensity in mid-winter is about 300 kcal/m$^2$.day and reaches about 5000 kcal/m$^2$.day on clear summer days.

The effect of the standard mixture of essential metals was studied in 5 different experiments with 3 typical combinations of climatic factors (Table 5).

TABLE 5 The Seasonal Distribution of the Dominant Species in Groundwater and the Species most Favored by the Essential Heavy Metal Standard Solution

| water temperature °C | average light intensity kcal/m$^2$.day | months | dominant species in groundwater | species stimulated by FeMe 1 x* |
|---|---|---|---|---|
| 10-16 | < 1500 | Oct-Nov | Melosira varians | Melosira varians |
| | | Dec | Melosira varians | Melosira varians |
| | > 2500 | June | Diatoma hiemale | Diatoma (?) Tribonema vulgare |
| | | July | Diatoma hiemale | Tribonema vulgare |
| > 18 | ∿ 1500 | Sep | Tribonema some Synedra ulna and Nitzschia ssp. | no effect |

(?) reaction uncertain * see Table 4

The qualitative and quantitative responce of algal biocenosis to the addition of the metal mixture is variable (Table 5 & 6). The result on the accumulation of the standing crop ranges from no effect to more than a doubling over the ground water control. The sociological effects were of 2 kinds.
- either a stimulation of the species which at the time is dominant in ground water so that no alteration in the sociology of the community can be observed
- or some of the subordinate species is stimulated in such a way that it gains in abundance what may then result in a noticable change of the composition of the biocenosis and possibly a conspicuous modification of its aspect.

The greatest effect is noticed in summer at high light intensities and relatively low water temperatures (between 10 and 15°C) when Diatoma hiemale is dominant.

In early fall when the temperature in the channels reaches the maximum between 18° and 20°C the vegetation is dominated by Tribonema vulgare and is accompanied by Synedra ulna. In addition there are several species of Nitzschia which grow predominantly as crusts on the walls of the channels. This vegetation did not respond to metals.

TABLE 6 The Modification of the Effect of the Essential Heavy Metals on Algal Benthos by Climatic Conditions

| exp | water temperature °C | light intensity kcal/m².day | age days | density of vegetation 1)4) | | | | | | biomass production | | | | |
|---|---|---|---|---|---|---|---|---|---|---|---|---|---|---|
| | | | | total | | filaments | | diatoms | | standing crop g odw/m² | | "productivity" 2) g odw/m².day | | F 3) |
| | | | | Co | FeMe | Co | FeMe | Co | FeMe | Co | FeMe | Co | FeMe | |
| Eu 14/1 | 10-16 | < 1500 | 0-20 | 4.0 | 5.7 | 0 | 0 | 5.7 | 5.7 | 8.8 ±2.3 | 11.6 ±3.0(8) | .44 | .58 | * |
| Eu 14/2 | | | 0-35 | 3.6 | 6.7 | 1.4 | 1.4 | 2.8 | 6.6 | 3.5 ±.9 | 14.2(4) ±2.0 | .10 | .41 | * |
| Eu 15 | | > 3000 | 0-23 | 7.1 | 8.0 | 0 | 1.5 | 7.0 | 7.7 | 42.6 ±3.8 | 60.6(6) ±4.6 | 1.85 | 2.64 | *** |
| Eu 16 | | | 0-17 | 7.0 | 8.5 | 0 | 6.0 | 7.0 | 4.5 | 37.5 ±6.8 | 55.4(8) ±5.2 | 2.21 | 3.26 | *** |
| Eu 17/1 | > 18 | ∿ 1500 | 0-20 | 5.3 | 5.3 | 5.3 | 5.3 | 1.5 | 1.5 | 13.2 ±2.6 | 12.9(12) ±2.6 | .66 | .65 | - |

1) see Table 2
2) accumulation of standing crop
3) F-test for significance of difference in Anovar (effect of chemistry) p for Null-hypothesis, $* \leq .05$, $** \leq .01$, $*** \leq .001$.
4) for dominant species see Table 5
in brackets: number of samples for each condition (distributed on 2 or 3 channels)
co: ground water without any additions. FeMe: see Table 4.

When Melosira varians appears in late fall and winter the light intensity and the production are low. The addition of the metal mixture results in some stimulation of the dominant species Melosira.

Clearly the reaction of the algal community to essential metals is modified by its sociology and by the physiological state of its species. We can expect that the dose response curves to various metal levels depend highly on the ambient conditions.

## 5. THE SEPARATION OF THE EFFECT OF IRON FROM THE EFFECT OF THE OTHER ELEMENTS

In view of the different modes of action of its components the observed effect of the metal mixture is likely to result from the superposition of separate reactions to the elements present.

Iron and the micronutrients Mn, Co, Cu, Mo and Zn were applied separately in a concentration 10 time higher than is present in the mixture FeMe. The supply of iron 1.0 μMol Fe/l (56 μg/l) does not affect the vegetation (Table 7). On the other hand the addition of the micronutrients as Me 10x completely suppresses the growth of the ground water dominant diatom Melosira varians; this diatom is on the other hand cleary stimulated by the mixture FeMe 1x. In the channels treated with Me 10x Stigeoclonium becames dominant. This alga is rarely met in channels fed with pure ground water.

The effect of Me 10x demonstrates how a fundamental change in the aspect of an algal community can occur without any significant change in the biomass formation of the community. As the further development of the biocenosis depends partly on the previous species composition some differences in productivity possibly could develop at a later date.

The appearance of Stigeoclonium in Me 10x and the complete disappearance of the diatoms and filamentous algae characteristic for FeMe 1x points out that the two metal mictures FeMe 1x and Me 10x trigger different physiological reactions. The clear depression of the algal growth at the start of the experiment with Me 10x suggests that the mixture has a toxic effect and that only the alga Stigeoclonium can thrive under the new conditions.

## 6. THE POSSIBLE ROLE OF AN INTERACTION BETWEEN PHOSPHORUS AND IRON

Iron as Fe(III) and phosphates form highly insoluble iron(3)phosphates, a fact which has been made use of by treatment plants to eliminate phosphorous from sewage. If the phosphate added to ground water channels in an attempt to test its role in eutrophication were to block some iron, then the observed lack of stimulation of algal growth by P-dosage might result from experimentally induced iron deficiency. In this case however we would expect occasional growth reduction when the phosphate supply is artificially increased. In the experiments performed in different seasons with various sociological compositions of the algal communities the increase of P-concentration from 15 μg P/l to 100-200 μg P/l had no stimulatory effect on the standing crop. Neither was any inhibition observed (Table 3). An interference of P with the availability of iron is therefore most unlikely.

On the other hand it might be argued that the natural concentrations of the trace elements in our ground water is so low, that some metal limits the growth and that the plants can therefore not take advantage of an increase of the P-concentration

TABLE 7 The Separation of the Effect of Iron from the Effect of the other Essential Heavy Metals

| condition | | Co | FeMe 1x | Me 10x | Fe 10x |
|---|---|---|---|---|---|
| Fe μMol/l(μg/l) | | | .1 (5.6) | - | 1 (56) |
| Mn μMol/l(μg/l) | | | .05 (2.8) | .5 (27.5) | |
| Co,Cu,Mo,Zn | | | .01 (∿ .6) | .1 (∿ 6) | |
| channels/condition | | 2 | 2 | 2 | 2 |
| | age days | | | | |
| standing crop | 35 | 3.5 | 14.2 | 3.9 | 3.8 |
| g odw/m$^2$ | n=4 | ± 9 | ± 2.0 | ± 1.9 | ± 1.6 |
| | 66 | 24.2 | 56.4 | 33.6 | 22.0 |
| | n=4 | ± 7.4 | ± 14.0 | ± 11.5 | ± 8.4 |
| significance 1) | | | ** | - | - |
| biomass accumulation g odw/m$^2$.day | 0-66 | .37 | .86 | .51 | .33 |
| sociology total | 15 | 4.3/4 | 5.7/6 | 1.3/1.2 | 4.3/4.3 |
| | 28 | 4/3.3 | 6.7/6.7 | 4.7/4 | 3.7/3.7 |
| | 34 | 4.5/3 | 6/7 | 4/3.5 | 3.5/3 |
| | 62 | 7/5 | 7/8 | 7/6.5 | 7.5/5.5 |
| filamentous algae | 15 | 1.3/1 | 1.7/1.3 | 1.3/1.2 | .3/.3 |
| | 28 | 2/.7 | 2/.7 | 3.3/4 | 1.3/.2 |
| | 34 | 2/1 | 1.5/- | 4/3.5 | 1/.5 |
| | 62 | 2/2 | 3/.5 | 6.6/6.5 | -/1 |
| diatoms | 15 | 3.3/3 | 5/6 | -/- | 4/4.3 |
| | 28 | 3/2.8 | 6.5/6.7 | -/- | 3/3.5 |
| | 34 | 3.5/2 | 6/7 | -/- | 3.5/2.5 |
| | 62 | 6/4.5 | 3.5/8 | -/- | 7.5/5.5 |

December 1972 - February 1973. Each condition 2 channels, recolonization of the entire channel, n= number of samples in each channel

1) t-test of the significance of the difference to groundwater channels ** $p < .01$.

sociology: density average along each of the 2 replicate channels.

by dosing. Only after the removal of the metal limitation, as might be caused by sewage, the potential P-limitation might be revealed.

That this kind of phosphorous limitation can not be serious is demonstrated by the fact that the addition of FeMe 1x at the most productive season stimulates growth in the presence of only 15 μg $PO_4$-P/l by about 50 % (Table 6; Fig. 2) or more over short periods.

The experiments therefore suggest that under these conditions there is no interaction between metals and phosphorous which results in growth limitation.

## 7. THE EFFECTS OF COPPER AND MANGANESE SUPPLIED SEPARATELY

The addition of .1 µMol Cu/l at a time when diatoms are favored in pure ground water results in an almost complete suppression of diatoms (except Nitzschia ssp); after some lag, the colonizable surface is taken over by Tribonema (Table 8). In a short burst the biomass accumulates faster than in ground water giving the impression that copper is stimulating production. The filaments are however partly contorted, signalling some damage which becomes manifest in a subsequent growth standstill, presumably as the result of progressive copper accumulation.

In the higher Cu-concentration of 1.0 µMol/l neither diatoms nor Tribonema nor any other filamentous algae appear; new growth is extremely reduced except for the formation of a crust of coccoid green algae.

TABLE 8 The Effect of Copper and Manganese on the Development of Algal Communities

| element | age days | n | | | copper | | manganese | |
|---|---|---|---|---|---|---|---|---|
| µMol/l | | | - | - | .1 | 1.0 | .5 | 5.0 |
| µg/l | | | - | - | 6.4 | 63.5 | 27.5 | 274.7 |
| standing crop g odw/m$^2$ | | | | | | | | |
| | 12 | 2 | 10.3±1.4 | 12.8±1.7 | 4.0±.5 | - | 15.6±2.0 | 7.0±2.2 |
| | 20 | 2 | 26.1±.4 | 24.0±1.0 | 7.1±.6 | - | 36.9±.4 | 29.9±5.5 |
| | 30 | 3 | 54.4±4.2 | 48.8±3.1 | 21.4±3.6 | 1.9±.9 | 45.9±4.0 | 42.2±3.7 |
| biomass accumulation g odw/m$^2$.day | 12-20 | | 1.98 | 1.40 | .39 | - | 2.66 | 2.86 |
| sociology total | 18 | 3 | 6 | 6 | 4 | - | 7 | 6 |
| | 32 | 3 | 7 | 7 | 4.7 | 1 | 8.3 | 8 |
| | 40 | 3 | 7 | 7 | 5 | .5 | 8 | 8 |
| filamentous algae | 18 | 3 | - | 1 | 4 | - | 7 | 6 |
| | 32 | 3 | .5 | .5 | 4.3 | - | 8 | 7.7 |
| | 40 | 3 | .5 | .5 | 5 | - | 6.6 | 6.7 |
| diatoms | 18 | 3 | 6 | 6 | - | - | 2 | 2 |
| | 32 | 3 | 7 | 7 | 1 | - | 2.3 | 2.3 |
| | 40 | 3 | 6.7 | 7 | .5 | - | 2.5 | 3.3 |
| frequent species | | | Synedra, Nitzschia, Melosira, Diatoma vulgare, Hormidium, Tribonema, Ulothrix | | Tribonema* Nitzschia | coccoid green algae | Tribonema, Hormidium, Ulothrix, Diat. vulgare, Synedra, Nitzschia | |

March-April 1974. Each condition one channel, recolonization of the upper reaches of the channels, n= number of samples.

underlined: species making up more than 50 % of the biomass (dominant).

* filaments often contorted

Copper not only heavily checks the development during colonization. If added in winter to well developped algal communities major shifts in sociology occur after a short time. With 1 μMol/l the old algal community composed of Melosira and Tribonema disappeared in the course of a few days leaving the channels practically bare; then some thin film of a resitant blue green alga became established.

The addition of .5 μMol/l manganese to ground water strongly reduces the characteristic development of the diatoms but does not eliminate them as copper does. Tribonema is stimulated and becomes the dominant species (Table 8). The higher concentration of 5 μMol Mn/l results in an initial reduction of growth, which is soon overcome by Tribonema. In contrast to the channels fed with copper no morphological damages of the algae are noticed despite concentrations being higher by a factor of 5.

In this experiment the lower concentrations of both elements first act on the diatoms, the algae with are at the time most favored in pure ground water. Whereas with Mn all the species typical for ground water survive and start to spread again after some time, the effect of 1 Cu is radical and only some Nitzschias are found in .1 μMol/l and none at all in 1 μMol/l.
Copper at .1 μmolar seems to reach the upper limits of its stimulating range for Tribonema and then becomes toxic when the alga has an opportunity to accumulate the metal. In contrast to copper the critical concentration for Mn appears considerably higher.

In the solution Me 10x (chapter 5, Table 7) .1 μMol Cu/l and .5 μMol Mn/l were supplied as in this experiment. However the filamentous species which took then over was not Tribonema but Stigeoclonium. Whether the effects produced by Cu and Mn alone are modified in Me 10x by the presence of other micronutrients or by some other factor, climate for example (the experiment with Me 10x was performed in December-February, with Cu and Mn alone in March-April) needs further clarification.

## 8. THE RECOVERY FROM METAL STRESS

As we have seen, the addition of metals frequently results in changes of the sociological composition of the algal community. How fast does a biocenosis shaped by metal stress revert to the composition typical for ground water when the supply of the metals is discontinued?

The changes in density of the major algal groups in 4 channels have been observed from October to March (Fig. 3). In the 2 ground water channels a relative stable cover dominated by Melosira varians is maintained throughout the winter. The addition of Fe 1x (0.1 μMol/l) is without effect; however the supply with Me 10x results in the disappearance of Melosira and the spread of Tribonema and Stigeoclonium, a regrouping comparable to what had been observed in recolonization experiments (chapter 5, Table 7).

At the age of 4 months, after being subjected for 2 months to Fe 10x, the two metal channels were returned to a pure ground water regime. After 40 days without metals the filamentous algae (now Tribonema alone) were still dominant. On the 160 th day of the experiment, after more than a month under identical chemical conditions, the general appearance of the vegetation in the once metal treated channel strongly contrasts with the permanent ground water control where filamentous algae are quite unimportant. The two conditions could unfortunately not be followed up longer; still they show a remarkable stability of the sociology induced by stress.

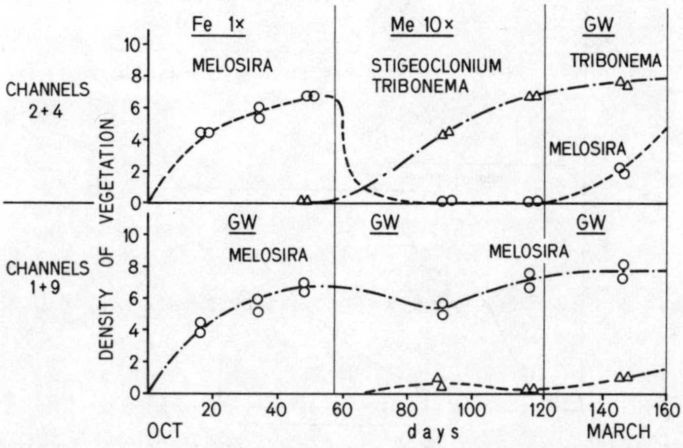


Fig. 3 Stress recovery of the metal-treated algal communities. Each condition 2 channels, average vegetation density over 50 m. (see Table 2)
o diatoms, Δ filamentous algae, GW ground water,
Fe 1x:.1 μMol Fe/l, Me 10x:Mn .5, Co, Cu, Mo, Zn .1 μMol/l

It is well possible that an analysis for metal of these algae with short generation time would not reveal their previous treatment with metals. A person not familiar with the history of the channels would then be at a loss to explain the differences in sociology between the channels. This phenomenon calls for some caution when we attempt to correlate the chemistry of natural rivers to biological observations in order to establish lists of indicator organisms.

## 9. OUTLOOK

Few aspects of river ecology demonstrate so emphatically the highly dynamic relationship between environmental and biotic parameters as does the study of the reaction of the ecosystem to essential metals. Here the two phenomena, eutrophication and toxicity, are intertwined to constitute a physiological reality still highly confusing to the ecologist, specially when measures to protect rivers from pollution are envisaged.

The response to a stepwise increased supply of an essential metal under otherwise not growth limiting conditions transgresses from a concentration range which increasingly furthers healthy development to a range where metabolic disturbances spread and finally result in the death of the organism (Fig. 4). It is important not to loose sight of the reversed sequence of reactions when we reduce a toxic load; this may lead to more growth what could be considered troublesome under some situations.

In naturally mixed populations some species will markedly differ in their dose-response characteristics. An increase in the metal concentration which eliminates species A (Fig. 4) may enhance the production of species B as we noted when Tribonema replaced Melosira upon addition of 6 μg Cu/l to the river water (Table 8). Climatic conditions modify the growth response of algae. The interaction between climate and metal concentration can be considered in the simplified system of Fig. 4 with the species A and B; A is assumed to grow in winter, B in summer and both A and B thrive together in fall. We can then distinguish 3 different reactions

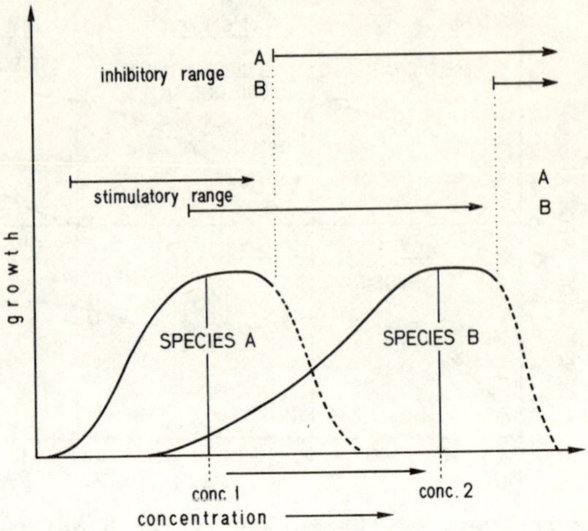


Fig. 4 The response of algae to increasing concentrations of an essential metal (for explanations see text).

to a shift in supply of an essential metal from the lower concentration conc 1 to the higher conc 2:
1) in winter: species A grows well in conc 1 but will be killed in conc 2
For climatic reasons however species B can not take advantage of the disappearance of A and the habitat remains free of vegetation; this is a clear case of inhibition. In the long run the habitat is open to the colonization by some other species which can occupy the new niche, as we observed for example in the spread of the blue-greens after the displacement of Melosira and Tribonema by 1 µMol Cu/l.
2) in summer: species A does not grow in summer and in conc 1 species B lacks the metal indispensible for its development. A shift from the lower to the higher conc 2 will stimulate production of B. This is a clear case of eutrophication.
3) in fall: the transition from conc 1 to conc 2 will cause a complete shift in sociology as species B will replace A. The overall biomass production however will remain unaltered. On can argue whether to consider this an eutrophication or an inhibition.

Analoguous situations will exist for each one of the other essential metals leaving us with a large array of possible reactions.

The reaction of a species to a set of physico-chemical conditions in a natural habitat is frequently different from its reaction to the same environmental conditions when bred in isolation as Ellenberg (1963) has pointed out. Nevertheless the knowledge of the physiological potential of algae in pure cultures will be usefull to understand their distribution in nature. Unfortunately running water species are difficult to study in pure cultures maintaining ecologically meaningfull conditions; therefore we are not yet able to compare physiological dose response curves of single benthic algal species with their behaviour in a benthic community.

Several investigators have pointed to interactions between metals such as synergistic effects whose mode of action are still open. Gächter (1976) has demonstrated that Hg, Cu, Cd, Zn and Pb are considerably more toxic to phytoplankton when given together than when applied singly (in concentrations from about $.5\times10^{-9}$ Mol/l for the most toxic Hg to $50x10^{-9}$ Mol/l for the least toxic Pb). Seasonal variations on the toxic effect of the metals are believed to result from changes in phytoplankton composition.

As far as the essential heavy metals are concerned their toxic action has received more attention than their growth stimulating effects; complex interactions with sociology are to be expected. Among the 4 essential elements Cu, Co, Mn and Zn, copper is by far the most toxic for the marin diatom Nitzschia closterium (Rosko and Rachlin, 1975). A growth inhibition of 50 % in a 4 day period ($EC_{50,96hrs}$) results from 33 µgCu/l (about .5 µMol/l); for the same inhibition by the other metals 271 µg Zn/l, 10200 µg Co and 25700 µg Mn/l are required. Unter similar condition however the same inhibition of Chlorella vulgaris the Zn-concentration needs to be about 10 times higher than with Nitzschia. Chelators will strongly modify the response; the addition of citric and amino acids increases tolerance towards Co, Mn and Zn by a factor of about 2, towards Cu by a factor of 7.

Chelators, either introduced from the terrestrial environment or by the river vegetation will affect the availability of metals and thus their eutrophying as well as inhibiting influences (Stumm, 1972). If organic pollution is mild their concentrations in the free water may be too low to be effective. However in dense vegetation reduced mass exchange leads to accumulation of excreted organic compounds which can locally modify the action of metals.

In simple laboratory systems with organisms of short generation times the adjustment to changes in environmental conditions is fast. In the more complex, less homogenous benthos where some organisms have longer generation times the manifestation of altered physiological conditions requires much more time, as can be seen for example in stress recovery (chapter 8). Past conditions can still find expression long after they have ceased to exist. In order to recognize the forces which shape a momentaneous situation one has to know the history of the habitat; in this sense a benthic community has a memory. In view of the manifold reactions of essential metals and the fact they can accumulate in a biotope these compounds could play a considerable role in shaping the memory of the river benthos.

REFERENCES

Bowen, H. J. M. (1966). Trace Elements in Biochemistry. Academic Press, London. pp. 241.

Eichenberger, E. (1972). Oekologische Untersuchungen an Modellfliessgewässern. III. Die jahreszeitlichen Veränderungen im Verhältnis von heterotropher zu phototropher Biomasse bei verschiedenen Abwasserbelastungen. Schweiz.Z.Hydrol., 34, 155-172.

Eichenberger, E. (1976). Biologische und ernährungsphysiologische Gesichtspunkte zur Ausbreitung des Flutenden Hahnenfusses im Oberrhein. Wasser, Energie, Luft, 68, 234-239.

Eichenberger, E. and A. Schlatter (1978). Effect of herbivorous insects on the production of benthic algal vegetation in outdoor channels. Verh. Internat. Verein. Limnol., 20 , 1806-1810.

Ellenberg, H. (1963). Vegetation Mitteleuropas mit den Alpen in kausaler, dynamischer und historischer Sicht. Ulmer, Stuttgart. pp. 943.

Eyster, C. (1964). Micronutrient Requirements for Green Plants, Especially Algae.In D.F. Jackson (Ed.). Algae and Man. Plenum Press, New York. pp 87-119.

Gächter, R. (1976). Untersuchungen über die Beeinflussung der planktischen Photosynthese durch anorganische Metallsalze im eutrophen Alpnachersee und der mesotrophen Horwer Bucht. Schweiz.Z.Hydrol., 38, 97-119.

Mitchell, D. S. (Ed.). (1974). Aquatic vegetation and its use and control. Unesco, Paris. pp. 135.

Rosko, J. J. and J. W. Rachlin (1975). The effect of copper, zinc, cobalt and manganese on the growth of the marine diatom Nitzschia closterium. Bull. Torrey Bot. Club, 102, 100-106.

Schwarz, D. (1965). Der Einfluss von Wirkstoffen auf das Wachstum und die Vermehrung von Algen (Literaturübersicht). Veröffentlichungen der Hydrologischen Forschungsabteilung der Dortmunder Stadtwerke AG. 8, 1-206.

Stumm, W. (1972). Die Rolle der Komplexbildung in natürlichen Gewässern und allfällige Beziehungen zur Eutrophierung. Gewässerschutz-Wasser-Abwasser, Heft 8, 57-87. Aachen.

Wuhrmann, K. and E. Eichenberger (1975). Experiments on the effect of inorganic enrichment of rivers on periphyton primary production. Verh. Internat. Verein. Limnol., 19, 2028-2034.

Acknowledgement: I would like to thank H.-J. Schälchli and his staff of the Experimental Station Tüffenwies for the maintenance of the channels, L. Marosvöglyi and G. Scherrer for their technical assistance in the execution of the experiments.

# Heavy Metal Pollution in Freshwater Ecosystems

**U. Förstner and F. Prosi**

*Institut für Sedimentforschung, Universität Heidelberg, Im Neuenheimer Feld 236, D-6900 Heidelberg, Federal Republic of Germany*

ABSTRACT

Geochemical and biological processes involving metal pollutants in contaminated rivers, lakes and dams are reviewed with particular emphasis on recent developments in the study of toxic metal species and their availability for organisms. Examples are given to demonstrate the usefullness of sediment analyses to assess the provenance, distribution, extent and also the possible hazards of metal pollutants in freshwater ecosystems. The mechanisms of immobilization and remobilization of metals in freshwater particulates and the question of the accumulation or depletion of heavy metals in the aquatic food chain are discussed in detail.

Keywords: heavy metal pollution, speciation of heavy metals, heavy metal transport, sediment, biocides, industrial effluents, restoration of lakes food chain, trophic levels.

## INTRODUCTION

Industrial, domestic and agricultural wastes are continuously being discharged to our rivers and lakes whereby certain pollutants are deposited and concentrate in the sediments. Such pollutants can endanger public health by being incorporated in the food chain, or being released into overlying water layers which serve as drinking water supplies. In this regard, heavy metals in particular can constitute a serious health hazard.

Attention has been primarily focussed on heavy metals as environmental pollutants over the last two decades, especially since the well-known instances where fishermen from the Japanese Minamata Bay and villagers from the Jintsu River died or became paralyzed from mercury and cadmium poisoning respectively.

Two factors contribute to the deleterious effects of heavy metals as environmental pollutants: heavy metals can not be destroyed through biological degradation as is the case with most organic pollutants; and heavy metals tend to accumulate in the environment - especially in the bottom sediments of rivers and lakes - by association with organic and inorganic matter through processes of adsorption, complex formation and chemical combination.

## METAL POLLUTION IN INLAND WATERS

Some heavy metals which, in normal concentrations are essential components of biochemical functions (Cu, Zn, Fe, Mn, Cr, Co), are toxic when present in higher concentrations: "The most important mechanism of toxic action is thought to be the poisoning of enzymes" (Bowen, 1966). The total extent of the heavy metal pollution is unknown, neither for the atmosphere nor for the aquatic environment. A first attempt to tackle this problem involves a comparison of the consumption of heavy metal with the natural concentration of these elements in the different spheres (lithosphere, pedosphere, hydrosphere, atmosphere). The ratio

$$\frac{\text{metal consumption (in tons/a)}}{\text{average metal content in a specific sphere (in g/ton)}}$$

is introduced as a measure of the relative pollution potential of each element in a certain sphere.

The "Index of Relative Pollution Potential" for some important industrial metals in the pedosphere is shown in Table 1. It becomes evident that the more "rare" (and more toxic) metals are enriched by one order of magnitude as compared with iron, manganese, chromium and nickel.

TABLE 1 World's Consumption of Heavy Metals in 1968; Metal Contents in Soils; Index of Relative Pollution Potential for the Pedosphere; and Technophility Index

| | Consumption x 1,000 t/y (from Sames, 1971) | Soils (ppm) (from Bowen, 1966) | Index of Relative Pollution Potential (Förstner and Müller, 1973) | Technophility Index (Nikiforova and Smirnova, 1975) |
|---|---|---|---|---|
| Iron | 400,000 | 38,000 | 1 | $5.3 \times 10^7$ |
| Manganese | 9,200 | 850 | 1 | $5.0 \times 10^7$ |
| Copper | 6,400 | 20 | 30 | $1.1 \times 10^9$ |
| Zinc | 4,600 | 50 | 10 | $5.4 \times 10^8$ |
| Lead | 3,500 | 10 | 35 | $1.6 \times 10^9$ |
| Chromium | 1,700 | 100 | 2 | $2.0 \times 10^8$ |
| Nickel | 493 | 40 | 1 | $9.0 \times 10^7$ |
| Tin | 232 | 10 | 2 | - |
| Cadmium | 15 | 0.06 | 25 | - |
| Mercury | 10 | 0.03 | 30 | $1.5 \times 10^9$ |

Technogenic migration of a metal and the degree of its utilization in the noosphere can also be estimated through its "Technophility Index" (Nikiforova and Smirnova, 1975), that is, the ratio of the annual output of a metal to its "Clark" (mean concentration in the earth's crust). The higher the TP of a metal, the more intensively it is involved in the technogenic migration: mercury and lead are characterized by high TP values of 1.5 and $1.6 \times 10^9$ respectively (Table 1, last column). The TP index varies through time and thus each metal can be characterized by its own rate of TP growth; the TP of lead, for example, has grown two and a half times from the beginning of the century up to the present time and will have further increased four and a half times by the year 2000.

## The Sources of Heavy Metal Pollution

In general, there are five sources of heavy metals in inland waters:

(1) Geological weathering. This is the source of "background levels". It is to be

expected that in areas characterized by metal-bearing formations, these metals will also occur in high levels in the water of the area.

(2) Industrial processing of ores and metals. During the processing of ores, metal-bearing dust particles are formed which may be only partially filtered out by air purification systems. Appreciable quantities of metals go to waste during chemical metal refinement processes (e.g. galvanizing and pickling) by way of heavy metal solutions which are often discharged without any reclamation measures.

(3) The use of metals and metal compounds. Examples are the use of chromium salts in tanneries, copper compounds as plant protection agents, and tetramethyl lead as an anti-knock agent. Examples of metal pollutants in industrial waste streams (sources "2" and "3") are listed in Table 2 (from Barnhart, 1978).

TABLE 2 Some Hazardous Materials in Industrial Waste Streams (Barnhart, 1978)

| Industry | As | Cd | Cr | Cu | Pb | Hg | Se | Zn |
|---|---|---|---|---|---|---|---|---|
| Mining and metallurgy | x | x | x | x | | x | x | x |
| Paints and dyes | | x | x | x | x | x | x | |
| Pesticides | x | | | | x | x | | x |
| Electrical and electronic | | | | x | x | x | x | |
| Cleaning and duplicating | x | | x | x | x | | x | |
| Electroplating and metal finishing | | | x | x | | | | x |
| Chemical manufacturing | | | x | x | | x | | |
| Explosives | x | | | x | x | x | | |
| Rubber and plastics | | | | | | x | | x |
| Batteries | | x | | | x | x | | x |
| Pharmaceuticals | x | | | | | x | | |
| Textiles | | | x | x | | | | |
| Petroleum and coal | x | | | | x | | | |
| Pulp and paper | | | | | | x | | |
| Leather | | | x | | | | | |

(4) Heavy metals in animal and human excretion. Heavy metals are present in human and animal food - zinc in particular at relatively high concentrations. These metals are concentrated in excretions and find their way mainly to the water environment. The adult human excretes between 7 and 20 mg of zinc per day.

(5) Leaching of metals from garbage and solid waste dumps. The contribution of this source to the heavy metal pollution of inland waters merits close attention. Mine dumps expecially can be a serious source of pollution.

## Toxic Metals[1]

Viewed from the standpoint of environmental pollution, metals may be classified according to three criteria: (i) non-critical, (ii) toxic but very insoluble or very

[1] This section was written by Dr. G.T.W. Wittmann, Pretoria University as part of a larger article in the book Förstner/Wittmann: Metal Pollution in Aquatic Environments, Springer-Verlag: Berlin-Heidelberg-New York, 1979 and is reproduced here with the kind permission of the author.

rare, and (iii) very toxic and relatively accessible. Such a classification has been made by Wood (1974) and is listed in Table 3. Here, special importance must be attached to the non-metals arsenic and selenium besides the heavy metals mercury, cadmium and lead.

TABLE 3 Classification of Metals According to Toxicity and Availability (from Wood, 1974)

| Non-Critical | | | Toxic but Very Insoluble or Very Rare | | Very Toxic and Relatively Accessible | | |
|---|---|---|---|---|---|---|---|
| Na | C | F | Ti | Ga | Be | As | Au |
| K | P | Li | Hf | La | Co | Se | Hg |
| Mg | Fe | Rb | Zr | Os | Ni | Te | Tl |
| Ca | S | Sr | W | Rh | Cu | Pd | Pb |
| H | Cl | Al | Nb | Ir | Zn | Ag | Sb |
| O | Br | Si | Ta | Ru | Sn | Cd | Bi |
| N | | | Re | Ba | | Pt | |

Water quality critera. Although it has been well established that many inorganic constituents enter inland waters from natural or man-made sources, their significance with regard to surface water quality depends on many interdependent factors. Not only is the abundance and widespread occurrence of a particular constituent of importance, but also its availability in the form of solubized species (see below). When setting permissable limits or ultimate goals for drinking water standards, cognizance must be taken of the bioaccumulation via the food chain. Moreover, it is imperative not only to impose limits which protect man's health on the basis of trace metal quantities in surface water from which potable water is extracted, but also to consider the environmental impact of these waters discharged to the environment. Such considerations involve the ecosystem as a whole; self-purification of river systems, biological treatment plants, the effects of trace metal enrichment on biological purification treatment, the effects on crustaceans, fish and ultimately on man. With regard to the different types of drinking water contaminants, the trace metals have received considerable attention in terms of their toxic effects during the past few years. Unfortunately, it has to be admitted that many basic questions regarding this group of elements still remain unanswered. As pointed out before, the question of chemical speciation poses on of the most difficult problems to be resolved by the chemist, pharmacologist and toxicologist, especially regarding synergistic effects encountered in natural waters. Drinking water standards have been proposed by various governmental bodies in accordance with toxicity data obtained from human clinical investigations and various other studies, auch as animal experiments. A brief summary is given in Table 4 compiled by Hattingh (1977) with additional data from the F.R.G. (Schöttler, 1977)

Concentrations of dissolved trace metals. Due to the variety of rock formations and the fluctuations in water transport (especially in rivers) it is relatively difficult to establish global mean values for the individual trace elements in inland waters. In order, however, to determine the extent of civilizational pollution, background values are most desirable, which, although not "absolute", serve as guidelines for water quality control and as indicators of the extent and possible consequences of contamination when exceeded. It is in this sense that the background value data listed in the first column of Table 5 are to understood. Most of the data was obtained from large inland waters, e.g. from the Amazon and Mississippi Rivers (2nd and 3rd columns in Table 5) where pollution either still low or where the civilisational inputs are diluted by the large water mass.

TABLE 4 Drinking Water Quality Criteria (after Hattingh, 1977; Schöttler, 1977). All concentrations in μg/l

| Parameter | USPHS (1962) | Japan (1968) | USSR (1970) | WHO European (1970) | WHO Intern. (1971) | SABS (1971) | NAS (1972) | Australia (1973) | EPA (1975) | FRG (1975) |
|---|---|---|---|---|---|---|---|---|---|---|
| Arsenic | 10 | 50 | 50 | 50 | 50 | 50 | 100 | 50 | 50 | 40 |
| Barium | 1000 | - | 4000 | 1000 | - | - | 1000 | 1000 | 1000 | - |
| Cadmium | 10 | - | 10 | 10 | 10 | 50 | 10 | 10 | 10 | 6 |
| Chromium | 50 | 50 | 100 | 50 | - | 50 | 50 | 50 | 50 | 50 |
| Copper | 1000 | 10000 | 100 | 50 | 50 | 1000 | 1000 | 10000 | - | - |
| Lead | 50 | 100 | 100 | 100 | 100 | 50 | 50 | 50 | 50 | 40 |
| Mercury | - | 1 | 5 | - | 1 | - | 2 | - | 2 | 4 |
| Selenium | 10 | - | 1 | 10 | 10 | - | 10 | 10 | 10 | 8 |
| Silver | 50 | - | - | - | - | - | - | 50 | 50 | - |
| Zinc | 5000 | 100 | 1000 | 5000 | 5000 | 5000 | 5000 | 5000 | - | 2000 |

TABLE 5 Metal Concentrations (in μg/l) in Inland Waters

| Metal | Background[a] | Amazon River[b] | Mississippi[e] | Lake Michigan[g] | Lake Constance[i] | Rhine River[k] | Mine effluent[l] |
|---|---|---|---|---|---|---|---|
| As | 1[g] | - | 3[f] | 1 | 3.7 | 13 | - |
| Cd | 0.07[c] | 0.07[c] | 0.1 | 0.3 | - | 5.5 | 52 |
| Co | 0.05[b] | 0.06[b] | - | 0.2 | - | 10 | 3300 |
| Cr | 0.5[e] | 2[b] | 0.5 | 1.7 | 5.7 | 33 | 4000 |
| Cu | 1[d] | 1[d] | 2 | 5 | 8.2 | 34 | 5400 |
| Hg | 0.01 | - | <0.1[f] | 0.03 | 0.18 | 0.65 | - |
| Ni | 0.3[b] | 0.27[b] | 1.5 | 3 | 5 | 20 | 6400 |
| Pb | 0.2[e] | - | 0.2 | 1.5[h] | - | 57 | 290 |
| Se | 0.1[g] | - | - | 0.08 | 1.3 | 6.5 | - |
| Zn | 10 | - | 10[f] | 16 | 37 | 330 | 26000 |

a) Förstner and Wittmann (1979, in press) from references b-g and others.
b) Gibbs (1977), c) Boyle and co-workers (1976), d) Boyle (1978), e) Trefry and Presley (1976), f) U.S. Geol. Survey, Baton Rouge (1972-1976), g) Copeland and Ayers (1975), h) Edgington and Robbins (1975), i) Quentin and Winkler (1974), mean value 1971-1973, k) Inst.f. Wasser-, Boden- und Lufthygiene, Bundesgesundheitsamt Berlin (unpublished data, Working Group "Metals", German Research Society, Internal Report for 1970-1973), l) Wittmann and Förstner (1977), example from West Wits Goldfields, R.S.A.

The example of the *Amazon River* shows that the various sub-systems ~~of large river~~ of large river systems may differ widely with respect to the concentrations of trace metals in both water and suspended matter. We have found from studies of Amazon "lake" sediments (Irion & Förstner, 1975) that deposits contributed by rivers from regions with heavily leached soils - "blackwater rivers" (e.g. Rio Negro) are significantly depleted of iron, manganese, zinc, copper and cobalt as compared to the "white water" river deposits, e.g. from Rio Solimoes. Recent analyses of copper in the Amazon River system by Boyle (1978) indicate a similar distribution for the dissolved species; the values of the black - or clear water samples center around 0.5 μg/l, whereas the Cu concentrations of the second group of rivers originated in the Andes, pre-Andes and southwestern Amazon lowlands range between 1.5 and 2.0 μg/l.

The investigations on the large rivers of relatively unpolluted areas, such as those on the *Mississippi River* have shown that with improved sample extraction, storage and handling procedures, most of the data are significantly below those values established as background data ten years ago (e.g. Turekian, 1969). Nonetheless, there is still a good deal of controversy as to the origin of the strong divergences in the metal contents sometimes found in inland waters which are even greater than those for seawater. It seems that the amount of suspended matter which, on the one hand can partly pass through the filter (usually 0.45 µm pore size) and on the other hand can adsorb metals from solution,, plays a considerable role for the metal concentrations in water analyses. The Mississippi River for example, showed however no significant seasonal changes in dissolved trace metals although the suspended matter concentrations in the Mississippi decreases from over 300 mg/l during periods of normal flow to 30 mg/l or less during the three month flow period (Trefry and Presley, 1976)

The size of the water body also seems to be an important factor, particularly in areas of stronger pollution such as in the *Lake Michigan* area. Compared to the background values, most of the metals are enriched by a factor of approximately three. This contamination is mainly due to atmospheric contaminants (Klein, 1975) as indicated by the high amounts of dissolved lead in water samples from Lake Michigan. Metal enrichments have also been observed in water samples taken from *Lake Constance* although none of these exceeded the standards set for drinking water. This is of importance since Lake Constance is the largest drinking water reservoir in Europe, providing water for large parts of southern Germany. In other water bodies in the Federal Republic of Germany however, metal concentrations have already reached or even surpassed the permitted limits. The analytical data for metals in the central and lowever sections of the *Rhine* taken between 1970 and 1973 indicate that lead, iron and manganese clearly exceed the acceptable maximum levels for raw and drinking water (WHO-European limits), (see data in Table 5 from the station Bimmen near the Dutch/German border). In some cases, cadmium, chromium, mercury and selenium are present in critical amounts. Approximately 50% of the nickel and arsenic is already approaching intolerable levels (Interim Report 1975, Working Group Metal, German Research Society).

The highest concentrations of trace metals have been observed in areas with *acid mine drainage*. One example is seen in Table 5, last column, from South African Goldfields, where the metal concentrations in solution are enriched by factors of $10^3$ to $10^5$ as compared to normal values. The serious effects of mine effluents on the water quality of rivers and lakes, as well as on the biotopes, particularly on fish populations have been known of for many years. One of the first descriptions of this problem is to be found in the Report of the 1868 River Pollution Commission in Britain which studied the grave damage caused by the dispersal of toxic metals from lead, zinc and arsenic mines in mid-Wales (cit. Lewin and co-workers, 1977):

"All these streams are turbid, whitened by the waste of the lead mines in their course; and flood waters in the case of all of them bring down poisonous slime which, spreading over the adjoining flats, either befoul or destroy grass, and thus injure cattle and horses grazing on the dirtied herbage, or, by killing the plants whose roots have held the land together, render the shores more liable to abrasion and destruction on the next occasion of high water."

In a study published in 1924 on the fauna of rivers in the Aberystwyth District of Cardiganshire, Kathleen Carpenter showed that despite legislation, several rivers in this area were completely devoid of fish, probably as a result of pollution by the local lead mines, even from those which had been previously closed. The major process affecting the lowering of pH values (down to pH 2 to 3) is the exposure of pyrite ($FeS_2$) and of other sulfide minerals to atmospheric oxygen and moisture, whereby the sulfidic component in pyrite is oxidized to sulfate ($SO_4^{2-}$) and acidity ($H^+$ ions) is generated. Bacterial activity, for example by Thiobacillus ferroxidans, can assist the oxidation of $Fe^{2+}$ (aq) in the presence of dissolved oxygen.

By applying the opposite process of sulfide formation, it has been proposed to limit heavy metal pollution in natural waters by the introduction of sewage together with heavy metal effluents into anaerobic settling ponds (Jackson, 1978).

## METAL SPECIES IN FRESHWATER

Until recently most environmental research on trace metals was based on an assessment of the total metal concentration. During the last years it has become increasingly evident that the environmental impact of a particular metal species may be more important than the total metal concentration. The toxicity of copper, for example, is strongly reduced with the formation of copper organic complexes (Davey and co-workers, 1974). Organic ligands, such as fulvid acids, NTA and EDTA, can inhibit the uptake of metals and may thus raise the toxic threshold (Sibley and Morgan, 1975; Andrew, 1976; Zitko, 1976; Brown, 1976). The experiments of Jackson and Morgan (1976) have shown that free ion activity is a good indicator of copper toxicity to phytoplankton.

### Metal Speciation

According to Guy and Chakrabarti (1977), three major questions must be answered in regard to the speciation of metal ions:
(i) Is the soluble metal present as a complexed species or as a simple aquated ion?
(ii) Is the species charged?
(iii) What is the size of the metal species?
Two analytical techniques have in particular been applied in metal ion speciation: anodic stripping voltammetry (ASV) and ultrafiltration. The first procedure divides the metal species into two categories - electroactive (aquous ions and "labile" complexes) and electroinactive (organic complexes and colloidal species). Ultrafiltration and dialysis is used to divide the metal species into different size fractions. The species that pass through the smallest pore size are generally taken to be free metal ions or small complexes.
A general schema of metal speciation - mainly based on the particle size fractions - has been given by Stumm and Bilinski (1973) and is reproduced in Table 6.
In practice, the first step applied in separation of particulate from soluble metals involves filtration through a 0.45 μm pore size membrane filter. The group of soluble metal ions can be subdivided into following categories: (i) simple aquated ions, (ii) metal ions complexed by organic ligands such as amino, fulvic and humic acid (analytical schemes are given by Florence and Batley, 1976 and Hart and Davies, 1978).

As a practical example, the distribution of manganese in the grain size spectrum in a sample from the Amazon is reproduced in Fig. 1 (from Gibbs, 1977). The size separation for the various fractions to be analyzed was accomplished by centrifugation of the fractions > 0.1 μm. For the finer solids and dissolved materials, a series of membranes of 300 Å, 100 Å, 30 Å, 10 Å and 5 Å were used. An additional phase differentiation of the solid particles - the methods are described below - indicates the decrease of the surficially bound metal content with increasing grain sizes, while the structurally bonded (more or less inertly incorporated) metal concentrations are fairly constant over the grain size spectrum from 0.1 μm to 100 μm. The concentrations of both solid phases rapidly decrease to zero at approximately 0.1 μm particle diameter. The left-hand modes of dissolved organics ( 0.003 - 0.02 μm) and dissolved inorganic ( < 0.008 μm) Mn constituents are sufficiently small in particle size to assure physical transport of the material with the water mass. Conversely, the majority of right-hand Mn modes (Fig. 1) are of sufficiently large particle size to be separated from the water mass.

Transport phases of trace elements. The source of trace metals in aquatic systems significantly determines their distribution ratio between the aqueous and solid phases. For example, the bulk of the detrital trace element particulates never leaves the solid phase from initial weathering to ultimate deposition. Similarly, metal dust particles (e.g. from smelters) and effluents containing heavy metals associated with inorganic and organic matter, undergo little or no change after be-

TABLE 6 Metal Species in Aquatic Systems
(Stumm and Bilinski, 1972)

| Metal Species | Range of diameters (µm) | Examples |
|---|---|---|
| Free aquated ions | | $Fe(H_2O)_6^{3+}$; $Cu(H_2O)_6^{2+}$ |
| Complex ionic entities | | $AsO_4^{3-}$, $UO_2^{2+}$, $VO_3^-$ |
| Inorganic ion-pairs and complexes | | $CuOH^+$, $CuCO_3^o$, $Pb(CO_3)_2^{2-}$ $AgSH^o$, $CdCl^+$, $Zn(OH)^-$ |
| Organic complexes, chelates and compounds | 0.001 | Me - $OOCR^{n+}$, $HgR_2$ 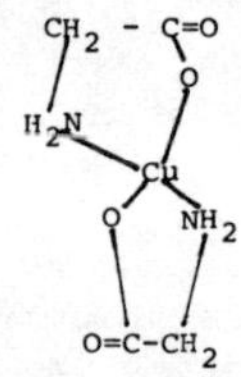 |
| Metals bound to high molecular weight organic materials | 0.01 | Me-humic/fulvic acid Polymers |
| Highly dispersed colloids | | FeOOH, Mn(IV) hydrous oxides |
| Metals sorbed on colloids | 0.1 | Me $\cdot aq^{n+}$, $Me_n(OH)_y$, $MeCO_3$, etc. on clays, FeOOH, organic |
| Precipitates, mineral particles, organic particles | | $ZnSiO_3$, $CuCO_3$, CdS in FeS, PbS |
| Metals present in live biota | | Metals in algae |

(Arrows spanning the size range: IN TRUE SOLUTION; DIALYSABLE; MEMBRANE FILTERABLE; FILTERABLE)

(Me=metal; R=alkyl)

ing discharged into a river. This is attributable to the average residence times (of the order of days or weeks), too short for the establishment of stable, dynamic equilibria between water and suspended material (Bowen, 1977).

Estimates on the world river transport of transition metals to oceans suggest that less than 3% is associated with dissolved species (Gibbs, 1977): iron 0.13%, cobalt 0.3%, nickel 0.5%, copper 1%, chromium 2.5%, manganese 3%. It should be noted, however that there are very strong differences, even in the less polluted systems; Gibbs (1977) gives the example of the Amazon system where the contributary Rio Negro carries very little sediment and therefore almost the entire load of transition metals is transported in the dissolved-complexed phase; when the Rio Negro reaches the main channel of the Amazon River, the sediment-related metal transport overwhelms the minor load carried in solution. A characteristic picture of heavy metal transport in less polluted systems in moderate climates is given by Trefry and Presley (1976) from the lower Mississippi River. Table 7 indicates, that only 11.1% (Cd) to less than 0.02% (Fe) of the heavy metal input to the Gulf of Mexico is transported in a dissolved-complexed state (arsenic as a potential metal toxin is discharged at approxi-

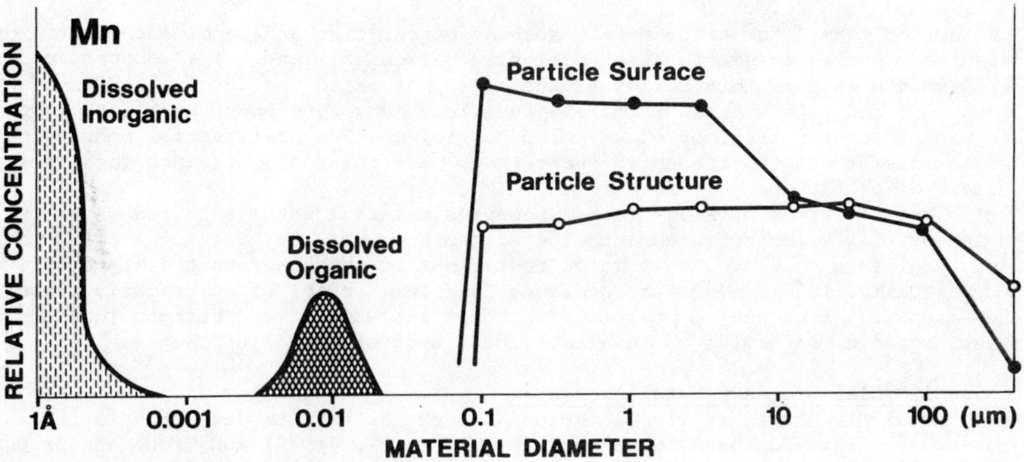


Fig. 1 Mn phases in the grain size spectrum samples from the Amazon River (after Gibbs, 1977)

TABLE 7 Percentage Particulate-Associated Metals of Total Metal Discharge (Solid + Aqueous)

| | Mississippi[a] | Polluted U.S. Rivers[b] | Polluted F.R.G. Rivers[c] | Rhine (Netherlands)[d] |
|---|---|---|---|---|
| Sodium | - | - | 0.5 | - |
| Calcium | - | - | 2.5 | - |
| Strontium | - | 21 | - | - |
| Boron | - | 30 | - | - |
| Cadmium | 88.9 | - | 30 | 45 |
| Zinc | 90.1 | 40 | 45 | 37 |
| Copper | 91.6 | 63 | 55 | 64 |
| Mercury | - | - | 59 | 56 |
| Chromium | 98.5 | 76 | 72 | 70 |
| Lead | 99.2 | 84 | 79 | 73 |
| Aluminum | - | 98 | 98 | - |
| Iron | 99.9 | 98 | 98 | - |

a) Trefry and Presley (1976)
b) Kopp and Kroner (1968)
c) Heinrichs (1975)
d) De Groot and co-workers (1973)

30% in solution.)

A similar sequence of the ratios of particulate and dissolved heavy metal phases has also been found for polluted systems, with the typical effects, however, that the dissolved discharges in polluted waters are significantly higher that in the less polluted systems, particularly for metals such as cadmium, zinc and copper.

Examples are given in Table 7 indicating the fractions of metals in particulate form as percentages of the total metal discarges from U.S. rivers (Kopp & Kroner, 1968), rivers in West Germany (Heinrichs, 1975) and from the Rhine River in the Netherlands (De Groot et al., 1973). The order of sequence of the mobility for a few selected metals is as follows: Alkali and alkali-earth metals are predominantly present in a dissolved form; for trace metals such as boron, zinc and cadmium the ratio of dissolved species to particulate species is between 2:1 and 1:1 ; copper, mercury, chromium and lead exhibit ratios of the solid phases to the aqueous phases of between 2:1 and 4:1; whereas iron, aluminum (and manganese under normal Eh conditions are almost totally transported as solid particles. The preferential bonding of heavy metals in effluents to the solid phase is of particular significance for tracing sources of pollution.

During reduced rates of river flow, suspended material settles to the river bed, becoming partially incorporated into the bottom sediment. By virtue of their composition, sediments conserve heavy metal contamination and "express the state of a water body" (Züllig, 1956). Vertical sediment profiles (cores) often uniquely preserve the historical sequence of pollution intensities; lateral distributions (quality profiles) serve to determine and evaluate local sources of pollution.

### Seasonal Cycles of Heavy Metal Transport

Heavy metal discharges in rivers may undergo characteristic developments in a one-year-cycle. This has been shown by Carpenter et al. (1975) and Troup and Bricker (1975) in their investigations on the temporal variability of metal transport by the *Susquehanna River* to the Chesapeake Bay. At first they found that the trace metal concentration correlated well with the amounts of solids discharged: the concentrations seemed to be highest in the spring and lowest in the summer and fall. This concurs with the above-mentioned findings of a major transport of heavy metals in associations with particulate matter. Upon closer study however, Mn, Ni, Zn and Co exhibited large concentrations in January, and Cu, Cr and Mn had concentration peaks in the late spring and early summer. When data are calculated for weight concentrations of metals in the solid fraction, it is found that all metals generally peak during December and January and secondary peaks occur for Co, Cr, Ni, Cu and Mn in July. This suggests that since decaying organic matter is abundant in the Susquehanna River during these two periods, the high concentrations may be the result of metal bonding to these phases. Grimshaw and co-workers performed studies on the *River Ystwyth* in mid-Wales, where strong metal pollution from former mining operations is still obvious. Their findings indicate that metal concentrations in solution are highest at low flows, suggesting a *dilution effect* (which has been found by many other investigators particularly from less polluted rivers but also from polluted rivers). For brief periods during the initial stages of storm run-off, there is a very significant increase in the metal concentrations in solutions which can be explained by a *flushing effect*. The authors note that summer convection storms in mining areas appear to be particularly significant for river biota.

Temporal variations in the metal distribution in three phases - biomass, allochthonous particulates and dissolved phase - were studied by Baccini (1976) from the *Lake of the Four Cantons* (Vielwaldstättersee). In the epilimnion the biogenic portion of the total particulate phase for copper is approximately the same as the portion of the allochthonous phase and can even be higher during periods of high organic production. The distribution of copper in the hypolimnion, however, indicates that the allochthonous portion dominates during periods of greatest sedimentation (May to September). These data show that copper (and zinc) introduced in soluble form is transported into the particulate matter through the agency of plankton and that the Cu from decomposed sedimentary plankton is partially returned into solution whereas the allochthonous particles are deposited on the lake bottom relatively unchanged.

## HEAVY METALS IN FRESHWATER SEDIMENTS

Pollution reconnaissance in stream and lake sediments is carried out using the standard practices of mineral exploration. As an introductory example, the heavy metal contamination of the Rhine River, one of the most heavily polluted rivers, is described.

Between Lake Constance and the North Sea, the Rhine flows through some of the most highly industrialized areas of Central Europe (Rhine-Main, Ruhr areas). In the Middle and Lower Rhine, some pollutants clearly exceed the permissible standards for drinking water; the most dangerous toxic metals, i.e. cadmium, mercury and selenium, are present in critical concentrations (see above). This poses an acute danger to the common practice of artificial ground water enrichment by bank filtration, should "breakthroughs" of elevated metal concentrations in the river water ever occur. Even the traditional methods of drinking water reclamation are not capable of eliminating the metals completely. It was of primary importance, therefore, to establish the extent and the source of the various metal pollutants in the Rhine River.

often seemed to be used for determination of h-metals

Sediment samples were collected during low water periods in the winter of 1971/1972 from the otherwise flooded banks of the Rhine River. In order to diminish grain size effects, the pelitic fractions ( < 2 µm ) were separated from the bulk sediment in settling tubes; chemical analyses were performed by atomic absorption spectroscopy.

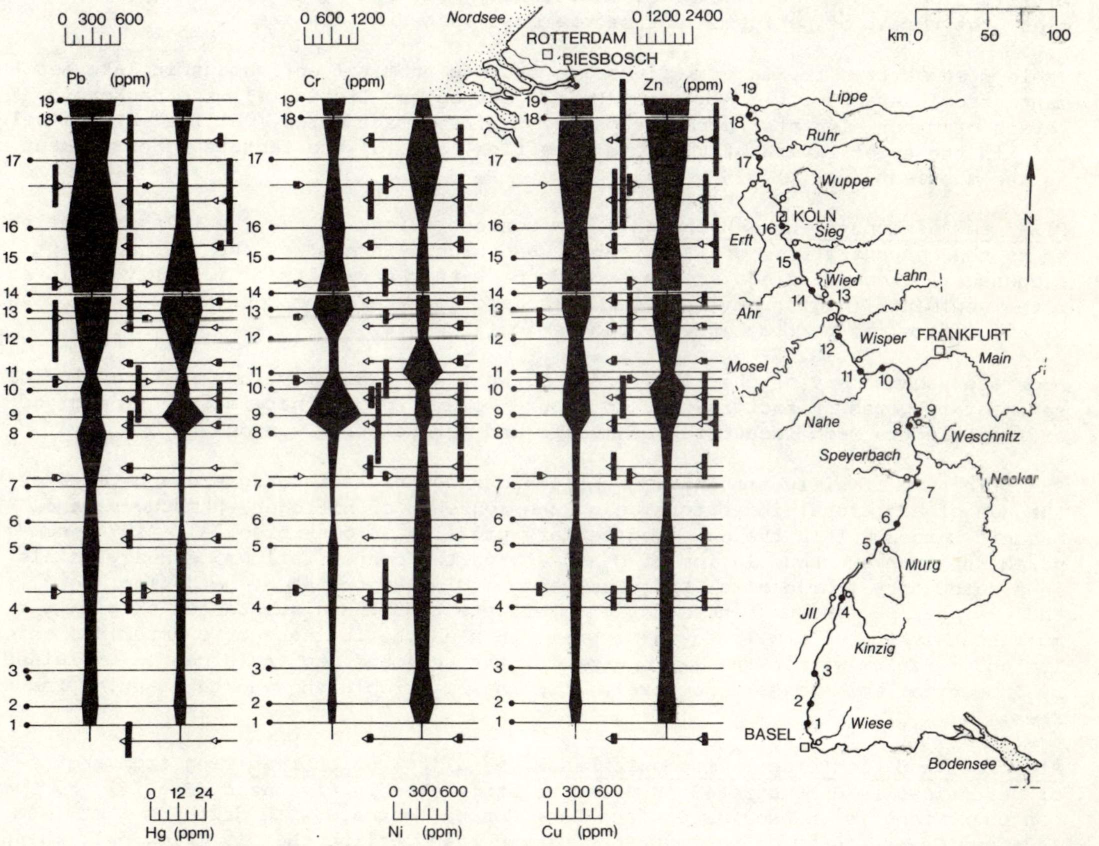


Fig. 2 Heavy metals in the pelitic fraction (<2 µm) of sediment samples from the Rhine River (from Banat et al., 1972a)

Fig. 2 shows the distribution of 6 selected heavy metals in sediments from 19 sampling locations between the Swiss/German and the German/Dutch borders. At Station

1, the Rhine River is already significantly contaminated by mercury, lead and cadmium, when compared with sediments in Lake Constance. A strong increase of chromium in the sediments is evident at Station 9, where the Weschnitz River, highly polluted by waste from tanning and other industries, flows into the Rhine. Between Stations 9 and 19 (Dutch border), all the elements investigated, with the exception of nickel and cobalt, are further enriched by a factor of two to three. In the Netherlands, an additional increase is evident at Biesbosch (De Groot et al., 1971) where the Rhine Estuary begins. The extent of heavy metal pollution in the Rhine sediments is strongest for metals with a high "Index of Relative Pollution Potential" (see above). As compared with uncontaminated clay-sized sediments, e.g. the "average shale", the following enrichment factors are found in sediments between Station 9 and 19: Cd (40), Hg (20), Pb (18), Zn (13), Cr (5); Ni and Co are only enriched by a factor of 2.5 and 1.5 respectively.

## Lake Sediments as Indicators of Metal Pollution

The significance of sediments in assessing the conditions in aquatic systems is best illustrated by examples from the lacustrine environment since freshwater lakes have been the center of important cultural developments ever since the earliest days of civilization. As a consequence of increased population and industrialization densities, the threat of pollution has become most acute in such areas.

Table 8 summarizes the major influencing factors on metal enrichment in lake sediments from characteristic examples (United States) by listing (1) the background levels of minor elements in the deeper part of sediment cores, (2) the maximum values in the upper layers of the core profiles, and (3) the factors of enrichment as the quotient (2):(1).

Mixed sewage inputs: Lake Michigan. Low concentrations of heavy metals are reflected by the concentrations of iron, cobalt and nickel in most of the lake sediment sequences. These findings are in accordance with the results from highly polluted river sediments, e.g. from the lower Rhine section where the contents of Fe, Co and Ni are influenced particularly by geochemical factors.

Moderate enhancement of the zinc, lead, mercury and cadmium values in Lake Michigan sedimentary cores by factors of 2 to about 7 seems to originate mainly from mixed effluent inputs from industrial, domestic and agricultural sources.

Specification of effluents into domestic and industrial discharges can be made by the use of additional indicators, e.g. the contents of nitrogen, phosphorus and organic carbon within the same sedimentary profiles. Förstner et al. (1974) compared the mode of accumulation of N and P with the concentrations of heavy metals in a dated core from central Lake Constance. Similar trends of both zinc /lead and nitrogen/phosphorus concentrations indicate that these substances originate simultaneously in the public sewage system, whereas the increase of cadmium and chromium and chromium within the upper layers of the sediment sequences can be explained by industrial emissions, repectively from the electroplating and the tanning industries.

Algicides and herbicides: Wisconsin Lakes. Metal pollution resulting from the use of pesticides is demonstrated by examples from lakes in Wisconsin (Fig. 3). *Copper* concentrations in the sediments from Lake Monona increase with depth to a maximum (434 ppm Cu at 15-20 cm sediment depth) and then decline sharply (Fig. 3a). These effects are a result of the intermittent treatment of the lakes with copper sulfate to control algal growth during 1918-1944. Since the rates of increase of copper concentrations in sediments from other lakes, e.g. Lake Michigan are generally low, local copper accumulations may, as a rule, be taken as indicators of previous application of copper for algal control. *Arsenic* values were not available from Lake Monona and were therefore taken from a study of Lake Minocqua in northeastern Wis-

TABLE 8 Distribution of Minor Elements in Sedimentary Profiles from Lake Michigan[a], Lake Monona/Wisconsin[b], Lake Washington[c], and Lake Erie[d]

| | Lake Michigen | | | Wisconson Lakes | | | Lake Washington | | | Lake Erie | | |
|---|---|---|---|---|---|---|---|---|---|---|---|---|
| | back-ground | max. value | *F* | back-ground | max. value | *F* | back-ground | max. value | *F* | back-ground | max. value | *F* |
| Zinc | 120 | 317 | *2.5* | 15 | 92 | *6* | 60 | 230 | *4* | 7 | 42 | *5* |
| Chromium | 77 | 85 | *1* | 7 | 49 | *7* | n.d. | | | 13 | 42 | *4.5* |
| Nickel | 54 | 44 | *1* | 34 | 50 | *1.5* | (iron: | | *1*) | 40 | 95 | *2.5* |
| Copper | 44 | 75 | *1.5* | 22 | 268 | *12* | 15 | 50 | *3* | 18 | 59 | *4* |
| Lead | 40 | 145 | *3.5* | 14 | 124 | *9* | 20 | 400 | *20* | n.d. | | |
| Arsenic | 11 | 22 | *2* | ( 2 | 51 | *25* ) | 10 | 200 | *20* | 0.6 | 3.2 | *5.5* |
| Mercury | 0.04 | 0.2 | *5* | 0.24 | 1.14 | *5* | 0.1 | 1.0 | *10* | 0.04 | 4.48 | *12* |
| Cadmium | n.d. | | | 2.5 | 4.6 | *2* | n.d. | | | 0.14 | 2.4 | *17* |

a) Ruch et al., 1970; Shimp et al., 1971; Kennedy et al. 1971; Frye and Shimp, 1972),
b) Syers et al., 1972; Shukla et al., 1972; Iskandar & Keeney, 1974
c) Barnes and Schell, 1973; Crecelius and Piper 1973; Schell, 1974; Crecelius, 1975;
c) Walters et al., 1975;
Data of background and maximum values in parts per million (ppm). F = factor of enrichment.

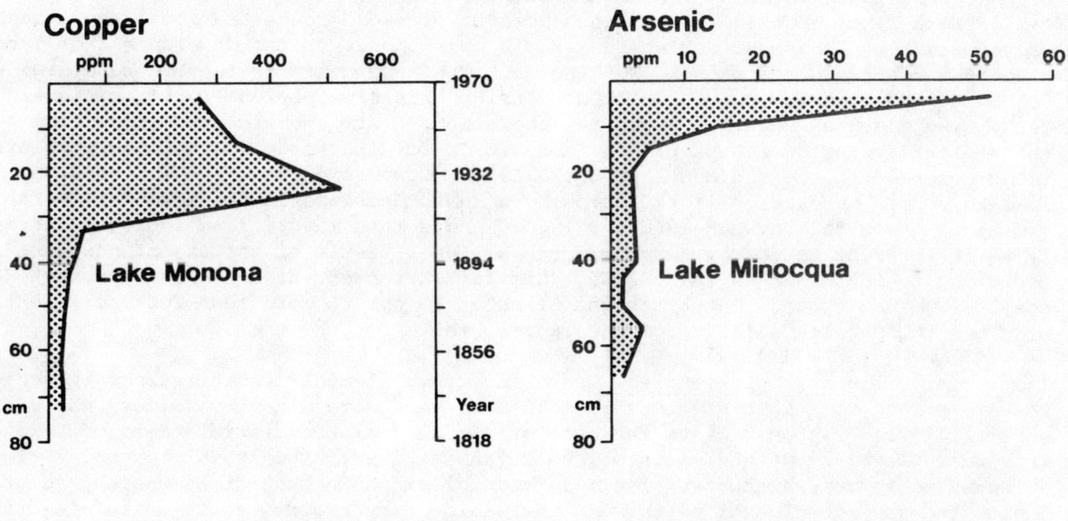


Fig. 3 Variation of copper (a) and arsenic (b) with sediment depth in lakes of Wisconsin (Iskandar and Keeney, 1974; Shukla et al., 1972)

consin(Fig.3b). The characteristic increase of arsenic in these sediments was explained by the fact that several Wisconsin lakes had been treated with sodium arsenite to reduce the population of noxious weeds. It is assumed that fertilizers, and above all, houshold detergents are responsible for the general enrichment of arsenic concentrations observed, for example, in the sediments from Lake Michigan.

could result in enrichment of river sediments.



Airborne particulate fallout: Lake Washington. Characteristic sources of metal pollution are constituted by smelters and by coal-fired power plants, as well as by heavy city traffic.

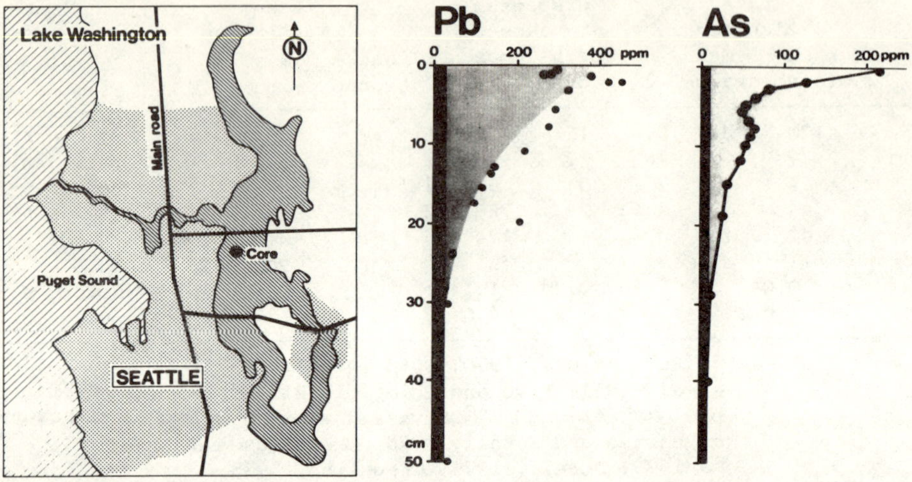


Fig. 4 Vertical distribution of Pb and As in sediments from Lake Washington (Crecelius and Piper, 1973)

An instructive example of atmospheric influences on metal concentrations in lacustrine deposits is given by Crecelius and Piper (1973) from sedimentary core studies of Lake Washington (Fig. 4): enchancement of *lead* concentrations from a background level of 20 ppm Pb during the 1880's corresponds to the development of land along the western slope of Lake Washington. The population of Seattle rose from 7,600 in 1880 to over 100,000 in 1900. During this same period, the Tacoma smelter, 50 km south of Seattle, began its operations. Atmospheric discharges from the smelter are held responsible for a 5- to 10-fold increase in the lead concentrtions observed in the sediments up until 1916. Conversion of the smelters from lead to copper production is reflected by a sharp decrease in the lead values in sediments deposited between 1916 and 1920. Above this layer, lead increases strongly to its present values of 400 ppm, probably as a result of the use of gasoline additives, which began in the 1920's. Similar enrichment of *arsenic* in present deposits from background concentrations of about 10 ppm to more than 200 ppm is attributed to atmospheric fallout from the Tacoma smelter.

Industrial effluents: Lake Erie. Trace and minor element concentrations in sedimentary cores from eight stations in Lake Erie were determined by Walters and coworkers (1973). Table 8 lists the data of a core from the central basin of Lake Erie near Cleveland at the mouth of the Cuyahoga River. The graphic example (Fig. 5) seems to be representative of a moderate pollution in that area, whereby according to the distribution of mercury in the sedimentary record, early cultural activity might have taken place beginning in about 1935. Major increases in the values of zinc, arsenic and copper during 1939-1955 reflect the general growth of industry during World War II and the Korean conflict. Strong enrichment of *chromium* occurred during the late 1940's, corresponding for the most part to the growth of the Cleveland electroplating industry. It is assumed that the establishment of chemical plants in the Cleveland area in 1949 caused a characteristic increase in the mercury concentrations: the major break in 1955 would than correspond to the opening of the Detrex chlorine-alkali plant near Ashtabula, Ohio.

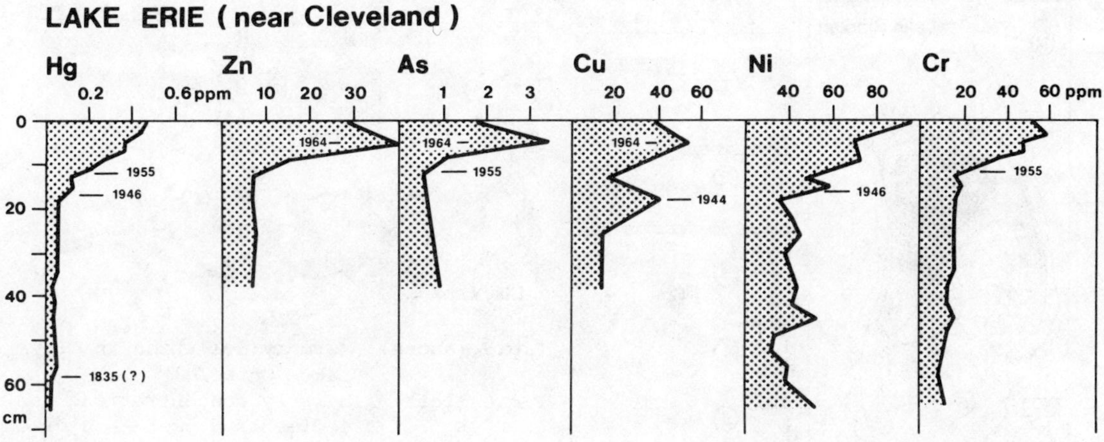


Fig. 5 Metal enrichment in the upper sediment layers of a core profile from Lake Erie (Walters et al.,1973)

Mercury in Polluted Freshwater Sediments.
Among the various metal pollutants, *mercury* received much attention with regard to its accumulation in sediments and subsequent incorporation into the aquatic food chain. Mercury and many of its compounds exhibit a distinct residue-forming tendency in aquatic ecosystems. Ionic mercury is quickly removed from solution and immobilized above its very low background concentrations. Microbial processes, however, are capable of converting originally inorganic mercury into more mobile and more toxic methylated forms. Studies carried out by Wood and co-workers (1968) and by Jensen and Jernelöv (1969) on the factors and mechanisms influencing the biosynthesis of mercury alkyls clearly indicated that the methylation rates are particularly affected by a good supply of nutrients and elevated water temperature. Bioturbation of the upper sediment layers by microorganisms, e.g. tubificids and mussels may prolong the release of methylated mercury from polluted bottom mud for decades. Mercury contamination from sediment and biota has been reported in aquatic systems far removed from characteristic industrial effluents. One likely source of mercury accumulations in remote areas is atmospheric particulate matter, either from industrial plants burning fossil fuel, or from the smelting of copper, lead and zinc ores. For example, Lake Saoseo and Lake Tuma in southern Switzerland, both situated at high altitudes, show evidence of atmospheric mercury contamination, which probably originates in the industrial complexes of northern Italy (Vernet and Thomas, 1972). Acute cases of mercury pollution arise mainly from direct industrial emissions. Several examples of mercury pollution by wastes from pulp mills and chlor-alkali plants were found in lakes in Sweden and North America:

One of the smaller but most polluted lakes in *Sweden* is Lake Björken, situated in in the central part of the country with an area of 1.8 km$^2$ and a mean water depth of 4 m. Maximum mercury concentrations of up to 50 ppm Hg were reported from surface sediment (Hasselrot, 1971). In the upper 10 cm of Lake Björken sediment, the median mercury concentration was about 1 ppm, but many of the top samples of core profiles from the central part of the lake contained more than 6 ppm of mercury (Fig. 6; Axelsson and Håkanson, 1973). The principal source of contamination are wastes from a pulp mill which are discharged just upstream from Lake Björken. During the period 1955 to 1965 about 352 kg phenylmercury was used (the equivalent of about 180 kg metallic Hg) and probably most of it reached the lake.

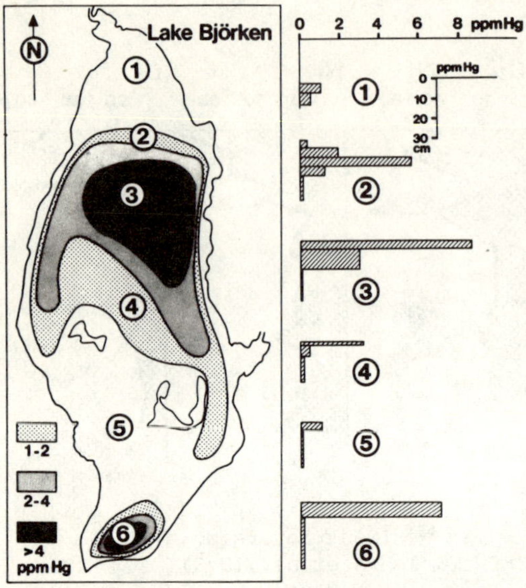


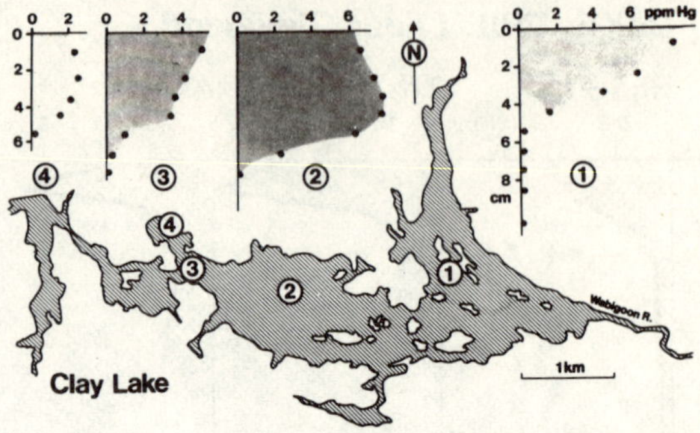


Fig.7 (above) Mercury pollution in Clay Lake (Ontario)

Fig.6 (left) Mercury concentrations in sediments from Lake Björken (Central Sweden)

One of the areas in *Canada* which is most heavily polluted with mercury is the Wabigoon River system in northwestern Ontario. Analyses of sediments, water and biota were performed by Armstrong and Hamilton (1973) in particular on samples from Clay Lake. This lake has a surface area of about 6 $km^2$, a maximum depth of 24 m and an average depth of 8 m. High concentrations of more than 8 ppm Hg were found in the upper 4-6 cm of the sediment cores, whereas the Hg concentration below 6 cm was usually less than 1 ppm; the background level in the sediments of Clay Lake is estimated at 0.1 ppm Hg (Fig. 7). From the median vales of 3.1 ppm Hg in the core sediments, the total amount of mercury in the top 6 cm was calculated at about 2,000 kg, equivalent to 20% of the mercury discharged during the first eight years of operation of a chlorine-alkali plant in Dryden situated approximately 200 km upstream from Clay Lake at the Wabigoon River. After stringent enforcement of legislation early in 1970, mercury discharges have been reduced to about 3% of the previous values. Nonetheless, any improvement would require that the existing deposits be covered by material with a low mercury content. The accumulation of the several centimeters of sediments needed might take several centuries. In 1972, however, the material being deposited in Clay Lake still had mercury levels of more than 2 ppm (Armstrong and Hamilton, 1973).

Restoration of mercury-contaminated lakes. Several methods for the restoration of mercury-contaminated water bodies have been suggested: (1) dredging of polluted sediments, (2) converting mercury to mercury sulfide with a low methylation rate (anaerobic conditions), (3) binding of Hg to silica or coprecipitation with hydrous Fe and Mn oxides (aerobic conditions), (4) increasing the pH to give volatile dimethyl mercury rather than monomethyl mercury, (5) isolating the polluted sediments from the water body by means of physical barriers, such as polymer film overlays, blankets, or plugs of waste wool, sand and gravel overlays (Jernelöv and Lann, 1973).

Recent data on the recovery of a river-reservoir system (North Fork of the Holston River in Southwestern Virginia; Turner and Lindberg, 1978) formerly subjected to high discharges of mercury by a chlor-alkali plant indicate that these processes are very slow and far from being complete; continued losses of mercury from mercury-contaminated solid wastes stored at the plant site are implicated in this slow recovery. In this sense, river systems highly polluted with mercury (for examples see Table 9) and cadmium are becoming increasingly dangerous to both aquatic life and the drinking water supply[2].

TABLE 9 Mercury Concentrations (in ppm) in Sediments from Polluted RIvers in North America and Europe (Examples)

| River | Max. Conc. | Source | Reference |
|---|---|---|---|
| **North America** | | | |
| Stream in Taylor Mts. (Alaska) | 10 ppm | mining industry | Clark et al. (1970) |
| Tennessee River (North Fork) | 32 ppm | chlor-alkali industry | Derryberry (1972) |
| (Wheeler Dam) | (550 ppm) | chlor-alkali industry | Lloyd et al. (1972) |
| Wabigoon (Ontario) | 56 ppm | pulp-wood mill (chlor-alkali production) | Buffa (1976) |
| St. Clair River | 60 ppm | chlor-alkalki industry | Cline et al. (1973) |
| | (1700 ppm) | | Ont. Water Comm. (1970) |
| Detroit River | 86 ppm | chlor-alkali industry | Turney (1971) |
| Wisconsin River | 792 ppm | chlor-alkali industry | Iskandar et al. (1972) |
| **Europe** | | | |
| Seine (France) | 8.4 ppm | industrial, municipal | Chesterikoff et al. (1973) |
| Blies/Saar (F.R.G.) | 9.6 ppm | industrial (?) | Becker (1976) |
| Gurk (Austria) | 23 ppm | chlor-alkali industry | Ebner & Gams ( 1975) |
| Kamniska Bistrica (Slovenia/Yugosl.) | 32 ppm | industrial (?) | Štern and Förstner (1976) |
| Elbe (F.R.G.) | 35 ppm | industrial, municipal | Müller and Förstner (1975) |
| Toce (Italy) | 46 ppm | industrial (?) | Gaggino et al. (1975) |
| Wupper (F.R.G.) | 47 ppm | industrial | Banat et al. (1972) |
| Rhône (Switzerland) | 55 ppm | chemical industry | Vernet and Johnstone (1974) |
| Sambre (Belgium) | 70 ppm | industrial | Billen et al. (1974) |
| Ume Älv (Sweden) | 102 ppm | wood-pulp mill | Hasselrot and Göth-berg (1974) |
| Ginsheimer Altrhein (F.R.G.) | 110 ppm | industrial waste | Laskowski et al.(1974) |
| Monte Amiata Rs. (Italy) | 288 ppm | mercury mining area | Batti et al. (1975) |
| Mt. Avala (Yugoslavia) | 6000 ppm | mercury mining area | Maksimovic and Dangic (1973) |

## Particle Associations of Trace Metals

The availability of trace metals for metabolic processes is closely related to their chemical species both on solution and in particulate matter. For this reason a thorough investigation into the behavior of trace metals in the dissolved state was undertaken (see above). In comparison, very little is known about the chemical association of metals in suspended materials and sediments, although the knowledge of characteristic bonding types may be the key to the detection of specific pollution

[2] A compilation of examples from cadmium-polluted aquatic systems has been given by Förstner(1979) in J.O.Nriagu's study "Biogeochemistry of Cadmium"(Ann Arbor Science Publishers.)

sources in the aquatic environment.

In spite of their different origins, environmentally related types of metal bonding are comparable to the natural forms:

(1) Heavy metals are transported and deposited as major, minor or trace constituents in the natural *rock detritus*, mostly in inert positions
(2) Slight alkalinity prevailing in normal surface waters results in the formation and *precipitation of hydroxides, carbonates, sulphides and phosphates* of heavy metals.
(3) *Sorption and cation exchange* particularly takes place on fine-grained substances with large surface areas. A generalized sequence of the capacity of the solids to sorb heavy metals was established by Guy and Chakrabarti (1975) in the order:

$MnO_2$ > humic acid > hydrous iron oxides > clay minerals

(4) Dissolved organic polymer is increasingly regarded as a carrier material in the transfer of metals from the mineral phase into the organic material. The younger, less condensed *humic acids* (fulvic acids) play an especially important role in the bonding of heavy metals in aquatic systems because of their large number of functional groups.
(5) *Co-precipitation of heavy metals with hydroxides and carbonates* appears to be an important mechanism controlling the metal concentrations in the aquatic environment (Lee, 1975; Jenne, 1976).

TABLE 10 Extraction Sequence for Metal Associations in Recent Sediments

| Extraction method | Major exchangeable phase | Reference |
|---|---|---|
| 1. $H_2O$ | easily soluble fraction | |
| 2. 0.2 N $BaCl_2$-triethanolamin pH= 8.1 | exchangeable cations | Jackson (1958) |
| 3. 0.1 N NaOH | humates, fulvates, phosphate | Volkov and Fomina (1974) |
| 4. Acidic cation exchanger | carbonates | Deurer et al.(1978) |
| 5. 0.1 N $NH_2OH \cdot HCl$ + 0.01 n $HNO_3$ | easily reducible fraction (Mn-oxide, amorph. Fe-oxide) | Chao (1972) |
| 6. 30% $H_2O_2$ + 1 N $NH_4OAc$ | organic residues + sulfides | Gupta and Chen (1975) |
| 7. 1 N $NH_2OH \cdot HCl$ + 25% acetic acid | moderately reducible fraction (hydrous Fe-oxides) | Chester and Hughes (1967) |
| 8. $HF/HClO_4$-digestion | inorganic residues | |

A compilation of several of these steps are given in Table 10 (from Förstner and Patchineelam, 1979). As an example, figure 8 shows the application of our extraction sequence to a representative sediment sample from the Lower Rhine River. Due to the large catchment area, all essential pollution may be considered to be represented in the sample under consideration. Starting with the metal contents in the detrital-mineral association, two groups of metals can be distinguished: Firstly iron, nickel, lead and cadmium with very low content of detrital mineral bonding. The same sequence has been obtained, when we established an order of man-made effects on the metal concentration of the Lower Rhine sediments, by comparing the data from recent material with analyses of ancient Rhine sediments obtained from drilling cores in the Cologne area. According to our calculations, cultural influences are relatively low for iron, nickel and cobalt, moderately high for chromium and copper, and very high for lead and zinc. Only two percent of the cadmium content in the recent sediments of the Lower Rhine originate from natural sources, 98% are derived from industrial and urban effluents as well as from agricultural and atmospheric pollution.

These and other investigations indicate that metal contaminants which are particularly enriched in sediments and soils, such as lead, zinc, cadmium and mercury, occupy the least stable sediment fractions and may thus be more easily remobilized and introduced into the food chain than others which are inertly bound in the residual sediment particles, e.g. nickel, cobalt, chromium and copper.

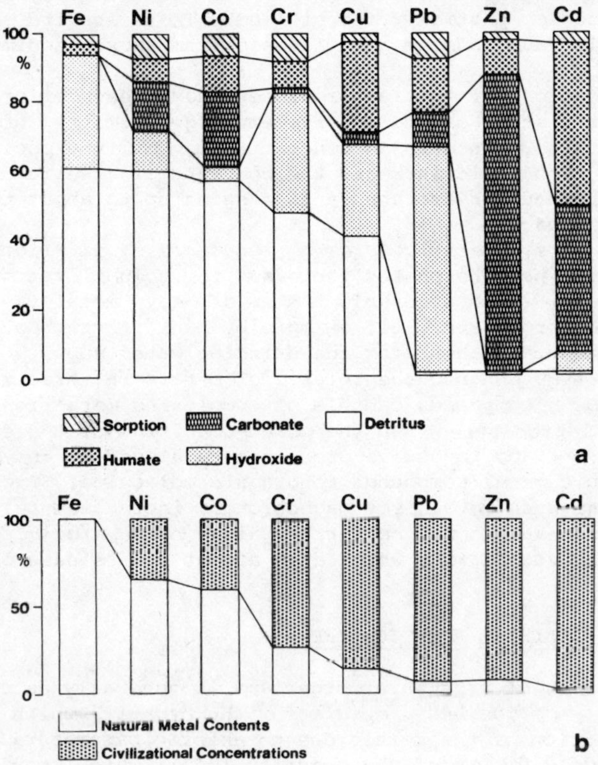


Fig. 8 Distribution of bonding types (a) and sources (b) of metals in Rhine sediment

The Release of Trace Metals from Freshwater Sediments

Heavy metals which are "immobilized" in the bottom sediments of rivers and lakes constitute a potential hazard to water quality since they may be released as a result of chemical changes in the aquatic milieu:

1) *Increased salinity* of the water body leads to competition between dissolved cations and adsorbed heavy metal ions and results in partial replacement of the latter. Such effects should particularly be expected in the *estuarine environment*. Apart from the highly complicated situation there, however, it has been demonstrated that the decrease in heavy metal concentrations at the river/sea interface is caused my mixing processes of polluted tiver particulates with relatively "clean" marine sediments rather than by solubilization and/or desorption.

(2) *A lowering of pH* leads to the dissolution of carbonate and hydroxide minerals also - as a result of hydrogen ion competition - to an increased desorption of

metal cations.

Long-term changes of the pH conditions have been observed from highly industrialized areas by atmospheric $SO_2$ emissions mainly in waters poor in bicarbonate ions. Acid mine drainage effects a 1,000 - 10,000-fold increase of iron, manganese, nickel, cobalt, copper and zinc (see above).

3) A change in the *redox conditions* is usually caused by the increased input of nutrients. Oxygen deficiency in the sediments leads to an initial dissolution of hydrated manganese oxide, followed by that of the analogous iron compound. Since these metals are readily soluble in their divalent states, any co-precipitates with metallic coatings become partially remobilized. There are indications that, for example, copper, zinc and cadmium are released from anoxic sediments into surface water.

The possible implications of the latter process have been depicted by Edington and Callender (1970) in a study of the minor element geochemistry of Lake Michigan ferromanganese nodules. Since these concretions contain unexpectedly large amounts of arsenic, dissolution would increase the contents of arsenic in the waters of Green Bay, where the nodules are abundant, ranging up to about twice the permissible amounts for drinking water.

4) The growing use of synthetic *complexing agents* (e.g. nitrilotriacetic acid) in detergents to replace polyphosphates increases the solubilization of heavy metals from aquatic sediment. A considerable number of heavy metal chelates are not destroyed during water processing, for example by bank filtration, and may be considered to be a potential danger for the drinking water supply.

5) *Microbial activities* enhance the release of metals in three main ways: (i) by formation of organic compounds capable of complexing metal ions; (ii) by influencing the physical properties and the pH-Eh-conditins of the environment; (iii) by means of oxidative and reductive processes catalyzed by enzymes, microorganisms can convert inorganic metal compounds to organic molecules. Organo-mercurial transformations show that such mechanisms may strongly increase metal toxicity.

6) The latter two mechanisms may be further aided by *physical effects such as erosion, dredging and bioturbation*, which also affect the release of pore solutions rich in metals.

## HEAVY METALS IN AQUATIC LIFE

The heavy metal burden of freshwater organisms is generally determined by:

1) *The availability of the heavy metals* for the organism which can be defined by the chemical speciation of the metals due to abiotic parameters.

2) *Intrinsic biologic factors* of the organism itself which regulate the net uptake of the metals and which are responsible for the *biologic half life* of the metal.

3) *The biological behavior* of the organism, which is expressed in its life cycle and life history, and by the *trophic level* in which the organism feeds.

### The Influence of Abiotic Factors on Toxicity and Uptake of Heavy Metals

The presence of heavy metals in ecosystems becomes dangerous for organisms when the concentration rises above the natural background in water, sediment and the food supply. In order to cause a significant heavy metal uptake from contaminated sources, the metal must be in a bioavailable chemical species. It is therefore quite obvious that those physico-chemical parameters which influence the metal availability play an important role for the toxicity and the bioaccumulation in organisms. Heavy metals easily enter biota in the form of active ions. In most cases, particulate heavy metal compounds are to a great extent excreted if they enter the organism, for example, through the digestive tract. Whereas trace metals in water not contaminated by heavy metals are in equilibrium between soluble and particulate forms, in polluted rivers either precipitation or solution sets in due to different physico-chemical composition of waste effluents and the river water. The availability of the metals for the organism on the various trophic levels is thereby decisively influenced. Various rates of contamination of an organism can be expected according to the metal's distribution coefficient in the water.

Water hardness. Lloyd (1965) has shown using *rainbow trout* that a high degree of water hardness can reduce the toxicity of lead, zinc and copper. Even with decreasing toxicity of any one metal, the metals' general order of toxicity remains the same (Tabata, 1969). In this respect, the influence of calcium hardness is greater than that of magnesium. Since both of the above investigations are concerned with the acute toxicity of metals, it is obvious thatby means of elimination or deactiviation of the metal ions, the toxicity of the metals will prove to be less than that expected from the original concentrations. This phenomenon can be explained through precipitation or co-precipitation with carbonates and the formation of less toxic inorganic heavy metal compounds such as aquo- and hydroxo- complexes. In an extensive experiment, Bell (1976) proved the reduction of the toxicity of Cu, Pb and Zn ions by increasing hardness ($CaCO_3$) (see Fig. 9). Similar slopes were found for Pb and Zn, whereby the detoxification effect of the subsequent addition of 1 µg $CaCO_3$ was higher in soft water than in hard water.
For copper, the most toxic of the three metals investigated, the slope was much steeper than for Zn and Pb, indicating the important influence of water hardness on copper toxicity. In order to attain the same high metal toxicity, however, only 2 µg/l Cu for each addition of 1 mg $CaCO_3$/l are necessary, whereas for zinc and lead 20 µg/l and 18 µg/l respectively, are required in soft water. At higher $CaCO_3$ levels a straight line relationship between hardness and the lethal threshold concentration of all three metals was found (probability level: $P > 99.9\%$)

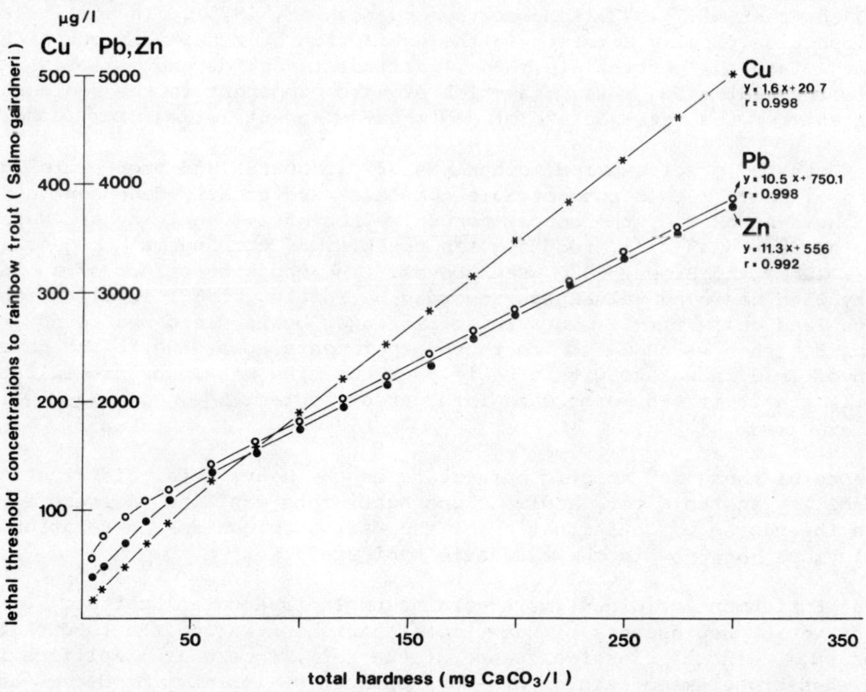


Fig. 9 Influence of water hardness on the lethal threshold concentration of *Cu*, *Zn* and *Pb* in *rainbow trout* (*Salmo gairdneri*). (Linear regression is calculated on 76 values for each metal (after Bell, 1976)

After addition of $Cu^{2+}$ ions, Shaw and Brown (1974) investigated the various copper compounds occurring in precipitation. At pH 8.0, dissolved $Cu^{2+}$ dropped to about 50% from the original solution (1.1 mg/l) within two hours. For other metals, Patchineelam (1975) also established carbonate precipitation or co-precipitation. Since metal uptake occurs for many organisms mainly through water, it can be assumed that through the elimination of some of the ion-active metals from the water, the uptake in the animal will likewise be reduced. Since especially in the proximity of waste water outfalls into freshwater, the carbonate and Fe-hydroxide precipitation mechanisms play an important role, an elimination of dissolved heavy metals in freshwater occurs directly in the outfall area itself. pH variations can either increase or reduce the effects of these mechanisms. Another mechanism of detoxifying heavy metal ions caused by water hardness ions is the competition of Ca and Mg ions with heavy metal ions for active sites in fish tissues (Zitko and Carson, 1976). This mechanism is most important in water with higher salinity like the mixing zone of estuaries.

Organic compounds. Organic substances in solution can remove metal ions from the water by the formation of organic compounds. This can occur as a complex formation, as a sorption of cations or a cation exchange. In laboratory experiments on *Daphnia* Tabata (1969) not only established a reduction of toxicity of copper through citrate and EDTA, but also found a reduced copper uptake caused by organic substances in oysters. The uptake rate with EDTA was even less than the control. Other investigations on the detoxification of heavy metals using organic complexing agents and freshwater organisms were conducted, e.g. by Sprague (1968) and by Shaw and Brown (1974). Complex metal compounds are therefore much less available to organisms, and have an important function in waters heavily polluted by organic substances (cf. Gächter et al., 1973; Ramamoorthy and Kushner, 1975). In still water bodies, this function finally consists in the extraction of metal ions out of the aqueous phase into the particulate phase (particulate solids and colloids) while settling to the sediment. This metal-rich organic component in the sediment is very important for the metal uptake of sediment-dependent benthic organisms.

pH Value. Both the precipitation mechanisms of carbonates and iron-hydroxides and the bonding of heavy metals onto organic complexes are greatly dependent upon pH-Eh conditions. At pH 8.0, the copper concentration of 1.1 mg/l in hard water sinks within two hours to 0.5 mg/l, reducing the toxicity of the "normal concentration" of 1.1 mg/l (Shaw and Brown, 1974, see above). The opposite effect of an increase in toxicity with basic pH values was recorded by Whitley (1968) in experiments conducted with lead on *tubificid worms* in small finger bowls; lead was found to be more toxic at pH 8.6 than at pH 6.5 . In these experiments, according to the chemical speciation of lead in waters with a basic pH vlaue, the metal was precipitated and consequently the tubificid worms came into direct contact with the metal on the bottom of the bowls.

The influence of these few abiotic parameters on the heavy metal distribution and the availability in the water, sediment and nutriments explains the resulting variations in the degree of contamination of the various organisms in relation to their ecological behavior in their aquatic habitats.

Heavy Metal Enrichment of Organisms According to Their Aquatic Habitat

In the last years, new aspects of heavy metal enrichment along the food chain have been revealed. In the classical sense of the term, food chain amplification is confirmed when organisms of a high bio-mass production (primary producers and primary sonsumers ) have lower metal concentrations than organisms with a small biomass (end consumers). For heavy metals, however, this classical definition is only valid for a few of them and only under certain circumstances. One of the best investigated metals in freshwater ecosystems is mercury (Ackerfors et al. 1970; Potter et al. 1975). Its high affinity to organic ligands due to its methylation by bacteria and the consequent turnover to higher organisms as organic mercury, causes a classical food chain enrichment in many cases. However, this normally only occurs in polluted waters, and in many investigations with normal or only slightly elevated mercury levels, no food chain amplification could be detected.

In field investigations dealing with heavy metal enrichment in organisms, it is imperative to group the organisms according to their habitat and their ecologic behavior i.e. feeding habits (phytophageous, carnivorous, omnivorous, filter feeding, sediment feeding, detritus feeding etc.), life cycle, life history, sessility and wandering. In addition, the physiologic response of various organisms towards metal pollution may be different with respect to organ distribution of the metal, synergistic or antagonistic effects of other metals on metal uptake, heavy metal resistance etc. (Merlini et al. 1971; Reay, 1972; Brown, 1977, Müller and Prosi, 1978). When all these factors are considered in respect to heavy metal amplification along the food chain, it becomes clear that, in many cases, elevated heavy metal concentrations in higher trophic levels do occur but not necessarily in the classical sense of food chain enrichment.

Heavy metal enrichment at different trophic levels. In contrast to the high mobility of organo-mercurials, the bioavailability of other metals is generally lower. Therefore, those aquatic habitats with the highest metal concentrations may become dangerous for their inhabitants.

In most cases of polluted water, the concentration gradient of heavy metals in aquatic ecosystems follows the order:

$$\text{sediment} \geq \text{suspended matter} > \text{water}$$

The fact that the lowest and highest concentrations are found in water and sediment respectively still does not allow any conclusions as to the toxicity of the metal load in these media, the reason being that the dissolved metal load in the water has a greater bioavailability than the total sediment load. The metal speciation in the sediment reveals that only a small amount of the total metal content is available to organisms. However, even small amounts may be toxic for benthic organisms and can be responsible for the bioaccumulation of heavy metals from the sediment.

In a study of a tributary of the Neckar River and the Elsenz River (F.R.G.), Prosi (1977) found high concentrations of Pb, Cd, Zn and Cu in the sediment, whereby in the flowing water the metal concentrations were even lower - except for Zn - than the river water standard (Tab.11). It must therefore assumed that these heavy metals which are introduced into the river via waste water either precipitate on reaching the river, or are already in particulate form and are deposited in the sediment. The high water hardness (6.1. mval/l), organic substances (average COD of 140 mg $K_2Cl_2O_7$/l) and a slightly basic pH of the river water contribute to these processes. The heavy metal concentrations in the organisms here reflect the frequency and the availability of heavy metals in the different habitats of the system. Accordingly, the concentrations decrease the farther the organism's habitat is away from the sediment (Fig. 10). Studies on the suspended particulates, sessile algae and higher sessile water plants from the *Elsenz River* confirm this. No food chain enrichment via the primary producers takes place (Heydt, 1977; Bibo, 1977; Table 12), despite the longer life cycles of the majority of the consumers *(leeches, fish)*. Only the benthic sediment- and detritus-feeding tubificids and isopods showed significant differences in the two sections of the river for Cd, Zn and Cu concentrations (see Fig. 10). The decrease of heavy metals with distance from sediment in organisms of the Elsenz River is quite obvious for *lead* and *cadmium* in the urban-influenced river section, whereas for *zinc* and *copper* and intermediate behavior can be observed, probably due to their function as essential metals.

TABLE 11 Heavy Metals in Sediments, Water and Biota of the Elsenz River System (All concentrations of sediments and organisms are dry mass related and are based on 13 sampling stations of the river) (Prosi, 1977)

| $\bar{x}$ and range | Pb | Cd | Zn | Cu |
|---|---|---|---|---|
| water ⌀ 0.45 µm, ppb | 1 | 0.5 | 25.5 (14.0-63.5) | 2.2 (0.8-5.5) |
| sediment < 2µm ppm | 273.1 (178.6-378.1) | 11.52 (1.34-61.92) | 1200 (650-2500) | 149.24 (70.99-683.93) |
| Tubificidae, ppm | 26.6 (5.0-80) | 1.98 (0.09-8.68) | 614.2 (104.0-2226.1) | 46.65 (19.43-131.9) |
| Asellus aquaticus, ppm | 6.80 (2.69-16.50) | 1.21 (0.47-2.97) | 181.9 (118.6-203.1) | 12.99 (8.64-19.91) |
| Herpobdella octoculata, ppm | 6.84 (2.70-13.58) | 0.62 (0.28-1.46) | 418.1 (291.2-640.5) | 23.07 (13.12-33.48) |
| fish (whole, Rutilus rutilus, Gasterosteus aculeatus, ppm | 1.0 | 0.1 | 70-178 | 10-26 |

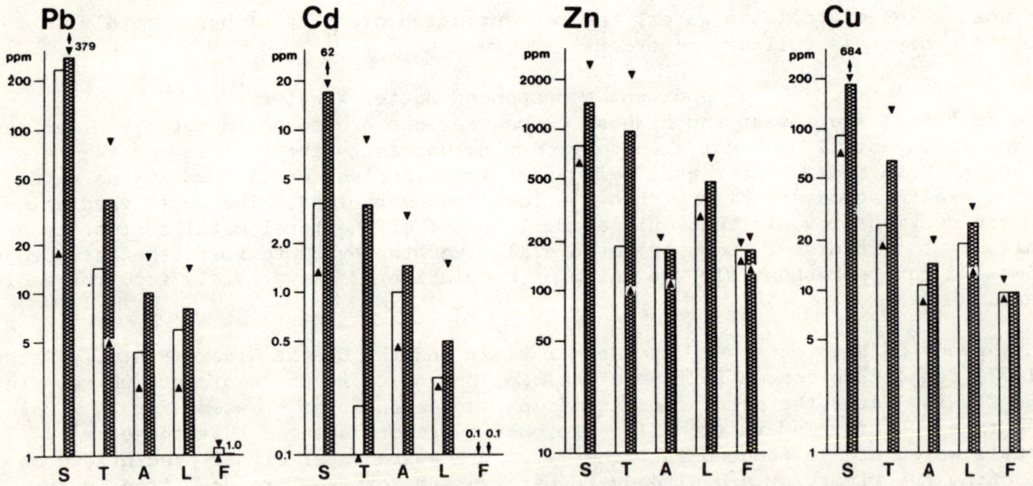


Fig. 10 Heavy metal concentrations (Pb, Cd, Zn, Cu) in a rural (white column) and an urban (grey column) river section of the *Elsenz River*. (All concentrations are dry mass related; arrows indicate minimum and maximum values; S = sediment < 2 µm; T = tubificids (*Tubifex tubifex; Limnodrilus hoffmeisteri*), A = isopods (*Asellus aquaticus*), L = leeches (*Herpobdella octoculata*), F = fish (*Rutilus rutilus, Gasterosteus aculeatus*). (after Prosi, 1977)

The high metal concentrations of the sediment-dependent tubificids evidence the uptake of heavy metals from the sediment, however, Prosi (1977) found no direct correlation with the total metal concentration in the sediment. Therefore, the partitioning of trace metals among different types of binding substrates (amino acids, humic acids, lipoproteins; bacteria and fungi; clay minerals and carbonates; interstitial water) determines the bioavailability of the metals; the relative concentrations of these different substrates are responsible for the total amount of bioavailable metal in the sediment. (cf. Gilbert et al., 1976; Schindler et al., 1977; Luoma and Bryan, 1978; Beveridge, 1978; Prosi, in prep.).

TABLE 12 Heavy Metals in Suspended Matter and Primary Producers of the Elsenz River (Data from Heydt, 1977; Bibo, 1977). (All data are dry mass related).

| $\bar{x}$ and range, ppm | Pb | Cd | Zn | Cu |
|---|---|---|---|---|
| suspended matter | 115.6 | 5.41 | 866.7 | 408.3 |
| | (13.8-429.8) | (0.35-28.14) | (420.1-2174.9) | (33.4-3562.2) |
| Cladophora rivularis | 28.9 | 0.45 | 139.2 | 4.88 |
| | (5.8-84.2) | (0.08-1.37) | (24.1-375.0) | (0.79-31.65) |
| Potamogeton pectinatus | 8.5 | 0.49 | 275.5 | 14.28 |
| | (3.9-16.1) | (0.07-1.82) | (141.8-1178.6) | (4.04-48.74) |

The activity of benthic organisms in respect to the turnover of heavy metals from the abiotic phase of the sediment to the biotic phase and an additional loss from the sediment was shown for Cd (Ueda et al., 1976) and for Zn (Renfro, 1973) on marine polychaetes. After 8 days 12% of the $^{115}Cd$ in the sediment was transferred to the worms and the $^{65}Zn$ loss in the sediment was 3 to 7 times higher than in sediment without polychaetes.
Thus, in polluted freshwater sediments where sediment-dependent and sediment-associated benthic food webs are predominant, similar effects can be expected. The possible release of heavy metals by bioturbation is indicated by the investigations of Wood (1975): The burrowing activities and faeces production by freshwater oligochaetes can set free large amounts of interstitial water containing high metal concentrations. Sediment particles from the deeper sediment layers of the reductive zone are transported to the oxidative surface of the sediment whereby further chemical speciation of the heavy metal in this sediment sets in.

All these factors imply that benthic organisms are better indicator organisms for heavy metal pollution in freshwater biota than fish or other organisms of higher trophic levels. However, even these can be affected by contaminated sediments. This has been proven by laboratory experiments by Deslisle and co-workers (1975). A significant uptake of *Pb, Cd, Zn* and *Cu* from sediment by *catfish* and *goldfish* could be confirmed, whereby the goldfish had a higher uptake than catfish. In another study, Gillespie (1972) established *mercury uptake in guppies* from contaminated sediments. However, a possible uptake from water was not investigated. In further field investigations on food chain enrichment and heavy metal monitoring it has been seen that the most reliable data was obtained on benthic organisms, including water plants (Dietz, 1972; Namminga and Wilhm, 1977; Enk and Mathis, 1977). A corresponding distribution pattern among organisms was found by Leland and McNurney (1974) for lead (Fig. 11). In the river system with a rural and urban influenced catchment area, they found even higher lead concentrations in omnivorous fish species than in predatory fish. The consequent decline of lead concentrations from benthic organisms to higher trophic levels shows the danger of intoxication of the benthos by sediment-bound trace metals (Brkovic-Popovic, 1977). Lead transport along the food web may take place but uptake via food seems to be predominantly dependent on the metal gradient of the different habitats of the various organisms. Even for the relatively mobile metal cadmium in littoral ecosystems, no significant enrichment in higher trophic levels was found (e.g. Ravera et al., 1973). Therefore, fish may be good indicators of an acute heavy metal toxicity in the water caused by the uptake of dissolved heavy metals whereas benthic sediment-dependent organisms such as sludgeworms and mussels are more suitable monitoring organisms for ecosystems contaminated by heavy metals. In recent investigations an analogous use of *freshwater mussels* to that of mussels in the marine environment have been performed by Merlini and co-

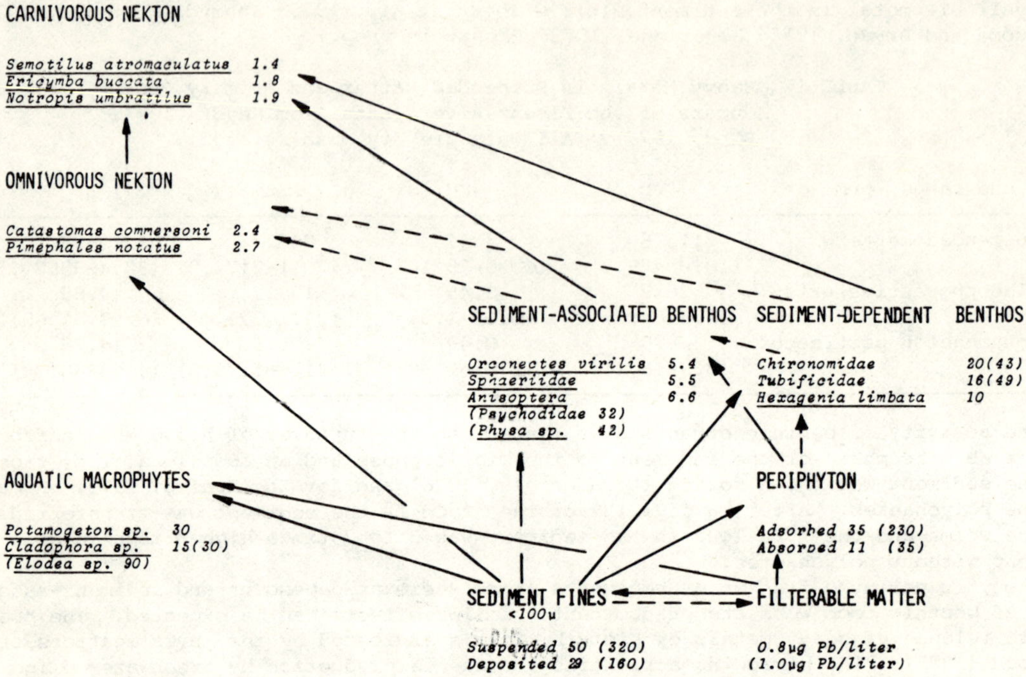


Fig. 11 Lead distribution in a river ecosystem of a rural area (urban area: numbers in brackets; concentrations in ppm) (according to Leland and McNurney, 1974).

workers (1977) and by Manly and George (1977).

The unexpectedly high metal contamination of organisms in low trophic levels of polluted freshwater ecosystems - which is contradictory to food chain enrichment theories - can be explained by the fact that most of these lower organisms feed directly on highly contaminated substances such as sediment and suspended matter caused by waste water and industrial wastes. Due to the ecologic impact of man on the environment, these organisms no longer feed in the lower trophic levels but in the highest trophic level possible. This change of trophic level can only be attributed to man's contribution to pollution in freshwater ecosystems.

Acknowledgements
We are grateful to U. Kästner for the graphics and our special thanks go to J. Broadbridge for translating, editing and patiently typing the manuscript.

## REFERENCES

Review articles and compilations on the present subject (examples):

Andrew, R.W., P.V. Hodson, and D.E. Konasewich (Eds.) (1976). *Toxicity to Biota of Metal Forms in Natural Water*. Intern. Joint Commision Great Lakes Research, Windsor, Ontario.

Förstner, U. and G. Müller (1974). *Schwermetalle in Flüssen und Seen als Ausdruck der Umweltverschmutzung*. Springer, Berlin-Heidelberg-New York.

*Heavy Metals*, (prepared by H.V. Leland and co-workers), Literature Reviews for the June Editions of the *Journal Water Pollution Control Federation*, 1974-1978 (Vol. 46-50).

Hemphill, D.D. (Ed.) (1967-1978). *Trace Substances in Environmental Health*. Vol. I to Vol. XI. University of Columbia, Missouri.

Krenkel, P.A. (Ed.) (1975). *Heavy Metals in the Aquatic Environment*. Pergamon Press, Oxford-New York.

*Proceedings of the International Conference on Heavy Metals in the Environment*, Toronto, Ontario, Canada, October 27-31, 1975. Vol. I (released 1976), Vol. II/1, Vol. II/2, Vol. III (1977). Institute for Environmental Studies, University of Toronto, Canada.

Singer, P.C. (Ed.) (1973). *Trace Metals and Metal-Organic Interactions in Natural Waters*. Ann Arbor Science Publ., Ann Arbor, Michigan.

Other References

Abernathy, A.R. and P.M. Cumbie (1977). Mercury accumulation by Large-Mouth Bass (*Micropterus salmoides*) in recently impounded reservoirs. *Bull.Environ.Contam.Toxicol.*, 17, 595-602.

Ackefors, H., G. Löfroth, and C.G. Rosen (1970). A survey of the mercury pollution problem in Sweden with special reference to fish. *Oceanogr.Mar.Biol.*, 8, 203-224.

Andrew, R.W. (1976). Toxicity relationships to copper forms in natural waters. In R.W. Andrew, P.V. Hodson, and D.E. Konasewich (Eds.), *Toxicity to Biota of Metal Forms in Natural Waters*. Int.Joint Comm., Windsor, Ontario. pp. 127-143.

Anon. (1972-1975). *Water Resources Data for Louisiana*. U.S. Geol.Surv., Water Resources Div., Baton Rouge. Cit. Trefry and Presley (1976).

Armstrong, F.A.J. and P.M. Hamilton (1973). Pathways of mercury in a polluted northwestern Ontario lake. In P.C. Singer (Ed.), *Trace Metals and Metal-Organic Interactions in Natural Waters*. Ann Arbor Science Publ., Ann Arbor, Michigan. pp. 131-156.

Axelsson, V. and L. Håkanson (1973). Kvicksilver i södra Vätterns, Ekolns och Björkens sediment. Univ. Uppsala, Sweden, *UNGI Rapport No. 25*.

Baccini, P. (1976). Untersuchungen über den Schwermetallhaushalt der Seen. *Schweiz. Z.Hydrol.*, 38, 121-128.

Banat, K,. U. Förstner, and G. Müller (1972a). Schwermetalle in den Sedimenten des Rheins. *Umschau Wiss.Technik*, 72, 192-193.

Banat, K,, U. Förstner, and G. Müller (1972b). Schwermetalle in Sedimenten von Donau, Rhein, Ems, Weser und Elbe im Bereich der Bundesrepublik Deutschland. *Naturwissenschaften*, 59, 525-528.

Barnes, R.S. and W.R. Schell (1973). Physical transport of trace metals in the Lake Washington watershed. In M.G. Curry and G.M. Gigliotti (Eds.), *Cycling and Control of Metals*, U.S.EPA, Cincinnati, Ohio. pp. 45-53.

Batley, G.E. and T.M. Florence (1976). A novel scheme for the classification of heavy metal species in natural waters. *Anal.Lett.*, 9, 379-388.

Batti, R., R. Magnaval, and E. Lanzola (1975). Methylmercury in river sediments. *Chemosphere*, 1, 13-14.

Becker, R. (1976). Geochemische Untersuchung der Sedimente des Flusses Blies, Saarland. *Naturwissenschaften*, 63, 144.

Bell, A.V. (1976). Waste controls at base metal mines. *Environ.Sci.Technol.*, 10, 130-135.

Beveridge, T.J. (1978). The interaction of metals in aqueous solution with bacterial cell walls from *Bacillus subtilis*. In W.E. Krumbein (Ed.), *Environmental Biogeochemistry and Geomicrobiology*, Vol. 3. pp. 975-985.

Bibo, J. (1977). *Schwermetalluntersuchungen an Wasser, Schwebstoffen, Aufwuchs und Cladophora rivularis der Elsenz*. Dipl.Arbeit Univ. Heidelberg, 117 p.

Billén, G., C. Joiris, and R. Wollast (1974). A bacterial methylmercury mineralizing activity in river sediments. *Water Res.*, 8, 219-225.

Boyle, E.A. (1978). Trace element geochemistry of the Amazon and its tributaries. *EOS*, 59, 276.
Bowen, H.J.M. (1966). *Trace Elements in Biochemistry*. Academic Press, London.
Bowen, H.J.M. (1976). Residence times of heavy metals in the environment. Proc. Int.Conf. on *Heavy Metals in the Environment*, Toronto, 1975, pp. 1-19.
Brannon, J.M., R.M.Engler, J.R. Rose, P.G. Hunt, and I. Smith (1976). Distribution of manganese, nickel, zinc, cadmium and arsenic in sediments and the standard elutriate. *Dredged Material Research Program*, U.S. Army Engineer Waterways Experiment Station, Vicksburg, Miss. Misc. Paper D-76-18. 39 p.
Brkovic-Popovic, M. (1977). Effects of heavy metals in survival and respiration rate of tubificid worms. *Environ.Pollut.*, 13, 93-98.
Brown, B.E. (1977). Uptake of copper and lead by a metal-tolerant Isopod *Asellus meridianus* Rac. *Freshwater Biol.*, 7, 235-244.
Brown, V.M. (1977). Aspects of heavy metal toxicity in freshwater. In R.W. Andrew, P.V. Hodson and D.E. Konasewich (Eds.) *Toxicity to Biota of Metal Forms in Natural Water*. Int.Joint Comm., Windsor, Ontario. pp. 59-75.
Buffa, L. (1976). *Review of environmental control of mercury in Japan*. Canada Environmental Protection Service, Rept. EPS 3-WP-76-7, 81 p.
Carpenter, J.H., W.L. Bradford, and V. Grant (1975). Processes affecting the composition of estuarine waters ($H_2CO_3$, Fe, Mn, Zn, Cu, Ni, Cr, Co, and Cd). In L.E. Cronin (Ed.), *Estuarine Research*, Vol. I, Academic Press, New York. pp. 137-152.
Carpenter, K.E. (1924) A study of the fauna of rivers polluted by lead mining in the Aberystwith District of Cardiganshire. *Ann.Appl.Biol.*, 11, 1-23.
Chao, L.L. (1972). Selective dissolution of manganese oxides from soils and sediments with acidified hydroxylamine hydrochloride. *Soil Sci. Soc.Am.Proc.*, 36, 764-768.
Chester, R., and M.J. Hughes (1967). A chemical technique for the separation of ferro-manganese minerals, carbonate minerals and adsorbed trace elements from pelagic sediments. *Chem.Geol.*, 2, 249-262.
Chesterikoff, A., A.M. Carru, B. Garban, D. Ollivon, and C. Chesterikoff (1973). La pollution ou la Basse-Seine par le mercure (du Pecq à Tancarville). *Centre Belge d'Etude et de Documentation des Eaux, Nos. 335-356*, pp. 269-276.
Clark, A., W.A. Condon, J.M. Hoare, and D.H. Sorg (1970). Analysis of rocks and stream sediment samples from the Taylor Mountains C-8 Quadrangle, Alaska. *U.S. Geol.Survey, Open-File Rept.*, 110 p.
Cline, J.T., J.B. Hillson, and S.B. Upchurch (1973). Mercury mobilization as an organic complex. Proc. *16th Conf.Great Lakes Res.* pp. 233-242.
Copeland, R.A. (1972). Mercury in the Lake Michigan environment. In R. Hartung and B.D. Dinman (Eds.), *Environmental Mercury Contamination*. Ann Arbor Science Publ., Ann Arbor, Michigan. pp. 71-76.
Crecelius, E.A. (1975). The geochemical cycle of arsenic in Lake Washington and its relation to other elements. *Limnol.Oceanogr.*, 20, 441-451.
Crecelius, E.A. and D.Z. Piper (1973). Particulate lead contamination recorded in sedimentary cores from Lake Washington. *Environ.Sci.Technol.*, 7, 1053-1055.
Davey, E.W., M.J. Morgan, and S.J. Erickson (1974). A biological measurement of the copper complexation capacity of seawater. *Limnol.Oceanogr.*, 19, 993-997.
Davies, P.H., J.P. Goettl jr., J.R. Sinley, and N.F. Smith (1976). Acute and chronic toxicity of lead to rainbow trout *Salmo gairdneri* in hard and soft water. *Water Res.*, 10, 199-206.
Derryberry, O.M. (1972). Investigation of mercury contamination in the Tennessee Valley region. In R. Hartung and B.D. Dinman (Eds.), *Environmental Mercury Contamination*. Ann Arbor Science Publ., Ann Arbor, Michigan. pp. 76-79.
Deslisle , C.E., B. Hummel, and K.G. Wheeland (1977). Uptake of heavy metals from sediment by fish. Proc.Int.Conf. on *Heavy Metals in the Environment*, Toronto, 1975, Vol. II/2, pp. 821-831.
Deurer, R., U. Förstner, and G. Schmoll (1978). Selective chemical extraction of carbonate-associated trace metals in recent lacustrine sediments. *Geochim.*

*Cosmochim. Acta,* 42, 425-427.

Dietz, F. (1972). Die Anreicherung von Schwermetallen in submersen Wasserpflanzen. *GWF-Wasser/Abwasser,* 113, 269-273.

Ebner, F., and H. Gams (1975). Schwermetalle in den Flüssen Glan und Gurk unter besonderer Berücksichtigung der Quecksilbergehalte. *Österr.Abwasser-Rdsch.,* 20, 51-53.

Edgington, D.H., and E. Callender (1970). Minor element geochemistry of Lake Michigan ferromanganese nodules. *Earth Planet.Sci.Lett.,* 8, 97-100.

Edgington, D.H., and J.A. Robbins (1976). Records of lead deposition in Lake Michigan sediments since 1800. *Environ.Sci.Technol.,* 10, 266-273.

Engler, R.M., J.M. Brannon, and J. Rose (1974). *A practical selective extraction procedure for sediment characterization.* Paper pres. at 168th Meeting American Chemical Society, Atlantic City, New Jersey.

Enk, M.D., and B.J. Mathis (1977). Distribution of cadmium and lead in a stream ecosystem. *Hydrobiologia,* 52, 153-158.

Förstner, U. (1976). Lake sediments as indicators of heavy-metal pollution. *Naturwissenschaften,* 63, 465-470.

Förstner, U. and G. Müller (1973). Heavy metal accumulation in river sediments: a response to environmental pollution. *Geoforum,* 14, 53-62.

Förstner, U., G. Müller, and G. Wagner (1974). Schwermetalle in Sedimenten des Bodensees, natürliche und zivilisatorische Anteile. *Naturwissenschaften,* 61, 270.

Frye, J.C., and N.F. Shimp (1973). Major, minor and trace elements in sediments of late Pleistocene Lake Saline compared with those in Lake Michigan sediments. *Environ.Geol.Note, Illinois Geol.Surv. No. 60,* 14 p.

Gächter, R., K.L.-S. Chan, and Y.K. Chau (1973). Complexing capacity of the nutrient medium and its relation to inhibition of algal photosynthesis by copper. *Schweiz.Z.Hydrol.,* 35, 252-261.

Gaggino, G.F., M. Gerletti, R. Marchetti, and A. Pennacchioni (1975). *Inquinamento da mercurio nelle aque interne italiane: il fiume Toce.* Instituto di Ricerca sulle Aque (CNR), Sez. di Idriobiologia Appl., Milan, Italy, 10 p.

Gibbs, R.J. (1977). Transport phases of transition metals in the Amazon and Yukon Rivers. *Geol.Soc.Am.Bull.,* 88, 829-943.

Gilbert, T.R., A.M. Clay, and D.A. Leighty (1976). Influence of the sediment/water interface on the aquatic chemistry of heavy metals. *Air Force Civil Engineering Center, Rept. AFCEC-TR-76-22,* 90 p.

Gillespie, D.C. (1972). Mobilization of mercury from sediments into guppies (*Poecilia reticulata*). *J.Fish.Res.Board Canada,* 29, 1035-1041.

Grimshaw, D.L., J. Lewin, and R. Fuge (1976). Seasonal and short-term variations in the concentration and supply of dissolved zinc to polluted aquatic environments. *Environ.Pollut.,* 11, 1-7.

De Groot, A.J., E. Allersma, and W. van Driel (1973). Zware metalen in fluviatile en marine ecosystemene. Symp.*Waterloopkunde in Dienst van Industrie en Milieu.* Publ. No. 110 N, Sect. 5.

De Groot, A.J., J.J.M. Goeij, and C. Zengers (1971). Contents and behavior of mercury as compared with other heavy metals in sediments from the Rivers Rhine and Ems. *Geol.Mijnbouw,* 50, 393-398.

Gupta, S.K., and K.Y. Chen (1975). Partioning of trace metals in selective chemical fractions of nearshore sediments. *Environ.Lett.,* 10, 129-158.

Guy, R.D., and C.L. Chakrabarti (1975). Distribution of metal ions between soluble and particulate forms. Abstr.Int.Conf.*Heavy metals in the Environment,* Toronto , 1975, D-29/30.

Guy, R.D., and C.L. Chakrabarti (1977). Analytical techniques for speciation of trace metals. Proc.Int.Conf. on *Heavy Metals in the Environment,* Toronto, 1975, Vol. I, pp. 275-294.

Hart, B.T., and S.H.R. Davies (1978). A study of the physico-chemical forms of trace metals in natural waters and wastewaters. *Caulfield Institute of Technology, Water Studies Centre, Techn. Rept. No. 6,* 188 p.

Hasselrot, T.B. (1972). Mercury in fish, water and bottom sediment. *Research Lab. Nat. Swedish Environ.Prot. Board. SNV Pm-239 ULV*, 922.

Hasselrot, T.B. and A. Göthberg (1974). The ways of transport of mercury to fish. Proc.Int.Conf. on *Persistent Chemicals in Aquatic Ecosystems*, Ottawa, May 1-3, 1974, III-37/41.

Hattingh, W.H.J. (1977). Reclaimed water: a health hazard? *Water S.A.*, 3, 104-112.

Heinrichs, H. (1975). *Die Untersuchung von Gesteinen und Gewässern auf Cd, Sb, Hg, Tl, Pb und Bi mit der flammenlosen Atom-Absorption Spektralphotometrie*. Dissertation Univ. Göttingen, 82 p.

Heydt, G. (1977). *Schwermetallgehalte von Wasser, Wasserpflanzen, Chironomiden und Mollusca der Elsenz*. Dipl.Arbeit Univ. Heidelberg, 143 p.

Irion, G. and U. Förstner, (1975). Chemismus und Mineralbestand amazonischer See-Tone. *Naturwissenschaften*, 62, 467.

Iskandar, I.K. and D.R. Keeney (1974). Concentration of heavy metals in sediment cores from selected Wisconsin lakes. *Environ.Sci.Technol.*, 8, 165-170.

Jackson, G.A. and J.J. Morgan (1975). Theory of chelator-metal ion interaction and phytoplankton growth. Unpubl. manuscript. Cit. Sibley and Morgan (1976).

Jackson, M.L. (1958). *Soil Chemical Analysis*. Prentice Hall, Englewood Cliffs, New York.

Jackson, T.A. (1978). The biogeochemistry of heavy metals in polluted lakes and streams at Flin Flon, Canada, and a proposed method for limiting heavy-metal pollution of natural waters. *Environ.Geol.*, 2, 173-189.

Jenne, E.A. (1976). Trace element sorption by sediment and soils - sites and processes. In W. Chappel and K. Petersen (Eds.). Symposium on *Molybdenum*, Vol. 2. Marcel Dekker, New York. pp. 425-553.

Jensen, S. and A. Jernelöv (1969). Biological methylation of mercury in aquatic organisms. *Nature*, 233, 753-754.

Jernelöv, A. and H. Lann (1973). Studies in Sweden on feasibility of some methods for restoration of mercury-contaminated bodies of water. *Environ.Sci.Technol.*, 7, 712-718.

Kennedy, E.J., R.R. Ruch, and N.F. Shimp (1971). Distribution of mercury in unconsolidated sediments from southern Lake Michigan. *Environ.Geol.Notes, Illinois Geol. Survey, No. 44*.

Klein, D.H. (1975). Fluxes, residence times and sources of some elements to Lake Michigan. *Water Air Soil Pollut.*, 4, 3-8.

Kopp, J.F. and R.C. Kroner (1968). *Trace Metals in Waters of the United States*. U.S. Fed.Water Pollut. Control Administration, Div.Pollut. Surveillance.

Laskowski, N., D. Pommerenke, A. Schäfer, and H.J. Tobschall (1974). Hohe Quecksilberkonzentration in Sedimenten des Ginsheimer Altrhein. *Naturwissenschaften*, 61, 681.

Lee, C.R., R.M. Engler, and J.L. Mahloch (1976). Land application of waste materials from dredging, construction and demolition processes. *Dredged Material Research Program*. U.S. Army Engineer Waterways Experiment Station, Vicksburg, Miss. Paper D-76-5, 42 p.

Lee, G.F. (1975). Role of hydrous metal oxides in the transport of heavy metal in the environment. In P.A. Krenkel (Ed.), *Heavy Metals in the Aquatic Environment*. Pergamon Press, Oxford. pp. 137-147.

Leland, H.V. and J.M. McNurney (1974). Lead transport in a river ecosystem. Proc. Int.Conf. on *Transport of Persistent Chemicals in Aquatic Ecosystems*. Ottawa, May 1-3, 1974.

Lewin, J., B.E. Davies, and P.J. Wolfenden (1977). Interactions between channel change and historic mining sediments. In K.J. Gregory (Ed.), *River Channel Changes*. Wiley and Sons, New York. pp. 353-367.

Lloyd, R. (1965). Factors that effect the tolerance of fish to heavy metal poisoning. Third Seminar *Biological Problems in Water Pollution*, U.S. Dept. of Health, Education and Welfare, 1962. p. 181.

Luoma, S.N. and G.W. Bryan (1978). Trace metal bioavailability: modelling chemical and biological interactions of sediment-bound zinc. Reprints of papers presented at the *176th National Meeting American Chemical Society*, Miami Beach, Florida, Sept. 10-15. pp. 413-414.

Maksimovic, Z. and A. Dangic (1973). Mercury mine at Mount Avala, a source of environmental pollution by mercury and arsenic. *Geol.Ann.Balk. Poluostrva* (Serbian), 38, 349-358.

Manahan, S.E. and M.J. Smith (1973). The importance of chelating agents. *Water Sewage Works*, Sept. 1973. pp. 102-106.

Manly, R. and W.O. George (1977). The occurrence of some heavy metals in populations of the freshwater mussel *Anadonta anatina* from the Thames. *Environ.Pollut.*, 14, 139-146.

Merlini, M., F. Argentesi, A. Brazelli, B. Oregioni, and G. Pozzi (1971). The effects of sublethal amounts of cadmium and mercury on the metabolism of zinc-65 by freshwater fish. Int.Symp. on *Radioecology Applied to the Protection of Man and his Environment*, Rome, Sept. 7-10, 1971.

Merlini, M., G. Cadario, and B. Oregioni (1977). The unionid mussel as a biogeochemical indicator of metal pollution. In W.E. Krumbein (Ed.), *Environmental Biogeochemistry and Geomicrobiology*, Vol. 3. Ann Arbor Science Publ., Ann Arbor, Michigan. pp. 955-965.

Müller, G. and U. Förstner (1976). Schwermetalle in den Sedimenten der Elbe bei Stade. Veränderungen seit 1973. *Naturwissenschaften*, 63, 242-243.

Müller, G. and F. Prosi (1978). Verteilung von Zink, Kupfer und Cadmium in verschiedenen Organen von Plötzen (*Rutilus rutilus L.*) aus Neckar und Elsenz. *Z. Naturforsch.*, 33c, 7-14.

Namminga, H. and J. Wilhm (1977). Heavy metals in water, sediments, and chironomids. *J.W.P.C.F.*, 49, 1725-1731.

Nikiforova, E.M. and R.S. Smirnova (1975). Metal technophility and lead technogenic anomalies. Int.Conf. on *Heavy Metals in the Environment*, Toronto, 1975, Abstr. C-94-96.

Patchineelam, S.R. and U. Förstner (1977). Bindungsformen von Schwermetallen in marinen Sedimenten. *Senckenberg.Marit.*, 9, 75-104.

Potter, L., D. Kidd, and D. Standiford (1975). Mercury levels in Lake Powell. Bioamplification of mercury in man-made desert reservoir. *Environ.Sci.Technol.*, 9, 41-46.

Prosi, F. (1977). *Schwermetallbelastung in den Sedimenten der Elsenz und ihre Auswirkung auf limnische Organismen.* Dissertation Univ. Heidelberg, 210 p.

Quentin, K.-E. and H.A. Winkler (1974). Vorkommen und Nachweis von anorganischen Schadstoffen im Oberflächen- und Grundwasser. *Zbl. Bakt. Hyg., I. Abt.Orig., B. 158*, 514-523.

Ravera, O., R. Gommes, and H. Muntau (1973). Cadmium distribution in aquatic environment and its effects on aquatic organisms. Conf. on *Problems of the Contamination of Man and his Environment by Mercury and Cadmium.* Comm.Eur. Luxembourg, pp. 317-331.

Reay, P.F. (1972). The accumulation of arsenic from arsenic-rich natural waters by aquatic plants. *J.Appl.Ecol.*, 9, 557-565.

Renfro, W.C. (1973). Transfer of zinc-65 from sediments by marine polychaete worms. *Mar.Biol.*, 21, 305-316.

Ruch, R.R., E.J. Kennedy, and N.F. Shimp (1970). Distribution of arsenic in unconsolidated sediments from southern Lake Michigan. *Environ.Geol. Notes, Illinois Geol.Surv., No. 37.*

Sames, C.-W. (1971). *Die Zukunft der Metalle.* Suhrkamp, Frankfurt.

Schell, W.R. (1974). Sedimentation rates and mean residence times of Pb and 210-Pb in Lake Washington, Puget Sound estuaries and a coastal region. Abstr.*Meet. Am.Soc.Limnol.Oceanogr.*, Seattle, June 23-27, 1974.

Schindler, J.E. and J.J. Alberts (1977). Behavior of mercury, chromium and cadmium in aquatic ecosystems. *U.S. EPA-600/3-77-023.*

Schöttler, U. (1977). *Ausbreitung und Eliminierung von Spurenmetallen bei Infiltration und Untergrundpassage.* Literaturstudie. Veröff. Inst. Wasserforschung

Dortmund, Nr. 27. 99 p.

Shaw, T.L. and V.M. Brown (1974). The toxicity of some forms of copper to Rainbow trout. *Water Res.*, 8, 377-382.

Shimp, N.F., J.A. Schleicher, R.R. Ruch, D.B. Heck, and H.V. Leland (1971). Trace element and organic carbon accumulation in the most recent sediments of southern Lake Michigan. *Environ.Geol.Notes, Illinois Geol.Surv., No. 41.*

Shukla, S.S., J.K. Syers, and D.E. Armstrong (1972). Arsenic interference in the determination of inorganic phosphate in lake sediments. *J.Environ.Qual.*, 1, 292-295.

Sibley, T.H. and J.J. Morgan (1976). Equilibrium speciation of trace metals in fresh water/sea water mixtures. Proc.Int.Conf. on *Heavy Metals in the Environment*, Toronto, 1975, Vol. I, pp. 319-338.

Sprague, J. (1968). Promising anti-pollutant: chelating agent NTA protects fish from copper and zinc. *Nature*, 220, 1345-1346.

Štern, J. and U. Förstner (1976). Heavy metals distribution in the sediments of the Sava basin in Slovenia. *Geologija*, 19, 259-274.

Stumm, W. and H. Bilinski (1972). Trace metals in natural waters. Difficulties of interpretation arising from our ignorance on their speciation. In S.H. Jenkins (Ed.), *Advances on Water Pollution Research.* Pergamon Press, New York. pp. 39-52.

Syers, J.K., I.K. Iskandar, and D.R. Keeney (1973). Distribution and background levels of mercury in sediment cores from selected Wisconsin lakes. *Water Air Soil Pollut.*, 2, 105-118.

Tabata, K. (1969). Studies on the toxicity of heavy metals to aquatic animals and the factors to decrease the toxicity. *Bull.Tokai Region Fish.Res.Lab.*, 58, 203-261.

Trefry, J.H. and B.J. Presley (1976). Heavy metal transport from the Mississippi River to the Gulf of Mexico. In H.L. Windom and R.A. Duce (Eds.), *Marine Pollution Transfer.* D.C. Heath, Lexington Books, pp. 39-76.

Troup, B.N. and O.P. Bricker (1975). Processes affecting the transport of materials from continents to oceans. In T.M. Church (Ed.), *Marine Chemsitry in the Coastal Environment*, ACS Symp.Ser., 18, 133-151.

Turekian, K.K. (1969). The oceans, streams and atmosphere. In K.H. Wedepohl (Ed.), *Handbook of Geochemistry*, Vol. 1. Springer-Verlag, New York.

Turner, R.R. and S.E. Lindberg (1978). Behavior and transport of mercury in river-reservoir system downstream of inactive chloralkali plant. *Environ.Sci.Technol.*, 12, 918-923.

Turney, W.G. (1971). Mercury pollution - Michigan's action program. *J.W.P.C.F.*, 43, 1427-1438.

Ueda, T., R. Nakamura, and Y. Suzuki (1976). Comparison of cadmium-115-accumulation from sediments and sea water by polychaete worms. *Bull.Jap.Soc. Sci.Fish.*, 42, 299-306.

Vernet, J.-P. and L.M. Johnstone (1974). Heavy metal pollutants in the sediments of the Rhône River. Proc.Int.Conf. on *Transport of Persistent Chemicals in Aquatic Ecosystems*, Ottawa, May 1-3, 1975.

Vernet, J.-P. and R.L. Thomas (1972). Levels of mercury in the sediments of some Swiss lakes including Lake Geneva and the Rhône River. *Eclogae Geol.Helv.*, 65/2, 293-306.

Volkov, I.I. and L.S. Fomina (1974). Influence of organic material and processes of sulfide formation on distribution of some trace elements in deep-water sediment of Black Sea. In E.T. Degens and D.A. Ross (Eds.), *The Black Sea, Geology, Chemistry and Biology. Am. Assoc-Petrol.Geol.Mem.*, 20, 456-476.

Walters, L.J., T.J. Wolery, and R.D. Myser (1974). Occurrence of As, Cd, Co, Cr, Cu, Fe, Hg, Ni, Sb and Zn in Lake Erie sediments. Proc. *17th Conf. Great Lakes Res.* pp. 219-234.

Whitley, L.S. (1968). The resistance of tubificid worms to three common pollutants. *Hydrobiologia*, 32, 193-205.

Wittmann, G.T.W. and U. Förstner (1977). Heavy metal enrichment in mine drainage.

III. The Klerksdorp, West Wits and Evander Goldfields. *S.Afr.J.Sci.*, 73, 53-57.

Wood, J.M. (1974). Biological cycles for toxic elements in the environment. *Science*, 183, 1049-1053.

Wood, J.M., F.S. Kennedy, and C.G. Rosen (1968). Synthesis of methylmercury compounds by extracts of a methanogenic bacterium. *Nature*, 220, 173-174.

Wood, L.W. (1975). Role of oligochaetes in the circulation of water and solutes across the mud-water interface. *Verh.InternVer.Limnol.*, 19, 1530-1533.

Zitko, V. (1976). Structure activity relationships and the toxicity of trace elements to aquatic biota. In R.W. Andrew, P.V. Hodson, and D.E. Konasewich (Eds.), *Toxicity to Biota of Metal Forms in Natural Water*. Int.Joint Comm., Windsor, Ontario, pp. 9-43.

Zitko, V. and W.G. Carson (1976). A mechanism of the effects of water hardness on the lethality of heavy metals to fish. *Chemosphere*, 5, 299-303.

Züllig, H. (1956). Sedimente als Ausdruck des Zustandes eines Gewässers. *Schweiz. Z.Hydrol.*, 18, 7-143.

# Radioactive Contamination of the Aquatic Food-chain and Consequent Risk to Man

A. A. Cigna

*CNEN - Dipartimento Radiazioni, Divisione Protezione Ambiente*

**ABSTRACT**

**The extreme caution adopted to keep any radioactive contamination within safe and severe limits implies that any component of the environment is protected from possible harmful effects, considering that man is among the organisms most sensitive to radiation. The general criteria of radiation protection lead to the fixation of the quantity of radioactive substances which can be discharged into an environment. This quantity is usually expressed in a so called "discharge formula". Many difficult problems are involved in the determination of such a formula: characteristics of environmental media, various phenomena which may modify the distribution of radioactivity and all factors which influence the exposure of man to radiation. All these factors must be evaluated against general requirement of public health and of economical feasibility. Only a fraction of the limiting environmental capacities will eventually be utilized for the effective discharge of radioactive wastes into the environment. Within this fraction the contribution of each radionuclide species is defined relative to the contribution of all other species.**

KEYWORDS

Radiation protection, Limiting capacity, Radionuclides, Wastes, Nuclear plants, Dose limits Radioecology.

INTRODUCTION

The term environmental pollution in its widest meaning implies a chemical, physical, or biological change of the environment as a result of the introduction of foreign substances into it. Therefore, defining "radioactive contamination" as that particular form of pollution due to radionuclides not present naturally in a certain environment, it is evident that any discharge, albeit a modest one, of radioactive substances can represent a definite contamination of that environment. Nevertheless, in practice, when evaluating a state of pollution (or of contamination) one cannot neglect the presence or absence of harmful effects of the above changes, meaning by harmful effects those effects which render the environment less suitable for any of those uses for which it would be suitable if it were in its natural state. With this more precise statement the concept of pollution (or of contamination) assumes a significance which is no longer strictly tied to naturalistic criteria, but rather to broadly economic criteria and to the concept of use. From this it follows that the total preservation of the natural characteristics of an environment is not the objective to pursue everywhere and always, in the struggle for its protection. In the case of protection from ionizing radiation the basic criterion consists, as is well known, not only in limiting the injury to persons but also in positively containing the risk of such damage within limits which are not statistically significant. Extreme caution, which is rarely encountered when conventional pollution is considered, of course implies that all the other components of the environment are protected from possible harmful changes, considering that man is among the organisms most sensitive to radiation.

## GENERAL CRITERIA OF RADIATION PROTECTION

The above mentioned criterion finds its expression in the condition that no one shall receive by total irradiation (internally and/or externally) doses greater than those established as dose limits by the basic standards for the protection of health against the hazards of ionizing radiation. If therefore the nature and the amount of the radioactive substances and the rate at which they are discharged into an environment are such that the dose limits are not exceeded, then one can exclude the occurence of "radioactive contamination" in the previously specified sense. For every environment therefore there exists a maximum quantity of radioactive substances which can be introduced in a given time, under specified conditions and in a specific manner, without there being "contamination", i.e. without producing a situation which can carry a statistically significant risk of harm to the population. The precise knowledge of such a quantity is thus indispensable in safeguarding public health; it must also be said, however, that an over-cautious estimate of the receptive capacity of an environment could cause unwarranted limitations on the discharges and therefore a limitation in the corresponding industrial or research activity, or else the adoption of unnecessary treatment of precautions, so that it is easily understood that in any case one ends up with a greater economic burden. The determination of the quantity of radioactive substances which can be discharged into an environment, in such a way as to balance the fundamental requirements of public health protection with those of general economics, presupposes knowledge of the chemical, physical and ecological characteristics of the receiving environment, and of the phenomena which affect the radionuclides after discharge and which can have an influence on their distribution in such an environment, and finally of the use which is made of that environment or of some parts of it. The nuclear industry should refuse to follow the bad example of the environmental disturbance still being produced by the inadequate siting of some non nuclear industrial plants. The siting of these plants sometime occurs notwithstanding local opposition (often reduced to silence by the prospect of advantages in the district employment situation or economy), on the basis of socioeconomic criteria and supported by rough public health guarantees, without detailing the health and ecological analysis as is always done in the case of nuclear plants. It is therefore necessary to withstand the temptation to imitate the procedures used for other industrial plants. Moreover, it is still necessary to act rather widely in the private and public industrial world to diffuse the principles and the cultural attitude behind the environmental safeguards or to make it clear that ecological problems are not less important than the prevention of diseases. Years were required to give the industrial world a complete health consciousness; years more will be necessary to give it an ecological awareness. Some narrow-minded attitudes, some procedural impatience or intolerance reveal the cultural immaturity of some sectors of the economic and industrial media as far as the environment is concerned. When industrial media accept the ecological attitude, it will be possible to include environmentalists among the decision makers for site proposals, together with technologists and economists. But this simple occurence will still require time and maturation to be generally accepted. Many debates still hindering the installation of some conventional factories, and sometimes also of nuclear plants, would come to an end if the aspect of environmental safeguard were kept in mind from the beginning of the selection of the site to be proposed to the competent authorities.

## THE CHARACTERISTICS OF THE RECEIVING ENVIRONMENT

When radioactive liquid wastes are introduced into a body of water they then disperse over an area (the receiving environment) with boundaries which are generally somewhat difficult to define. On the other hand the problem of the location of the area boundaries is less important than it may seem at first. The critical area however must be identified with great accuracy and this will be defined later. Immediately after discharge the wastes meet a water-course or a lake (aquatic environment); such an environment is defined here as the primary receiving environemnt. Later the radioactive contamination can be transferred to the land environment (plants and/or animals) which one may call the secondary receiving environment Obviously, from the point of view of propagation to man, the radionuclides can come either directly from the primary or from the secondary environment, as will be explained further on. Nevertheless the subdivision referred to above has advantages, since the interaction between the liquid wastes and the receiving environment takes place mainly in the aquatic environment. The primary environment (aquatic) consists principally of the receiving body of water and its sediments and also of the aquatic flora and fauna. Given then the possibility of the exchange of water between surface water and subterranean water, one considers a further subdivision represented by the adjacent aquifer(s). From the receiving body of water as such, man can undergo both external and internal exposure to radiation; if the water is used only for drinking or in food, then the external hazards predominate over the external hazards, the latter being linked with an indirect use of the body of water such as navigation and recreational activities(bathing, water skiing, etc.). Another important subdivision of the primary environment consists, as has been said, of the sediments at the bottom of the body of water. The risks associated with the areas in the vicinity of beaches of the materials originating from the sediments on the bottom (in dredging and filling) involve only an external exposure. The importance of the sediments is due also to the interchanges between them and the water and aquatic organisms. The transfer of the radioactive contamination from one to the other of these subdivisions can reach quite high values because of the concentration of the contamination in the sediments themselves. The aquatic flora and fauna are the other two subdivisions taken into consideration, obviously on the hypothesis that the animals and/or plants present in the body of water are abundant and the species are edible. Certain plants of the aquatic flora can be eaten directly by man (for example cress) and therefore can act as a source of internal irradiation. In any case the aquatic flora produce food for the successive trophic levels which make up the aquatic fauna. In their turn the edible species of these expose man to internal irradiation. The characteristics of the aquatic environment can be subdivided into the following groups:

Physical parameters (flowrate, volume, temperature, density, water stratification, nature of bottom).

Chemical and chemical-physical parameters (suspended solid content, pH values, oxygen concentration, salinity).

Biological parameters (biotic communities).

Each of the above mentioned characteristics will be briefly illustrated from the point of view of a possible interaction with the released radioactivity.

Flowrate

Is one of the most important parameters of the water courses and the factors on which this parameter depends are well known, in particular the meteorological factors. For the purposes of the present investigation, the knowledge of the average annual flowrate over many years and also of the average flowrate for each individual month acquires a particular importance, as will be seen later.

Volume

If the receiving waters are not flowing (lakes, reservoirs, etc.) then obviously one does not speak of the flowrate but rather of the volume of the whole body of water in question. Consider for example the case of a lake with thermal stratification, where the useful volume, as far as dispersion of the radionuclides is concerned, is at most the volume in which the circulatory motion occurs, i.e.

that of the epilimnion (see "Water Stratification").

Temperature

It is well known that this parameters depends on the absorption of solar radiation by the water and on the heat exchange with the atmosphere and with the bottom and is therefore variable within wide limits in time and space. The temperature exercises a direct influence of many of the chemical reactions which take place in the receiving water mass. In particular the rate at which the oxygen is consumed in the oxidation processes which take place in the hypolimnion (see "Water stratification") can double or triple following an increase of temperature of some ten degrees. Temperature is evidently important therefore in the processes of absorption of radioactive substances by aquatic organisms.

Density

For surface and ground waters the density values lie within fairly narrow limits, although in certain cases (brackish water) there is a substantial difference from the fresh-water values.

Water stratification

In certain types of aquatic environment water stratification phenomena can occur which are due to differences in density. The stratum which separates two strata of different density is called a pinocline. With stratification of thermal origin, the stratum which has the most rapid drop of temperature is called the thermocline or the metalimnion. This stratum separates the water above it, known as epilimnion, where there is not thermal stratification or if there is it is not important, from the water below it, which extends to the lake bottom and is known as the hypolimnion.

Nature of the bottom

In most cases the original bed of rock beneath the mass of water is overlaid with a stratum of sediment, the thickness of which can vary widely and the physical, chemical and biological characteristics of which are closely related to the processes of production of the sediment itself (deposition of inorganic particles transported by the water or of dead vegetable or animal organic material).

Suspended solids content

Surface water often contains considerable quantities of finely divided particles in suspension. It is known that such particles can differ in origin and chemical composition and also in physical-chemical behaviour. The concentration of these materials in water can change within very wide limits either from one mass of water to another or within the same mass of water at different times or places.

pH values

The value of this parameter depends in the first place on the carbon dioxide content. In general the pH value is lower when the concentration of carbon dioxide is higher. The pH value is a particularly distinctive factor in the biological processes of an aquatic environment.

Dissolved oxygen

The oxygen balance in a mass of water involves two fundamental constituents.

a) The oxygen which is dissolved from the atmosphere (reaeration) and that envolved by the vital metabolism of aquatic plants containing chlorophyll.

b) The oxygen consumed by the aerobic biochemical activity which decomposes organic matter, and more generally the oxygen consumed by all the oxidizable substances present. Considering that the metabolism carbon dioxide is complementary oxygen, in the aquatic environments where biogenic phenomena occur, there is a well defined inverse correlation between oxygen and carbon dioxide.

Salinity

The quantity and nature of substances dissolved in a body of water depend on various factors and in particular on the solvent power of the water, on the degree of solubility of the minerals of which the rocks are composed, on the duration and extent of contact with the rocks, and on the temperature. Evidently the salinity must also be considered in relation to possible domestic or industrial discharges which take place into the body of water in question. It will be shown in detail further on that the dissolved substances may give rise to side-effects which interfere with the effluents.

Biotic communities

A biotic community is defined as the totality of the animals and plants which exist in a particular environment. In the present case it is necessary to consider those communities in the receiving environment which are able to intervene significantly in the transport and diffusion processes of the radioactive contamination. In practice one studies the communities which directly furnish food suitable for human consuption (fish, watercress, etc.) or else those aquatic plants etc. which although not edible can constitute an intermediate link in the food chain. In order to study the fate of discharged radionuclides it is necessary to determine both the quality and the quantity of the flora and fauna present. These studies essentially concern the behaviour of radionuclides in the environment i.e. their eco-cycles, and aim at defining the limiting capacity of the environment. The limiting capacity of the environment for a radionuclide can be defined as the maximum annual quantity of that radionuclide which can be discharged in certain defined conditions into a particular sector of the environment in such a way that the resulting dose commitment received via the critical pathways does not exceed the annual dose limit for individual members of the public as laid down in the basic standards of the Commission of the European Communities. By "in certain defined conditions" is understood the complete set of conditions (chemical characteristics of the effluents, the mode of discharge, etc.) which characterise the release of the radioactive residues of the environment. Similarly by "a particular sector of the environment" is implied a limitation of the concept of environmental capacity to a reduced area of the surroundings which is directly affected by the discharge and where the critical pathways are situated. The " dose commitment" resulting from a particular discharge is defined as an integral for an infinite period of the intensity of the average dose received by individuals of the critical group of the exposed population, a definition analogous to that of UNSCEAR (1966) for the world population. When the discharges contain several radionuclides some distribution has to be made so that the sum of the contributions of each radionuclide does not exceed the annual dose limit already laid down. Furthermore, it should be emphasized that the actual discharges do not correspond to the limiting capacity of the environment already defined but only a fraction of this which can be called the stipulated capacity of the environment. The value of the latter is established in each particular case by the competent authorities responsible for radiation protection on the basis of the characteristics of the surroundings affected, the technical aspects of the discharge and the requirements of radiation protection. It is appropriate to remember that as far as radioactive contamination is concerned man is the most sensitive organism and, given the structure of the environments considered, is always regarded, in practice, as the limiting factor.

## PHENOMENA CAPABLE OF INFLUENCING THE DISTRIBUTION OF RADIONUCLIDES WITHIN THE RECEIVING ENVIRONMENT

The factors which have been listed and described in the previous section, as has already been mentioned, can enter into various combinations, thus determining the manner of propagation of the radionuclides to man. Evidently not all the phenomena mentioned here are of equal importance everywhere, but form a spectrum of values which can be different from one case to another. Most often the principal phenomena are those connected with water movement, sedimentation and suspension of particulate substances, the absorption of radionuclides by materials present in the receiving body of water, uptake by plants and/or animals, and of course radioactive decay. Nevertheless in principle one cannot exclude the possibility that because of some particular local situation a factor which is usually of secondary importance may become one of primary importance. This shows that it is desiderable to obtain complete information on all the phenomena which can theoretically take place in the environment after the discharge; thus case by case one can fix attention on those which most influence the distribution and the subsequent routes of the radionuclides. Another comment concerns the concrete possibility of appraising in numerical terms the phenomena which we are speaking. It is easily seen that analytical information of such a type would allow one to proceed with the calculation of the concentrations of radionuclides in the various homogeneous components of the environment in a manner which is direct and scientifically rigorous. Neverthless, although today they is a great mass of data and reports on the various phenomena capable of modifying the distribution af the radioactivity in the environment, the complexity and variability in time and space of such phenomena still do not allow one to make such a sophisticated calculation. But, apart from the obvious consideration that, with the development of methods of investigation and research, the direct quantitative evaluation of the above mentioned phenomena will become steadily easier, it must be said explicitly that for practical purposes one can dispense with a detailed knowledge of each phenomena and substitute for it a treatment of a more empirical type. Particular emphasis should be given to phenomena capable of increasing the concentration of a radionuclide. The most important ones are described here:

Sedimentation and suspension

It has been indicated that even the particulate pollutants or those associated with particles suspended in water can be transported or dispersed within a current according to the kinetic energy of the water and the size and weight of the pollutants themselves. The transporting capability of a water-course and more generally of a current, as is well known, is bigger for a greater velocity and density of water and for finer and lighter materials. If however the transporting capability of the current lessens, then the pollutants in suspension tend to form a sediment on the bed, undergoing the same fate as the rest of the detritus trasported by the flow and clouding the water. In general therefore a clarification of the water is a sign of a purification as regards radioactive contamination but at the same time means a progressive accumulation of radioactive substances in the sediments, a circumstance which should not be underrated, having regard to the possible presence of detritus-eating fauna or certain aquatic flora or the use of dredged silt for land reclamation. It must be remembered that the deposition on the bed can also be of a temporary nature in the sense that the water transporting capability, being dependent on variable factors such as flowrate and velocity, is itself variable and may therefore later be more than that needed to remove and maintain in movement the solids already settled (suspension). The final deposition of such solids will happen at a point farther downstream, depending on the conditions of the current to which the greater flowrate and velocity of the water correspond. The importance of this is obvious when one considers that the processes of sedimentation and suspension greatly influence the behaviour of the radioactive substances discharged into a body of water. In this regard however the literature provides data which can be extremely useful. The preceding considerations are substantially valid for moving water. For lakes, which generally behave like settling tanks, the particulate material reaches the bottom and remains there indefinitely, there being no renewal of the sediments as is brought about in rivers by the action of floods.

Ion exchange. Elution

It is well known that many natural materials possess ions or groups, mainly cationic, replaceable by other ions or groups. The natural ions exchangers are of either the organic or the inorganic type; the latter can be divided into two categories, clays and zeolites. From the point of view being considered here, the clays (montmorillonite, kaolinite, illite, chlorite) are the most interesting, being very widespread in nature and hence often present in the receiving water;

but in certain geological situations, the zeolites (clinoptilolite, tufa etc.) can also have an important role. If the radionuclides present in the wastes are cationic, they can exchange with the argillaceous and/or organic materials transported by moving water or otherwise present in the receiving body, for example in the sediments on the bottom of the banks. The chemical nature of the radionuclides has an influence in that the selectivity of the exchanging materials is greater for certain ions than for others, according to certain scales. For some important bivalent ions, the scale is

$Ba^{++}$ $Sr^{++}$ $Ca^{++}$ $Mg^{++}$

and for monovalent ions

$Cs^{+}$ $Rb^{+}$ $K^{+}$ $Na^{+}$ $Li^{+}$

By the general laws which govern the equilibrium of exchange such scales are inverted, for solutions of suitable concentration, which in principle can allow the elution of exchanging materials and the return into solution of the radionuclides. The characteristics of exchange which need to be known to try to forecast the variation in time and space of the contamination introduced into a particular hydrogeological environment are as usual those of the exchange capability exchange kinetics, and the ability of any exchange material, present both in suspension and in the sediments on the bottom and on the banks, to remove from solution an ion present in trace amounts. This last property is best expressed by means of the coefficient of distribution, defined as the ratio of the ion concentration in the solid phase (q) and that in the liquid phase (c) under equilibrium conditions. The exchange mechanism of a particular material with respect to a certain ion depends not only on the chemical nature of that ion, as previously stated, but also of course on the nature of that material, the particle size, the pH value, the concentration of other ions (especially those chemically similar to the absorbed ion) and also the time in contact. As with all absorption phenomena, ion exchange is also reversible; the absorbed ions can in theory become eluted, if for example, as was stated before, the salinity of the receiving water exceeds certain levels or if the pH value falls below certain values. In practice, however, such conditions are rarely realised except in cases such as the sudden introduction of industrial waste or water rich in fertilizers.

Adsorption

The phenomena of adsorption can occur between the radionuclides present in solution in the wastes and substances transported by the receiving water or contained in the sediments on the bottom or the banks. The effect diminishes as the surface area of the material lessens and with an increase in temperature. Carbon and silica are the substances for which the adsorption phenomena can occupy an important place in the phenomenological complex which determines the fate of the radionuclides introduced into the environment. Metallic particles possibly discharged into the environment by metallurgical industries upstream can also cause intense adsorption processes.

Colloidal particles can also undergo adsorption phenomena, although in such a case the mechanism seems to be different from that for solutes. The adsorption process is nearly always irreversible.

Coprecipitation

If the radionuclides are in a cationic form they are liable to coprecipitate as a result of flocculations in the receiving body of water, if for example this has a high iron or aluminium content and for some reason there has been a rise in the pH value. This process is very effective in regard to trivalent cations and those of higher valencies, and therefore to the alpha emitters and in particular to plutonium. Plainly, other cases are possible: for example one can think of the possibility of effluent rich in phosphates (such as the effluent resulting from the brewing of beer) being poured into the receiving water upstream of the radioactive discharge. If the radioactive wastes originate from a water supply which is fairly hard and if they have a high pH value, then the meeting of the wastes and the receiving water precipitates the calcium as a basic phosphate. The precipitate has a flocculent consistency and acts as a carrier for many of radionuclides present in the liquidin a cationic form, but is not efficient as regard anions and radiocolloids (among the latter may be mentioned caesium, zirconium, scandium, ruthenium and plutonium). These two examples show, among other things, the importance of an accurate survey of the discharges of neighbouring industries.

Ingestion by animals (including man)

As a result of their metabolism and feeding habits, it is known that certain kinds of animals can concentrate radionuclides in their bodies, often by many orders of magnitude. If therefore such animals take part in a certain food chain they can furnish man with food with a high concentration of radioactivity. If not, they can exert an inhibiting action which slows the progress of the radionuclides in the environment.

Absorption by plants

The same considerations as were developed above for animals are valid for this phenomenon. It is necessaty to add that root absorption is spoken of here; leaf uptake, partly because of the special characteristics of the process, is discussed in the following sub-section.

Surface adsorption by animals and plants

This form of intake can make a significant contribution to the radioactive contamination of organisms, considered as a whole. As a rule surface adsorption is more important for plants than for animals. In the case of irrigation by rain for example, leaf uptake can represent a critical pathway for the transfer of contamination for those radionuclides which, like Cs 137, are retained by the soil or for which the time required for intake through the roots is too long compared with their short half-life ( I 131 for example). In the same way as has been mentioned before, this process may or may not reduce the total quantity of radionuclides available to man according to whether the organisms under consideration do not take part or take part in some human food chain.

Sedimentation of dead aquatic plants and animals

When aquatic animals and plants die, the radioactive contamination contained within them is transferred to the sediment on the bottom. In certain cases this process can be responsible for a considerable contribution to the accumulation of contamination within the sediments, for example when the life of the organisms is fairly short and their number high. The radionuclides in the sediments can pass back into solution and if the rate at which they do so is high enough an equilibrium can be established between the concentration in the sediment and that in the water. None of these processes can be easily evaluated quantitatively but must be considered on the basis of the particular characteristics of each single environment.

## CRITICAL PARAMETERS

Descriptions were given in the preceding section of the characteristics of the environment influencing the general distribution of the radionuclides. A detailed study allows one to specify the critical elements in the environment (critical groups, critical pathways, critical areas). The critical group refers to that group of the population most exposed to injury from radiation; the critical pathways are those by which this group can most easily suffer injury. Finally the critical area will be the geographical area involved in the radionuclide propagation phenomena along the critical pathways previously mentioned. As has been pointed out in a publication by the Commission of European Community (1970) the term critical is used here not to introduce some idea of danger but to give a basis wich allows decisions to be taken in the field of health protection. It is first af all necessary in each case to delimit the effective receiving environment; for this purpose we identify all the ecosystems (various kinds of plants cultivated, animals reared etc.) which are dependent directly or indirectly on the body of water into which the discharge is made. For this purpose a general knowledge of the overall character of the environmental situation will be sufficient. At present accuracy of the quantitative knowledge of the processes which link the various subdivisions of the receiving environment is in general somewhat low. But is is likely that in future the amount of experience acquired in this field will allow an increasingly objective evaluation of the problem, thus reducing the uncertainty associated with the theoretical suppositions. The next step is to carry out a phytosociological survey in conjunction with a social-economic enquiry. A deep and analytical knowledge will be obtained both of the individual components of each ecosystem involved and of the composition and feeding and living habits of the populations. Such components can be identified in terrestrial environments by direct enquiry in situ or by using survey maps and information obtained from agricultural firms, syndicates etc. Thus a chart may be drawn of the environment involved, indicating the areas covered by various colonies and the forms of use of the body of water and soil. Such a survey will be concerned with the species which are present in such numbers and density as to allow practical and easy measurement of the radioactive contamination. For example, in cornfields, lucerne, various specialized cultivations, etc. the survey of the species will be limited specifically to the principal one and possibly some others (also easily evaluated) concerned from the point of view of a high concentration factor. Besides the identification of the species, quantitative determinations (productivity) of each one will be necessary, distinguishing between the percentage destined for local consumption and that destined for markets outside the area. It will be necessary then to establish the distribution area, since, the more limited this is, the greater the risk arising from any possible contamination, for the human population. Analogous procedures should be followed for qualitative and quantitative surveys of livestock characteristics. In certain cases where a component of the food chain causes a "bottleneck" in the propagation of radiocontamination, because it discriminates against the contamination itself for example, the study of the component must be particularly deep. On the other hand the opposite case must be considered, that is, when a component of the food chain has a particularly high concentration factor. These data will be supplemented by an investigation into the prevailing consumption by the population involved and into its feeding and living habits. Qualitative and as far as possible quantitative surveys should be carried out analogously for aquatic environments on the animal and plant species present and the collective form and the interactions between the various species should be established. Having determined the food consumption and more generally the living habits of the particular population groups related in any way to the receiving environment, the incorporations and/or exposures that result for each of these groups from a unit injection of a given radionuclide can be calculated on the basis of the previously determined concentrations of the radionuclide. The group which turns out to be by far the highest among all the incorporations and exposures is obviously the critical group for the radionuclide involved. It is possible, even probable, that a value will not always be found which is clearly above all the others, or that more than one group will have a value of exposure of irradiation which is equally high, or is of the same order of magnitude. In such a case critical groups are referred to (for that particular radionuclide). It would be illusory to establish a sequence for values which were not clearly distinguishable one from another (that is, differing by at least an order of magnitude), having regard to the uncertainty associated with the evaluation of many of the parameters used to calculate these values. The critical group (or groups), once identified, will lead to an unequivocal determinanion of the corresponding critical pathway, i.e. that pathway, among all those possible, which contributes predominantly to the total exposure of the group. The same considerations as those developed for the identification of the critical group are valid also in this phase, and there may be more than on critical pathway, again with reference to the single radionuclide. Finally the critical area will be the geographical region across which each of the critical pathways mentioned above extends. From these definitions of critical pathway and area it is evident that they will be related not so much to the definition of the quantity of radioactive substances dischargeable, as to the surveillance and control operations. These operations may be limited to those parts of the whole receiving environment in which the most important propagation phenomena of the radionuclides are located. All the above refers to a single radionuclide, as has been stated repeatedly. Plainly, when it is foreseen that more than one radionuclide is to be discharged, it will be necessary to repeat the whole procedure described above for each of them. As a result of this, it could be found that different radionuclides result in critical groups which are completely different or which are partly or wholly coincident. Nevertheless, as will be seen in the following section, such differing possibilities have no effect on the derivation of the discharge formula, although they could extend the surveillance and control operations. If neither edible flora nor edible fauna are present in the aquatic environment and if contaminated water is not used for irrigation downstream of the point at which the radioactive wastes are discharged, then it is not necessary to carry out an ecological investigation, and the maximum permissible concentrations for drinking water can be taken as the limiting factor.

## DOSE EVALUATION AND DISCHARGE LIMITS

Considerations developed in the previous section have shown how in practice the critical group or groups can be identified for a particular radionuolide. Such a group has the maximum incorporation and/or exposure following the injection into the environment of a unit discharge in unit time. A simple ratio will now allow the maximum dischargeable quantity ($Q_{max}$) to be calculated, such that the resultant incorporation and/or exposure is equal to the maximum allowed for a year by the regulations for protection from ionizing radiation:

$$Q_{max} = \frac{\text{Annual maximum permissible incorporation}}{\text{Incorporation corresponding to } Q=1}$$

or, in the case of external irradiation,

$$Q_{max} = \frac{\text{Annual dose limit}}{\text{Dose corresponding to } Q = 1}$$

The value $Q_{max}$ is said to be the discharge limit for the radionuclide being considered. Clearly the condition which must be satisfied for the discharge of that single radionuclide will be

$$\frac{q}{Q_{max}} \leqslant 1,$$

where q represents the quantity effectively discharged. In the more general case in which there may be more than on radionuclide discharged the above mentioned condition obviously will be changed to

$$\sum_{r=1}^{n} \frac{q_r}{(Q_{max})_r} = \frac{q_1}{(Q_{max})_1} + \frac{q_2}{(Q_{max})_2} + \ldots + \frac{q_n}{(Q_{max})_n} \leqslant 1.$$

This inequality represents the discharge formula for the receiving environment and the type of discharge which is being considered. It is easily seen from the structure of this formula that this ensures the safety of each individual critical group in regard to all the radionuclides concerned in the discharge, even those for which another group may be critical. On the basis of the definition of the discharge limit, $(Q_{max})r$ represents the maximum quantity of a particular radionuclide which may be continuosly discharge throughout the year. Consequently, the quantity effectively discharged $q_r$ must also be understood to be continuously introduced into the environment. Nevertheless, it is possible to carry out instantaneous discharge of not more than one twelfth of $\sum_{r=1}^{n} q_r$ in a single monthly operation. If several nuclear installations or centres are located within the same receiving environment, all the radionuclides associated with the waste from such installations are taken into account in the determination of the discharge formula. In this case the above mentioned formula is an overall one and for practical purposes needs to be divided into as many partial formulae as there are installations involved. To do this one must proceed to apportion the various discharge limits on the basis of the requirements of installation themselves. Before concluding this work it is desirable to develop certain considerations on an important aspect of the whole problem. It has been seen from the beginning that the whole treatment has been carried out taking into account the fact that each radionuclide in an environment behaves in a manner which is dependent on its chemical characteristics, and therefore in general in a manner which is different from all the others. It is therefore not permissible to group chemically different radionuclides on the basis of their emission characteristics (for example, beta emitters). Nevertheless indiscriminate measurements of radionuclides (for example, total beta) can be still usefully carried out for contamination levels which are much lower than those permissible, when all the contamination can be considered to be due to the most restrictive radionuclide. In this case, if the indiscriminate total measurement furnishes a lower result than that permitted for the most restrictive radionuclide, it will be not necessary to carry out any further measurement. As the contamination level approximates to the maximum permitted, one must proceed to specific measurements, starting with the most restrictive nuclide and continuing in order of decreasing restrictivity. When the contamination levels allow, it is an obvious advantage to be able to carry out indiscriminate total measurements.

## REFERENCES

Commission of the European Community (1970). Principles and general methods of establishing the limiting radiological capacity of a hydrobiological system. EURATOM-C.I.D., Bruxelles.

UNSCEAR (1966). United Nations Scientific Committee on the Effects of Atomic Radiation. General Assembly, Official Records, 19th session, Suppl. no. 14 A/6314.

# Biological Effects of Ionizing Radiations in Aquatic Environment

O. Ravera

*Department of Physical and Natural Sciences, Commission of the European Communities, Joint Research Centre - Ispra Establishment I-21020 Ispra (Va.), Italy*

ABSTRACT

The natural and artificial sources of radiation were listed and the inadequate information about the biological effects due to the H-3 (radioisotope discharged by nuclear plants) was emphasized. On the basis of the present information on the environmental radioactivity (natural plus artificial) the effects on the aquatic populations were discussed. The different importance of the genetical and somatic damages was a matter of speculation based on the rich information from laboratory experiments and the scarce ones concerning natural populations. Environmental conditions (e.g. temperature), species, strain, age, sex and behaviour of the organism were taken into account as factors interferring with the radiation effect. The importance of the recovery was related to the physiological properties of the species and, particularly, to the demographic characteristics (e.g. intrinsic rate of natural increase) of the irradiated populations. From the present knowledge it seems that the level of environmental contamination by radioisotopes is too low for producing noticeable effects on aquatic biota.

KEYWORDS

Aquatic populations, biological effects, ionizing radiations, natural dose-rate, radioactive fall-out, radioactive wastes.

Plants and animals have always received a certain dose of ionizing radiation due to cosmic rays, radioactive minerals and natural radioisotopes present in the environment and in their bodies (Table 1). This means that, below a certain threshold, ionizing radiations are not able to damage organisms. According to certain authors, the "spontaneous" mutations are produced by ionizing radiations and, consequently, they should be the cause of biological evolution.

The natural dose-rate measured in different areas of the Earth is very different according to their lithological, geographical and ecological characteristics (Table 2). The hazard of high doses of ionizing radiations was well known even before the Second World War, but the consequences of the nuclear explosions in Japan during this war, have made the public afraid of radiation effects and radioactive pollution (Table 3).

Actually, the use of the atomic bomb in 1945 and the following 336 nuclear tests have caused a sudden increase in the natural level of radioactivity, particularly in northern temperate latitudes. This increase, for the fall-out contamination, was worldwide and all foodchains, man included, were contaminated. It has been calculated that the mean dose due to fall-out absorbed by a man living in a temperate zone was equal to about 2.4% of the total dose due to the natural level of radioactivity (Table 4). On August 5, 1963, the United States, the Soviet Union and the United Kingdom agreed to end nuclear tests in the atmosphere, seas and space. As a consequence of this agreement, the level of fall-out radioactivity has continuously decreased with time. Until now the Moscow Treaty (1963), signed by more than 100 countries, has been respected and the nuclear tests of France and China have not been so important as to influence significantly the fall-out decrease.

In addition to fall-out, there are, as a consequence of peaceful uses of nuclear energy, other sources producing radioactive wastes: 1) reprocessing of reactor fuel; 2) mines of radioactive ores; 3) enrichment installations for radioactive minerals; 4) research and power plant reactors; 5) nuclear ships; 6) industrial firms, hospitals and research laboratories using radioisotopes; 7) engineering applications of atomic energy.

The wastes, produced by these sources, contain different radioisotopes and are discharged into the atmosphere, seas and continental water.

TABLE 1 Mean Annual Dose, Calculated for the Gonads, due to Different Sources of Natural Radioactivity (UK). The Total Dose Varies from 85 to 106 mR (in: Medical Research Council, 1960)

| Source | Dose (mR) |
|---|---|
| Cosmic rays | 24 |
| Soil and rocks radioactivity | 36 - 57 |
| Radon in the air | 1 |
| $^{40}K$ in the body | 21 |
| $^{14}C$ in the body | 1 |
| Radon and decay products in the body | 2 |

Fall-out contamination is extended to the whole biosphere, whereas radioactive discharges always concern small areas. For the present low level of fall-out radioactivity, the principal problem seems to be the wastes produced by nuclear installations for the production of electric power. The importance of these wastes will increase with the future development of nuclear power industry. For example, $^{3}H$ and $^{85}Kr$, discharged by nuclear plants, may become, in the near future, an important problem of radioprotection. Unfortunately, there is inadequate in-

TABLE 2 Mean Values of Environmental Radioactivity Calculated for Different Altitudes and for Different Types of Substrate (in: Alexander, 1965)

| Altitude (msl) | Dose from cosmic rays (mR/year) | Total dose including cosmic rays (mR/year) | | |
|---|---|---|---|---|
| | | Sea | Granite | Sedimentary rocks |
| 0 | 33 | 53 | 143 | 76 |
| 1500 | 40 | - | 150 | 83 |
| 3000 | 80 | - | 190 | 123 |
| 4600 | 160 | - | 270 | 203 |
| 6100 | 300 | - | 414 | 347 |

TABLE 3 Number and Percentage of Humans Wounded and Dead, Calculated on the Total Number of Inhabitants after the Nuclear Explosion at Hiroshima and Nagasaki (in: Oughterson and S. Warren, 1956)

| Town | Number of inhabitants | Number of wounded | % wounded | Number of dead | % dead |
|---|---|---|---|---|---|
| Hiroshima | 255,200 | 72,200 | 28.3 | 64,600 | 25.3 |
| Nagasaki | 195,300 | 25,000 | 12.8 | 39,000 | 20.0 |

TABLE 4 Average Genetically Effective Dose to the Gonads of Population of Great Britain. Dose Rates to the Population from Natural Background and Man-Made Sources (in: Jenkin, 1964)

| Source of radiation | Dose rate as percentage of natural background (%) | Date of calculation |
|---|---|---|
| Natural background | 100.0 | - |
| Diagnostic radiology | 19.0 | 1957 |
| Luminous clocks and watches | 1.0 | 1956 |
| Occupational exposure | 0.5 | 1959 |
| Other causes | 0.1 | 1956 |
| Fall-out | 2.4 | 1955-1959 |

formation about the effects of $^3H$ on organisms and no data about the accumulation of krypton in aquatic biota.

Long and mean half-life radioisotopes are persistent pollutants. Indeed, radioactive decay may be considered a process of degradation because by this process the hazard of radioactive material decreases with time. Radioactive decay is a physical process with a rate completely independent from environmental characteristics whereas degradability is an effect of biological activity and of chemical and physical processes and, consequently, its rate is strictly dependent on the conditions of the environment. In other words, the physical half-life of radioisotopes is constant, whereas that of other pollutants (e.g. several organic substances) is variable. As a consequence, if the persistent radioactive pollutants (radionuclides with half-lives of years) may be separated in an arbitrary way, from those which are not persistent (radionuclides with half-lives of minutes or seconds), it is difficult to know the degradation rate of certain non-radioactive pollutants present in natural environment.

The aim of international and national rules is to prevent the increase of environmental radioactivity and, more precisely, to prevent it exceeding the mean level of natural dose by a factor of two. Although we may suppose that the same rules, applied to the control of radioactive pollution, on the basis of the protection of man, should also preserve the whole community of the ecosystem, the radiation effects on the environment cannot be neglected when considering the existing and especially the future disposal of radioactive wastes into the environment.

The importance of the protection of ecosystems has been emphasized at several meetings and in many scientific papers. In the beginning it was a general rule that the amount of radioactive wastes discharged in an environment should be lower than its dilution capacity, that is than its volume. Subsequently, in order to control more accurately the level of pollution of an ecosystem, the concept of "environmental radiological capacity" was established, that is the maximum amount of radioactive substances that can be discharged into an ecosystem without producing a contamination that could present a danger to man. This concept is based on the relation between the amount of radioactivity taken in with food and that present in the environment in which food is produced. For example, to find a relationship between the radioactivity in food produced in an environment contaminated by radioactive effluents, this general equation may be adopted:

$$(Qa)_i = (Qr)_i \cdot f(K_{H_i}, K_{P_i}, K_{B_i}) \cdot e^{-\lambda_i \Delta t}$$

where: $(Qa)_i$ = amount of radioactivity due to the radioisotope $\underline{i}$ in a certain food $\underline{a}$,

$(Qr)_i$ = amount of radioactivity due to the radioisotope $\underline{i}$ in an effluent $\underline{r}$,

$K_{H_i}, K_{P_i}, K_{B_i}$ = parameters expressing the hydrological, physico-chemical and biological processes acting on the dilution and/or concentration of the radioisotope $\underline{i}$ in the environment,

$e^{-\lambda_i \Delta t}$ = radioactive decay of the radioisotope $\underline{i}$ during the period of its transfer from the effluent to man.

Unfortunately, this concept is not applied to other types of pollution. To adopt the concept of "radiological capacity" in radioprotection the characteristics of the environment and the properties of the radioisotopes discharged in it and their fate through the different compartments of the ecosystem must be known. In other words, radioecological studies must be carried out before permitting any discharge of radioactive material into the environment.

Radioecology is that part of applied ecology concerned with the radioactive pollution of ecosystems.

The fundamental topics of radioecology are the following: 1) distribution of radionuclides in the environment; 2) concentration and turnover rate of radionuclides in organisms; 3) radiation dose rate absorbed by organisms and somatic and genetic effects of ionizing radiations at the level of populations and communities.

This paper concerns one aspect of radioecology that is the biological effects of ionizing radiations.

The total dose rate absorbed by an aquatic organism which is the sum of the dose rates from incorporated radionuclides (internal dose) plus the dose rate from radionuclides present in the water and the sediment is closely related to the behaviour of the organism. For example, the dose rate from water assumes a comparatively great importance for planktonic and nectonic organisms, whereas the dose rate from the sediment influences benthonic organisms. In addition, animals burrowing into the sediments receive from them about double the dose rate as those living on the interface between water and sediment. For example, Folsom and Harley (1957) estimate that organisms living in deep sea sediments receive from 40 to 620 millirads/year in relation to the type of sediment. The dose rate from incorporated radionuclides depends on the amount and distribution of the radionuclides in the organism, the type of radiation of the emitters (alpha, beta and gamma) and the size and shape of the organism.

The absorption coefficient is high for alpha-particles, lower for beta and very low for gamma radiation. As a consequence, if the size of the organism is small (for example, the phytoplanktonic algae) a relevant part of the alpha and beta radiation, emitted by the incorporated radionuclides, is dispersed in the environment and gamma emitters give a very negligible contribution to the total dose rate. Larger organisms, such as worms, molluscs, crustaceans and fish absorb most of the radiation emitted from incorporated alpha and beta radionuclides and a fraction (more or less important depending on the size of the organism) of the gamma radiation.

The dose rate absorbed from radionuclides in the water is, generally, very low in comparison with that from incorporated radionuclides and the latter is equal to or lower than that from sediment. This is true for natural and fall-out radioactivity and for environments contaminated by radioactive wastes.

The dose rate absorbed from fall-out radionuclides varies widely with the environment but the information on this subject is very scarce. It seems that $^{90}Sr$ and $^{137}Cs$ are the most important sources of the dose rate absorbed by the freshwater organisms.

There are few data on direct measurements of the dose rate actually absorbed by aquatic organisms. This information may be obtained by means of thermoluminescent dosimeters attached to the external surface of the body of an aquatic organism, but, unfortunately, this method has been applied only in a few cases. Since direct measurements are practically non-existent, some researchers have tried to calculate the dose rate absorbed by aquatic organisms in relation to their size. The values listed in Table 5 have been reported from the "Technical Report Series No. 172", IAEA (1976).

It is well known that dangerous and lethal effects on plants and animals are produced by high doses of radiation. A very large range of sensitivity to ionizing radiation has been found for different taxonomic groups. In Table 6 are reported the LD-50 (lethal dose for the 50% of the organisms exposed, at the 30th day after treatment) for some terrestrial and aquatic animals. Data on the resistance of some marine animals to ionizing radiations are reported in Table 7. Radiobiological research has been carried out on almost all the animal taxa, from Protozoa to Chordata, but most of the information concerns only a few species.

The organisms have been chosen, obviously, for the facility with which they can be reared and because their physiological and pathological characteristics are similar to those of man. For this reason, the animals most often used are the ones traditionally bred in the laboratory: mouse, rat, guinea-pig and rabbit.

Aside from the scientific interest of knowing the effects of irradiation on a greatest possible number of species, it is useful to get a more complete information on those invertebrates that have a considerable importance in the biosphere and, then, indirectly to man.

Bacteria (and particularly their spores), fungi and blue green algae are more resistant than other plants and animals. Algae are more tolerant than higher plants and animals. Invertebrates have a very high tolerance to ionizing radiations as compared with that of vertebrates. Photosynthesis and other functions of the algae are less severely influenced than reproduction. Cell growth without division produces "giant cells". These abnormally large cells due to the arrest of reproduction, have also been observed in algal cultures contaminated with low doses of heavy metals (e.g., Premazzi and co-workers, in press).

From the results of our experiments on Biomphalaria glabrata it seems that the radioresistance of this fresh-water snail as in other taxa, for instance the mammals, varies not only with the species but also with the variety; indeed, the "normal" animals appeared more radioresistant than the "albino". Actually, this is true only for the adult stage, while the radiosensitivity of the embryo is not significantly different between the two varieties. The adult of the "normal variety is easily distinguishable from the "albino" because its mantle is dark as a result of accumulation of melanic pigments which are absent in the "albino" variety. The embryos of both varieties do not differ in mantle which is always very light. It could be that the higher radioresistance of the "normal" variety, which is evident only for the adults, is an effect of some biochemical mechanism involved in

TABLE 5 Summary of Dose Rates (urad.$h^{-1}$) to Freshwater Organisms from Environmental Radioactivity (in: IAEA, 1976)

| Source | Phytoplankton 1 m depth; 1 m from river bed | Zooplankton 1 m depth; 1 m from river bed | Mollusca 2 m depth on river bed | Crustacea 2 m depth on river bed | Fish 1 m depth; 1 m from riverbed | Fish 2 m depth on river bed |
|---|---|---|---|---|---|---|
| Natural Background | | | | | | |
| Cosmic radiation | 2.7 | 2.7 | 2.2 | 2.2 | 2.7 | 2.2 |
| Internal activity | NDA | NDA | NDA | NDA | 3.6-4.8 | 3.6-4.8 |
| Water activity | $7.4x10^{-3}$-6.2 | $9.8x10^{-4}$-0.85 | $4.2x10^{-4}$-0.35 | $4.2x10^{-4}$-0.35 | $8.3x10^{-4}$-0.70 | $4.2x10^{-4}$-0.35 |
| Sediment activity | - | - | 1.5-16.0 | 1.5-16.0 | - | 1.5-16.0 |
| | - | - | 1.6-21.0 | 1.6-21.0 | - | 1.6-21.0 |
| Total(*) | 2.7-8.9 | 2.7-3.6 | 3.7-18.6 | 3.7-18.6 | 6.3-8.2 | 7.3-23.4 |
| Fall-out | | | | | | |
| Internal activity | NDA | NDA | $1.6x10^{-2}$-0.15 | NDA | 0.12-26.4 | 0.12-26.4 |
| Water activity | 0.18-0.53 | 0.13-0.40 | $6.2x10^{-2}$-0.13 | $6.2x10^{-2}$-0.13 | 0.12-0.26 | $6.2x10^{-2}$-0.13 |
| Sediment activity | - | - | 0.16-5.8 | 0.16-5.8 | - | 0.16-5.8 |
| | - | - | $9.8x10^{-2}$-3.5 | $9.8x10^{-2}$-3.5 | - | $9.8x10^{-2}$-3.5 |
| Total(*) | 0.18-0.53 | 0.13-0.40 | 0.24-6.1 | 0.22-5.9 | 0.24-26.7 | 0.34-32.3 |
| Waste Disposal | | | | | | |
| Columbia River | | | | | | |
| Internal activity | 195 | $1.5x10^{4}$ | $2.4x10^{4}$ | - | $4.8-2.1x10^{3}$ | $4.8-2.1x10^{3}$ |
| Water activity | 0.55-2.6 | 0.55-2.4 | 0.27-1.1 | 0.27-1.1 | 0.54-2.3 | 0.27-1.1 |
| Sediment activity | - | - | 860 | 860 | - | 860 |
| | - | - | 29 | 29 | - | 29 |
| Total(*) | 196-198 | $1.5x10^{4}$ | $2.5x10^{4}$ | 860 | $5.3-2.1x10^{3}$ | $870-3.0x10^{3}$ |
| River Danube | | | | | | |
| Internal activity | - | - | - | - | $6.0x10^{-3}$-0.13 | $6.0x10^{-3}$-0.13 |
| Water activity | - | - | - | - | $6.3x10^{-4}$ | $3.2x10^{-4}$ |
| Sediment activity | - | - | - | - | - | $6.7x10^{-2}$-2.23 |
| | - | - | - | - | - | - |
| Total(*) | - | - | - | - | $7.0x10^{-3}$-0.13 | $7.3x10^{-2}$-2.36 |

(*) Excluding the contribution due to beta-radiation from the sediment

TABLE 6 Lethal Dose (LD-50) Expressed in R and Calculated at the 30th Day for Different Animals Exposed to X-rays (in: Alexander, 1965)

| Animal | LD-50 |
|---|---|
| Guinea pig | 250 |
| Dog | 335 |
| Pig | 350 |
| Goat | 350 |
| Monkey | 600 |
| Burro | 580-780 |
| Man(*) | 650 |
| Mouse | 500-700 |
| Rat | 550-800 |
| Rabbit | 750 |
| Chicken | 600-800 |
| Goldfish | 670 |
| Frog | 700 |
| Tortoise | 1500 |
| Newt | 3000 |
| Snail | 8000-20000 |

(*) Various authors have calculated LD-50 for man ranging from 400 to 500 R. This difference is probably due to the age and health conditions of the examined subject.

the synthesis of the melanic pigments (Ravera and co-workers, 1969).

Different tissues of the same individual have a different resistance to the same dose. Germ and embryonic cells, which have a very high mitotic activity, are very sensitive to radiation compared with the somatic cells of the same species. In addition, embryos of some species (e.g. *Rana*, *Fundulus*, *Onchorhyncus*, *Drosophila*, *Lymnaea* and *Physa*) increase their resistance to ionizing radiation with age.

As an example of approach to evaluate the influence of radiation at population level, we have measured the value of the radiosensitivity of a fresh-water snail (*Physa acuta*, Drap.) at all stages of its development from egg to adult (Ravera, 1966). From our results it is clear that high doses may reduce the life-span of the adult (e.g. doses higher than 10,000 R) but comparatively low irradiation increases embryo mortality and kills germ cells (Figs. 1,2). Egg viability is decreased by irradiating the adult with an acute dose of 2,000 R, and about 1,000 R increases the mortality of the embryo at 4-blastomers to 50%. The irradiation-induced delay in development, which is particularly evident during the younger stages, may increase the frequency of death through other causes, such as predation, parasitism, algal coating of the capsule, etc. Because all the individuals

TABLE 7 Relative Tolerance of Some Marine Animals to Ionizing Radiation (in: Chipman, 1972)

| Group and Species | LD-50 (30 d) in Röntgen | Authors |
|---|---|---|
| Fishes (juvenile or postlarval) | | |
| Micropogon undulatus | 1,050 | White and Angelovic (1966) |
| Fundulus heteroclitus | 1,120 | " " " " |
| Mugil cephalus | 1,450 | " " " " |
| Lagodon rhomboides | 3,000 | " " " " |
| Eucinostomus sp. | 3,500 | " " " " |
| Paralichthys lethostigma | 5,550 | " " " " |
| Crustaceans | | |
| Palaemonetes pugio | 1,500 | Rees (1962) |
| Gammarus salinus | 1,700 | Hoppenheit (1969) |
| Gammarus zaddachi | 1,700 | " " |
| Gammarus duebeni (females) | 3,500 | " " |
| " " (males) | 3,900 | " " |
| Uca pugnax | 8,000 | Rees (1962) |
| Callinectes sapidus | 56,600 | Engel (1967) |
| Calliopius laeviusculus | 10,000* | Bonham and Palumbo (1951) |
| Allorchestis angustus | 10,000* | " " " " |
| Artemia salina | 14,450** | White and co-workers (1966) |
| Echinoderms | | |
| Arbacia punctulata | 38,900 | White and Angelovic (1966) |
| Molluscs | | |
| Thais lamellosa | 20,000° | Bonham and Palumbo (1951) |
| Urosalpinx cinerea | 38,000° | Hargis and co-workers (1957) |
| Nassarius obsoletus | 37,500 | White and Angelovic (1966) |
| Crassostrea virginica | 99,000 | Price (1965) |
| Mercenaria mercenaria | 109,000 | " " |

*LD-50 5 weeks; **LD-50 25 days for mixed adult population (13,000 rads); ° LD-50 30 days calculated by White and Angelovic (1966)

were exposed to very high doses, it seems improbable that the surviving embryos have greater radioresistance than the ones that died. The reason why some irradiated individuals live and others die is probably due to chance. Radioresistance of development stages has also been studied in some salmonoid species (e.g. Wadley and co-workers, 1971; Welander, 1954; Welander and co-workers, 1948). Radioresistance does not necessarily imply a greater resistance to other environmental stresses. Thus, irradiation, which increases mortality and reduces the size of the population, does not select the fittest individuals in the environment

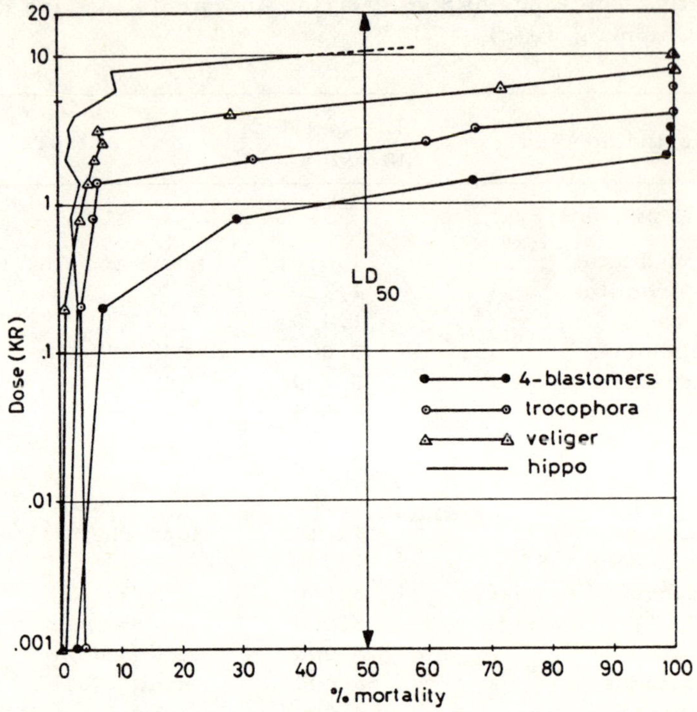


Fig. 1. Relation between dose and mortality of embryo. The broken segment of the curve for hippo indicates the LD-50 at this stage (in: Ravera, 1966)

in which they live. From a radioecological point of view, the recovery of germ tissues is particularly interesting. The development of mature functional germ cells, following a period of temporary sterility, permits the survivors, irradiated with doses up to 10,000 R, to spawn fertile eggs and, consequently, to repair the damage produced in the irradiated population (Table 8).

The effects of irradiation on a population cannot be independent of environmental characteristics. For instance, it is generally recognized that the damage induced by irradiation is less if the animals are kept at a low temperature soon after irradiation. The reason for this protective effect seems due to the reduction of metabolic processes with decreasing temperature (Gros and co-workers, 1957). We have observed that temperature amplifies the effects of irradiation on the reproduction of Physa; for instance, increasing the temperature from 20°C to 30°C results in a marked decrease of fecundity (Ravera, 1967). The duration of embryonic development, considered as a whole as well as for each stage, is increased by lowering the temperature. For instance, eggs of Physa acuta develop twice as slowly at 20°C as at 30°C. Thus, higher temperature, if it amplifies the effects of irradiation, reduces at the same time the period of higher radiosensitivity (embryonic stage). Since, as shown in a previous work (Ravera, 1967), the reproductive cycle depends more upon the season than upon the temperature and since radiosensitivity decreases with increasing age, damages produced in a population will be strongly influenced by the season in which irradiation is delivered.

In conclusion, our data give a general picture of the effects of X-irradiation on the demographic characteristics of Physa acuta. However, to extrapolate our results to natural population it would be necessary to study the following points: (1) effects of chronic irradiation on demography; (2) role played by the most important ecological factors (e.g. temperature, oxygen-concentration, food) on the radiosensitivity of the population; (3) influence of irradiation on successive generations.

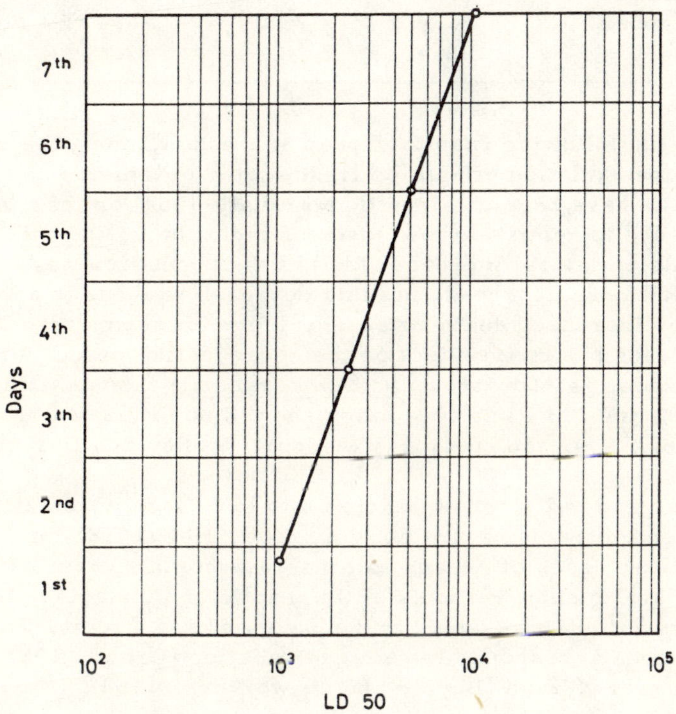


Fig. 2. Correlation between the log LD-50 and the age at which embryos were irradiated (in: Ravera, 1966)

To know and predict the biological effects produced by the discharge of radioactive waste into the environment, we need more information on the chronic effects from doses so low as to be comparable with those measured in natural ecosystems.

In recent years a very rich information on the biological effects produced by ionizing radiation has been collected. The literature dealing with the effects on aquatic organisms is also abundant, but most of the studies have been carried out with high and acute doses.

Ionizing radiation produces "somatic effects" on organisms and "genetic effects" on the progeny of irradiated individuals. The exposure may be acute or chronic, the latter being continuous or fractioned. In the natural environment the organisms are commonly submitted to long-term continuous exposure at low doses.

TABLE 8 Viability Expressed as Percent of Hatchers (in: Ravera, 1966)

| Month | Control | 2 kR | 10 kR |
|---|---|---|---|
| October | 97.0 | 40.9 | 0.0 |
| November | 95.2 | 51.1 | 2.1 |
| December | 86.9 | 55.7 | 56.4 |
| January | 98.4 | 87.4 | 87.6 |

As an example, the following results chosen at random among the rich information concerning the radiation effects on fresh-water organisms are reported. Interesting results have been obtained by Marshall (1962) on the effects due to long-term exposure to relatively low doses of radiation on Daphnia pulex. The decrease of the rate of natural increase (RNI) of the population was inversely correlated with the total dose absorbed, but this decrease was due to a reduced birth rate and not to an increased death rate. Therefore, it seems that the exposure to chronic low doses has more effect on the reproductive processes than on the longevity of Daphnia. Rainbow trout (Salmo gairdnerii) exposed to an acute dose of X-rays (21, 83 and 203 R) at eyed stage showed no difference in survival, growth or fecundity from the control (Welander, Wadley and Dysart, 1971).

Some organisms, generally, show reactions to low doses of ionizing radiation, but the perception mechanisms are not yet known. Hug (1958) and Born (1960) observed that at low doses of X- and gamma-rays some species of terrestrial and fresh-water snails show reflexes of the mantle and tentacles. Hug (1960) described evident reactions in mussel, anthozoans and sea-urchin. The Planaria Dugesia dorotocephala responds to gamma-radiation (Brown, 1963) and Daphnia magna may perceive X-rays (Bayler and co-workers, 1958).

It seems that 100 Rad (acute dose) increases the death rate of a population, but this dose is not sufficient to eliminate the whole population. Chronic radiation of the same dose, generally, produces a smaller effect.

At a dose of one Rad/day the biological effects are very difficult to detect (IAEA, 1976).

The protection of man is exerted at individual levels as well as at the population level. Indeed, great value is attributed to each member of human society. In fact, today a significant part of the population (at least in developed countries) is maintained in the population by means of medical care. The same criteria cannot be adopted for the protection of animal and plant populations and, consequently, the value of a single individual varies with the size and turnover-time of the population to which it belongs.

As a consequence the knowledge of the effects of pollution (radioactive and not radioactive) at population level is the fundamental background for protection of the biota.

Quantitative information on the effects produced by radiation on the natural population of aquatic organisms is very scarce when compared to the data concerning the effects on individuals tested under laboratory conditions. In addition, there is a lack of data about the mechanisms regulating the population density in freshwater environment. Most of the studies on this subject have been carried out on commercially exploited populations of marine fishes, but these investigations were undertaken from the point of view of fishing and not of radiation effects.

From the effects on morphological structures, physiological functions, enzymatic systems and chromosomes, produced from ionizing radiations, it is very difficult to foresee the ultimate consequences at population level. It is nearly impossible to extrapolate from the effects on individuals those produced by the same dose on the population to which the same individuals belong.

Blaylock (1966) examined for 5 years the larvae of Chironomus tentans (Insecta) living in contaminated sediments of White Oak Creek and White Oak Lake (Tennessee). The larvae, exposed to dose rates 1000 times (230-240 R/year) the background of the area, showed an increase in the frequency of chromosomal aberrations. This relevant genetic damage could not have any effects on the population because the individuals bearing aberrant chromosomes were eliminated by natural selection.

Indeed there are some properties of the population not displayed by single individuals; for example, birth rate, spatial and type of distribution, density.

A population is not a simple aggregate of individuals of the same species, chosen at random, but may be considered as a sample of individuals for a given species differentiated by selecting mechanisms and then adapted to survive in the environment in which it lives.

Relevant modifications of the properties of the ecosystem (in hydrological and climatic characteristics as well as for immigration or emigration of one or more species) influence the population which reacts to the new situation. As a result, a group of individuals, with a relatively high adaptive value to the new situation, is selected from the population and the others are eliminated. At the beginning the population size is reduced and if this reduction is compatible with the survival of the population, it may reach and surpass its initial density after a certain time. In conclusion, whatever the nature of the stress, if it is not too strong, a part of the population, numerically more or less important, survives.

Ionizing radiations may thus be considered as an additional stress that may or may not be tolerated by a population which will react by means of the same defence mechanisms that populations have against natural stress. The survival of a part of the population despite the stress from radiation is an important point of discussion and its keyword is "a part". In the first place, it is impossible to fix a value or a range to "a part", because "ceteris paribus" it varies with the species population and the ecosystem. Now, we will discuss only the probability of survival of populations living in a contaminated environment.

Species with high fecundity may support a large reduction in the size of their populations without being eliminated, whereas a relatively small numerical reduction may cause the extinction of populations belonging to a species with low fecundity. As a consequence, the effects produced by the same dose rate, eliminating the same percentage of individuals from populations of the same size, but with different rates of fecundity, will be very different. The number of young individuals recruited at each generation depends on: (1) the number of eggs (or newborn) laid and; (2) the viability of eggs and larvae, that is their survival rate. In species with high fecundity the number of eggs laid by the female is very large (for example, insects) and often one female spawns several times during life (for example fish). It is known that if there is no limiting factor the numerical growth of a population with high fecundity follows an exponential curve. After a certain period of numerical growth, the exponential curve is soon broken and changed into the more general curve of population growth, the logistic curve. This change is due to two types of factors: external, such as predators, diseases, parasites; and internal, such as intraspecific competition. Examples of exponential curves are given by laboratory cultures under constant conditions and by the immigration of populations into an environment with optimal conditions but with negligible or no competition (e. g. the growth of rabbit populations in Australia, the diffusion of the macrophyte Elodea canadensi in European fresh-waters).

Theoretically, only two newborns from the total number produced by a couple of animals have to survive to maintain the size of the population in time. The attempts to establish a relation between the number of parents and the size of the recruitment in species with high fecundity have had little success. For species with high fecundity it seems that the size of the recruitment is, within a certain range, independent of the number of seasonally mature individuals. If the survival of the newborn depends on the amount of food available, the size of the recruitment will vary more with food production than with the number of parents. The annual variations of the recruitment of the species of fishes feeding on zooplankton is, probably, an example of this mortality independent of the number of newborn. If this is true, the recruitment of these fishes depends on the recruitment of zooplankters eaten by them, and this is controlled by the productivity of the species of algae used as food by the zooplankton. In the abovementioned considerations we have not taken into account the effects of radiation on adults but only those on the eggs and larvae, which are the most sensitive stages for several species of aquatic organisms.

If the number of eggs or newborns is reduced for any reason (e. g. climatic stress, dose rate) the competition for food will be less strong and the probability that the survivors will attain sexual maturity will increase. In this case the mortality of the eggs and newborns due to dose rate absorbs a part of the mortality caused by the competition for food and, therefore, the size of the recruitment cannot be seriously affected.

Data from studies carried out on some species of marine fishes (IAEA, 1976), commercially exploited molluscs (Hancock, 1973) and Daphnia (Marshall, 1967) support this hypothesis.

It is not known what the minimum value of birth rate to maintain the size of population constant in time is, but it is evident that it varies with the strength of the selection exerted by means of competition and physical factors of the environment.

It is obvious that, if mortality is independent of the density of the young cohort, the total mortality will be the result of natural mortality plus the mortality produced by radiation. In this case, and when the mortality due to radiation exceeds that arising from the population density, it is less probable that the population will maintain its size. As a consequence, if this situation is continued for a sufficient number of generations (e.g. in a chronically contaminated ecosystem), the population will be eliminated from the environment.

There is a relation between the number of parents and that of newborns for species with low fecundity (e.g. whales) and this relation will be closed with the decreasing of fertility. In these species, natural selection is not very severe on the newborns and also a small decrease of their number caused by radiation or other agents may greatly reduce the survival of the population.

Previously, the probable effects of radiation on the recruitment of young animals have been considered; now, the probable influence of radiation on the recruited individuals will be discussed. An example may be given by some species of fish with high fecundity and heavily exploited (IAEA, 1976). The annual rate of fishing for these species is about 50% on all year-classes. In addition, the decrease in population due to natural death is about 10-20%. As a result, the total percentage of individuals eliminated annually from the population reaches 60-70%. This reduction does not significantly affect population size in successive years. As a consequence, it means that this reduction is compatible with the survival of the population but a higher level of exploitation of the stock would, probably, eliminate the species. If a species has low fecundity the reduction of the population compatible with its survival would obviously be smaller.

Conclusions similar to those reported for marine fish have been reported for a fresh-water Cladoceran (*Daphnia pulicaria*), which increases its medium life expectancy with increasing exploitation. Results from experiments carried out by Hoppenheit (1975) on a marine copepod (*Tisbe holothuriae*) show that the frequency of egg-sack production, the mean number of eggs per sack and the viability of the young ones, increase with increasing exploitation. The same author (1978) states that the decrease of population density of *T. holothuriae* after the addition of 148 or 222 ppb $Cd^{++}$ is lower at higher exploitation rates.

From these considerations we conclude that the density-dependent regulating mechanisms control the size of the population, which react to changes in its environment. As a result a population may survive and overcome reductions of its size down to a given level, fixed by its capacity of recovery. If this capacity cannot balance the effects of the reduction, the population is eliminated.

Applying these considerations to the radiation effects we may predict that dose

rate will wipe out those populations with a capacity of recovery lower than the number of individuals eliminated directly or indirectly by the radiation. Since the capacity of recovery varies with the species, the same dose rate will affect some species more strongly than others. This selection is not a result of the different sensitivity of different species to a given dose rate, but is due to demographic characteristics of the population.
Therefore, it is easy to demonstrate that knowledge on the effects of radiation at an individual level is an insufficient basis for the prediction of the effects of pollution at population level.

It seems that the complexity of the community is an important factor in the effects of radiation. The effects of radiation "ceteris paribus" seem to be smaller for very complicated food-webs than for simpler ones. Indeed, if in a community with a high degree of "diversity" one species with low resistance to pollutants is eliminated, its interactions with other species may be more easily substituted by other interactions. As a consequence, the structure and stability of the community, normally, is not changed. In contrast, in a more simple community the elimination of one species may create a series of variations in the whole structure of the community.

In different fresh-water communities with the same dose rate, the effects will be more dangerous in those with a lower "diversity" and degree of stability: that is in eutrophicated environments in comparison with others with a lower level of trophy. For this reason, the same dose rate has a greater influence on fresh-water communities than on marine ones; indeed, the number of species in marine ecosystems is very high when compared with the number of fresh-water species. In addition, the marine ecosystem is favoured by its high capacity of dilution (resulting from its great volume, as well as from the currents and tides) and by the great possibility for organisms to emigrate from contaminated zones to non-contaminated ones. Indeed, the open sea, up to now, has been unaffected by chemical pollution and the large modifications which have occurred are produced only by climatic changes (IAEA, 1976).

It is known that in natural populations a large genetic variability is stored in the heterozygous state and some of these genetic characteristics are harmful in the homozygous conditions. The degree of variability is controlled by various factors, for example, the size of population and the environmental characteristics. A high variability allows the population to adapt to the environmental variations, to extend its distribution area and to have several advantages in its competition with other populations with lower variability. Its fitness is the result of selection, exerted by the ecosystem on the genotypes. The level of fitness of a single individual or population is demonstrated by its capacity to survive and reproduce compared to other individuals and populations. Ionizing radiations (such as several pollutants) give rise to mutations and these,by increasing the natural rate of mutation, decrease its fitness. In natural populations the frequency of unfavourable genetic characteristics is lowered by selective processes, whereas favourable ones attain a high frequency. It is true that artificially induced mutations are frequently detrimental, but selective mechanisms exert an influence on these mutations also. If the detrimental effects produced by ionizing radiations are evident on a single generation, they are called short-term genetic effects which produce sterility and death. As a consequence, these effects, if their frequency is high, may reduce the size of the population but through mortality and sterility eliminate abnormal individuals and prevent the accumulation of unfavourable genetic characteristics within the population. Long-term effects are more dangerous at population level because they occur in several successive generations in the homozygous state as malformations and "genetic diseases". These effects are produced by the increase of deleterious recessive mutations and become evident with the increase of recessive alleles in successive generations. These considerations are generally accepted by researchers involved in this field, but some do not agree with this point of view (e.g. Crenshaw, 1965; Wallace, 1957). These authors have shown that, in some cases, a small increase of the mutation rate and variability may increase the fitness of populations of Drosophila (Diptera) and Tribolium (Coleoptera).

In spite of the paucity of information, on the basis of the results of the studies carried out so far, until now it seems that the present level of environmental contamination is too low to produce serious effects on aquatic biota.

## REFERENCES

Alexander, P. (1965). Atomic Radiation and Life. Penguin Books Ltd., Harmondsworth.

Blaylock, B. G. (1966). Cytogenetic study of natural populations of Chironomus inhabiting an area contaminated by radioactive waste. In Disposal of Radioactive Wastes into Seas, Ocean and Surface Waters (Proc. Symp., Vienna). IAEA, Vienna, p 835.

Bonham, K. and R. Palumbo (1951). Effects of X-rays on snails, Crustacea and algae. Growth, 15, 155-188.

Born, W. (1960). Zur Auslösung von Reflexen bei Schnecken durch Röntgen- und Alphastrahlen. Strahlentherapie, 112, 634-636.

Brown, F.A., Jr. (1963). An orientation response to weak gamma-radiation. Biol. Bull. mar. biol. Lab. Woods Hole, 125, 206-225.

Chipman, W.A. (1972). Ionizing radiation. In O. Kinne (Ed.) Marine ecology, Vol. 1, John Wiley and Sons, London. pp. 1579-1657.

Commissione delle Comunità Europee (1969). Principi e metodologia generale per la determinazione della capacità radiologica limite di una rete idrologica. Commissione delle Comunità Europee, Direzione Protezione Sanitaria, Bruxelles.

Engel, D.W. (1967). Effects of single and continuous exposures to -radiation on the survival and growth of the blue crab, Callinectes sapidus. Radiat. Res., 32, 685-691.

Folsom, T.R. and J.H. Harley (1957). Comparison of some natural radiations received by selected organisms. In The Effects of Atomic Radiations on Oceanography and Fisheries, U.S. National Academy of Sciences. (Publs. natn. Res. Coun., Washington, 551, 28-33).

Gros, C.M. and J. Bloch (1957). Influence de la température sur la survie d'un poisson, Carassius carassius (L.), après irradiation totale. C.R. Soc. Biol., 151, 602.

Hancock, D.A. (1973). The relationship between stock and recruitment in exploited Vertebrates. Rapp. P.-v. Réun. Cons. Int. Explor. Mer.,

164, 113.

Hargis, W.J., M.F. Arrighi, R.W. Ramsey, and R. Williams (1957). Some effects of high frequency X-rays on the oyster drill, Urosalpinx cinera. Proc. natn. Shellfish. Ass., 47, 68-72.

Hoppenheit, M. (1969). Strahlenbiologische Untersuchungen an Gammariden (Crustacea, Amphipoda). Helgoländer wiss. Meeresunters., 19, 163-204.

Hoppenheit, M. (1978). On the dynamics of exploited populations of Tisbe holothuriae (Copepoda, Harpacticoida). IV. Response to perturbation by temporarily elevated mortality. Helgoländer wiss. Meeresunters., 31, 295-297.

Hug, O.D. (1958). Die Auslösung von Fuhlerreflexen bei Schnecken durch Röntgen- und Alphastrahlen, Strahlentherapie, 106, 155-156.

Hug, O.D. (1960). Reflex-like responses of lower animals and mammalian organs to ionizing radiation. In Immediate and Low Level Effects of Ionizing Radiations. Int. J. Radiat. Biol. (spec. Suppl.), 1960, 217-224.

International Atomic Energy Agency (1976). Effects of Ionizing Radiations on Aquatic Organisms and Ecosystems, Technical Report Series No. 172, IAEA, Vienna.

Jenkins, E.N. (1964). An Introduction to Radioactivity, Butterworths, London.

Marshall, J.S. (1962). The effects of continuous gamma radiation on the intrinsic rate of natural increase of Daphnia pulex. Ecology, 43, 598-607.

Marshall, J.S. (1967). Radiation stress in exploited Daphnia populations, Limnol. Oceanogr., 12, 154

Medical Research Council, H. M. S. O. (1956). The Hazards to Man of Nuclear and Allied Radiations, I. Report.

Medical Research Council, H. M. S. O. (1960). The Hazards to Man of Nuclear and Allied Radiations, II. Report.

Oughterson, A.W. and S. Warren (1956). Medical Effects of the Atomic Bomb in Japan, McGraw-Hill, New York.

Premazzi, G., R. Bertone, A. Freddi and O. Ravera (1978). Combined effects of heavy metals and chelating substances on Selenastrum cultures. In Proc. Seminar Ecological tests relevant to the implementation of proposed regulations concerning environmental chemicals: evalua- and research needs, Berlin, Dec. 7-9, 1977 (in press).

Price, T.J. (1965). Accumulation and retention of radionuclides and the effects of external radiation on molluscs; LD-50 determinations of oysters and clams. In Annual Report, Radiobiological Laboratory, U.S. Fish & Wildlife Service (= Circular 204), 1963, 13-14.

Ravera, O. (1966). Effects of X-irradiation on various stages of the life cycle of Physa acuta, Draparnaud, a freshwater Gastropod. In Disposal of Radioactive Wastes into Seas, Oceans and Surface Waters. IAEA, Vienna, pp. 799-808.

Ravera, O. (1967). The effects of X-rays on the demographic characteristics of Physa acuta (Gastropoda: Basommatophora). Malacologia, 5, 95-109.

Ravera, O. and R. de Bernardi (1969). Effects of gamma and X-irradiation on the variety normal and albino of Biomphalaria glabrata, Say (Gastropoda, Pulmonata). In Actes du Symposium International de Radioecologie. Centre d'Etude Nucléaires de Cadarache, pp. 1137-1150.

Reefs, G.H. (1962). Effects of gamma radiation on two decapod crustaceans, Palaemonetes pugio and Uca pugnax. Chesapeake Sci., 3, 29-34.

Wadley, G. W. and A. D. Welander (1971). X-rays and temperature: combined effects on mortality and growth of salmon embryos. Trans. Amer. Fish. Soc., 100, 267-275.

Welander, A. D. (1954). Some effects of X-irradiation of different embryonic stages of the trout (Salmo gairdnerii). Growth, 28, 227-255

Welander, A. D., R. F. Donaldson, K. Bonham and A. H. Seymour (1948). The effects of röntgen rays on the embryos and larvae of the chinook salmon. Growth, 12, 203-242.

Welander, A. D., G. W. Wadley and D. K. Dysart. (1971). Growth and fecundity of rainbow trout (Salmo gairdneri) exposed to single sublethal doses of X-rays during the eyed embryo stage. J. Fish. Res. Bd. of Canada, 28, 1181-1184.

White, J. C., Jr. and J. W. Angelovic (1966). Tolerance of several marine species to $^{60}Co$ irradiation. Chesapeake Sci., 7, 36-39.

White, J. C., Jr., J. W. Angelovic, D. W. Engel and E. M. Davis (1966). Interactions of radiation, salinity and temperature on estuarine organisms. In Annual Report to the Atomic Energy Commission, August 1, 1966. Radiobiology laboratory, U. S. Fish & Wildlife Service, Beaufort, N. C. pp. 76-93.

# Methodology of Waste Water Treatment

**F. Malz**

*Essen, F.R.G.*

ABSTRACT

The paper will cover chemical and technical aspects of the treatment of domestic sewage and of industrial effluents, the combined treatment of domestic and industrial wastes and the problems connected on the interface between air pollution control and sewage treatment. In connection with domestic treatment plants the common used types are discussed. Special attention is given to the elimination of phosphorus. The interaction of the reuse of water and the application of unit operations to treat industrial wastes is shown in connection with the univers model of the cyclus of water in natur. The air pollution control by using wet technics leads to new problems in waste treatment. Examples to find solutions are demonstrated.

Keywords: domestic sewage treatment, industrial effluent treatment, air pollution control.

## INTRODUCTION

In this paper three topics
- treatment of domestic sewage
- treatment of industrial effluents
- interface air pollution control and sewage treatment

are covered to indicate the necessary correlations in this matter and to highlight, by means of a few concrete examples, the state of the art and the trend of future developments in the field of waste water control, more especially from chemico-technical points of view.

Natural waters are to be maintained or conditioned so as to preserve ecological equilibria, serve to health and wellbeing of people, and ensure a satisfactory water supply for the inhabitants and industry.

The aims of natural water purification are:
- preserving the state of as yet unpolluted waters
- improving the state of already polluted waters
- protecting waters against dangers and harm arising out of the storage, transport and use of substances liable to endanger water.

The objects of the management of existing waters are:
- increasing the amount of usable water on offer

- comprehensive control of river areas
- inclusion of environment-related social, economico-political and ecological factors in planning and decisionmaking processes.

The objects also include the intensification of research and training. The projected aims refer to the development of scientific foundations of preventative and protective measures for the supply and treatment of natural and waste water, and the improvement of technical measurement and control methods.

TREATMENT OF DOMESTIC SEWAGE

We shall distinguish between the forms of treatment of communal sewage on the one hand and industrial waste waters on the other, although the application of certain basic operations is possible and successful in both areas. The joint treatment of domestic sewage and industrial effluents in regions of large industrial agglomerations is also likely to be of special interst because this sort of combination will become more marked in future types of settlement.

The technology in the field of communal waste water treatment continues to be governed by classical methods. In a flow-sheet the purification is divided into
- grit removal
- preliminary treatment
- biological stage
- final clarification
- sludge digestion
- drying beds

The choice of treatment processes in the biological stages, i.e. whether trickling filters or activated sludge is used, which many years ago was clearly related to the number of inhabitants in the community, has become much more flexible. In the past trickling filter installations were generally used for smaller communities and activated sludge installations for major cities. In a number of cases the combination of trickling filters with secondary aeration as the polishing stage has proved eminently useful in a number of sewage plants for medium sized towns.

Primary sludge and excess activated sludge is treated anaerobically in digestion tanks which work exceptionally reliably if the material is well circulated at a controlled temperature of about 32 - 35 °C. In many cases the energy of the digester-gas can be utilised to cheapen at least the power costs of the whole of the treatment process. The digested sludge is mainly dewatered on drying beds. Storing the sludge still continuies to be a special problem. Work goes on in many places on either incinerating the sludge (e.g. together with domestic refuse) or on using it in agricultural areas as part of raw materials recycling as soil conditioners or fertilisers, always provided the sludge is mainly derived from communal sewage and contains no undue proportion of trade sludge and heavy metals.
Performance characteristics for the biological purification of communal waste waters are fixed by the standard values for waste water are fixed by the standard values for waste water purification methods. They specify that discharged, purified waste water must,

in accordance with the present state of the art, have values below 25 mg/l in the 5-day BOD and it must (after 2 h) not contain more than o,3 ml/l settleable matter. Well maintained sewage plants which are not overloaded generally have a 5-day BOD in the region of 15 mg/l, even when their retention times in the activation tank are about 1 1/2 - 2 h. In most parts of West Germany this purification is effective enough as far as the 5-day BOD is concerned. To meet an average of 2o mg BOD/l the ATV-guidelines recommend a BOD load of 1,o kg BOD/$m^3$ . d, a sludge load of o,3 kg BOD/kg TS and a load in volume of 5 $m^3/m^3$ . d.

The construction and operating costs of communal sewage plants are greatly affected by local factors. In the building costs it is necessary to observe that roughly 25 % of the overall costs are attributable to waste water and 75 % to the sewerage system. The costs of operation itself are mainly determined by the size of the plant, the number of inhabitants it serves or the amount of incoming waste water and the required degree of purification. The trend shows that centralisation in larger units leads to lower costs.

The incorporation of unit operations of advanced treatment in West Germany is connected with the solution of the eutrophication problem. The increasing load of inorganic nutrient salts in our lakes and impounding reservoirs has in many cases led to rapid and progressive eutrophication resulting in the mass development of algae associated with vegetable colouration, and that for several months every year this even involved the occurrence of anaerobic conditions at the bottom of lakes up to shallow water particularly for potable water supply but also for recreation, and this should be prevented. The aim of these efforts must therefore be the limitation of the supply of nutrients so that the equilibrium between production and decomposition processes in lakes and reservoirs is substantially maintained in accordance with the morphological conditions of these waters, and also that the long-term process of normal changes in the lakes shall not turn into accelerated eutrophication through the avoidable and excessive introduction of communal, industrial and agricultural waste waters.

The maeasures against progressive eutrophication consist in keeping the concentration of phosphorus in lakes and reservoirs as low as possible. This is best achieved by substantially inhibiting the supply of phosphorus to the lake. These considerations also apply to stagnant and very slowly running waters. However, in most rivers the light rather than the nutrients is the more likely limiting factor: The light may be affected by the turbidity of mineral matter churned up by turbulent water. Another limiting factor may be harmful substances discharged into the river. Here a given nutrient content, which would enable a certain number of algae to develop, might be desirable because they would favour the decomposition of organic substances and promote self-purification. It is therefore considered unnecessary to take special steps to eliminate phosphorus in such river catchment aereas.

Measures to limit eutrophication should be controlled and confined to those areas where such problems actually exist, i.e. mainly the catchment areas of lakes and reservoirs.

The phosphorus load is mainly derived from:
- definite sources
  communal and industrial waste waters
  waste waters from agricultural establishments
- disseminated sources
  erosion waters and elutriation, especially from areas in agricultural use

Before starting restorative measures, it is important to determine by controlled studies which of these two phosphorus sources may be predominant in a given catchment area. The countermeasures must then be taken in the light of the results of these studies.

If some stagnant water is especially burdened by the waste waters of riparian communities while there is only a limited nutrient load from the sources in the wider catchment area which we have called 'disseminated' and which can therefore be tolerated, there is the possibility of keeping the waste water away by building a sewer along the edge or possibly an intercepting sewer which would not only keep away the nutrients but also all other constituents of the waste water. This is most advantageous from the sanitary point of view, quite apart from a subtantial limitation of an otherwise rapidly progressing eutrophication. The effectiveness of these measures can be deduced from the example of the Bavarian lakes. If, however, the predominant part of the phosphorus load is derived from the wastewaters of localities at a greater distance from the water, phosphorus elimination is advisable as part of mechanico-biological waste water treatment. Today there are chemical precipitation methods for the purpose, with which the residual phosphorus content in the effluent from the treatment plant can be reduced to 1 - 2 mg P/l. Admittedly this presupposes the inclusion of appropriate measures to eliminate waste waters of agricultural establishments in the catchment area.

Another conceivable measure would be to reduce the phosphorus content in communal waste waters, one half of which consists of human wastes and the other of detergents. Assuming a complete replacement of the phosphates in the detergents by suitable substitutes, which in the present state of the art is admittedly not yet practical, the current phosphorus load can only be reduced by about one third.

It must not be overlooked that the substitute would be bound to introduce a quantity of organic matter in the detergent which would be equivalent to the phosphorus both in its action and by weight, and this will lead to a further increase in the concentration of organic matter in waste waters.

The phosphorus content in the waste water would still be so high, even after the replacement of 1oo % of the phosphorus in detergents, that chemical treatment of waste water as part of the sewage treatment would continue to be necessary in the catchment areas of lakes and impounding reservoirs where the major part of the phosphorus derives from communities. Under normal treatment conditions only 25 - 3o % of P can be removed.

If the overwhelming part of the phosphorus in a catchment area derives from the sources we call 'disseminated', efforts must be made to see that agricultural cultivation techniques are adapted to the demands of water protection and that measures of agricultural and forestry cultivation will largely counteract the erosion of those areas which are not agriculturally utilised. Pasture management, too, may be the cause of considerable nutrient loads when animals use the brooks as watering places. This must be prevented by installing separate watering places for them.

Sometimes it may be technically and economically advisable - subject to the catchment area being of the proper type - to pretreat the main inflow chemically in such a way that it contains a residual maximum of o,o1 g/l total phosphorus. It is necessary to reduce the residual phosphorus concentration to this extent because this inflow will provide by far the major part of the phosphorus burden for the lake or reservoir and will therefore determine its phosphorus content. Even though it is possible to deal with and reduce the amount of phosphorus introduced from disseminated sources, it must not be overlooked that an elimination rate of about 95 % can hardly be bettered. Under these circumstances the phosphorus concentration of the incoming water must not be greater than about o,2 mg/l.

These explanations show that controlled measures of phosphorus elimination are considered appropriate where they are necessary, i.e. in the catchment areas of lakes and reservoirs and very slowrunning waters. Here, depending on existing conditions, the chemical elimination of phosphorus during the treatment of waste water and the construction of intercepting sewers and riparian collectors should be paramount. An effort could also be made to replace phosphates in detergents if suitable substitutes are available. If the major part in the catchment area derives from disseminated phosphorus sources, controlled measures to reduce the phosphorus content in the incoming waters by chemical methods may be mentioned.

In West Germany there are at present over 1oo treatment plants to eliminate P in conjunction with full biological sewage treatment.

## TREATMENT OF INDUSTRIAL EFFLUENTS

In the industrial manufacturing of a product x raw materials, water and air are used. After n process stages, the finished end-product is ready to go to the consumer. However, the overall process also produces exhaust air and effluent. These latter contain finite and changing quantities of the raw materials used, of the intermediate and by-products and also the end-product. The frequency of distribution and concentration of these substances depends upon the production process and is determined by the physical-chemical properties of the components, as for example volatility, density, solubility etc. This means that the components are present in the exhaust air of effluent, in solid, liquid, dissolved or gaseous form.

On the effluent side, there are four principal possible situations where effluent discharge is concerned:

- the effluent is completely purified alone, in accordance with regional requirements;
- the effluent is discharged without being clarified for a combined treatment

- the effluent is pretreated to a certain degree, for subsequent combined treatment along with other industrial or domestic effluents;
- the effluent is treated inside the industry in the form of subflows, to an extent that it is maintained partially in circulation.

On the exhaust air side, again we find four principal possible situations where emission is concerned:
- the exhaust air is released to the atmosphere directly, without any treatment;
- the exhaust air is partially pretreated and maintained in circulation;
- the exhaust air is treated using dry processes;
- the exhaust air is treated using wet processes.

In the latter case, we are confronted with a direct combination between exhaust air purification and effluent treatment.

## INDUSTRIAL EFFLUENTS

In comparison with the relatively uniform composition of domestic effluents, industrial effluents present us with an extraordinary variety of concentrations and compositions. It is possible, of course, to make a general distinction between industries producing primarily organic pollution of their effluents (e.g. foodstuffs factories, breweries, chemical industries) or primarily inorganic pollution (e.g. iron founding and metal industries) or again mixed pollution (e.g. coking plants, large-scale chemical plants). However, the mere fact that even in the same production unit the effluent quantity and composition varies radically in accordance with the nature of the initial products, the production processes and the reaction conditions, shows how complex is the overall problem we face and how differentiated the solutions to this problem must be. Each industrial effluent problem must be examined and assessed independently through an individual situation analysis.

Principally, the aforementioned four starting points will confront us:
- untreated discharge;
- full treatment;
- partial treatment as a precursor to common purification along with other effluents;
- purification inside the plant, for circulatory processes.

In addition, irrespective of these starting considerations, account must be taken of the point of action of the effluent purification measures, i.e.
- treatment of the collected total discharge;
- intervention in the production process, with integration of basic unit operations on site as it were, in the course of production.

As far as the basic unit operations are concerned, we have available to us biological, physical and chemical reactions and mechanisms. Biological purification of industrial effluents on their own or in combination with domestic sewage is applied for example to effluents from the food stuff industry, chemical industry etc. if the organic material is biodegradable.

In addition to the basic biological operation used to treat effluent, we can also draw from the range of basic physical and chemical unit operations in order to deal with the undissolved substances, these unit operations includ sedimentation, filtration, flocculation, adsorption and flotation. Dissolved substances can for example be dealt with by the following basic unit operations: adsorption, extraction, reversible osmosis, distillation. There are combined chemical and physical mechanisms which operate without any chemical change in the eliminated material. Basic unit operations primarily at the chemical level, which can be applied to dissolve substances, include neutralisation, precipitation, oxidisation, reduction, ion exchange. By an appropriate combination of these basic unit operations, including the ones performed at the biological level, it is possible to deal with the various kinds of industrial effluents.

The extent to which the corresponding treatment of industrial effluent must be pursued and from what points of view the direct discharge of an effluent should be discussed, is determined by the site situation analysis which covers both the characteristic properties of the effluent and also the features of the receiving water. It is superfluous to point out that the introduction of treated effluent into a river constitutes a concrete form of utilisation of the river in the context of the utilisation equilibrium between water extraction and water return. The effluent will have to be assessed from the points of view of composition, concentration and quantity, and on the basic of these, using figures describing the performance of the individual basic operations, the specific treatment combinations will be determined. The clarifying process which results will produce a specific purification effect. This effect, in turn, must be critically analysed in terms of what is technically and economically attainable and also in terms of what is required. The requisite purification effect of the treatment measures applied to industrial effluents will vary from case to case, however. It is determined first of all by the nature of the receiving water (stream, river, lake, dam, estuary, sea), by its water flow in relation to the input quantities, by the prior pollution (natural, artificial, controllable, diffuse, limit-exceeding) and ultimately by the way in which the water into which the treated effluents is being discharged, is utilised (drinking water supplies, industrial water, cooling water, agriculture, fisheries, leisure activities).

The situation analysis of course should also tie in the technical and economic possiblities offered by drinking water and utility water sorces, to the endeavours being made towards effluent purification, in order to optimiese the working partnership between on the one hand drinking water and service water supplies and on the other hand effluent purification, and in order also to avoid interference. If the measures taken in the context of industrial effluent treatment are discussed from these standpoints which have been developed in relation to the regional situations, then inevitably questions of the validity and viability of general standards, as related to concentration, ariese; this aspect becomes the more important if industrial effluent purification is performed not simply be treating the entire collected discharge, but instead by action directly at the point of development of sewage in the production process.

Although it was the normal procedure hitherto, to collect the discharges from the individual production stages and then to treat them commonly, nowadays it is becoming more and more the norm to pick up individual sub-flows for clearing, this with a view to recycling too. The treatment of the overall discharge after the collection of the pretreated sub-flow, then becomes a terminal high-purification operation. A prerequisite here, however, is that the sub-flows must be split into unfouled cooling water, lightly polluted rinsing water, polluted effluent and concentrates. This splitting presents no problem in new plants; in older plants, however, especially in the chemical industry, this splitting of the sub-flows is an extremely difficult and cost-intensive operation. Again, the time involved has to be taken into account because these measures must be introduced without too severely restricting production. Frequently, progress of the semi-technical trials and planning of the final clearing phase, can only be achieved after a point at which the separate effluents are available. It is self-evident, therefore, that the total costs frequently involve a higher proportion for these channelling works, than is required for the clearing installations themselves.

The 5o % limit is generally exceeded. This splitting of the individual effluent flows, however, is the precondition for successful and economic solution of effluent problems. The effects of treating sub-flows, associated with the goal of achieving a circulatory water system, on volume and concentration in the overall discharge, can be proven very readily.

A quantity of water $V_N$ is introduced into a production process with one-time utilisation of the water, approximately the same quantity is discharged from the process as effluent; it contains a pollution level $q_1$ specific to the particular production operation. Before introduction into the receiving water, the collected effluent must be purified in a common clarifying operation. The performance of the clarifying process chosen, will yield an elimination rate of $\Delta E$ and a concrete discharge concentration of $q_2$. From the clarifying plant the treated effluent carries with it a load content of $F_1 = V_N \cdot q_2 = V_N \cdot q_1 \cdot \Delta E$. In a clarifying plant operating on the collected effluents, the water volume can not be reduced unless basic unit operations such as, for example distillation or reversible osmosis are incorporated. A further reduction in the load leaving the plant can only be effected by increasing $\Delta E$.

We know, however, that every basic operation and every combination of such operations, is subject to a specific technical and economic level on performance. An increase in the clarifying effect is accordingly only possible by the incorporation of basic operations which produce a progressively more differentiating effect, e.g. additional precipitation and flocculation, adsorption on active carbon or, ultimately, distillation.

Alongside this primarily quality-determining range of action of effluent purification, which is restricted to clarification and "on-going purification" of the total discharge, there is a further range of application concerned with a quantity-determining effect. By intervention in the production process itself, we can influence the total discharge effluent volume by arranging that whilst maintaining the same useful quantities, individual sub-flows are separated off and after specific effluent treatment, recycled either for the same purpose or to other production stages, thus achieving multiple utilisation.

If $V_N$ is the water volume required for operations and $V_K$ the volume in circulation, then the final clarification plant will only receive a volume of $V_E = V_N - V_K$. This also means, however, that only this lesser volume, need be introduced in the form of a fresh water replenishment. For the same $\Delta E$, a load will now be released to the outfall which is calculated as $F_2 = (V_N - V_K)\ q_2$ or $F = (V_N - V_K) \cdot q_1 \cdot \Delta E$. At an elimination rate of around 9o %, taking the case of one-time transit (for the performing of useful operations) through the system, the load becomes $F_1 = 1 \cdot q_1 \cdot 0{,}1 = 0{,}1\ q_1$ and in the event of a 9o % circulation $F_2 = 0{,}1 \cdot q_1 \cdot 0{,}1 = 0{,}01\ q_1$.

This simple system analysis very clearly points up the following facts:

- an increase in the $\Delta E$ figure in the final clarification operation, from 9o to 99 % is accompanied by an exponential increase in cost and is very often out of the question anyway because of factors associated with the reaction processes; limiting factors are for example the solubility product in precipitation.
- the water in circulation can be afforded a wider tolerance in terms of the requisite quality, depending upon the process section to which it is recycled or into which it is introduced. The recycling rate can be made more elastic.
- the volume of effluent in the final clarification stage becomes correspondingly smaller. The units can be given smaller dimensions and it frequently turns out that the physico-chemical treatment of concentrated effluents can be controlled substantially better.

The fact that these systemanalytical considerations can in fact be implemented in practice, has been evident for years from the water economy of electroplating plants etc. By introducing ion exchangers into the process and purifying the rinsing water, it was possible to maintain around 9o % of the water in circulation.

It is the separation into cooling water and production effluent, however, which produces the first decisive reduction in the burden on the final clarification plant by its avoidance of unnecessary dilution and occupation of volume in the settling tanks by the cooling water which simply carries heat away with ist. On the other hand, this separation is the precondition for the creation of an intensive circulatory system, both in the cooling water and in the process water contexts. Taking the total industrial water requirement, in the German Federal Republic 72 % is used for cooling purposes and 23 % for process water. The remaining 5 % is used as boiler feed water. 66 % of these quantities of water is in circulation, and of this, cooling water demands make up around 8o % to form the dominant proportion. The following are some specific figures on the utilisation of cooling water and process water in circulatory systems: water in coal mining operations 88 %, refineries 75 %, iron and steel industry 65 %, paper and cardboard industry 5o %, chemical industry 4o %, textile industry 1o %. However, we are noticing increasing utilisation of circulatory systems where process water is concerned. The motivation is to be found not only in the need to recycle water for environmental protection reasons, but also in commercial interest in recycling raw materials which can be recovered from effluents although this too is beneficial from the environmental protection point of view and should be promoted.

The situation as far as the direct introduction of the treated effluent is concerned is determined first of all by the siting of the plant; it should also be investigated however, in cases where it is the discharge of unpolluted water or saline pit water, which is involved, i.e. effluents which when subjected to common clarifying with other effluents from the plant itself or outside effluents, e.g. in a biological clarification plant, experience not additional elimination but simply introduce a hydraulic burden as it were. This also applies to effluents which do not contain concentrations any greater than the expected discharge resulting from the common treatment, e.g. effluents from large metallurgical plants. The inorganic pollution remaining after treatment inside plants, generally salts occurring in the form of secondary pollutants and resulting from basic chemical operations such as neutralisation, precipitation etc., are usually in a truly dissolved form and can no longer be removed in passage through the common clarification plant; instead, due to the volume which they occupy in the settling tanks, they interfere with the purification of the other effluents. Short-period disturbances in the clarification plants of these factories, can in particular disturb municipal clarification plants of biological kind over a long period of time so that ultimately the influence of the outfall by the discharge from the municipal clarification plant, is greater that which would be caused by a brief shock pollution as it were, in the outfall itself.

The common treatment of industrial effluent with other effluents and domestic effluents, will generally be performed using biological processes. Although, under certain circumstances, other physico-chemical basic operations may be included into the clarifying system, producing additional effects, the reaction which ultimately determines the clarifying process is that of bacterial metabolism in the aerobic process, i.e. active sludge or percolating filter, or in the anaerobic process, using sludge digestion. However, bacterial metabolism processes can only include organic components which can be biologically broken down. Furthermore, the breakdown or decomposition rate depends upon a series of ecological factors, e.g. temperature, pH value, dissolved oxygen, levelling of the nutrient balance, etc. Shock loadings in terms of quantity of water, composition and concentration, must be avoided; toxic and blocking substances must fundamentally be isolated from biological systems. However, even if the organic substances in industrial effluents are fundamentally non-toxic but can only be broken down slowly, for economic reasons the technically attainable retention times in the activating tanks and percolating filters, are subject to a certain limit. These few facts suffice to show that it is necessary, in the common treatment of domestic and industrial effluents, to study each situation independently before taking any decision about whether to combine the two and about the degree of pretreatment necessary.

Common clarifying also has advantages. It is to be recommended in the situation the industrial effluents contain organic, decomposable pollutants, e.g. effluents from the foodstuffs industry, factories making preserves and meat-processing plants, certain areas of the organic chemical industry, coking plants. Many of these effluents do not have a compensated nutrient balance. In particular, the corresponding proportion of phosphorus and nitrogen is absent and they are frequently sterile. Mixing these with domestic effluents means a steady, polyvalent inoculation, with in most cases the introduction of the requisite phosphorus and nitrogen into the nutrient substrate, i.e. a mixed effluent. If the volume of domestic effluent predominates, then fluctuations on concentration and quantity are compensated for, otherwise it is necessary so to control the discharge of industrial effluent, which generally experiences major fluctuations in the course of a day, that no surge effect is produced in the municipal clarification plant. This requisite compensation is furthermore necessary in respect of the final clarification operations within the factory itself, when biological processes are being used.

Common effluent treatment is subject to certain limitations, which means that industrial effluents have to be pretreated. The object of this pretreatment is first of all to prevent disturbances in the channelling system, corrosion of structures and structural elements of the clarification units, and to prevent the clarifying processes themselves from being impeded. The pretreatment of industrial effluent will accordingly be concerned with the following problems: cooling the effluent to temperatures below 3o ℃ in order to prevent degassing, adjusting the pH value 6,5 to 9, separating off sludges down to a volume of o,5 to 1 ml/l, precipitating out floating sludge and oils, separating off volatile solvents, especially hydrocarbons, major breakdown of emulsions, oxidisation of cyanides, retention of non-ferrous metals, and separating off of hydrogen sulphide and mercaptans.

A certain proportion of iron ions or aluminium ions can have a positive flocculation effect during common treatment, whilst the heavy metals such as zinc, copper, cadmium, chromium have a toxic or blocking action on biological purification processes, depending upon their concentration. The activated sludge or percolating filter layer begins to be affected at heavy metal concentrations in the range from 1 to 2 mg/l. Anaerobic sludge digestion is particularly seriously jeopardised by heavy metals because they become enriched in the sludge from the preliminary clarifying phase and in the activated sludge, and enter the digestion tank in concentrated form. We can commence from the premise that around 8o to 9o % of the heavy metals are retained by secondarily active flocculation and adsorption mechanisms, in a normal biological clarification plant. Of the organic compounds, chlorinated hydrocarbons have a particularly strong blocking or toxic action on the aerobic and anaerobic stage. Sludge digestion, in particular, is jeopardised again. Chlorinated hydrocarbons in concentration of around 1 mg/l of sludge can seriously disturb the anaerobic processes. The disturbing of a digestion tank is a critical matter to the extent that because of the relatively slow rate of duplication of the anaerobic cells, purification and processing can last up to several weeks and the rapidly decomposing sludge has to be dumped elsewhere, something which virtually always leads to nuisance from smells. The inflow of heavy metals and chlorinated hydrocarbons into a common biological clarification operation, must therefore be carefully controlled. The discharge of sludges from factories into the common sewage system must be avoided if the formation of sludge banks is to be prevented. Likewise, sulphides, mercaptans, cyanides and highly volatile substances which in particular when in contact with acidic effluent or hot water, tend to degas, must be largely pretreated within the factory itself in order to avoid nuisance from smells on the way to the clarification plant.

Where common treatment is concerned, therefore, industrial effluent must comply with specific quality criteria which are not of course so stringent as they would be were discharge direct into an outfall involved. However, even where combination for common treatment purposes, is concerned, specific requirements must be met in order to guarantee satisfactory results from the clarifying operation during common treatment. The discharge conditions must be determind for each specific case and depend upon the method of production, the size of the factory, the percentage quantities and capacity of the common clarification plant, the nature and size of the feeder network and the kind of mixing with other effluents. The collection of effluents for common treatment, also requires continuous monitoring of discharges. In addition to a continuous inspection routine, it is a good idea where the main inputs are concerned, to install automatic measuring and monitoring equipment to check discharge paremeters. The collection of average samples and simultaneous observation of effluent toxicity in the form of a continuous fish test, have proved excellent. In the case of disturbances, the source can then more readily be detected.

Common treatment also requires that the costs be appropriated shared between the partners. These costs can be calculated on the basis of the effluent quantity and concentration. If the municipal clarification plant operates in accordance with a biological process, then the following parameters are relevant: sedimentable substances, in order to be able to calculate the capacity required for preliminary cleaning and sludge treatment, $BSB_5$ and COD in the sedimented sample in order to be able to calculate the load on the aerobic stage.

Im May 1977, the Emscher mouth biological clarification plant, went into service. In this plant, the Emscher industrial river, which draws its water from an area of 77o square kilometres, is treated. Connected to this river there are about 2 5oo ooo head of population and a like number of industrial equivalents comprised of industries such as for example, coal mining, coking plants, iron and steel production, chemical plants, petrochemical plants, foodstuffs industries, breweries, fertiliser plants, etc. etc. The industrial and communal are combined in a approximately 37o km long open channel system. At a mean water flow of around 16 $m^3/s$, the Emscher consists for more than 9o % of effluent. The Emscher mouth clarification plant is designed in the mechanical stage for a maximum of 3o $m^3/s$ and in the biological stage for a maximum of 2o $m^3/s$. Only in about 15 days over a 1o year average is there a statistical excess over the maximum flow.

The Emscher mouth clarification plant consists of the following treatment stages:

- 6o mm coarse rake
- 2o mm fine rake

- 7.6oo m³ sand trap with 12 m/h surface loading at 2o m³/h
- rectangular preliminary clarification tank with 16o.ooo m³ capacity; depth to length ration 1:3o, 1,3 m/h surface charge = 2,2 hours dwell time at 2o m³/h
- pumping station to pump the mechanically purified effluent into the activation stage, which is beyond the high water level, the station having a pumping capacity of 3 x 15 m³/h
- 144.ooo m³ activated tank with rotor aeration. Dwell time 2 hours at 2o m³ per hour
- 18o.ooo m³ secondary settling tank, dwell time 2,5 hours at 2o m³/h
- turbine station in order to recover energy from the lifting of the preliminarily clarified effluent; approximately 7oo KW.

The mechanical stage is already exhibiting very good clarification performance; an elimination rate of better than 5o % in relation to organic substances, is estimated. The efficiency in terms of precipitable substances, has been measured as being in excess of 95 %. With introduction of the biological stage into service, as tests in the pilot plant showed, $BSB_5$ discharges of around 25 mg/l is achieved. In particular, however, a decisive factor where the exploitation of the Rhine in Holland is concerned, has been a reduction in the phenols to <1 mg/l and in the odour-producing agents of >95 %.

From the mechanical clarification stage sludge quantities of around 5ooo m³ or 3oo metric tonnes of dry substance, per day, are produced. Together with the 9ooo m³ or 45o tonnes from the Emscher clarification plant at Karnap (This has been operating since 1928), these sludge quantities are dewatered in chamber filter presses and burnt. After the biological stage has been put into service, an additional 25 ooo m³ or 15o tonnes of dry substance, in the form of surplus sludge, will have to be pretreated by thermal conditioning, so that it can be dewatered along with the other sludge fractions, in the filter process.

The costs involved in the building of the overall plant, will be more than 3oo million DM. The operating costs should be around 25 million DM per year. This clarification plant represents a significant step in the treatment of effluent from this industrial area and concurbation; the beginnings of outfall regulations and clarification measures on a co-operative basis, extend right back to the year 19o4 when the Emscher Association was founded.

These measures, the Emscher plant and other plants are exemplary of the environmental protection effort on the Rhine; they are a unit part of the plurality of projects which are required in order to achieve the projected target of the environmental protection programme drawn up by the Federal Government, in accordance with which by 1985 both communal and industrial effluent will for more than 9o % by subjected to treatment having a purification efficiency of better than 9o %.

## EXHAUST AIR TREATMENT BY WET METHODS

As an introduction to this part of the paper a practical example concerned with the effluent exhaust air interface will demonstrate the mutual relationships.

Aluminium smelting plants are powerful sources of fluoride emission. The fluoride comes from the cryolite and the calcium fluoride added in the operation of molten electrolysis of the aluminium oxide. During the smelting operation, gaseous fluoride is liberated. To protect the work force, the air is drawn off from the smelting shops and scrubbed in order to protect the invironment. The effluent from the exhaust air scrubbing operation can contain fluoride in concentrations exceeding 5o mg/l and solids at a level of around 3oo mg/l. The effluent shows a clearly acidic reaction and the pH value is around 3,5. It is not possible to discharge this kind of effluent without appropriate pretreatment.

The problem was dealt with in the following way. In assiciation with calcium, fluorine form calcium fluoride which is insoluble in water and which can be separated off as a sludge. To precipitate the fluorine and to adjust the pH value to 6,o, lime milk was added to the exhaust air scrubbing water in a reactor vessel. Simultaneous dosing with iron-III-chloride, resulted in the production of flocculated iron oxide hydrate and this binds the fine calcium fluoride suspension, by an absorptive action. Adsorptive precipitation is promoted by the addition of a polyelectrolyte. For purposes of continuous operation, it was found to be a good idea to recycle a small fraction of the precipitated sludge, to the reaction chamber in order to act as a seed for crystallisation. The sludge discharged from the settling tank is treated in a concentrator and subsequently dewatered on a dry vacuum filter. The sludge is transported in containers to a dump provided for special waste products.

The process arrived at in resolving this problem, not only made it possible to keep fluoride out of the main outfall but also resulted in the scrubbing water udsed to clean the exhaust air being maintained in a closed circuit so that only the evaporation losses and the water lost with the sludge residual moisture, a total of about 5 % of the circulating quantity, have to be compensated by the addition of fresh water.

This example shows that where the site analysis of an industrial operation is concerned, in dealing with the effluent side it is essential also to discuss all matters pertaining to the exhaust air side, to the extent that wet processes are used at any rate.

Amongst the various methods of exhaust air treatment, wet methods involving scrubbing operations to wash out solid, liquid or gaseous constituents from gases, are well-known and to some extent have come into widespread usage.

The composition of the incident scrubbing water is determined by the composition of the gas flow being treated. The proportion of solids and also of liquid and dissolved materials, may be both organic and inorganic in nature. As far as solids are concerned, these can be separated from the scrubbing water by sedimentation or, in special cases, by flotation. At this point, we must ask ourselves the question of whether the solids separated off can be directly reused or used elewhere. If this is not the case, then the next operation is concentration and possible mechanical dewatering in filtering equipment. If the solids cannot be used directly, then they must be properly dumped in order to minimise the risk of ground water contamination.

Organic solids from the exhaust gas scrubbing operation, should as far as possible be supplied to a system which burns them off in a process involving high terminal temperatures. In the context of raw materials recycling, consideration should be given not merely to whether the dump should be secured from the point of view of water quality, but also to whether, looking ahead to later utilisation, the dump should be graded in accordance with material right away. The dump, especially, with all the problems and risks which it involves, must be looked at very critically where the handling of sludges coming from wet dust, is concerned.

The precipitation of inorganic components from the gas phase, is associated with a mechanism of dissolution in the scrubbing water. In order to achieve a substantially closed circiut the system can have integrated into it basic chemical unit-operations sewage purification. The inorganic components are frequently subjected to a precipitation phase, an equilibrium state developing in the circulatory system and the precipitated material being discharged and corresponding quantities of sludge effluent being removed from the circuit as the earlier indicated example pertaining to the scrubbing of exhaust air from smelting shops, showed. In dealing with $SO_2$ emissions, in order to bind the $SO_2$ as well as dust and fluorine fractions, a diluted lime suspension is used as the scrubbing solution. The calcium sulphite initially occuring in the scrubbing operation, in a largely undissolved state and in the form of a semi-hydrate, is realtively rapidly oxidised in aqueous solution, by the atmospheric oxyten so that in the sludge effluent a sulphite-sulphate ration of 1:4o is found. In the course of an examination over a period of several quard, made through observation wells beneath a dump containing this sludge, not deterioration in the ground water was observed, but hardness increased.

Exhaust air treatment operations are frequently also intended to deal with components which produce odours. Examples in this context include problems of exhaust from fish meal factories or carcase processing plants. In such instances, prior to a wet scrubbing operation, burning should be carried out and in this context it is necessary to achieve combustion temperatures well in excess of 6oo °C. Some substances require temperatures in excess of 1ooo °C.

For dust extraction, wet, mechanical precipitation techniques are used in many branches of industry and elimination rates of around 9o % are claimed. More recent technical developments make it possible to deal not only with particles over a grain size range of about o,o5 to 1oo microns, but also with gases. The elementary mechanism is the collision between water droplet and particles. The degree of elimination is the better the higher the water comsumption.

The range of application of wet dust extraction processes, nowadays extends from the woodworking, glassmaking, metalprocessing, ceramic and plastics industries, where it is used to extract dust from machinery, etc., up to loading, screening and vibratory plants, crushing and milling plants, producing coke dust, coal, limestone and stone dust, sandblasting plants, mixer machines used in road building and ore-sintering plants. Scrubbing plants are to be found in exhaust gas purification applications in the iron founding industry, in the treatment of the vrown smoke produced by Siemens-Martin furnaces, in the chemical industry, and in the high-grade purification of impellant-propellant gases. Cascade scrubbing plants are used to treat exhaust gases from converter-cupola furnaces, lime kilns, soda rotary furnaces and cellulose machines. Wet methods operating on the rotary atomiser principle, are in operation in the treatment of exhaust gases in paint-processing factories and the chemical industry, and are used to precipitate acidic vapours, chemical vapours, oil mists etc. This far from complete listing is sufficient to give some idea of the wide range of waste gases treated and of the resultant precipitated materials as well as of the degrees of differentiation and complexity in the composition of the effluents coming from the exhaust air processing phase.

As far as our common interest "clean air and clean water" is concerned, a critical factor is that with increasing degree of precipitation the water throughput increases too. This circumstance, and the overall measures in the water quality economy, are taken into account, working initially from the fresh water price, to the extent that in many cases the wet precipitation units are directly connected to a settling tank where the dust is precipitated and the water places in circulation. In sophisticated systems, up to 9o % recycling of the water is achieved. In smaller units, the direct coupling of the precipitation unit to the settling section is an obvious avenue, the sludge being removed manually or mechanically. In plants equipped with several precipitation units, on occasion a central settling system may be an economic proposition. Central sludge collection and concentration is a good idea in large-scale plants, especially if the particles are difficult to precipitate and if materials which tend to form a floating sludge are involved. Separation from the aqueous phase can then be substantially better controlled and the circuit water pH-regulated to control corrosion, by the introcuction of specific measures, e.g. the addition of aids to flocculation, or again flotation.

The avenues used to achieve solutions to these problems cannot be generalised either at the effluent side or at the exhaust air side. The best approach can only be determind from a thorough analysis of the particular situation. In order to avoid any problem shift between the effluent and exhaust air aspects, it would appear to be necessary in future to work closely at this interface.

# PREFACE

Modern benthic foraminifera are ubiquitous in all marine environments, and planktic species occur in all of the world oceans. Their geologic records span hundreds of millions of years, and they have left an abundant fossil record. Their abundance, distribution and fossil record make them ideal for studies of paleoecology, evolution, biostratigraphy and so on. All of this usefulness has not gone unnoticed. The 1976 edition of the Directory of Paleontologists of the World lists about 1,900 foraminiferal workers. The activity in the field produces a vast literature and a great deal of speculation. Consequently, a researcher in, say, ecology often has only a vague idea of what is going on in Mesozoic biostratigraphy, and vice-versa. The situation for non-foraminiferal workers is much worse, and we imagine getting together a lecture or two on foraminifera for a paleontology class must present a difficult problem.

In organizing this short course, we have tried to cover most of the field. All authors were urged to make their contributions as general as possible. Like you, we are also reading these contributions for the first time. We hope each chapter will inform scientists from different research areas about the "state of the art" elsewhere, and, at the same time, not tell them much more than they ever wanted to know.

Martin A. Buzas
Barun K. Sen Gupta

Special editorial assistance by Melody Branch, Deborah Love and Joyce Burke of the University of Tennessee and by Ann Broadhead is gratefully acknowledged.

ISBN-0-910249-05-9
University of Tennessee, Knoxville publication R01-1040-27-001-83

CONTENTS

Program . . . . . . . . . . . . . . . . . . . . . . . . . . inside front cover

Preface . . . . . . . . . . . . . . . . . . . . . . . . . . . . . ii

Participants . . . . . . . . . . . . . . . . . . . . . . . . . . iv

Biology/Paleobiology of Foraminifera . . . . . . . . . . . . . . 1
by Jere H. Lipps

Classification of the Foraminiferida . . . . . . . . . . . . . . 22
by Alfred R. Loeblich and Helen Tappan

Ecology of Benthic Foraminifera . . . . . . . . . . . . . . . . 37
by Barun K. Sen Gupta

Biology of Planktonic Foraminifera . . . . . . . . . . . . . . . 51
by Allan W. H. Bé

Biogeography of Modern Benthic Foraminifera . . . . . . . . . . . 90
by Martin A. Buzas and Stephen J. Culver

Cenozoic Benthic Foraminifera - Case Histories of Paleoceanographic and Sea-Level Changes . . . . . . . . . 107
by Kenneth G. Miller

Cenozoic Planktonic Foraminifera: a Paleobiogeographic Summary . . . . . . . . . . . . . . . . . 127
by Richard K. Olsson

Mesozoic Planktic Foraminifera . . . . . . . . . . . . . . . . 148
by William E. Frerichs

Paleozoic Foraminifera - Fusulinids . . . . . . . . . . . . . . 163
by Charles A. Ross

North American Paleozoic Agglutinate Foraminifera . . . . . . . . . . . . . . . . . . . . . . . . 177
by James E. Conkin and Barbara M. Conkin

References . . . . . . . . . . . . . . . . . . . . . . . . . . . 192

## PARTICIPANTS

Allan W. H. Bé
Lamont-Doherty Geological Observatory
Columbia University
Palisades, New York 10964

Martin A. Buzas
Department of Paleobiology
Smithsonian Institution
Washington, D.C. 20560

Barbara M. Conkin
Jefferson Community College
Louisville, Kentucky 40202

James E. Conkin
Department of Geology
University of Louisville
Louisville, Kentucky 40292

Stephen J. Culver
Department of Geophysical Sciences
Old Dominion University
Norfolk, Virginia 23508

William E. Frerichs
Department of Geology and Geophysics
University of Wyoming
Laramie, Wyoming 82071

Jere H. Lipps
Geology Department
University of California, Davis
Davis, California 95616

Alfred R. Loeblich, Jr.
Department of Earth and Space Sciences
University of California, Los Angeles
Los Angeles, California 90024

Kenneth G. Miller
Woods Hole Oceanographic Institution/ Massachusetts Institute of Technology Joint Program in Oceanography
Woods Hole, Massachusetts 02543

Richard K. Olsson
Department of Geological Sciences
Rutgers University
New Brunswick, New Jersey 08903

Charles A. Ross
Gulf Oil Exploration and Production Company
P.O. Box 36506
Houston, Texas 77236

Barun K. Sen Gupta
Department of Geology
Louisiana State University
Baton Rouge, Louisiana 70803

Helen Tappan
Department of Earth and Space Sciences
University of California, Los Angeles
Los Angeles, California 90024

# Biology/Paleobiology of Foraminifera

Jere H. Lipps
Department of Geology
University of California, Davis, California 95616

## INTRODUCTION

Foraminiferida are an order of rhizopod protozoans characterized by the presence of granuloreticulopods and a test. Together with the Order Athalamida, they are the only Rhizopoda with anastomozing pseudopodia that contain granules (Fig. 1). The difference between the Athalamida and Foraminiferida is arbitrary because some foraminifera are known to leave their tests and move about, later returning to them (Marszalek, 1969), and some Athalamida attach foreign particles to the cell surface.

Foraminifera are recorded in the geologic record from the Ordovician, and possibly from the Cambrian and late Precambrian, to the Recent. The order is rich both in species and numbers throughout most of the Phanerozoic, with over 40,000 species now described. About 4,000 species are estimated to live today (Tappan, 1971). They occur in all marine environments from the deep-sea to lagoons and estuaries, and a few species are even recorded from brackish and fresh waters (Boltovskoy and Lena, 1971; Resig, 1974). Fewer than a hundredth of the living species (less than 40) are planktonic, although these are commonly very abundant.

Foraminifera must be considered one of the most significant animal groups living today because they constitute a major portion of the biomass in many marine ecosystems, including the deep sea which covers almost 70% of the earth's surface (Hessler, 1974; Lipps, in press). They are probably the most important fossil group as well, if for no other reason than their extreme usefulness in solving geologic problems and in finding oil (Lipps, 1981). In spite of their importance, their biology is generally ignored in marine biologic studies. They are not considered by these biologists nor are they commonly included in meiofaunal studies, although most foraminifera fall in that size range. Examine any marine biology or ecology textbook, and little will be found concerning them. Thus, our knowledge about their biology is scanty indeed. Most of it is based on laboratory cultures and anecdotal accounts of foraminifera observed in other studies. For example, of the 4,000 living species, we know a little about the the reproductive biology of less than 20 species. And among those 20, there is considerable diversity. What remains to be discovered when we study species living in environments other than the very shallow photic zone?

This paper provides a brief summary of current knowledge of known foraminiferal biology and of anecdotal observations that are logical yet unconfirmed. Most paleobiology of foraminifera remains to be inferred and is a virtually unexplored field (see Lipps, 1981).

## CYTOLOGY

Foraminifera possess all the same cell structures and organelles (Fig. 1) that do most single cells, but there are some unique aspects as well (Anderson and Bé, 1978). Foraminifera differ in chiefly two respects: their ganuloreticulopodia and their possession of single nucleus at one stage and numerous nuclei at another of their life cycles.

The granuloreticulopodia consist of very fine extensions of protoplasm which contain granules of protein and mitochondria (Fig. 1, 2). These objects move along the pseudopodium in two directions at the same time (Fig. 3). This observation has lead to controversy concerning the nature of these pseudopodia. Jahn and Rinaldi (1959) suggested that a shear plane divided the pseudopodia into two streams moving in each direction (Fig. 4). No suggestion of such a shear plane could be detected in electron and optical microscopic studies of the pseudopodia of *Iridia diaphana* (Marszalek, 1969). Grell (1973) alternatively suggested that the pseudopodia consisted of numerous functionally independent threadlike units, which carried the granules and food particles. As one thread retracts, others extend so that the overall form of the pseudopodium remains the same. This suggestion is supported by electron microscropy showing the threadlike structures (Wohlfarth-Bottermann, 1964). The mechanism of pseudopodial movement in foraminifera is still not clear, but it is complex (Anderson and Bé, 1978).

The pseudopodia of foraminifera function in several ways. They are involved in food capture, digestion, locomotion, test construction, and possibly defense.

The dimorphic nuclear cycle is related to reproduction (Grell, 1973). Individuals reproducing sexually (gamonts) have a single nucleus and give rise to gametes. Multinucleate individuals develop through divisions of the single nucleus in zygotes formed by gamete fusion. The number of nuclei seems fixed, and all are produced before growth of the asexual individuals (agamonts). In some species, the agamonts are homokaryotic, with all the nuclei identical, and in others, they are heterokaryotic, with generative and somatic nuclei of different kinds. Only the generative nuclei participate in meiosis and the passing of genetic information, while the somatic ones take part in cell metabolism and disintegrate upon reproduction (Grell, 1973).

## REPRODUCTIVE BIOLOGY

Reproduction in foraminifera is quite variable (Table 1). Some species appear to reproduce only by asexual means, and others undergo an alternation of generations during their life cycle, which is haplodiplont, like most plants, in contrast to the haplont and diplont cycles that are more common among other protists and animals (Fig. 5). In the haplodiplont life cycle, meiosis is intermediate, wherein the cells producing gametes (the gamonts) are haploid and the asexually-reproducing cells (agamonts) are diploid.

Table 1. Reproductive modes in foraminifera.

| ASEXUAL | SEXUAL |
|---|---|
| Budding | Gametogamy |
| Multiple fission | Gamontogamy |
| | Autogamy |

Most species of foraminifera are dimorphic as a result of the alternation of generations. The initial chamber (proloculus) of specimens resulting from gametic fusion is generally smaller (microspheric) than those formed by cell division (megalospheric). The microspheric tests result from the smaller initial biomass of the two fused gametes, whereas megalospheric tests start with a larger amount of protoplasm as a result of cell division. These differences can be observed readily in Recent and fossil specimens (Fig. 6). The asexual, or megalospheric, tests are the most common, although the proportion of each varies with the seasons, at least in temperate areas. Myers (1943) reported that in *Rosalina bulloides* (as *Tretomphalus*) the ratio of microspheric to megalospheric varied from 1:340 in the spring to 1:24 in the winter. The asexual megalospheric stage has tests that are dominantly left coiling, at least in some species.

*Asexual reproduction.*--Foraminifera are reported to form new individuals by budding in some simple agglutinated species (*Bathysiphon*) (Loeblich and Tappan, 1964). How such a process fits into the normal haplodiplont life cycle is not clear. Asexual reproductions are normally part of the life cycle, and involve cell division (Fig. 7). Commonly the asexual reproducing test may differ morphologically from the sexual stage (Fig. 8). Asexual reproduction becomes the dominant mode of increasing population size when environmental conditions, usually associated with increased nutrient supply, are favorable. This accounts also for the seasonal variation in microspheric to megalospheric ratios noted above. Some species, however, are known only to reproduce asexually. Successive asexual generations may also alternate in generation length, as in *Bolivina doniezi* where generation length alternates between about 2 months and about 3 months.

Sexual reproduction.--The evolutionary potential of the foraminiferal life cycle varies with the sexual reproductive mode. Different species of foraminifera may reproduce sexually by the formation and fusion of free-swimming gametes (gametogamy), the fusion of gametes or gametic nuclei produced by the same gamont (autogamy), or the mating of gamonts which then exchange gametes or gametic nuclei (gamontogamy). Some species and/or populations apparently never undergo a sexual cycle, either through repeated cell divisions or, in rare cases, budding. The proportion of each mode among foraminifera is uncertain.

A common problem among foraminifera is the conservation of gametes. Because foraminifera use their own cytoplasm to form gametes, the number formed is limited and reproductive modes or habits that waste them are apparently uncommon. Foraminifera may have identical gametes (isogametes) or differentiated gametes (anisogametes) depending on the species. The known gametes among the foraminifera are biflagellate, triflagellate and amoeboid. These types may also be related to reproductive mode in an as yet undetermined way.

Gametogamous species, including some planktonic ones, release their gametes freely into the water column (Fig. 9). Characteristic of this type is Rosalina bulloides d'Orbigny (also placed in Tretomphalus). This species differs in one respect: it releases its numerous gametes at the air-water interface, a proceedure I interpret to conserve gametes by concentrating them on one surface. Other species may adopt a different behavior at gametogenesis; for example, planktonic foraminifera may sink in the water column before gametogenesis. Others, like Myxotheca arenilega, simply liberate great numbers of gametes directly into the water near where the adults live. In Rosalina globularis (Fig. 9), the diploid zygote grows into a benthonic trochospiral form, which reproduces asexually and is usually attached to the substrate (Delaca and Lipps, 1972). Meiosis occurs during one of these divisions followed perhaps by still more asexual divisions. When the water temperature reaches 18°C or greater, at least in culture (Sliter, 1965), gamonts are formed which form a diatom-debris cyst. The gamont then grows a float chamber, breaks free of its cyst, floats to near the air-water interface, and releases its gametes. The gametes fuse and settle to form a zygote. Although Myers (1943) reported that the alternation between the agamontic (asexual) generation and the gamontic generation proceeded in an orderly manner, Sliter (1965) showed in his cultures that the agamonts could continue to produce more agamonts (a process termed apogamogony) until the temperature became warmer than 18°C (Fig. 10). Not until then was the gamont with its float chamber formed. In the sea off southern California, R. globularis with float chambers may be found, but off northern California where the water never reaches more than about 16°C, specimens with float chambers have never been observed. This observation supports Sliter's suggestion that water temperature is critical to sexual reproduction and that asexual populations may exist that never undergo gamete exchange.

Gamontogamous species mate in a process in which two or more individuals may come together and exchange gametes. Gamontogamous species, of course, insure fertilization of gametes and the initial survival of young to a several chambered stage. In Patellina corrugata Williamson (Fig. 11), gamonts come together anastomozing their pseudopodia and pulling their tests together. They form a cyst of debris, undergo several asexual cell divisions and finally meiosis. These amoeboid isogametes fertilize one another within the cyst. The zygotes grow rapidly to two and a half whorls before they escape from the cyst.

A specialized form of gamontogamy, called plastogamy, occurs in species of Glabratella and related genera. In this case (Fig. 12), two or more gamonts come together apertural sides appressed and secrete an organic membrane between them (Myers, 1940; Lipps and Erskian, 1969). After the pseudopodia are retracted and the membrane is secreted, the apertural faces of the tests are dissolved forming a single so-called brood chamber. Gametes are formed and they are fertilized within the brood chamber, where the zygote grows to about three chambers. The young are liberated by degrading the membrane holding the parent tests together (Lipps and Erskian, 1969).

Autogamous species form gametes or gametic nuclei that then fertilize each other (Fig. 13). There is no exchange of genetic material from other individuals, and genetic variation is not introduced through this mechanism at all.

In general these reproductive strategies tend to increase fitness to the immediate environment and to decrease genetic variability through selection on the asexual generations. Genetic variation is introduced by sexual reproductions, and even in autogamous species, variability is introduced through meiosis and fusion. The variability introduced in these manners could easily be eliminated later except that reproductions are relatively rapid compared with most environmental changes. Thus foraminiferal populations may benefit from both immediate fitness and genetic variability, without large costs in storing the variability required for environmental plasticity over periods of time longer than single generations.

## FEEDING BIOLOGY

Food.--The feeding biology of foraminifera has been studied chiefly for species from the shallow photic zone. These few species have been observed in cultures eating food, mostly algae, presented to them. As a result, foraminifera are generally and erroneously considered to be herbivores, although in uncontrolled cultures foraminifera occasionally have been observed to take food other than plants (Loeblich and Tappan, 1964; Boltovskoy and Wright, 1976), and in deep, offshore water they must eat other materials as well. In addition, most planktonic foraminifera are known to eat both animal and plant food in nature, with the possible exception of Hastigerina pelagica (Anderson et al., 1979). Foraminifera probably eat as wide a variety of food types as any other major group of animals.

The photic zone foraminifera eat many different kinds of plant materials and bacteria (Lee, 1974, 1980). Most species tested show a preference for certain species of algae over others that occur with them in their natural habitats (Anderson et al., 1979; Lee, 1980), and bacteria seem to be required in their diets (Lee, 1980). At least one species is reported to take up dissolved organic matter directly from the water and sediment through its cell surface (Delaca, Karl and Lipps, 1981), and other species have been observed to capture and consume small animals, including other species of foraminifera, eggs and fecal matter (see Loeblich and Tappan, 1964; Murray, 1973; Lee, 1974, 1980; Boltovskoy and Wright, 1976, for summaries).

The food items of foraminifera are for the most part small. Organisms much larger than the foraminifera or those moving vigorously can not apparently be successfully captured because the pseudopodia are not sufficiently strong. Older reports that foraminiferal pseudopodia possess the ability to stun or kill prey on contact are not now accepted (Murray, 1973).

The quantity and quality of food available to foraminifera affects other aspects of their biology, such as reproduction, growth, size, distribution, and abundance. In environments where food supply may fluctuate seasonally, foraminifera show marked changes in standing crop and reproductive mode. Asexual populations dominate when food supply increases and decrease proportionally when nutrient supply decreases. In artificially eutrophicated environments, such as near sewer outfalls, certain species may increase in abundance whereas others become locally extinct (Bandy, Ingle and Resig, 1964, 1965a, 1965b; Bates and Spencer, 1979).

Species of foraminifera also compete among themselves in culture for food, and probably in nature as well and with other organisms. Experiments with several photic zone foraminifera show that these species take advantage of differing food supplies, both in terms of composition and abundance (Muller, 1975). Some species respond to changes in these factors immediately and are later outcompeted by other species better adapted to take advantage of lower food levels.

Trophic mechanisms.--Foraminifera obtain their nutrition in as many ways as most other groups of animals. Although few observations have confirmed these feeding mechanisms, they are logically inferred from life habits. Even in species where food capture has been observed, questions remain whether or not these items are actually digested. Table 2 lists the various mechanisms; these are discussed in greater detail in Lipps (in press).

Table 2. Trophic mechanisms in foraminifera. From Lipps (in press).

1. Uptake of dissolved organic material (DOM).
2. Herbivory (active and passive).
3. Carnivory (active and passive).
4. Omnivory (active and passive).
5. Suspension feeding.
6. Scavenging (epifaunal and infaunal).
7. Parasitism.
8. Cannibalism.
9. Symbiosis (ectosymbiosis and endosymbiosis).

Uptake of DOM.--Two species of agglutinated foraminifera are known to utilize dissolved organic matter. The two species are erect arborescent forms rooted in sediment. They take up DOM through all cell surfaces, but chiefly through the root system buried in mud. Eleven amino acids (Delaca, in press) and glucose (Delaca, Karl and Lipps, 1981) enter their metabolic pathways in this manner. These species live in a seasonally oligotrophic situation where particulate material may be nearly absent in the water most of the year. These species are also capable of capturing particles with short pseudopodia extending from the arborescent test above the substrate (Delaca, Lipps and Hessler, 1980) when particulate material is available. Many other species probably also utilize DOM, especially those living in oligotrophic environments such as the deep sea, coral reefs, and parts of the water column.

Herbivory.--Foraminifera may graze actively across substrata harvesting algae, bacteria and yeasts as they go, or they may be attached either permanently or temporarily and passively harvest plants with their pseudopodia. *Rosalina globularis* will do either depending on the density of food items. Sliter (1965) found that as food decreased in culture, this species began to move in search of more, but in many natural situations the species likely remains on one site where its test may even grow in conformity with its substrate (Delaca and Lipps, 1972). This species, then, is an example of both an active and passive herbivore. Passive herbivores must commonly extend their pseudopodia for long distances from the test. In two of these species, the pseudopodia are covered with an agglutinated layer formed by the foraminifera presumably to protect the pseudopodia from predation and other physical disruptions. Many species are reported to be active herbivores when in culture (Murray, 1973; Haynes, 1981). These species move with their pseudopodia both extended foreward and trailing behind, and they gather food items as they go. They may stop when food enough has been gathered, digest it externally, and leave a "food" or "feeding" cyst behind consisting of undigestible materials.

Carnivory.--Although widely reported among foraminifera (Sandon, 1932; Boltovskoy and Wright, 1976; Murray, 1973), carnivory has not been studied in detail except for planktonic species. Foraminifera apparently do not stalk their prey but rely on movements of the prey to bring them in contact with the pseudopodia. After contact is made, the prey apparently expires from exhaustion and is brought near the test for digestion. One foraminiferan, Spiculosiphon radiata (Christiansen, 1964), feeds on at least thirteen other species of foraminifera. Carnivores could also be passive, waiting at one site for prey, or active. A successful carnivore must deploy its pseudopodia in a manner suitable for encountering its prey, without exposing them to predators and turbulence that would disrupt them.

Omnivory.--Many, maybe most, foraminifera probably eat both plant and animal matter. They should function both as passive and as active omnivores. Perhaps many of the photic zone foraminifera studied in algal enriched cultures would also take animal food or detritus.

Suspension feeding.--Foraminifera that have tests with apertures elevated above the substrate are probably suspension feeders. The most obvious of these have arborescent tests anchored in or attached to the substratum (Fig. 14). From the elevated aperture, pseudopodia may stream into the water column. These pseudopodia need not form a sieve or filter, but may also capture food particles through direct interception, inertial impaction, motile particle deposition or gravitational deposition (Lipps, in press; Rubinstein and Koehl, 1977). These non-sieving methods utilize one or two pseudopodial threads to capture food. Pseudopodia seem to be easily broken by currents or turbulence when free in the water, and they may be strengthened by the support of secreted spines, as in spinose planktonic foraminifera, or foreign objects along which the pseudopodia may extend. All suspension-feeding foraminifera are passive, as they have no way to move water into their capture structures. They most likely rely on the currents, gravity and movements of the prey. Astrorhiza limicola extends pseudopodia into sandy or silty sediment where it captures interstitial organisms that move into them. This mechanism may be considered "interstitial suspension feeding" (Lipps, in press).

Detrital and bacterial scavenging.--Most foraminifera found in soft sediments likely are scavenging on detritus and bacteria as they live not on the surface but in the top one or two centimeters. These foraminifera force their way through the sediment, and may leave burrows behind them. The force required to move through fine-grained sediment is proportional to the cross-sectional area of the test; elongated tests that reduce this area but conserve the volume necessary for organelles are common among foraminifera from soft sediment. Scavengers, of course, may also occur on hard substrata.

Parasitism.--Parasitism by foraminifera of other foraminifera has been reported (LeCalvez, 1947; Todd, 1965; Banner, 1971; Boltovskoy and Wright, 1976), but only in a few cases is the parasite clearly consuming parts of its host. Fissurina marginata captures granules from the pseudopodial network of Discorbis mediterranensis (LeCalvez, 1947).

*Cannibalism*.--Some foraminifera consume their own gametes in the initial stages of growth. For example, zygotes of *Glabratella ornatissima* were reported by Myers (1940, as *Discorbis*) to eat the remaining gametes during plastogamy. I am unaware of mature individuals of the same species eating each other.

## SYMBIOSIS

A number of foraminifera, particularly larger and planktonic species, harbor algal symbionts in their cytoplasm. The symbiotic relationship affects many aspects of the foraminiferal biology besides nutrition; hence it is treated separately here. A variety of algae may be symbiotic with foraminifera. Diatoms, chlorophytes, rhodophytes, dinoflagellates, and even isolated algal chloroplasts have been reported as endosymbionts (Lee, 1980; Lopez, 1979). Dinoflagellates are known as ectosymbionts on planktonic foraminifera, diurnally migrating along the spines, and a diatom is known to be an ectosymbiont on large, benthic species (Lee, 1980). These symbioses appear to be widespread among foraminifera today, and the abundance of large foraminifera in the fossil record indicates that it has been a long-enduring relationship as well. It has probably played an important role in the evolution of larger foraminifera (Hallock, 1982).

*Algal symbionts and test morphology*.--The algal symbionts may be distributed throughout the cytoplasm of the foraminiferan, as in *Amphisorus hemprichii*, or they may live only in certain places in the test interior, as in *Amphistegina lessonii* where they are found only in the upper portions of the cytoplasm (McEnery and Lee, 1981). The presence of the symbionts may result in modifications to the calcareous test. These include small spaces within the test walls that harbor algae (Hansen and Dalberg, 1979), pits on the interior of the chamber walls in which the algal cells reside, much as eggs in an egg carton (Hansen and Reiss, 1972; McEnery and Lee, 1981), chamberlets and other similar test complexities (Ross and Ross, 1978). Hallock (1982) suggested that chamberlets and other partitions strengthen the thin-walled test while allowing it to remain transparent for algal photosynthesis.

The test wall structure may also be modified for algal symbiosis in certain species (Haynes, 1965, 1981). The walls of foraminifera include various crystal arrangements, such as rods, stacks of platelets, and a radial crystal orientation. These arrangements and a reduction in the amount of organic matter in the test wall may aid light transmission to the symbionts. Such special microstructure has been found in parts of the test where symbionts reside (Ross and Ross, 1978).

The overall test shape seems related to symbiosis. The larger foraminifera, in general, have three test forms: flattened, discoidal and fusiform. These shapes tend to increase the surface area to volume ratio, a morphology that has been suggested to allow a greater area over which symbionts may receive sunlight. Even in the same species and different species in the same genus, test form varies depending on the amount of light received (Hallock, 1979; Hallock and Hansen, 1979). More oblate and thicker tests are found in species inhabiting shallow water than in those found deeper, suggesting that light availability controls test morphology through symbiotic interactions (Hallock, 1979).

Nutrition.--Foraminifera may obtain all or part of their nutrition requirements from their endosymbionts (McEnery and Lee, 1981). The extrametabolites of the symbionts may be utilized, and in some species the symbionts themselves may be digested. These nutrient sources provide an energetic advantage to the foraminifera (Hallock, 1981b) living in low nutrient tropical waters. Symbionts are an important nutritional resource for larger foraminifera.

Test calcification and isotopic composition.--Test calcification rate is directly affected by symbiont photosynthesis over certain ranges of light intensity (Duguay and Taylor, 1978). Indeed, the symbionts may permit the foraminifera to grow to macroscopic sizes (Lee, 1980). On some reefs, foraminifera contribute as much $CaCO_3$ to reef construction as do corals, coralline algae and macrobenthos (Hallock, 1981c, 1982).

The stable isotopic composition of foraminiferal tests is also directly affected by algal symbiont photosynthesis. $^{18}O$ and $^{13}C$ are depleted in foraminiferal tests relative to sea water, and the depletion is a function of increased photosynthesis (Buchardt and Hansen, 1977; Williams and others, 1981; Wefer, Killingley and Lutze, 1981). Thus stable isotopic studies of foraminiferal tests for paleoclimatology and paleooceanography must first determine if the foraminifera contained symbionts and what the effects may have been.

## PREDATION

Foraminifera constitute a significant trophic resource in most marine environments because of their large biomass. Many animals are known to ingest foraminifera, either accidentally while consuming other resources or selectively (Table 3), although the trophic role of foraminifera in marine ecosystems is still only generally understood (Lipps and Valentine, 1970). Predation likely has major taphonomic implications (Lipps, in press), as tests that have been attacked or passed through animal guts are more subject to dissolution and other diagenetic destructive processes (Douglas, 1973).

Table 3. Reported predators of foraminifera (from Lipps, in press).

| INCIDENTAL CONSUMERS | SELECTIVE CONSUMERS |
|---|---|
| Flatworms | Foraminifera |
| Polychaetes | Nematodes |
| Chitons | Polychaetes |
| Gastropods | Gastropods |
| Nudibranchs | Scaphopods |
| Ammonites | Crustaceans |
| Bivalves | |
| Crustaceans | |
| Holothuroids | |
| Asteroids | |
| Ophiuroids | |
| Echinoids | |
| Crinoids | |
| Tunicates | |
| Fishes | |

Incidental consumers.--The incidental consumers include most detritus-feeding animals and herbivores feeding particularly on filamentous algae. Nearly all living phyla take foraminifera in this way (Lipps, in press), and at least one fossil occurence is known where an ammonite was found with foraminifera and ostracodes in its crop (Lehmann, 1971). Because the incidental consumers may reach high standing crops and because they frequently work the substrata, they may have major impact on foraminiferal populations. Where detrital-feeding megazoans are present, the foraminiferal standing crop may be reduced considerably (Christiansen, 1958; Buzas, 1978, 1982; Douglas, 1981).

Selective predators.--Selective predators of foraminifera are small, for individual foraminifera do not provide much energy. Some of these predators select foraminifera in preference to other prey and some select among different species of foraminifera. There are two kinds of predators--large species that take foraminifera only as juveniles when they are small and then switch to other food later in life (Brand and Lipps, 1982; Hickman and Lipps, in press), and species that take foraminifera as adults.

*Olivella biplicata*, a shallow-water gastropod, takes foraminifera routinely as juveniles (Hickman and Lipps, in press), but not in later life (Fig. 15). Similarly, Brand and Lipps (1982) found that another gastropod *Philine alata* and a polychaete from Antarctica did likewise. Foraminifera may thus be a common food resource for juvenile metazoans.

Selective feeders as adults include a variety of opisthobranch gastropods (Lipps, in press), the scaphopod *Dentalium* (Bilyard, 1974), and one foraminiferan (Christiansen, 1964). These are all small animals themselves. Foraminifera may in fact be selected by a wider variety of small metazoans than previously observed. Some of the reports indicate that these predators are capable of selecting even among different species of foraminifera (Bilyard, 1974; Shonman and Nybakken, 1978).

Another apparently selective predator or predators makes holes in the tests of calcareous foraminifera (Sliter, 1971; Douglas, 1973). The holes are straight-sided, sometimes aborted, and commonly numerous on single tests. They have been attributed to nematodes, gastropods or unidentified organisms. Sliter (1975) effectively used the presence or absence of such holes to determine whether or not anoxic conditions existed in Cretaceous deposits in southern California.

Parasites.--Foraminifera have been reported to be parasitized by other foraminifera, sporozoans, gregarinids, nematodes, amoebae, and fungus (see Boltovskoy and Wright, 1976). Nyholm (1962), for example, showed that sporozoans inhabited the cytoplasm of *Gypsina* eventually consuming it and causing the death of the foraminiferan. Some of the other reported cases of parasitism may be actually different sorts of relationships.

## DISPERSAL

Dispersal mechanisms in foraminifera have not been well studied but may be of biological and paleontological importance. Table 4 lists most of the dispersal mechanisms; most of them have not been studied in detail.

Table 4. Dispersal mechanisms in foraminifera.

Pseudopodial locomotion. All vagile foraminifera are capable of moving but not for long distances. Rates of movement are generally on the order of a few cm/day.

Planktonic habit. Planktonic species and at least one benthonic species with a planktonic float chamber are capable of freely floating in the water column. Gametes and zygotes may also remain in the water column. These may be dispersed by currents over wide ranges.

Current transport. In addition to species living in the water column, benthonic species that are lifted from their substrata have been found floating long distances from shore.

Turbulent transport. Where water turbulence is great, foraminifera may be dislodged from their substrata and be carried in suspension or on foam.

Bottom traction. Foraminifera living on the sea floor at various depths may be moved long distances by bottom currents.

Table 4. Continued.

Turbidity currents. Many shallow-water species have been reported in deeper waters, having been carried there by turbid water flow.

Rafting. Attached species and species inhabiting debris capable of floating may be carried long distances.

Hitch-hiking. Foraminifera may be transported by other organisms, including man, accidentally. Birds for example are suspected of carrying foraminifera to inland bodies of saline water where they may live.

Gut transport. Foraminifera are consumed by larger mobile animals and they may pass through the guts without ill affects. Fish, for example, could consume foraminifera along with their other food and release them in another location upon defecation.

Ice abrasion and rafting. In high latitudes, ice may be grounded and trap sediment containing foraminifera. These foraminifera could live long times and be deposited great distances from their point of origination upon melting of the ice.

---

## FUTURE WORK

Foraminiferal biology is in general not well known. Because foraminifera are reasonably hardy and can be kept in cultures even far from the ocean, they are easier to study than commonly realized. Natural history observations in uncontrolled cultures are valuable and easy to make. Caution must be taken in the interpretation of results, however, because culture results may differ from nature.

Paleontologically important topics needing attention include: genetic studies, competition with other foraminiferal species and organisms, predation, food especially of non-photic zone species, symbiosis, and processes and biochemistry of test secretion. Even the topics reviewed briefly here require much additional study.

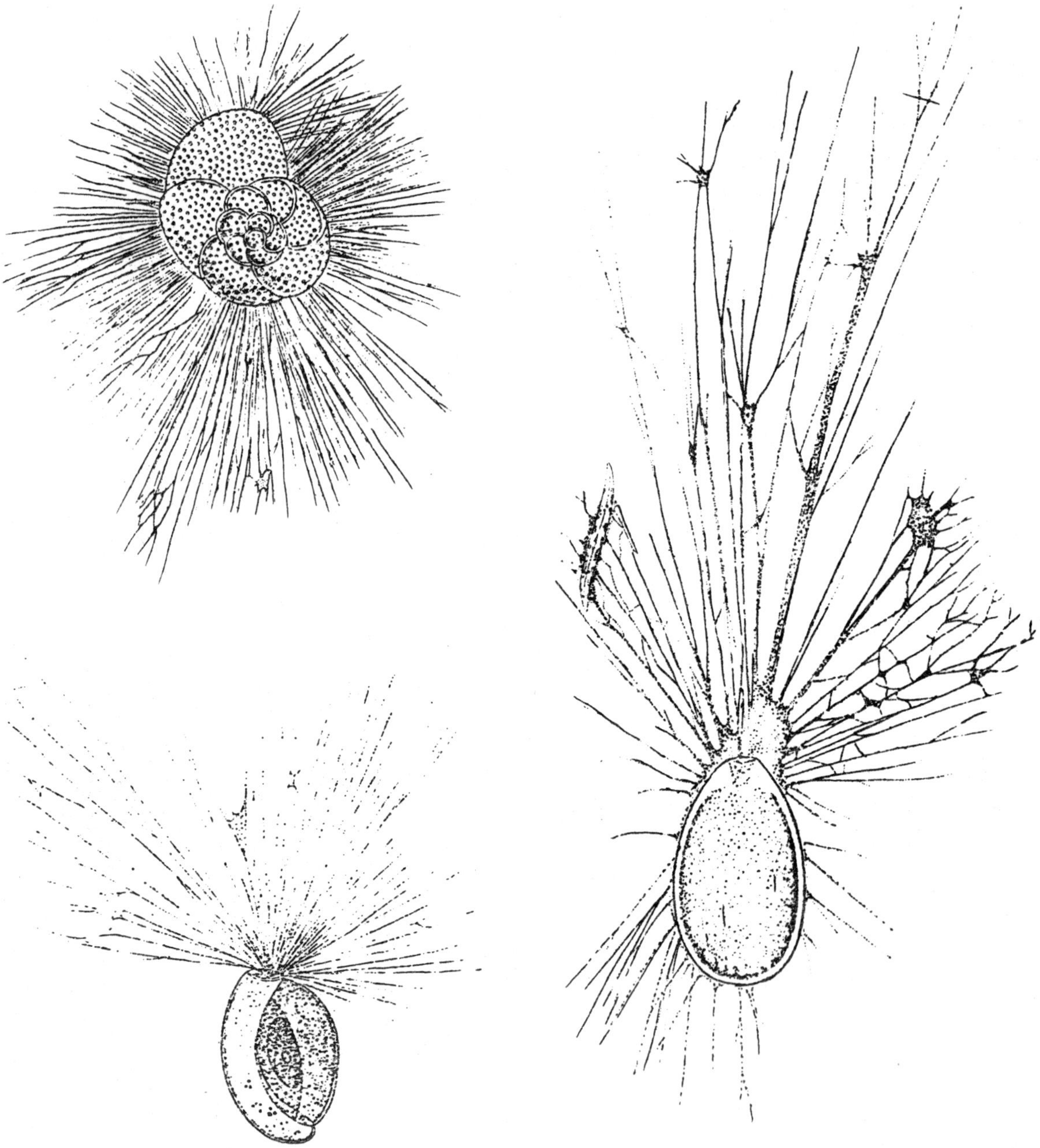

FIG. 1--Living foraminifera. Upper left: Rosalina, lower left: Miliolina, right: Allogromia. From Carpenter, Parker and Jones, 1862.

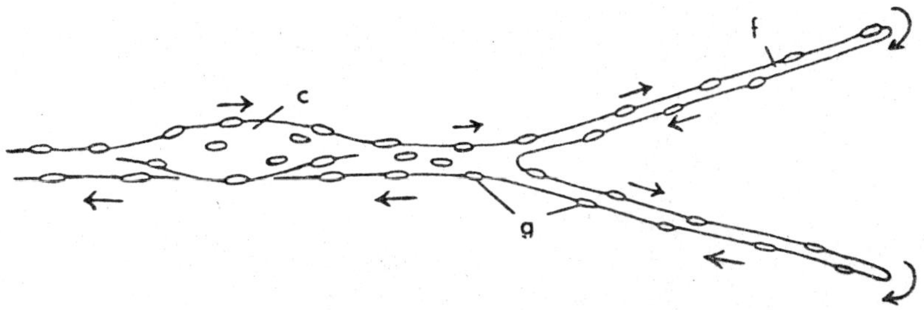


FIG. 2--The distal end of a pseudopodium of Allogromia laticollaris, showing the general shape, the granules (g) and their directions of movement (arrows), a small cytoplasmic mass (c), all of which are attached to actively moving filaments (f). From Jahn and Rinaldi, 1959.

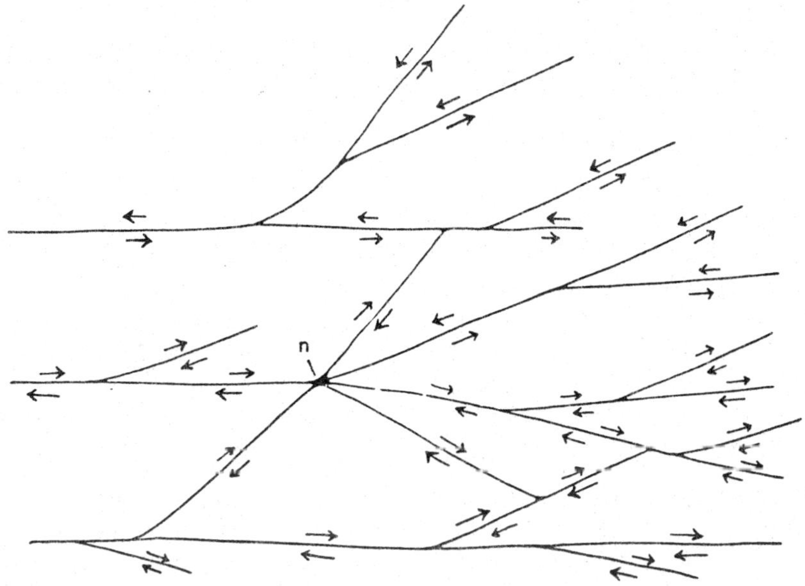


FIG. 3--A portion of reticulopodia of Allogromia laticollaris showing the direction of streaming (arrows). From Jahn and Rinaldi, 1959.

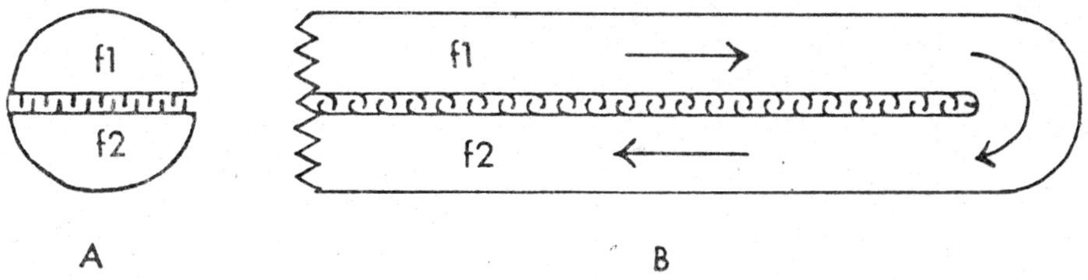


FIG. 4--The Jahn and Rinaldi (1959) model for active shear in foraminiferal pseudopodial movements. The moving protoplasm is semi-cylindrical, turned back upon itself at the tips. A. Cross section. B. Side view. From Jahn and Rinaldi, 1959.

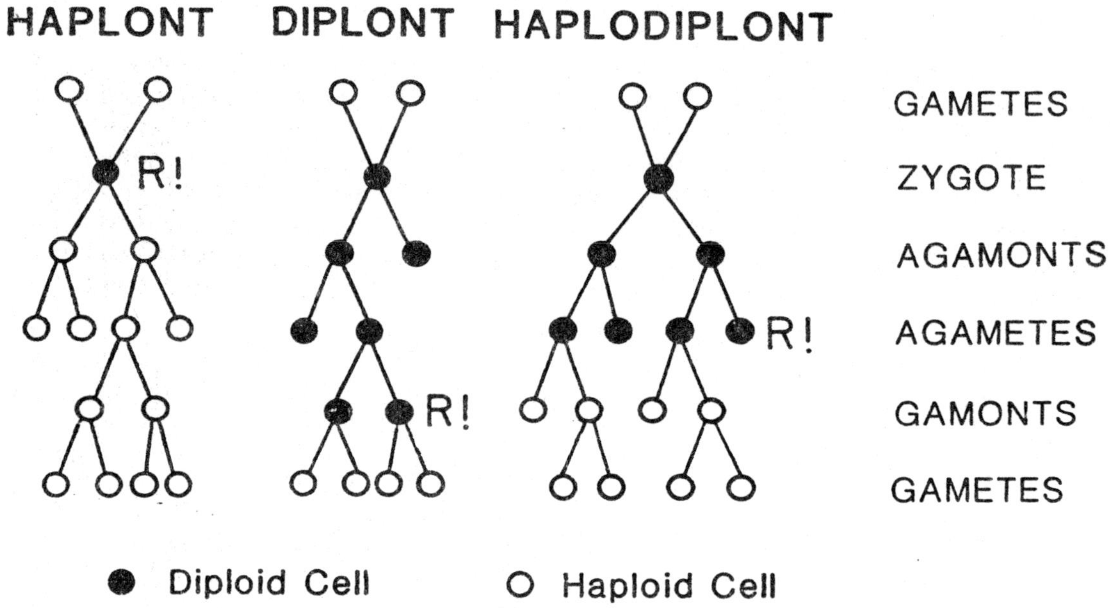


FIG. 5--Comparison of different modes of alternation of generations. In the haplont life cycle, possessed chiefly by unicellular algae, the majority of the cycle is haploid, with a diploid zygote immediately undergoing meiosis. In the diplont life cycle, possessed by most metazoans, ciliates and heliozoans, all cells are diploid with only haploid gametes. In the haplodiplont life cycle, possessed by foraminifera and most plants, meiosis is intermediate. Modified from Grell, 1973.

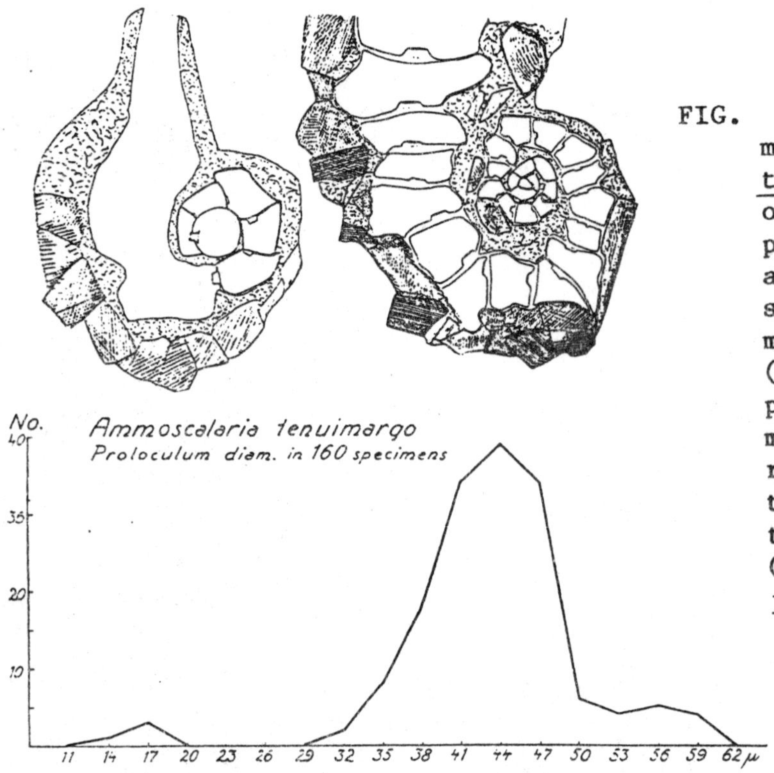


FIG. 6--Dimorphism in test morphology of Ammoscalaria tenuimargo from the Skagerak off Norway and Sweden produced by sexual and asexual reproduction. Cross sections of the megalospheric (asexual) test (upper left) show a larger proloculum than the microspheric test (upper right). The megalospheric tests are more abundant than the microspheric tests (lower). From Hogland, 1947.

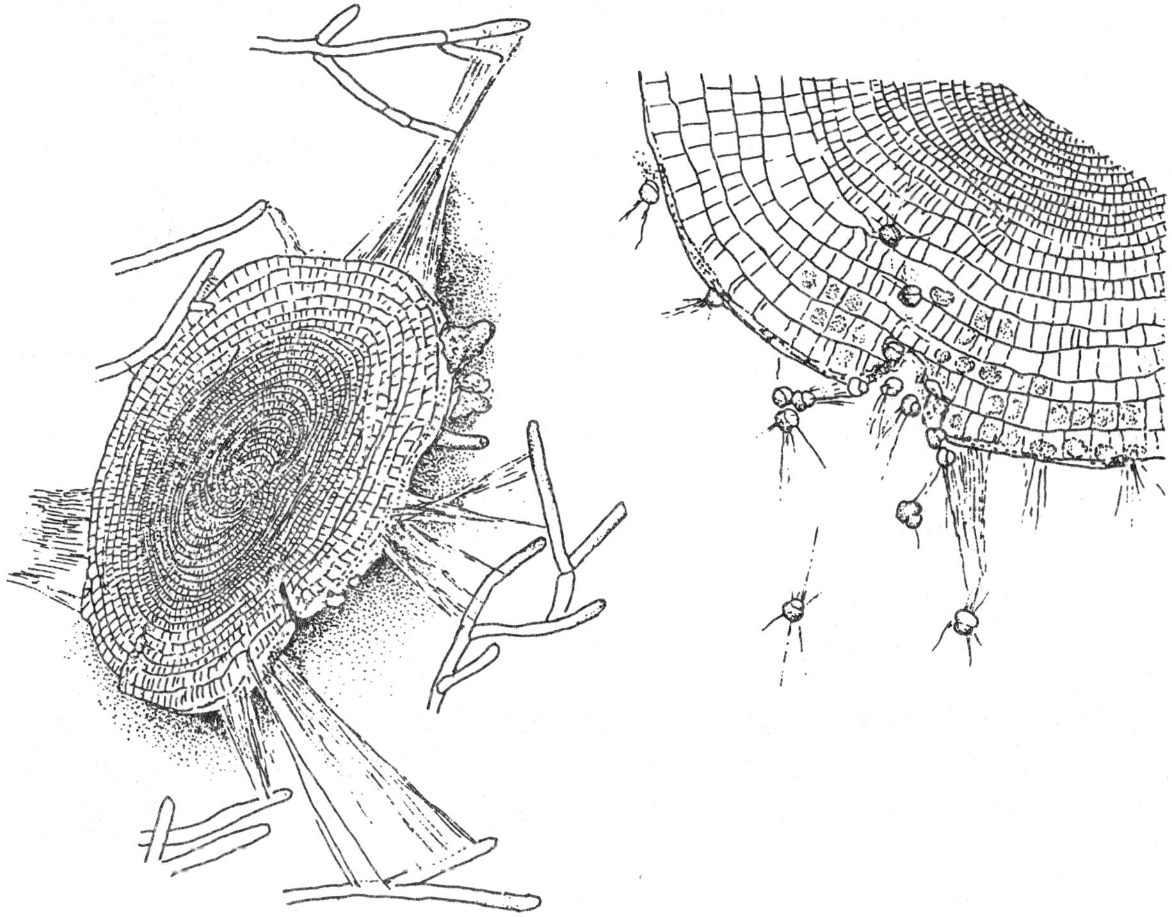

FIG. 7--Asexual reproduction in *Cyclorbiculina compressa*. Left: Live specimen with reproductive chambers, filling with cytoplasm and forming embryos by multiple fission. Right: Partial destruction of the reproductive chambers and escape of the bi-chambered embryos. From Lutze and Wefer, 1980.

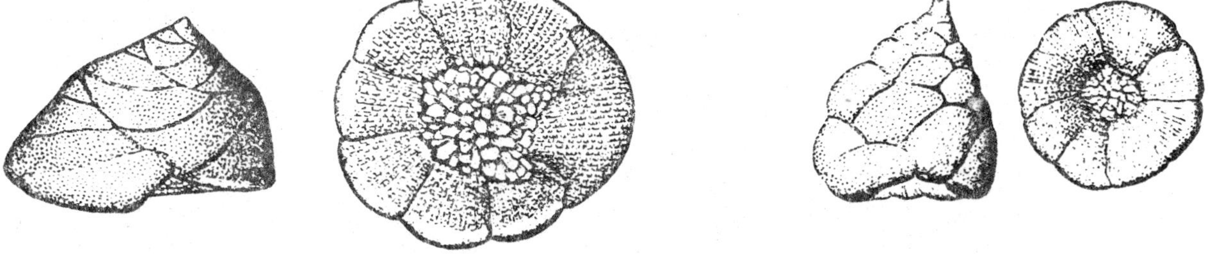

FIG. 8.--Dimorphism in *Discorbis patelliformis* corresponding with asexual agamonts (left two figures) and sexual gamonts (right two figures). From LeCalvez, 1938.

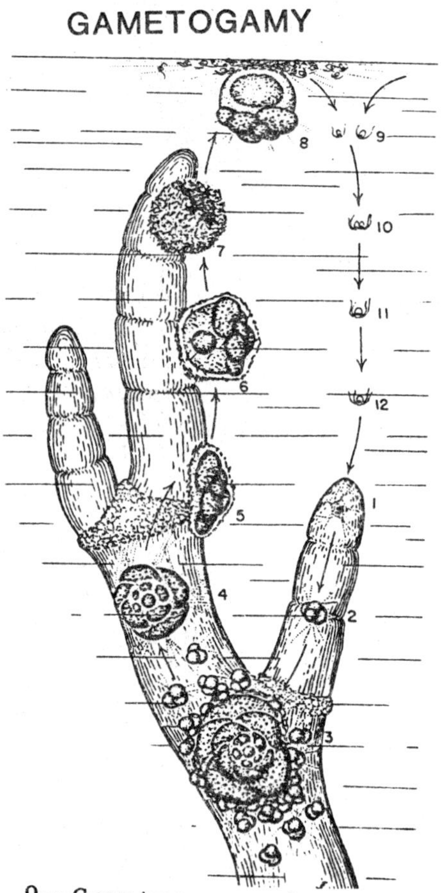


FIG. 9--Gametogamy in Rosalina globularis. In this species the gamont forms a special float chamber while encysted. Gametes are released at the air-water interface. From Myers, 1943.

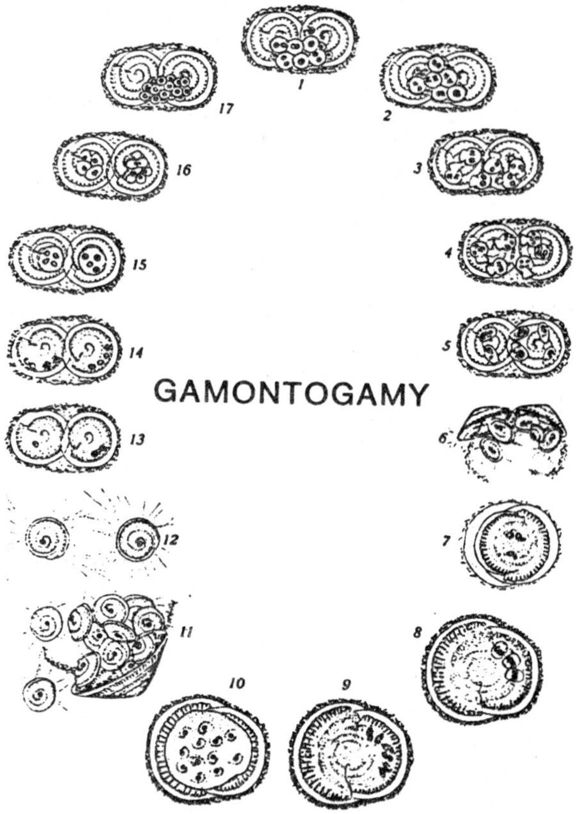


Fig. 11--Gamontogamy in Patellina corrugata. In this species two or more individuals come together and form a cyst in which to exchange gametes. From Myers, 1935.

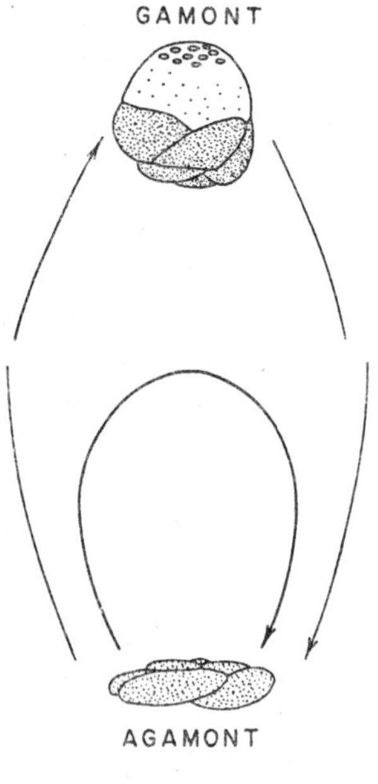


FIG. 10--Complete life cycle of Rosalina globularis. The agamontic generation (asexual reproductions) may be repeated numerous times until certain external conditions (temperature greater than 18°C?) prevail. The gamont (sexual stage) then forms and floats away to release gametes that settle as new agamonts. Modified from Sliter, 1965.

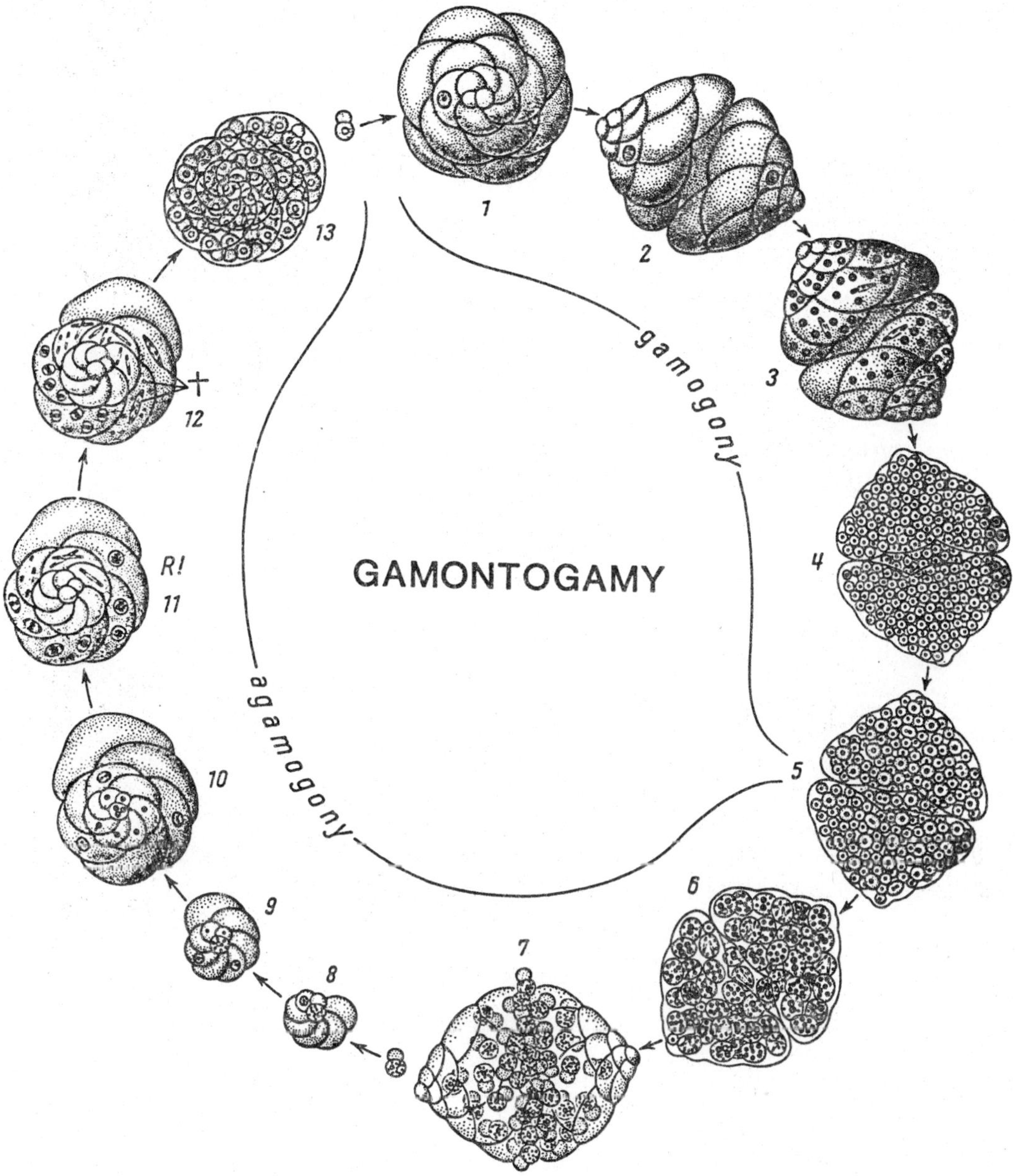


FIG. 12--Gamontogamy in Glabratella sulcata. In this genus, two or more gamonts (1) may join together with a membrane (2), a process given the special name of plastogamy, form gametes (3), dissolve the apertural sides of the tests (4) with fusion of gametes following (5). A further cell division takes place (6) with liberation of juvenile agamonts (7), which grow (8-10) to adults. The adult agamont undergoes meiosis (11, 12), forming individuals that grow into gamonts (13, 1). From Grell, 1979.

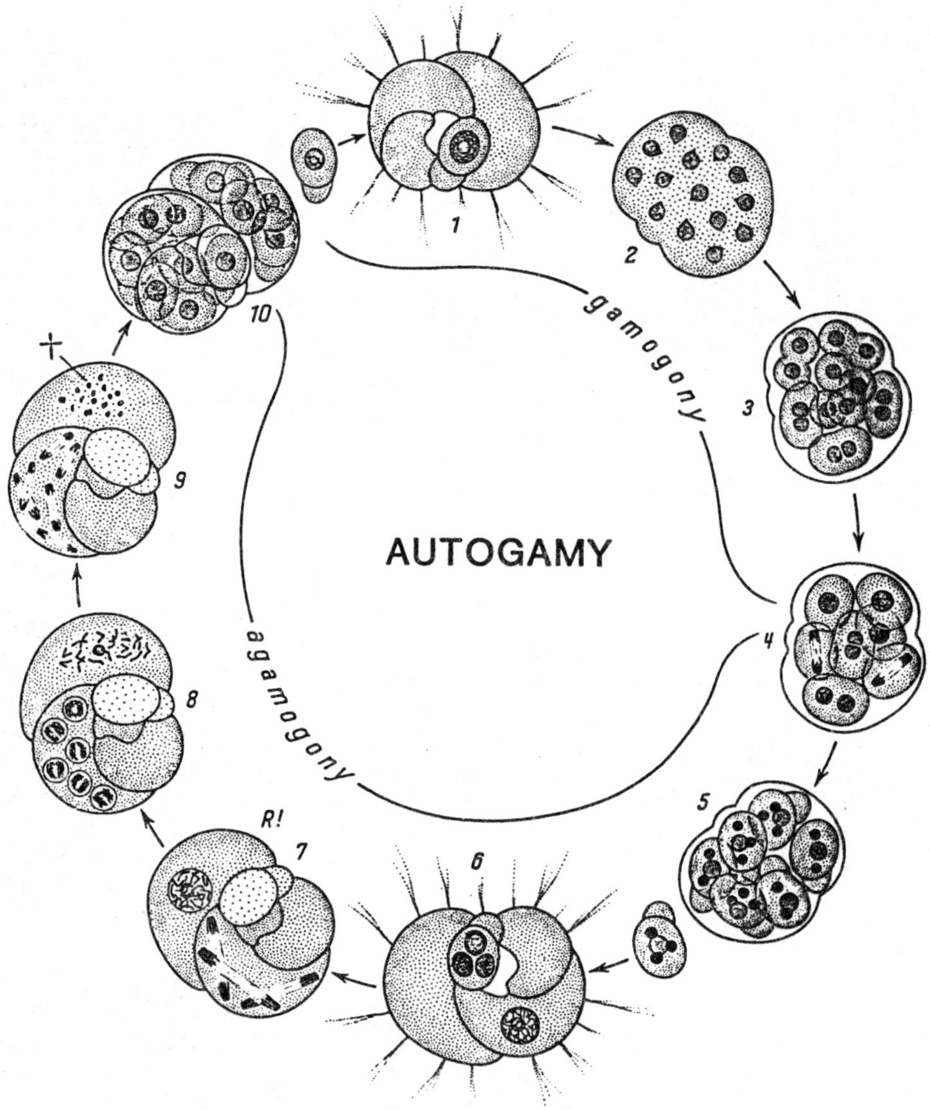


FIG. 13--Autogamy in Rotaliella roscoffensis. A gamont (1) forms gametic nuclei (2) which fuse (3) and undergo cell division (4) forming young agamonts (5) that grow to adults (6). These agamonts divide by meiosis (7-9) to form haploid cells (10). The gametes are formed by a single individual and fertilize one another within the parent test. From Grell, 1979.

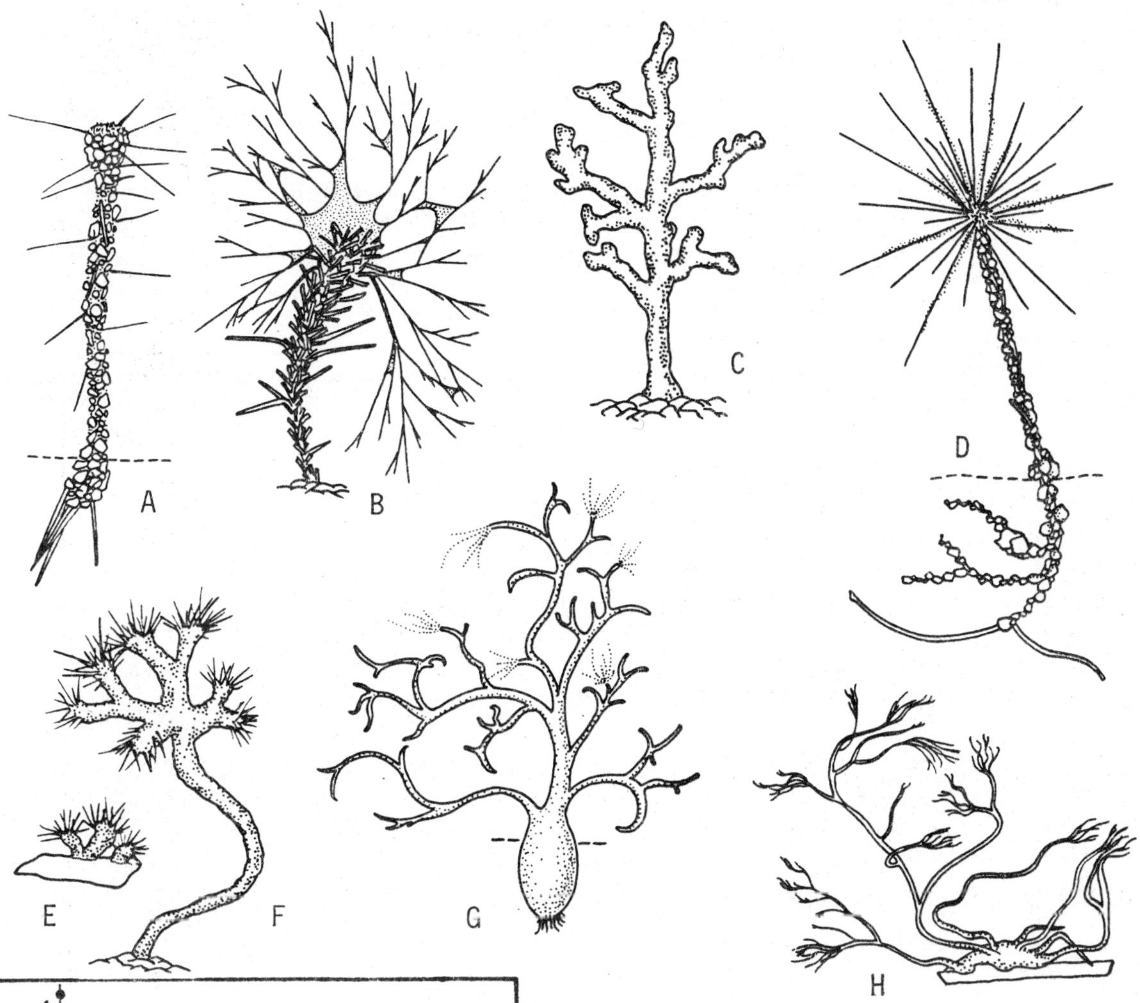


FIG. 14--Possible suspension feeding foraminifera. A. Jaculella Brady. B, E, F. Halyphysema Bowerbank. C. Dendrophrya Wright. D. Marsipella Norman. G. Pelosina Brady. H. Saccodendron Rhumbler.

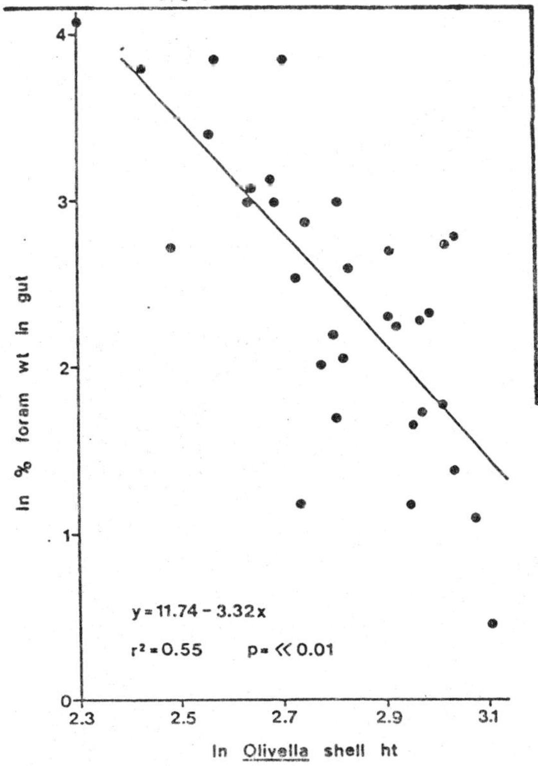


FIG. 15--Relationship between the gastropod Olivella shell height and foraminiferal weight in gut contents. Smaller Olivella ingest more foraminifera than larger ones. From Hickman and Lipps, in press.

# Classification of the Foraminiferida

ALFRED R. LOEBLICH, JR. AND HELEN TAPPAN
Department of Earth and Space Sciences
University of California Los Angeles
Los Angeles, California, 90024

## INTRODUCTION

A systematic classification generally reflects accurately the state of knowledge of a group. As information concerning the group increases, both similarities and differences become more apparent. New characters then may be utilized to develop a more complex classification, that better serves to indicate presumed relationships. Assuming that organisms have evolved through time, the more detailed classification may also prove more valuable to the biostratigrapher.

The earliest classifications that included the foraminifers regarded some as cephalopods, some as corals, and others as worms. Later most were included in the Cephalopodes Foraminifères, until Dujardin (1835) proved their Protozoan nature. In his earliest classification, d'Orbigny (1826) divided the group into five families. As more information was obtained, within 25 years the classification had tripled in complexity. Schultze (1854) recognized 10 families and 8 subfamilies, and 30 years later Brady (1884) recognized 10 families and 28 subfamilies. In 1927, Cushman recognized 45 families and 63 subfamilies, and the detail of classification has continued to increase (table 1). The most detailed to date is that of Saidova (1981), who utilized 517 suprageneric categories for the Cenozoic benthic foraminifers alone. Obviously, consideration of the Parathuramminacea, Endothyracea, Fusulinacea, and the many other genera restricted to the Paleozoic and Mesozoic would have required that this classification be still more lengthy.

Since de Blainville's publication (1825), the foraminifera have generally been regarded as representing an order, although at first within the Class Cephalopoda. D'Orbigny (1839) later elevated the foraminifera to class level, Delage and Hérouard (1896), Hofker (1951), Rauzer-Chernousova and Fursenko (1959), Grigyalis (1978), and Haynes (1981) considered the group as a subclass, but most publications include them as an order of the Rhizopoda (Cushman, 1927, 1940; Galloway, 1933; Chapman and Parr, 1936; Glaessner, 1945; Sigal in Piveteau, 1952; Pokorný, 1958; Loeblich and Tappan, 1964). Similarly, Protozoologists have generally regarded the foraminifera as an order (Grell, 1973; Levine et al., 1980), whereas the Russian school has generally placed the foraminifera as a subclass or class.

Table 1. Number of suprageneric categories used in foraminiferal classification

| | Class (Subcl.) | Order (Subor.) | Super-family | Family | Super-subfam. | Sub-family | Suprag. total |
|---|---|---|---|---|---|---|---|
| de Blainville, 1825 | | | | 9 | | | 9 |
| d'Orbigny, 1826 | | | | 5 | | | 5 |
| Schultze, 1854 | | | | 10 | | 8 | 18 |
| Reuss, 1862 | | | | 21 | | 7 | 28 |
| Brady, 1884 | | | | 10 | | 28 | 38 |
| Cushman, 1927 | | | | 45 | | 63 | 108 |
| Galloway, 1933 | | | | 35 | | 60 | 95 |
| Chapman, Parr, 1936 | | | 3 | 33 | | 84 | 120 |
| Glaessner, 1945 | | | 7 | 37 | | 38 | 82 |
| Sigal: Piveteau, 1952 | | (3) | 8 | 62 | | 50 | 123 |
| Rauzer, Fursenko, '59 | | 13 | 14 | 72 | | 90 | 189 |
| Loeblich, Tappan, '64 | | (5) | 17 | 94 | | 128 | 244 |
| Haynes, 1981 | | (2) | 32 | 101 | | ? | 144 |
| Saidova, 1981 | 3(5) | 19(16) | 50 | 118 | 39 | 267 | 517 |
| Loeblich, Tappan,'82 | | 1(10) | 42 | 220 | | 231 | 503 |

Most recently, Saidova (1981) considered the foraminifera as a subphylum including three classes. Nevertheless, the highest category devoted to this group must still fit within the general framework of protozoan classification. The most recent all-inclusive suprageneric classification of the Protozoa, developed by the Committee on Systematics and Evolution of the Society of Protozoologists (Levine et al., 1980), considers the Protozoa as a subkingdom of the Protista, with the following hierarchy relating to the foraminifera.

Kingdom Protista
Subkingdom Protozoa
Phylum Sarcomastigophora Honigberg & Balamuth, 1963
Subphylum Sarcodina Schmarda, 1871
Superclass Rhizopoda von Siebold, 1845
Class Granuloreticulosea de Saedeleer, 1934
Order Foraminiferida

This higher category subdivision was followed in the recent volumes "Synopsis and Classification of Living Organisms", hence we used ordinal status for the classification of living foraminifera therein (Tappan & Loeblich, 1982). No greater hierarchical status seems necessary for developing as detailed a classification as desired.

Of greater interest, perhaps, is the nature of the subdivisions of the foraminifera (regardless of the level assigned to these). The classifications cited in the table above are those that have been most inclusive. Others have appeared in the last 20 years that concerned only living taxa (Tappan & Loeblich, 1982), Cenozoic ones (Saidova, 1981), or only Paleozoic taxa (Miklukho-Maklay, 1963). Other reclassifications have covered only selected groups; among many that could be cited, only a few are listed here. Some have covered agglutinated taxa (Bermúdez & Rivero, 1963), or certain groups of these (Balakhmatova, 1972; Banner, 1966, 1970; Banner & Pereira, 1981; Barnard & Banner, 1980; Moullade, 1965; Tendal & Hessler, 1977). The fusulinids were revised by F. & G. Kahler (1966, 1967), Rozovskaya (1969, 1975), Toriyama (1960) and Zhang et al. (1981); the archaediscids by Pirlet & Conil (1973); involutinids and spirillinids by Hohenegger & Piller (1977), and the former also by Zaninetti (1976); and the miliolids by Anglada & Randrianasolo (1971), and Brönnimann & Zaninetti in Brönnimann et al. (1971). Many calcareous perforate taxa were reclassified by Reiss (1963), the Nodosariacea by Putrya (1970), the Discorbacea by Tappan & Loeblich (1966), and the planktonic foraminifera by Blow (1979), El Naggar (1971), Lipps (1964, 1966), Longoria & Gamper (1975), McGowran (1968), Moullade (1964), Pessagno (1967), and Saito et al. (1976, 1981).

This general activity in systematic classification does not indicate only continued subdivision or a reshuffling of old data; suprageneric revisions have resulted from the accumulation of new information as to wall composition (e.g., nature of the Barkerinidae, Involutinidae, Duostominidae), imperforate or canaliculate walls (as in the Spiroplectamminidae, Textulariidae, Ataxophragmiidae), development of lamellar structure (Lagenina), surface texture and spinosity (Globigerinacea), secondary septa, internal pillars, subepidermal partitions, buttresses (many agglutinated taxa); true monolamellar or bilamellar septa (Lagenina, Rotaliina), the crystallographic distinction of hyaline radial and hyaline oblique walls (formerly termed hyaline granular) and intermediates between these (Cibicididae, Elphidiidae). Other taxonomic changes have resulted from restudy of original material of poorly known type species of classical genera (Anomalina, Nonion).

Another major factor stimulating reclassification has been the still increasing number of known foraminiferal genera. When the Treatise of Invertebrate Paleontology on the Foraminifera was sent to press in 1962, some 1192 genera were recognized and 1267 included as synonyms of these. New information has since resulted in the reinstatement of many of the latter as distinct, and many additional genera have been described (fig. 1). Some of these probably are synonyms, and a few are homonyms. However, this continually increasing number of taxa, in combination with the new morphological and structural data has resulted in the necessity of a more detailed classification, as well as the reassignment of certain genera from one family to another. As of this writing, a total of 3558 generic taxa have been proposed for the foraminifera, nearly as many having appeared since the Treatise as were regarded valid therein! Some are from geologic horizons that were less well known earlier (e.g., Triassic), some from different facies than had been well studied (e.g., the deep sea), and many have resulted from more detailed methods of study, ranging from thin-sectioning for study of wall structures to electron microscopy.

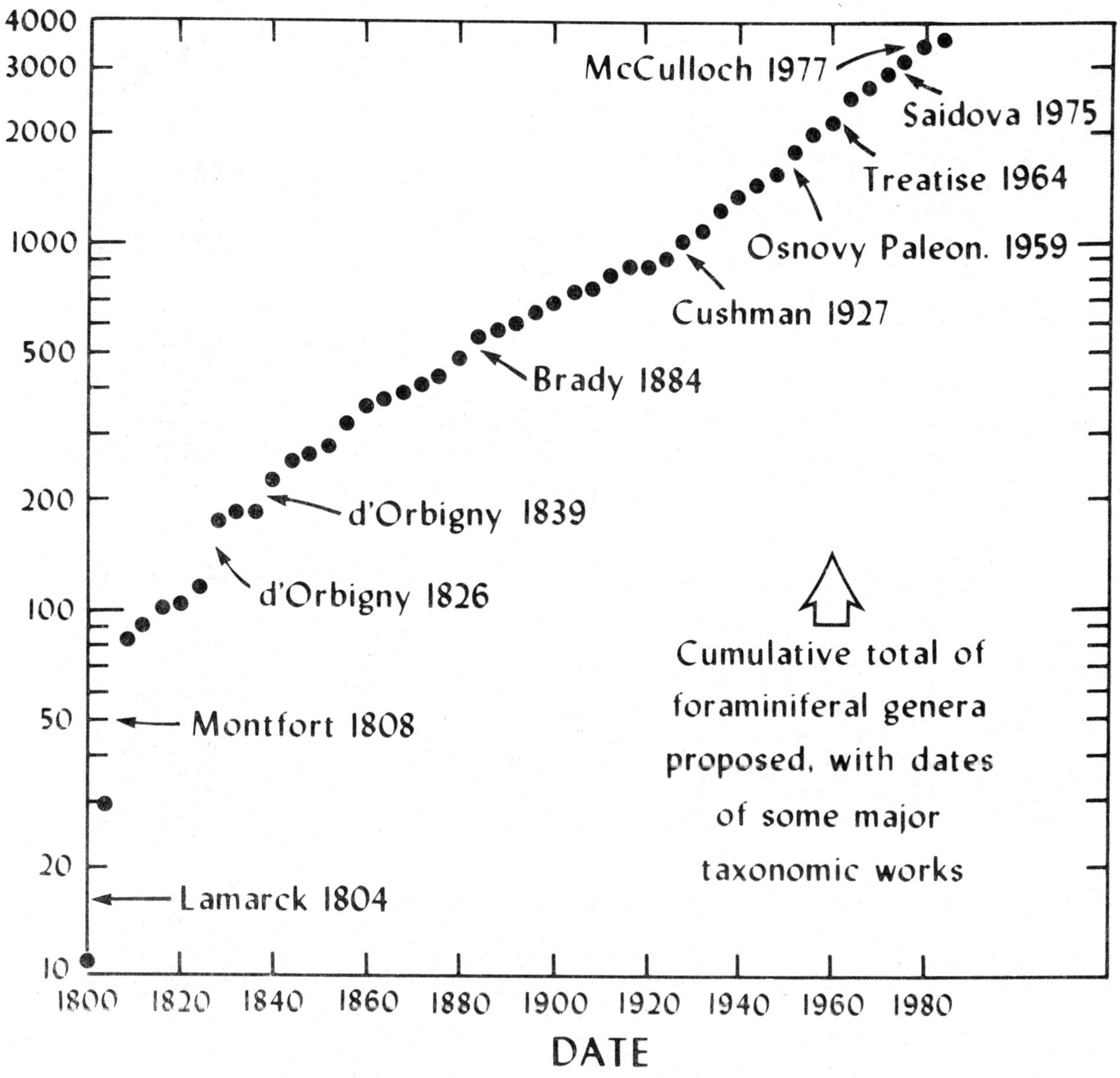


Fig. 1. Genera proposed through time; semi-logarithmic scale.

Our general philosophy regards the test composition and method of formation as of greatest importance (delimiting suborders), the single or multichambered nature of the test, presence or absence of wall perforations, canaliculi, canal systems, alveoli, simple or subdivided chamber interiors, and mode of chamber addition of next importance, and therefore separating the superfamilies.

The suprageneric classification that follows is only a reflection of the current state of knowledge and as new information accumulates in the future will also require modification, like all those that have preceded it.

## PROPOSED CLASSIFICATION

Order Foraminiferida Zborzewski, 1834
Suborder Allogromiina Loeblich & Tappan, 1961
Test membranous or proteinaceous, may have ferruginous encrustations or small quantity of agglutinated particles.

Superfamily Lagynacea Schultze, 1854
Family Lagynidae Schultze, 1854
Family Allogromiidae Rhumbler, 1904
Family Phthanotrochidae Arnold, 19788

Suborder Textulariina Delage & Hérouard, 1896
Test agglutinated, foreign matter held in organic or mineralized ground mass.

Superfamily Astrorhizacea Brady, 1881
Family Astrorhizidae Brady, 1881
Subfamily Astrorhizinae Brady, 1881
Subfamily Vanhoeffenellinae Saidova, 1981
Family Rhizamminidae Rhumbler, 1895
Subfamily Rhizammininae Rhumbler, 1895
Subfamily Dendrophryninae Vyalov, 1966
Subfamily Dendrophryinae Eimer & Fickert, 1899
Family Notodendrodidae Delaca, Lipps & Hessler, 1980
Family Hippocrepinidae Rhumbler, 1895
Family Hyperamminoididae Loeblich & Tappan, 1982
Family Schizamminidae Norvang, 1961
Family Stegnamminidae Moreman, 1930
Family Saccamminidae Brady, 1884
Subfamily Psammosphaerinae Eimer & Fickert, 1899
Subfamily Saccammininae Brady, 1884
Subfamily Thurammininae Miklukho-Maklay, 1963
Family Hemisphaeramminidae Loeblich & Tappan, 1961
Subfamily Hemisphaerammininae Loeblich & Tappan, 1961
Subfamily Crithionininae Hofker, 1972
Subfamily Oryctoderminae Saidova, 1981
Family Diffusilinidae Loeblich & Tappan, 1964
Superfamily Komokiacea Tendal & Hessler, 1977
Family Komokiidae Tendal & Hessler, 1977
Family Baculellidae Tendal & Hessler, 1977
Superfamily Ammodiscacea Reuss, 1862
Family Ammodiscidae Reuss, 1862
Subfamily Ammodiscinae Reuss, 1862
Subfamily Ammovolummininae Chernykh, 1967
Subfamily Tolypammininae Cushman, 1928
Subfamily Usbekistaniinae Vyalov, 1968
Subfamily Turritellellinae Saidova, 1981
Superfamily Rzehakinacea Cushman, 1933
Family Rzehakinidae Cushman, 1933
Superfamily Hormosinacea Haeckel, 1894
Family Aschemonellidae Eimer & Fickert, 1899

Subfamily Aschemocellinae Vyalov, 1966
Subfamily Aschemonellinae Eimer & Fickert, 1899
Family Hormosinidae Haeckel, 1894
Subfamily Reophacinae Cushman, 1910
Subfamily Cuneatinae Loeblich & Tappan, 1982
Subfamily Hormosininae Haeckel, 1894
Subfamily Nodosininae Saidova, 1981
Family Cribratinidae Loeblich & Tappan, 1964
Superfamily Lituolacea de Blainville, 1825
Family Oxinoxisidae Vyalov, 1968
Family Haplophragmoididae Maync, 1952
Subfamily Haplophragmoidinae Maync, 1952
Subfamily Discamminninae Mikhalevich, 1980
Family Haplophragmiidae Eimer & Fickert, 1899
Family Sphaeramminidae Cushman, 1933
Family Lituolidae de Blainville, 1825
Subfamily Lituolinae de Blainville, 1825
Subfamily Ammomarginulininae Podobina, 1978
Subfamily Flabellammininae Podobina, 1978
Family Ammosphaeroidinidae Cushman, 1927
Subfamily Ammosphaeroidininae Cushman, 1927
Subfamily Recurvoidinae Alekseychik-Mitskevich, 1973
Family Placopsilinidae Rhumbler, 1913
Family Haddoniidae Saidova, 1981
Superfamily Loftusiacea Brady, 1884
Family Barkerinidae Smout, 1956
Family Lituoliporidae Gušić & Velić, 1978
Family Cyclamminidae Marie, 1941
Subfamily Mesoendothyrinae Voloshinova, 1958
Subfamily Alveolophragmiinae Saidova, 1981
Subfamily Hemicyclammininae Banner, 1966
Subfamily Choffatellinae Maync, 1958
Subfamily Cyclammininae Marie, 1941
Family Biokovinidae Gušić, 1977
Family Spirocyclinidae Munier-Chalmas, 1887
Family Loftusiidae Brady, 1884
Family Coscinophragmatidae Thalmann, 1951
Superfamily Spiroplectamminacea Cushman, 1927
Family Spiroplectamminidae Cushman, 1927
Subfamily Spiroplectammininae Cushman, 1927
Subfamily Spirotextulariinae Saidova, 1975
Subfamily Morulaeplectinae Saidova, 1981
Superfamily Trochamminacea Schwager, 1877
Family Trochamminidae Schwager, 1877
Subfamily Trochammininae Schwager, 1877
Subfamily Rotaliammininae Saidova, 1981
Subfamily Conotrochammininae Saidova, 1981
Subfamily Jadammininae Saidova, 1981
Subfamily Tritaxinae Mikhalevich, 1972
Subfamily Polystomammininae Brönnimann & Beurlen, 1977
Subfamily Remaneicinae Loeblich & Tappan, 1964
Family Nezzazatidae Hamaoui & Saint Marc, 1970

Subfamily Nezzazatinae Hamaoui & Saint Marc, 1970
Subfamily Coxitinae Hamaoui & Saint Marc, 1970
Superfamily Verneuilinacea Cushman, 1911
Family Textulariopsidae Loeblich & Tappan, 1982
Family Verneuilinidae Cushman, 1911
Subfamily Verneuilininae Cushman, 1911
Subfamily Dorothiinae Balakhmatova, 1972
Subfamily Spiroplectinatinae Cushman, 1928
Family Plectorecurvoididae Loeblich & Tappan, 1954
Family Pseudobolivinidae Wiesner, 1931
Family Nouriidae Chapman & Parr, 1936
Family Plectotrochamminidae Saidova, 1981
Superfamily Ataxophragmiacea Schwager, 1877
Family Ataxophragmiidae Schwager, 1877
Family Pavonitinidae Loeblich & Tappan, 1961
Family Textulariellidae Grönhagen & Luterbacher, 1966
Family Pfenderinidae Smout & Sugden, 1962
Subfamily Pfenderininae Smout & Sugden, 1962
Subfamily Kurnubiinae Redmond, 1964
Family Coskinolinidae Moullade, 1965
Superfamily Textulariacea Ehrenberg, 1839
Family Textulariidae Ehrenberg, 1839
Family Valvulinidae Berthelin, 1880
Family Tawitawiidae Loeblich & Tappan, 1961
Family Globotextulariidae Cushman, 1927
Family Chrysalidinidae Neagu, 1968
Superfamily Orbitolinacea Martin, 1890
Family Dicyclinidae Loeblich & Tappan, 1964
Subfamily Cyclolininae Loeblich & Tappan, 1964
Subfamily Dicyclininae Loeblich & Tappan, 1964
Family Orbitolinidae Martin, 1890
Subfamily Dictyoconinae Moullade, 1965
Subfamily Orbitolininae Martin, 1890

Suborder Fusulinina Wedekind, 1937

Wall of homogeneously microgranular calcite, of tiny tightly packed equidimensional subangular crystals, a few μm in diameter. Advanced forms with differentiation of 2 or more wall layers.

Superfamily Parathuramminacea E. V. Bykova, 1955
Family Parathuramminidae E. V. Bykova, 1955
Superfamily Earlandiacea Cummings, 1955
Family Earlandiidae Cummings, 1955
Family Pseudoammodiscidae Conil & Lys in Conil & Pirlet, 1970
Superfamily Archaediscacea Cushman, 1928
Family Archaediscidae Cushman, 1928
Subfamily Planoarchaediscinae Mamet, 1975
Subfamily Archaediscinae Cushman, 1928
Subfamily Asteroarchaediscinae Miklukho-Maklay, 1957
Family Lasiodiscidae Reytlinger, 1956
Superfamily Moravamminacea Pokorný, 1951
Family Caligellidae Reytlinger, 1959

- Family Moravamminidae Pokorný, 1951
- Superfamily Nodosinellacea Rhumbler, 1895
  - Family Tuberitinidae Miklukho-Maklay, 1958
  - Family Ptychocladiidae Elias, 1950
  - Family Paratikhinellidae Loeblich & Tappan, 1982
  - Family Nodosinellidae Rhumbler, 1895
  - Family Colaniellidae Fursenko, 1959
- Superfamily Palaeotextulariacea Galloway, 1933
  - Family Palaeotextulariidae Galloway, 1933
    - Subfamily Palaeotextulariinae Galloway, 1933
    - Subfamily Pseudopalmulinae E. V. Bykova in Rauzer-Chernousova & Fursenko, 1959
  - Family Semitextulariidae Pokorný, 1956
  - Family Biseriamminidae Chernysheva, 1941
- Superfamily Tetrataxacea Galloway, 1933
  - Family Pseudotaxidae Mamet, 1974
  - Family Tetrataxidae Galloway, 1933
  - Family Valvulinellidae Loeblich & Tappan, 1982
  - Family Abadehellidae Loeblich & Tappan, 1982
  - Family Tournayellidae Dain, 1953
    - Subfamily Tournayellinae Dain, 1953
    - Subfamily Forschiinae Dain, 1953
    - Subfamily Septabrunsiininae Conil & Lys, 1977
    - Subfamily Lituotubellinae Miklukho-Maklay, 1963
    - Subfamily Chernyshinellinae Reytlinger, 1958
  - Family Endothyridae Brady, 1884
    - Subfamily Endothyrinae Brady, 1884
    - Subfamily Haplophragmellinae Reytlinger, 1959
    - Subfamily Endothyranopsinae Reytlinger, 1958
  - Family Bradyinidae Reytlinger, 1950
  - Family Loeblichiidae Cummings, 1955
- Superfamily Fusulinacea von Moeller, 1878
  - Family Ozawainellidae Thompson & Foster, 1937
    - Subfamily Ozawainellinae Thompson & Foster, 1937
    - Subfamily Pseudostaffellinae Putrya, 1956
  - Family Schubertellidae Skinner, 1931
    - Subfamily Schubertellinae Skinner, 1931
    - Subfamily Boultoniinae Skinner & Wilde, 1954
  - Family Fusulinidae von Moeller, 1878
    - Subfamily Fusulinellinae Staff & Wedekind, 1910
    - Subfamily Fusulininae von Moeller, 1878
    - Subfamily Eofusulininae Rauzer-Chernousova & Rozovskaya, 1955
    - Subfamily Wedekindellininae F. Kahler & G. Kahler, 1966
  - Family Schwagerinidae Dunbar & Henbest, 1930
    - Subfamily Polydiexodininae Miklukho-Maklay, 1953
    - Subfamily Chusenellinae F. Kahler & G. Kahler, 1966
    - Subfamily Schwagerininae Dunbar & Henbest, 1930
    - Subfamily Pseudoschwagerininae Chang, 1963
  - Family Verbeekinidae Staff & Wedekind, 1910
    - Subfamily Verbeekininae Staff & Wedekind, 1910
    - Subfamily Misellininae Miklukho-Maklay, 1958
    - Subfamily Pseudodoliolininae Leven, 1963

Subfamily Kahlerininae Leven, 1963
Subfamily Cheniinae F. Kahler & G. Kahler, 1966
Family Neoschwagerinidae Dunbar, 1948
Subfamily Neoschwagerininae Dunbar, 1948
Subfamily Sumatrininae Silvestri, 1933
Subfamily Thailandininae Toriyama & Kanmera, 1968
Family Staffellidae Miklukho-Maklay, 1949
Subfamily Pseudoendothyrinae Mamet, 1970
Subfamily Staffellinae Miklukho-Maklay, 1949

Suborder Involutinina Hohenegger & Piller, 1977

Proloculus followed by enrolled tubular second chamber, wall calcareous, perforate, radiate, probably originally aragonitic, but may be recrystallized to homogeneous microgranular structure; wall thickened or with pillar-like structures in umbilical region of one or both sides.

Family Involutinidae Bütschli, 1880
Family Trocholinidae Kristan-Tollman, 1963

Suborder Miliolina Delage & Hérouard, 1896

Test of porcellaneous calcite, commonly with organic lining, may have added adventitious material; usually imperforate in post-embryonic stage. May have flexostyle (spiral passage) between proloculus and later chambers.

Superfamily Cornuspiracea Schultze, 1854
Family Cornuspiridae Schultze, 1854
Subfamily Cornuspirinae Schultze, 1854
Subfamily Cornuspiroidinae Saidova, 1981
Subfamily Hemigordiopsinae Nikitina, 1969
Subfamily Meandropsininae Saidova, 1981
Subfamily Calcivertellinae Loeblich & Tappan, 1964
Family Fischerinidae Millett, 1898
Subfamily Fischerininae Millett, 1898
Subfamily Fischerinellinae Saidova, 1981
Subfamily Zoyaellinae Saidova, 1981
Subfamily Glomulininae Saidova, 1981
Subfamily Wiesnerellinae Saidova, 1981
Family Nubeculariidae Jones, 1875
Subfamily Nodobaculariinae Cushman, 1927
Subfamily Nubeculariinae Jones, 1875
Family Ophthalmidiidae Wiesner, 1920
Family Spiroloculinidae Wiesner, 1920
Family Discospirinidae Wiesner, 1931
Superfamily Miliolacea Ehrenberg, 1839
Family Squamulinidae Haeckel, 1894
Family Miliolidae Ehrenberg, 1839
Subfamily Miliolinae Ehrenberg, 1839
Subfamily Hauerininae Schwager, 1876
Subfamily Miliolinellinae Vella, 1957
Subfamily Sigmoilinitinae Łuczkowska, 1974

Subfamily Tubinellinae Rhumbler, 1906
Family Riveroinidae Saidova, 1981
Family Fabulariidae Ehrenberg, 1839
Superfamily Orbitolitacea Gray, 1840
Family Milioliporidae Brönnimann & Zaninetti, 1971
Family Peneroplidae Schultze, 1854
Family Meandropsinidae Henson, 1948
Family Rhapydioninidae Keijzer, 1945
Subfamily Praerhapydionininae Hamaoui & Fourcade, 1973
Subfamily Rhapydionininae Keijzer, 1945
Subfamily Crateritinae Saidova, 1981
Family Orbitolitidae Gray, 1840
Subfamily Archaiasinae Cushman, 1927
Subfamily Orbitolitinae Gray, 1840
Subfamily Fusarchaiasinae Saidova, 1981
Family Keramosphaeridae Brady, 1884
Family Alveolinidae Ehrenberg, 1839

Suborder Silicoloculinina Resig, Lowenstam, Echols & Weiner, 1980
Wall imperforate, of secreted opaline silica.

Family Silicoloculinidae Resig, Lowenstam, Echols, & Weiner, 1980

Suborder Spirillinina Hohenegger & Piller, 1975

Wall of calcite, optically a single crystal or few to a mosaic of crystals; a-axis preferred orientation along axis of coiling and c-axis parallel umbilical surface (Patellina); may have pseudopores or micropores filled with organic matter and closed by sieve plates. Wall formed by accretion at edge, not with organic template formed by pseudopodia.

Family Spirillinidae Reuss, 1862
Family Placentulinidae Kasimova, Poroshina, & Geodakchan, 1980
Family Patellinidae Rhumbler, 1906

Suborder Lagenina Delage & Hérouard, 1896

Wall of monolamellar optically radiate and structurally radiate calcite (crystal c-axes perpendicular to surface), crystal units enveloped by organic membrane.

Family Geinitzinidae Bozorgnia, 1973
Family Eocristellariidae Loeblich & Tappan, 1982
Family Nodosariidae Ehrenberg, 1839
Subfamily Nodosariinae Ehrenberg, 1839
Subfamily Lingulininae Loeblich & Tappan, 1961
Subfamily Plectofrondiculariinae Cushman, 1927
Family Vaginulinidae Reuss, 1860
Subfamily Lenticulininae Chapman, Parr & Collins, 1934
Subfamily Marginulininae Wedekind, 1937
Subfamily Vaginulininae Reuss, 1860
Subfamily Lingulinopsinae Loeblich & Tappan, 1962
Family Polymorphinidae d'Orbigny, 1839

Subfamily Polymorphininae d'Orbigny, 1839
Subfamily Webbinellinae Rhumbler, 1904
Subfamily Ramulininae Brady, 1884
Family Glandulinidae Reuss, 1860
Subfamily Glandulininae Reuss, 1860
Subfamily Entolingulininae Saidova, 1981
Subfamily Seabrookiinae Cushman, 1927
Subfamily Oolininae Loeblich & Tappan, 1961

Suborder Rotaliina Delage & Hérouard, 1896

Hyaline, perforate lamellar calcareous tests of aragonite (orthorhombic) or calcite (hexagonal crystal form).

Superfamily Duostominacea Brotzen, 1963
Family Asymmetrinidae Brotzen, 1963
Family Duostominidae Brotzen, 1963
Family Oberhauserellidae Fuchs, 1970
Superfamily Robertinacea Reuss, 1850
Family Conorboididae Thalmann, 1952
Family Ceratobuliminidae Cushman, 1927
Subfamily Ceratobulimininae Cushman, 1927
Subfamily Reinholdellinae Seiglie & Bermúdez, 1965
Family Robertinidae Reuss, 1850
Subfamily Robertininae Reuss, 1850
Subfamily Alliatininae McGowran, 1966
Subfamily Ungulatellinae Seiglie, 1964
Subfamily Cassidulinitinae Saidova, 1981
Family Epistominidae Wedekind, 1937
Subfamily Epistomininae Wedekind, 1937
Subfamily Epistominoidinae Saidova, 1981
Family Mississippinidae Saidova, 1981
Subfamily Stomatorbininae Saidova, 1981
Subfamily Mississippininae Saidova, 1981
Superfamily Pleurostomellacea Reuss, 1860
Family Pleurostomellidae Reuss, 1860
Subfamily Pleurostomellinae Reuss, 1860
Subfamily Wheelerellinae Petters, 1954
Subfamily Cribropleurostomellinae Owen, 1971
Superfamily Buliminacea Jones, 1875
Family Fursenkoinidae Loeblich & Tappan, 1961
Subfamily Fursenkoininae Loeblich & Tappan, 1961
Subfamily Sigmavirgulininae Saidova, 1981
Family Bolivinidae Glaessner, 1937
Family Bolivinoididae Loeblich & Tappan, 1982
Family Bolivinitidae Cushman, 1927
Family Loxostomatidae Loeblich & Tappan, 1962
Family Annulopatellinidae Loeblich & Tappan, 1964
Family Bolivinellidae Hayward, 1980
Family Trimosinidae Saidova, 1981
Family Pavoninidae Eimer & Fickert, 1899
Subfamily Pavonininae Eimer & Fickert, 1899
Subfamily Reussellinae Cushman, 1933

Family Turrilinidae Cushman, 1927
Family Praebuliminidae Loeblich & Tappan, 1982
Family Caucasinidae N. K. Bykova in Rauzer-Chernousova & Fursenko, 1959
Subfamily Baggatellinae N. K. Bykova in Rauzer-Chernousova & Fursenko, 1959
Subfamily Caucasininae N. K. Bykova in Rauzer-Chernousova & Fursenko, 1959
Family Lacosteinidae Sigal in Piveteau, 1952
Family Eouvigerinidae Cushman, 1927
Family Stilostomellidae Finlay, 1947
Family Buliminidae Jones, 1875
Subfamily Stainforthiinae Saidova, 1982
Subfamily Buliminiinae Jones, 1875
Family Buliminellidae Hofker, 1951
Family Virgulinellidae Loeblich & Tappan, 1982
Family Siphogenerinoididae Saidova, 1981
Subfamily Siphogenerinoidinae Saidova, 1981
Subfamily Siphogenerininae Loeblich & Tappan, 1982
Family Uvigerinidae Cushman, 1913
Subfamily Uvigerininae Cushman, 1913
Subfamily Trifarininae Srinivasan, 1966
Family Millettiidae Saidova, 1981
Family Sphaeroidinidae Cushman, 1927
Family Tremachoridae Lipps & K. E. Lipps, 1967
Family Delosinidae Parr, 1950
Superfamily Cassidulinacea d'Orbigny, 1839
Family Islandiellidae Loeblich & Tappan, 1964
Subfamily Islandiellinae Loeblich & Tappan, 1964
Subfamily Reissiinae Saidova, 1981
Subfamily Orthoplectinae Loeblich & Tappan, 1982
Family Cassidulinidae d'Orbigny, 1839
Subfamily Ehrenbergininae Cushman, 1927
Subfamily Cassidulininae d'Orbigny, 1839
Superfamily Heterohelicacea Cushman, 1927
Family Heterohelicidae Cushman, 1927
Subfamily Guembelitriinae Gallitelli, 1957
Subfamily Heterohelicinae Cushman, 1927
Subfamily Gublerininae Aliyulla, 1977
Subfamily Pseudoguembelinidae Aliyulla, 1977
Family Chiloguembelinidae Reiss, 1963
Superfamily Planomalinacea Bolli, Loeblich, & Tappan, 1957
Family Globigerinelloididae Longoria, 1974
Subfamily Globigerinelloidinae Longoria, 1974
Subfamily Eohastigerinellinae Loeblich & Tappan, 1982
Family Planomalinidae Bolli, Loeblich & Tappan, 1957
Family Schackoinidae Pokorný, 1958
Superfamily Rotaliporacea Sigal, 1958
Family Hedbergellidae Loeblich & Tappan, 1961
Subfamily Hedbergellinae Loeblich & Tappan, 1961
Subfamily Praeglobotruncaninae Loeblich & Tappan, 1982
Family Favusellidae Longoria, 1974

Subfamily Caucasellinae Longoria, 1974
Subfamily Favusellinae Longoria, 1974
Family Rotaliporidae Sigal, 1958
Subfamily Ticinellinae Longoria, 1974
Subfamily Rotaliporinae Sigal, 1958
Superfamily Globotruncanacea Brotzen, 1942
Family Globotruncanidae Brotzen, 1942
Subfamily Globotruncaninae Brotzen, 1942
Subfamily Globotruncanellinae Maslakova, 1964
Subfamily Abathomphalinae Pessagno, 1967
Family Rugoglobigerinidae Subbotina in Rauzer-Chernousova & Fursenko, 1959
Superfamily Hantkeninacea Cushman, 1927
Family Hantkeninidae Cushman, 1927
Family Cassigerinellidae Bolli, Loeblich & Tappan, 1957
Superfamily Globigerinacea Carpenter, Parker & Jones, 1862
Family Globigerinidae Carpenter, Parker & Jones, 1862
Family Hastigerinidae Bolli, Loeblich & Tappan, 1957
Superfamily Globorotaliacea Cushman, 1927
Family Eoglobigerinidae Blow, 1979
Family Candeinidae Cushman, 1927
Family Globorotaliidae Cushman, 1927
Subfamily Globorotaliinae Cushman, 1927
Subfamily Truncorotaloidinae Loeblich & Tappan, 1961
Family Turborotalitidae Hofker, 1976
Family Catapsydracidae Bolli, Loeblich & Tappan, 1957
Superfamily Discorbacea Ehrenberg, 1838
Family Bagginidae Cushman, 1927
Subfamily Conorbininae Hofker, 1954
Subfamily Serovaininae Sliter, 1968
Subfamily Baggininae Cushman, 1927
Family Pegidiidae Heron-Allen & Earland, 1928
Family Eponididae Hofker, 1951
Subfamily Eponidinae Hofker, 1951
Subfamily Alabaminellinae Saidova, 1981
Subfamily Rectoeponidinae Saidova, 1981
Subfamily Cribroeponidinae Shchedrina, 1964
Family Bueningiidae Saidova, 1981
Family Bronnimanniidae Loeblich & Tappan, 1982
Family Laticarininidae Hofker, 1951
Family Pseudoparrellidae Voloshinova, 1952
Subfamily Pseudoparrellinae Voloshinova, 1952
Subfamily Stetsoniinae Saidova, 1981
Subfamily Concavellinae Saidova, 1981
Family Discorbidae Ehrenberg, 1838
Subfamily Discorbinae Ehrenberg, 1838
Subfamily Rosalininae Reiss, 1963
Family Asterigerinatidae Reiss, 1963
Family Glabratellidae Loeblich & Tappan, 1964
Family Buliminoididae Seiglie, 1970
Family Rotaliellidae Loeblich & Tappan, 1964
Family Epistomariidae Hofker, 1954

Subfamily Epistomariinae Hofker, 1954
Subfamily Eponidellinae Seiglie & Bermúdez, 1965
Subfamily Nuttallidinae Saidova, 1981
Subfamily Alfredininae S. K. Singh & Kalia, 1972
Family Siphoninidae Cushman, 1927
Subfamily Siphonininae Cushman, 1927
Subfamily Siphonidinae Saidova, 1981
Subfamily Siphoninoidinae Loeblich & Tappan, 1982
Family Asterigerinidae d'Orbigny, 1839
Family Amphisteginidae Cushman, 1927
Family Boreloididae Reiss, 1963
Family Lepidocyclinidae Scheffen, 1932
Subfamily Lepidocyclininae Scheffen, 1932
Subfamily Helicolepidininae Tan, 1936
Family Planulinidae Bermúdez, 1952
Subfamily Planulininae Bermúdez, 1952
Subfamily Caribeanellinae Saidova, 1981
Family Cibicididae Cushman, 1927
Subfamily Cibicidinae Cushman, 1927
Subfamily Falsocibicidinae Saidova, 1981
Subfamily Stichocibicidinae Saidova, 1981
Subfamily Annulocibicidinae Saidova, 1981
Family Cymbaloporidae Cushman, 1927
Subfamily Cymbaloporinae Cushman, 1927
Subfamily Fabianiinae Deloffre & Hamaoui, 1973
Subfamily Halkyardiinae Saidova, 1981
Family Planorbulinidae Schwager, 1877
Family Acervulinidae Schultze, 1854
Subfamily Acervulininae Schultze, 1854
Subfamily Gypsininae Silvestri, 1905
Family Linderinidae Freudenthal, 1969
Family Homotrematidae Cushman, 1927
Subfamily Carpenteriinae Saidova, 1981
Subfamily Rupertininae Loeblich & Tappan, 1961
Subfamily Victoriellinae Chapman & Crespin, 1930
Subfamily Homotrematinae Cushman, 1927
Superfamily Orbitoidacea Schwager, 1876
Family Orbitoididae Schwager, 1876
Subfamily Orbitoidinae Schwager, 1876
Subfamily Omphalocyclinae Vaughan in Cushman, 1928
Family Hellenocyclinidae Freudenthal, 1969
Superfamily Rotaliacea Ehrenberg, 1839
Family Rotaliidae Ehrenberg, 1839
Subfamily Cuvillierininae Loeblich & Tappan, 1964
Subfamily Pararotaliinae Reiss, 1963
Subfamily Rotaliinae Ehrenberg, 1839
Subfamily Chapmanininae Thalmann, 1938
Family Calcarinidae Schwager, 1876
Family Elphidiidae Galloway, 1933
Subfamily Elphidiinae Galloway, 1933
Subfamily Notorotaliinae Hornibrook, 1961
Subfamily Faujasininae Bermúdez, 1952

Subfamily Parrellininae Saidova, 1981
Family Lepidorbitoididae Vaughan in Cushman, 1933
Family Pseudorbitoididae Rutten, 1935
Subfamily Pseudorbitoidinae Rutten, 1935
Subfamily Vaughanininae Macgillavry, 1963
Family Nummulitidae de Blainville, 1825
Subfamily Nummulitinae de Blainville, 1925
Subfamily Cycloclypeinae Bütschli, 1880
Family Discocyclinidae Galloway, 1928
Family Orbitoclypeidae Brönnimann, 1946
Family Miogypsinidae Vaughan in Cushman, 1928
Superfamily Chilostomellacea Brady, 1881
Family Chilostomellidae Brady, 1881
Family Quadrimorphinidae Saidova, 1981
Family Globorotalitidae Loeblich & Tappan, 1982
Family Spirotectidae Saidova, 1981
Family Parrelloididae Hofker, 1956
Family Alabaminidae Hofker, 1951
Subfamily Gyroidininae Saidova, 1981
Subfamily Alabamininae Hofker, 1951
Family Oridorsalidae Loeblich & Tappan, 1982
Family Osangulariidae Loeblich & Tappan, 1964
Family Gavellinellidae Hofker, 1956
Family Karreriidae Saidova, 1981
Family Trichohyalidae Saidova, 1981
Family Almaenidae Myatlyuk in Rauzer-Chernousova & Fursenko, 1959
Subfamily Almaeninae Myatlyuk in Rauzer-Chernousova & Fursenko, 1959
Subfamily Anomalinellinae Saidova, 1981
Superfamily Nonionacea Schultze, 1854
Family Nonionidae Schultze, 1854
Subfamily Nonioninae Schultze, 1854
Subfamily Pulleniinae Schwager, 1877
Subfamily Spirotectininae Saidova, 1981
Family Melonidae Chapman, Parr & Collins, 1934

Suborder Carterinina Loeblich & Tappan, 1981

Trochospiral test of secreted spicules, each a single elongated crystal of calcite, held in organic matrix and groundmass of smaller spicules.

Family Carterinidae Loeblich & Tappan, 1955

# Ecology of Benthic Foraminifera

Barun K. Sen Gupta
Department of Geology
Louisiana State University
Baton Rouge, Louisiana 70803

## ABSTRACT

Benthic foraminifera live in a wide variety of marine habitats, from marshes to abyssal plains. Both epifaunal and infaunal adaptations are present. The distribution of species is controlled by a complex of abiotic and biotic variables of the environment, usually resulting in a depth-zonation of assemblages on continental margins.

The composition of the substrate may have a direct or indirect influence on the constitution of the assemblage. Laboratory studies and field observations demonstrate the effects of temperature and salinity variations, particularly on nearshore species. Many calcareous species have algal or chloroplast symbionts; their distribution is partly controlled by light tolerance.

Benthic foraminifera are omnivorous and feed mainly by the pseudopodia. Large supply of nutrients, as in areas of upwelling or water mixing, may give rise to large blooms of opportunistic species.

Marine marshes of diverse geographic locations support some common species. Deep bathyal or abyssal species are widespread, but dominance patterns are apparently associated with identifiable water masses, although causal relations are unclear.

## INTRODUCTION

The foraminifera constitute an "order" with both benthic and planktonic adaptations. The vast majority of foraminiferal taxa, however, are benthic. They include all tectinous, agglutinated, calcareous microgranular, and calcareous porcelaneous forms, and the bulk of the calcareous hyaline forms--in all 109 out of 119 families in Loeblich and Tappan's revised classification (1974, 1981). Modern members of this diverse group are among the most widely distributed marine organisms, and usually the most abundant shelled benthic meiofauna. Environmental information on these species, particularly that on depth, temperature, and substrate, has been reported in the literature for a long time, but serious efforts to understand aspects of benthic foraminiferal ecology began in the 1930's (Myers & Cole, 1957).

## DEPTH-ZONATION

Following the pioneering work of F.B. Phleger, F.L. Parker, and O.L. Bandy in the late 1940's and the 1950's, numerous species or assemblage distribution studies have been undertaken on continental shelves and slopes with the focus on depth-zonation; some of these studies have included the deep ocean floor. Many of these investigations had an ultimate paleoecologic and geologic goal: interpretation of past environments from the distribution of fossil benthic foraminifera, using the bathymetric patterns seen for their modern representatives or analogs. It was, of course, understood that water depth is not a primary environmental factor controlling the distribution of modern foraminifera, but only a "location in three-dimensional space" (Gibson, 1968). The existence of foraminiferal depth-zonation is related to the fact that some or all of the factors that affect local distributions of benthic foraminifera, such as substrate grain-size, water temperature, salinity, light, pressure, dissolved oxygen, nutrients, and even biotic factors, may have a relation to the depth gradient.

In any case, much of the ecological literature on modern benthic foraminifera deals with depth-zones proposed for small or large parts of the sea floor by various authors, and in one well-known case, by a committee (Albers et al., 1966). The earlier studies (see Sen Gupta, 1977, for review) were based on the observed depth ranges of selected stenobathic species, and often on the common depth range of several of these species -- a procedure similar to the determination of the age of a stratigraphic unit by using the common geological range of a group of fossils. Later, methods of numerical analysis were used, the data generally being the relative abundance (percentage) of species in the total or living assemblage, or simply the presence or absence of species. The most popular techniques have been different forms of cluster and factor analyses. Other techniques, such as canonical variate analysis (Buzas, 1967) or recurrent group analysis (Sen Gupta & Hayes, 1979), have also been used.

The classic area for the study of foraminiferal depth-distribution has been the Pacific continental margin of North America. The earliest of these studies in this region was by Natland (1933) who recognized "five distinct faunal assemblages in close proximity" in the channel between Long Beach and Santa Catalina Island, California. These assemblages were regarded as constituting "faunal- or life-zones" with characteristic bottom-water temperatures and depths. The zones and some of their abundant species were as follows: Zone I- nearshore, brackish-water lagoon: *Ammonia beccarii*; Zone II- depth range 4-38 m: *Nonion scapha* and *Elphidium articulatum*; Zone III- depth range 38-275 m: species of *Cassidulina*; Zone IV- depth range 274-1981± m: species of *Bolivina*, *Cassidulina cushmani*, *Globobulimina pacifica*, *Uvigerina peregrina*; Zone V- depth range 1981±-2542± m: *Bulimina rostrata*, *Pullenia bulloides*. In the same general area, McGlasson (1959), using relative abundances of benthic foraminiferal species, identified three depth zones (0-37, 38-74, and 75-183+m) around Santa Catalina Island. The compositions of living and dead assemblages were found to be "essentially different" in each of these zones.

Expansion of depth-distribution studies to other areas of the Pacific coast resulted in the publication of several local zonal schemes. However, for a long stretch of the shoreline, from Central America to northern California, a distinct fauna, dominated by *Ammonia beccarii* and species of *Elphidium* and *Quinqueloculina*, was recognized between the tide line and a depth of 20-30 m (Smith, 1964).

Overall, the foraminiferal depth-zones most commonly identified on continental shelves correspond to the inner, middle, and outer parts of the shelves. The inner or nearshore zone is generally limited to a depth less than 30 m, in many areas less than 20 m. The bathymetric limits of the middle and outer shelf zones are highly variable, primarily because of the variation in the depth of the shelf-edge. Multiple foraminiferal zones have been reported from shelves whose outer limits range between 50 and 200 m in depth. The shelf-edge, however, does not necessarily constitute a conspicuous faunal boundary. For well-studied continental shelves, generalizations can be made regarding dominant or relatively abundant species within particular biogeographic provinces for each of the depth-zones, in spite of special patterns of local distribution. For some latitudinally extensive shelf areas, however, conflicting schemes have been proposed by different workers. Similar assemblages are likely to occupy comparable locations on different shelves, with respect to the shoreline and the shelf-edge, but the habitat depths may vary, because of change in bottom-water conditions.

The distribution of some stenobathic species, with varying depth limits in different areas, transgresses many biogeographic provinces. Most of these species, e.g., *Ammonia beccarii*, *Elphidium excavatum*, and *Quinqueloculina seminulum*, usually characterize the inner and middle shelf zones. Indeed, the nearshore zone in most areas is inhabited by an *Ammonia-Elphidium* assemblage. *Elphidium excavatum*, however, is also adapted to much deeper waters in some areas. Recent studies have demonstrated its existence, and even dominance, on parts of the western North American continental slope (Schafer & Cole, 1982). The widespread species of the outer shelf zone, e.g., *Uvigerina peregrina*, *Cassidulina subglobosa* and *Bulimina marginata*, attain their highest abundances in bathyal or abyssal waters.

Benthic foraminiferal species diversity, expressed by various indices, increases from the shoreline to the shelf-edge and beyond (e.g., Walton, 1964; Betjeman, 1969; Schnitker, 1971; Gibson & Buzas, 1973), and at least in one area (Sen Gupta & Kilbourne, 1974), shows a marked change at the seaward limit of the inshore part. Depth-zone boundaries, however, are generally not distinguishable on the basis of diversity changes.

When attention is turned from depth-zonation to environmental parameters usually measured in marine ecological investigations, few broad patterns of benthic foraminiferal distribution emerge. Many significant local studies, however, have been made that demonstrate the direct or indirect influence of these factors on the geographic extent and abundance variations of species. Much of the following discussion on the better-known of these factors relates to nearshore foraminfera, because the available data deal mainly with them.

## SUBSTRATE

Benthic foraminifera include both epifauna and infauna. Thus, the characters of the substrate may exert significant influence on the distribution of some species. Many examples of this influence have been reported in the literature, although the exact cause and effect may be difficult to understand. In terms of ecological relations, a substrate includes, besides inorganic particles or shell-debris of various sizes, shapes, and compositions, organic particles (e.g., plant material, fecal pellets, and detritus), and interstitial water and air (Gerlach, 1972). The amount of available nutrients often depends on the substrate; the effect of one may be indistinguishable from that of the other.

For benthic organisms in general, it is generally understood that the most significant properties of the substrate relate to its suitability for movement, the chemistry of interstitial water (including availability of oxygen), the size and density of pore spaces, and the amount and nature of decomposable organic material (Gerlach, 1972). All of these properties should be important for benthic foraminifera, because, contrary to earlier belief, they are not confined to the substrate at the sediment-water interface or to a layer about 1 cm thick. Various studies have demonstrated that they survive, often in large numbers, to sediment depths of 9-10 cm, and in some cases, even to 15-16 cm (e.g., Boltovskoy, 1966; Schafer, 1971; Buzas, 1974). Furthermore, for some species, the population size in a unit volume of sediment may not change significantly from the sediment-water interface to a layer a few centimeters below. For instance, *Ammobaculites exiguus* maintains its abundance in Rhode River, Maryland, down to a sediment depth of 10 cm (Buzas, 1974), and the living foraminifera of the coasts of Nova Scotia and Prince Edward Island, Canada, may not show any significant variation in numerical density from the 0-1 cm layer to the 5-9 cm layer (Schafer, 1971). The population size of *Fursenkoina fusiformis*, a dominant species off the Northumberland coast of England, does not undergo any major change within a sediment depth of 11 cm, but that of *Ammonia batavus*, another member of the same assemblage, shows a sharp drop below a depth of 1 or 2 cm, (Collinson, 1980). In Deseado Creek, Argentina, Boltovsky (1966) found a progressive decline in foraminiferal density and diversity from the surface to a depth of 16 cm. The deepest occurence of living foraminfera (species of *Elphidium*) in this study was observed in coarse sand and the shallowest in clayey silt; this leads to the possibility that availability of oxygen may be a limiting factor in the vertical distribution of living foraminifera in sediments. Matera and Lee (1972) suggest that feeding habits of species place an additional restriction on their vertical distribution; they found abundant *Elphidium incertum* in the upper 3 cm of a marsh sediment which was finer than the layer immediately below (3-8 cm), where *Trochammina inflata* was abundant.

Reproductive and feeding activities, rather than mere presence, of benthic foraminifera within sediments have been demonstrated by Frankel (1972, 1975) in short cores from a Connecticut estuary. Individuals (of uncertain taxonomy) undergoing gamogony were found in nearshore sands from 3 to 34 mm (mainly, from 3 to 9 mm) below the sediment-water interface. Furthermore, pseudopodia-bearing individuals of *Ammonia beccarii*,

Protelphidium tisburyense, and Miliammina fusca were identified in these sediments down to a depth of 55 mm. Both the pseudopodia and the protoplasm within the tests contained food, mainly algae and bacteria.

Correlations between foraminiferal biotopes and types of substrate have been reported in numerous studies, but unequivocal substrate preferences of species, other than those of attached forms, are hard to demonstrate. For instance, the calcareous assemblage of the "sand facies" and the agglutinated assemblage of the "mud facies" on the shelf of the southern Gulf of Maine exist elsewhere on other substrates in comparable water depths (Phleger, 1952, 1960). On the other hand, there are reports that a benthic foraminiferal assemblage may occur in different depth-zones where the substrates are similar. In Gulf of Panama, the typical outer shelf fauna, dominated by Hanzawaia nitidula, Uvigerina juncea, or Cassidulina minuta, lives on a sandy substrate between water depths of 65 and 100 m. The inner shelf fauna, dominated by Bulimina denudata, lives between 30 and 65 m, on clayey mud. However, in areas shallower than 65 m that have a sandy bottom, the assemblage is similar to that of the outer shelf (Golik and Phleger, 1977). Over much of the Florida-Hatteras slope, Bolivina lowmani is the dominant species at depths greater than 100 m. However, the species attains sizeable populations, even dominance, on parts of the adjacent continental shelf where the substrate, as on the slope, has a high proportion of silt and clay. A significant positive correlation between the abundance of B. lowmani and the proportion of finer clastics in the sediment of this region is established by regression analysis (Sen Gupta & Strickert, 1982).

In nearshore turbulent areas, foraminifera living on sandy bottoms have to cope with frequent displacement and burial. Some species, e.g., those of Buccella, Buliminella, and Elphidium, are well-adapted to this environment, and apparently prefer it to nearby non-turbulent bottoms. The adaptation to the turbulent environment may give them a competitive edge they do not have in other environments (Lankford and Phleger, 1973). Many species of rocky substrates have special adaptive features. Benthic foraminifera living on shallow water rocky or weedy substrates protect themselves from surf action by cementation (e.g., Acervulina and Marginopora) or pseudopodial attachment (e.g., Spirillina, Patellina, Rosalina) to the substrate, or by finding shelter in crevices or weeds (e.g., Calcarina) or in epiphytic algal growth (e.g., Peneroplis) (Murray, 1973).

The combined effect of depth and substrate on the distribution of some soritids and nummulitids has been reported from the Gulf of Elat, Red Sea (Hottinger, 1977). Amphisorus hemprichii and Sorites orbiculus live on Halophila leaves and algae between 0 and 20 m. Heterostegina compressa prefers hard bottoms between 20 and 70 m. Operculina ammonoides and Heterocyclina tuberculata live on soft bottoms, the former between 30 and 150 m, and the latter between 70 and 150 m. The dependence of attached foraminifera on plants in a very different environment is illustrated by Planorbulina acervalis and Rosalina globularis. These 2 species are permanent members of the Sargassum (pelagic gulfweed) community in waters near Bermuda; long-distance transport by rafting is one result of this adaptation (Spindler, 1980).

Some of the best examples of overall substrate control on species or assemblage distribution come from habitats associated with coral reefs. In St. Lucia, West Indies, the assemblage associated with reef patches is characterized by an abundance of Amphistegina gibbosa, Rotorbinella rosea, Sorites marginalis, and Rosalina floridensis, whereas the assemblage of an adjacent fine substrate is dominated by Ammonia beccarii and Buliminella elegantissima (Sen Gupta & Schafer, 1973). Similar assemblages are present in comparable environments in other Caribbean areas (e.g., Bandy, 1964; Cebulski, 1969). Using foraminiferal families, Brasier (1975) has characterized 5 assemblages in the reefs and shoals around Barbuda, West Indies. These are: (1) back-reef, fine sediment or seagrass assemblage, dominated by miliolids, soritids, rotaliids, and discorbids, (2) inter-reef, sandy substrate assemblage, dominated by miliolids, soritids, rotaliids, and discorbids, (3) fore-reef, talus slope assemblage, dominated by amphisteginids, miliolids, cassidulinids and textulariids, (4) weed surface assemblage, dominated by miliolids, discorbids, soritids, and planorbulinids, with abundance of attached forms, and (5) hard substrate assemblage, dominated by amphisteginids, homotrematids, cibicidids, miliolids, and discorbids. The diversity is high, except for the second assemblage.

The preference of typical attached species, e.g., Cibicides lobatulus, to sandy, gravelly, or rocky bottoms (e.g., Phleger, 1952; Parker, 1952; Sen Gupta, 1971) is well-known. The exact nature of the attachment may be species-specific. For instance, 3 common attached species of the New Jersey outer continental shelf show 3 different modes of attachment to quartz grains. The attachment of Elphidium subarticum is by an organic film, that of Webbinella concava is by a calcitic flange growing from the side of the test, and that of Vasiglobulina reticulata is by multiple spines connecting the test to a calcite lamina which is cemented to the substrate (Poag, 1982).

Finally, the composition of aggulinated tests depends on the nature of substrate particles. Beyond this relationship, a connection between the gross shape of the test and the grain-size of the substrate has been suggested for Textilina pseudogramen of the Adriatic Sea (Haake, 1977). The tests, for some unknown reason, are broader in individuals that live on a coarser substrate.

## TEMPERATURE

Temperature is a key property of the marine environment; it strongly affects many other physical and chemical properties and many biological processes. Marine microorganisns, such as the foraminifera, are poikilothermic, that is, their body temperature is very close to that of the surrounding water. Furthermore, the small mass of these organisms responds very quickly even to small changes in the ambient temperature. These responses, however, are partially dependent on changes that may take place simultaneously in other properties of the environment, such as light, salinity, pressure, and dissolved gases (Kinne, 1970).

The relative effects of different environmental factors on benthic foraminifera are hard to evaluate. It is, however, the consensus that temperature is the single most important physical factor influencing the distribution of species or assemblages (Lee, 1974).

The overall temperature range in open oceans is about -2° to 30°C and that of all marine (including brackish) waters is about -2° to 43°C (Kinne, 1963). Species of benthic foraminifera are present in all temperature regimes within these ranges. Their tolerances, however, are highly variable. Stenothermal species have been reported from different parts of the ocean floor, including the deep sea. On the other hand, many coastal species are eurythermal; Elphidium excavatum forma clavatum and Ammonia beccarii are found in inshore waters of many latitudinal biogeographic provinces.

Culture studies of a small number of species have resulted in the determination of their temperature limits for survival, growth, and reproduction (Boltovskoy & Wright, 1976). These limits, however, are likely to be different in natural habitats, where other environmental factors would have a concomitant effect on the species. Temperature ranges, including optimum values, under natural conditions have been reported for a few species by various authors (Boltovskoy & Wright, 1976), but the data are not particularly illuminating.

The large scale distribution of benthic foraminiferal assemblages in well-known biogeographic zones can be clearly related to temperature ranges, particularly to seasonal ranges that affect reproduction. Empirical relations between foraminiferal zones and bottom-water temperatures have been reported in the literature ever since Norton (1930) published a 15-sample survey of the distribution of benthic foraminifera in Florida and West Indies and suggested a zonation based on both depth and temperature. Natland (1933), in his classic study of the foraminiferal zonation of southern California (see section on depth-zonation), recorded the bottom-water temperatures. For the open marine environments, the ranges were as follows: Zone II (4-38 m)- about 13°-21°C; Zone III (38-274 m)- about 9°-13°C; Zone IV (274-1981± m)- about 4°-9°C; Zone V (1981± -2542±m)- about 2°-4°C. In McGlasson's (1959) comparable scheme of zonation around Santa Catalina Island, the temperatures reported were about 11°-21°C between 0 and 37 m, about 10°-14°C between 38 and 74 m, and about 8°-11°C between 75 and 183+ m. Literature on such temperature zonation in various parts of the world ocean does not reveal any overall patterns. A relation, however, may exist between local depth-distributions of benthic foraminifera and the vertical thermal structure of oceans. Phleger (1960) emphasized the relation between the amount of thermal variation within a given water layer and the nature of benthic foraminifera affected by this layer, and drew attention to the contrasts among the seasonal layer (great temperature variation), the permanent thermocline (little variation), and the deep water below about 1,000 m (uniform condition). Golik & Phleger (1977) concluded that the best correlation between an environmental factor and the distribution of benthic foraminifera on the outer continental shelf (65-200 m) of the Gulf of Panama was given by the temperature. The shallow water boundary of the fauna matched the base of the seasonal thermocline; the deep water boundary was immediately above the intersection between the steep continental slope and the permanent thermocline, which marks a zone of abrupt change in bottom-water temperature and one of oxygen minimum.

In some areas, the shelf-edge may undergo high seasonal fluctuations in bottom-water temperature because of the special nature of the slope water. For example, the waters of the outermost part of the Georgia continental shelf are frequently affected by the meanderings of the Gulf Stream; the middle part of the shelf shows the least seasonal variation in bottom-water temperature. The distinct foraminiferal assemblages recognized for these 2 parts of the shelf are apparently related to different thermal regimes (Sen Gupta & Kilbourne, 1976). The faunal break between the outer shelf and slope off Long Island, New York, is related to the "confluence of cold, lower salinity shelf waters with the warmer, more saline waters of the slope" (Gevirtz, Park & Friedman, 1971).

## SALINITY

In discussing the role of salinity in marine ecology, Kinne (1971) wrote: "The concept of salinity has been coined by physical oceanographers. It is of limited use to the marine ecologist unless its biologically significant components are realized and separately assessed." In practice, however, the separate influences of the ionic constituents of the ocean salt on organisms are extremely difficult to assess in the natural environment; the ecologist is usually resigned to studying the effect of the total salinity and its variation. A change in salinity may have a direct effect on the well-being of a marine organism, or it may have an indirect but significant effect on the organism by modifying the composition of its ecosystem, that is, by affecting the distribution of other species (Kinne, 1971). Responses of benthic foraminifera to salinity variations differ greatly. Euryhaline species, adjusted to salinity changes of 10‰ to 30‰ , have high cellular osmotic tolerances.

Many nearshore benthic foraminifera are euryhaline. The salinity tolerances of a few of these species have been determined in the laboratory. Feeding, growth, and reproduction are retarded, and finally halted, beyond certain threshold salinities, before lethal limits are reached (Bradshaw, 1955, 1961; Murray, 1963). In nature, salinity tolerances of marine organisms decline as other significant factors, such as temperature or concentration of dissolved gases, increase or decrease beyond optimal values. Benthic fauna are protected to varying degrees from drastic salinity changes in the waters above by the dampening effect of the substrate (Kinne, 1971).

Some euryhaline species are also eurythermal, and have wide geographic distributions. One such species is the agglutinated <u>Ammotium cassis</u>, common in hyposaline (<32‰) lagoons with a temperature range of 1°-26°C (Murray, 1973). In the non-tidal Patos Lagoon of southern Brazil, the species is adjusted to salinity variations between 0 and 28‰ (Closs, 1963). On the Swedish west coast, it exists in waters whose salinity may be as low as 10‰, but the optimum salinity is higher (Olsson, 1976). In the Baltic Sea, it prefers the region of saline water incursion from the North Sea, where the salinity range is 17-29‰. In this region, there has been a slight increase in salinity in recent years, resulting in an increased abundance of A. cassis and Reophax aduncus (Lutze, 1965).

Even in a low salinity environment, multiple benthic foraminiferal assemblages may exist in response to small but significant salinity differences. The calcareous foraminifera of Lake Pontchartrain, Louisiana, provide a good example (Otvos, 1978). The bottom-water salinities in the lake generally range between 1 and 5‰, occasionally reaching an unusual high of 12-19‰ in a few localities. In the eastern part of the lake, there is a predictable dominance of agglutinated species. In the rest of the lake, however, biotopes dominated by agglutinated species alternate with those dominated by calcareous ones. The most abundant calcareous species is Ammonia beccarii; Elphidium gunteri, E. galvestonense, and Nonion depressulum matagordanum are also common. The overall abundance of calcareous forms is most unusual for this environment. Otvos concluded that the forms are migrants from the Gulf of Mexico that reached the lake through a canal. A support for this idea comes from the fact that calcareous species are absent in the eastern part of the lake, which has the highest salinity, but is separated from the open Gulf. The dominant calcareous species of Lake Ponchartrain, listed above, are typically found in brackish, nearshore waters of the Gulf. The lake, however, has very small populations of some other species (e.g., Buliminella elegantissima, Cribroelphidium poeyanum, and Milionella circularis) that are characteristic of open marine environments in this region and "probably could not survive the lower salinities for extended periods of time" (Otvos, 1978).

The quick effect of salinity change on a brackish water assemblage has been demonstrated by Lamb (1972), on the basis of data recorded from the eastern Mississippi Sound by Phleger (1954) and Anderson (1968). In this area, there has been a complete turnover from an agglutinated (Ammobaculites-Miliammina) to a calcareous (Elphidium-Ammonia) fauna between 1954 and 1967, in response to a salinity increase of less than 10‰. A probable cause for this increase may be the diversion of some fresh water from the Sound, as a consequence of the building of a bridge system and the sedimentation that followed.

Other than the tectinous group (Allogromiina); benthic foraminifera are extremely rare in fresh water (salinity<5‰). A few species have been reported from northern Argentina (Boltovskoy & Lena, 1971) and southern Brazil (Closs & Medeiros, 1965). According to Boltovsky and Lena (1971), the species fully adjusted to fresh water are Protelphidium tisburyense, Miliammina fusca, Psammosphaera sp., and Trochammina sp. In addition, there are some species that normally live in brackish water, but can tolerate fresh water for a considerable time. These are Ammonia beccarii parkinsoniana, Reophax arcticus, Elphidium excavatum, Haplophragmium wilberti, and Ovammina sp.

## LIGHT

The protoplasm of some calcareous benthic foraminifera contains algal or chloroplast symbionts. Sustenance and growth of these foraminifera depend on the intensity of the available light and the exposure time. Laboratory studies of Hawaiian *Amphistegina* demonstrate a niche separation of 2 very similar species because of photoinhibition (Hallock, 1981a). *A. lobifera*, which prefers very shallow (<3 m) waters, requires higher light intensities for reproduction in cultures, compared to *A. lessonii*, which lives in deeper waters, has thinner test walls, and does not reproduce in well-lighted near-surface waters. *Heterostegina depressa* from the same region is also adapted to low intensities of light (Rottger, 1976).

In Baltic Sea shoals, where temperature and salinity change little, the reproduction of *Elphidium excavatum* forma *clavatum* is independent of the variation in dissolved oxygen, but is related to the availability of light. The species has symbiotic algae, and its main reproductive periods coincide with the times of maximum extension of the photic zone (Wefer, 1976).

Thus, in natural habitats, species with algal or chloroplast symbionts may be depth-zoned because of difference in light tolerance. Depth, of course, is not the only factor affecting light penetration in the sea; water turbidity, primary production, geographic location, and atmospheric conditions also play their roles.

An area where the overall foraminiferal zonation is strongly influenced by the degree of light penetration is the Gulf of Elat, Red Sea, where the euphotic zone is at least 150 m deep (Reiss, 1977). This 1830 m deep gulf is warm (20-26°C) and highly saline (>42‰), but has no significant vertical variation in temperature, salinity, and pressure. Benthic foraminifera are abundant, diverse, and depth-zoned. The upper zone (0-50m) is dominated by porcelaneous forms--species of *Borelis*, *Sorites*, *Amphisorus*, *Spirolina*, and *Peneroplis*. *Amphistegina* and *Operculina* are common down to a depth of 150 m. Below 150 m, smaller benthic species, e.g., those of *Cassidulina*, *Lamarckina*, *Stomatorbina*, *Gyroidina*, and *Uvigerina*, dominate. Except for some larger foraminifera studied by Hottinger (1977) (see section on substrate), the distribution of the species is little influenced by the nature of the substrate. Light penetration is apparently the significant factor controlling the species distribution. Reiss (1977) noted that algal symbionts are present in all porcelaneous and larger hyaline forms of the Gulf, but not in all smaller hyaline forms. Thus, the distribution of light in the Gulf of Elat affects the depth-zonation of the benthic foraminifera in some unknown way, besides controlling the photosynthetic symbionts.

## NUTRIENTS

Foraminifers are heterotrophic organisms, and capture food with their reticulate pseudopodia. Bacteria, diatoms and other protistans, and various small metazoans are consumed by benthic foraminifera as food, by intake through the aperture. Furthermore, liquids and macromolecules may be assimilated by invaginations of the cell wall (Sheehan & Banner, 1972). *Notodendrodes antarctikos* and another arborescent agglutinated species are even capable of utilizing dissolved organic carbon directly from sea water (Delcaca, Karl & Lipps, 1981). Some species are selective feeders; in laboratory cultures, their growth rates depend on the kinds and amounts of diatoms and flagellates in their diet (e.g., Bradshaw, 1955; Lee et al., 1961). Also, bacteria may be an essential component of the diet (Muller and Lee, 1969).

In species with algal or chloroplast symbionts, the development of the calcareous test may depend on their presence; specifically, oxygen released by the symbionts may be of use. Furthermore, they may even provide nutrition; Heterostegina depressa can grow in the absence of external food when symbionts are present (Rottger, 1973). Overall, the spectrum of food used by benthic foraminifera is wide: "dissolved and colloidal organic molecules; organic rich grains (fungal and algal coated and including fecal pellets); particulate organic detritus; live plants; live animals; dead organisms. This success is due to the versatility of the pseudopodial network" (Haynes, 1981).

Predation by benthic foraminifera is demonstrated by the action of the aggulatinated species, Astrorhiza limicola, off the Northumberland coast (Buchanan and Hedley, 1960). The population size is estimated to be about 50 individuals per square meter. The pseudopodia may extend to 6 or 7 cm from the test in order to capture and digest various organisms, including copepods, nematodes, amphipods, full-grown cumaceans, caprellids 2-3 cm long, and small sea urchins. Thorson (1966) comments: "If we use the figure given above of roughly 50 specimens of Astrorhiza per sq. m in the sandy-bottoms where it predominates and take the "radius" of its pseudopodia as roughly 6 cm, this will mean that the pseudopodia of all Astrorhizas inside 1 sq. m of bottom may together be active over an area of about 5,000 sq. cm or half the total bottom area. This area, accordingly, will be a "danger zone" for many other animals, among them the newly settled young of the macrofauna."

Lipps (1975) identified 4 main groups of foraminifera in terms of their feeding strategies: suspension feeders, detrital scavengers, herbivores, and carnivores. The species of these groups, however, are not narrowly adapted to special types of food and may mostly be opportunistic omnivorous feeders.

Large standing crops of benthic foraminifera are related to nutrient enrichments. On the southern part of the Grand Banks of Newfoundland, for example, unusually large assemblages--90,000 or more individuals per square meter for a sediment thickness of 1 cm -- are found in an area characterized by the presence of Slope Water, produced by a mixing of Central Atlantic and Labrador Waters. This mixed water mass has a high fertility, because of an increase in mineral nutrients or other biologically active substances (Sen Gupta, 1971). Phleger (1964) noted that several known examples of conspicuous standing crops of benthic foraminifera, as in Baja California Lagoons or Mississippi deltas, are related to high organic productivity or increased supply of plant nutrients and trace minerals.

Areas of upwelling are nutrient-enriched, and support large standing crops of benthic foraminifera. One such area lies of San Diego, California (Uchio, 1960). Another area is on the continental slope off northern

Florida where periodic upwelling has resulted in a high organic carbon and hydrocarbon content. The locality supports one of the highest recorded densities (over 3,000 individuals in 3 ml of sediment) of benthic foraminifera in the world ocean. Furthermore, an opportunistic species, Bolivina subaenariensis, generally rare in the western North Atlantic, is a major constituent of the assemblage (Sen Gupta, Lee & May, 1981).

The foraminiferal biotopes of Broken Bay, southeastern Australia, are apparently related to availability of nutrients (Albani & Johnson, 1975). The area is a drowned river valley, with water depths of 10-15 m. Albani and Johnson were able to distinguish 3 high-nutrient (fluvial, fluviomarine, and marine) biotopes from a low-nutrient fluvial one.

## OTHER FACTORS

Among other abiotic factors of the environment that influence the distribution of benthic foraminifera, the pH and oxygen content of sea water have received some attention. The effect of low pH is seen mainly in the stress produced on calcareous species in the secretion of $CaCo_3$; the effect of high pH is unknown. A depletion of oxygen may lead to a reduction in species diversity, and to an increase in the populations of some species, particularly those of the Bolivina group (Boltovskoy and Wright, 1976). On the continental slope and deep ocean floor, sharp changes in diversity and assemblage composition have been correlated with the presence of oxygen-minimum layers (e.g., Golik & Phleger, 1977; Ingle, Keller & Kolpack, 1980).

Predation by various organisms, particularly by grazers and detrital feeders, may have a serious effect on the distribution and standing crop of benthic foraminifera. Some small predators (5-10 mm size) may feed almost exclusively on foraminifers (Brand & Lipps, 1982).

Within any foraminiferal biotope, the distribution of living populations is essentially patchy. Observations in littoral zones have shown that enormous blooms of benthic foraminifera may occur next to an area virtually barren of living individuals. This patchiness may be due to factors such as grazing by consumers, but may also be an ecological strategy, leading to a diversification of habitats and limitation of competition (Lee, 1974).

## SPECIAL ASSEMBLAGES

At the two ends of the spectrum of benthic foraminiferal habitats, namely, marine marshes and the deep sea, the assemblages have pronounced special features. Neither the marsh fauna nor the deep sea fauna is uniform, but they have characteristics that can be recognized in diverse latitudes.

Marsh fauna. -- Marsh environments are complex, and in some respects, highly variable, but they have some common features that can be observed in diverse geographic locations. These include prevention of extreme temperature changes by shade trees, retention of substrate moisture by algal mats, presence of a muddy silt substrate that is rich in organic matter and poor in oxygen, and the existence of a low diversity benthic foraminiferal assemblage in the surface oxidized layer (Murray, 1971a). Most marsh samples contain 4-5 species of benthic foraminifera, but more than 15 species have also been reported (Boltovskoy & Wright, 1976). Miliammina fusca, Trochammina inflata, and Jadammina polystoma are found in most marshes; Ammotium salsum and Arenoparrella mexicana are also cosmopolitan species (Phleger, 1970; Murray, 1971a). These are all agglutinated forms.

Deep sea fauna. -- The deep sea environment, compared to continental margins, shows very small variations in physical and chemical properties. The bottom-waters are dark and under great pressure. Over extensive regions, there is little change in salinity, oxygen content, and temperature, which, below a depth of 2,000 m, is everywhere less than 4°C. The sediments are mainly derived from shells of planktonic microorganisms; the food supply is low, because of the great vertical distance from the zone of primary production (Sanders and Hessler, 1969).

The major water masses of the deep sea have a density-dependent distribution, although their temperature and salinity differences are small. For instance, in the Guiana Basin of the South Atlantic Ocean, parcels of the Antarctic Bottom Water (AABW) and the North Atlantic Deep Water (NADW) may have virtually the same salinity (about 34.9‰) and only small differences in temperature. Typical depths and temperatures of these parcels (Dietrich et al., 1980) are given below.

| | Depth | Temperature |
|---|---|---|
| Upper NADW | 1,500 m | 4.5°C |
| Middle NADW | 2,000 m | 3.5°C |
| Lower NADW | 3,000 m | 2.8°C |
| AABW | 4,500 m | 2.0°C |

Aspects of assemblage composition of deep sea benthic foraminifera and their relationships to water masses have been the subject of several recent studies (see Schnitker, 1980c, and Douglas & Woodruff, 1981, for review). In general, the common species are found in many basins. However, changing patterns of dominance have led to the suggestion that different assemblages are associated with different water masses. The best-known patterns are those for AABW and NADW. The faunas associated with AABW are dominated by Nuttallides umbonifera; those associated with lower NADW are dominated by several species, including Epistominella exigua and Cibicidoides wuellerstorfi (Streeter, 1973; Schnitker, 1979b). Furthermore, U. peregrina is reported to be associated with old, oxygen-depleted (<5ml/1) water occurring between depths of 2,000 and 4,000 m, and to be absent from the well-oxygenated core of NADW.

Although water masses are defined by temperature and salinity, the small variations in these properties at depths below 2,000 m are unlikely to exert a major influence on the distribution of benthic foraminifera. Biotic ecological factors for these depths are poorly known, and the effect of some significant physical and chemical factors, such as oxygen availability, is not fully understood. Even the preference of Uvigerina peregrina for oxygen-depleted waters is not unequivocal. A recent study of the abundance of the species on the western North American continental slope and rise (Miller and Lohmann, 1982) shows a correlation with the amount of silt and organic carbon in the substrate, but not with the amount of dissolved oxygen in the overlying water. Thus, the distribution of U. peregrina in this region may be related to an oxygen depletion in the substrate, rather than in the water.

# Biology of Planktonic Foraminifera

Allan W.H. Bé
Lamont-Doherty Geological Observatory
of Columbia University, Palisades, N.Y. 10964

## INTRODUCTION

The history of investigations of planktonic foraminifera leading to their current, wide applications in biostratigraphic correlation of Mesozoic and Cenozoic marine sediments can be traced to the discovery by Owen (1867) of the floating habit of certain foraminifera, later confirmed by Brady (1884), and the recognition by Murray and Renard (1891) that their shells constitute a major portion of deep-sea sediments over large regions of the seafloor. The utilitarian value of planktonic foraminifera in paleoecological analysis is also widely accepted today. Fortunately for students of paleoecology, many species of planktonic foraminifera are still extant and live in large numbers in all oceanic regions. Their areal, vertical and seasonal occurrences have been resolved to such a degree that a global synthesis of their distribution patterns is now available (e.g. Bradshaw, 1959; Belyaeva, 1964; Bé and Tolderlund, 1971).

Modern species provide information on present-day environmental conditions, which form the basis for the underlying assumptions in paleoecological analysis of fossil planktonic foraminiferal assemblages. When these "modern analogues" become progressively rarer, as the number of fossil species with living representatives decreases with increasing geologic age, our ability to resolve paleoecological problems also diminishes in accuracy.

Most propitious for foraminiferal paleoecologists is the availability of extant species for scrutiny and experimentation in the laboratory -- an opportunity that is denied their colleagues studying fossil groups without living descendants. In this paper, we shall review our present knowledge of the life processes of planktonic foraminifera, based on recently published field observations and laboratory experiments. Their fine structure will not be discussed, as this aspect was treated by Anderson and Bé (1978).

## FIELD OBSERVATIONS AND COLLECTING

In the past the conventional method of collecting planktonic foraminifera with fine-meshed plankton nets towed from a slowly moving vessel invariably caused damage to their long, thin spines and other delicate structures and may account partly for the failure in previous attempts to maintain them in the laboratory. This deficiency has now been surmounted by the relatively recent technique of collecting planktonic foraminifera by scuba-diving (Alldredge and Jones, 1973; Bé et al., 1977). Although the shells of planktonic foraminifera are generally less than 1 mm in diameter, the long rhizopodia and slender spines of some species extend several millimeters in length, thus increasing their overall diameter manifold. They are rendered visible by this relatively large size, exceeding that of many other microzooplankters, and their long spines that reflect light at a certain angle, so that they can then be collected with a small jar by a scuba diver. Foraminifera obtained in this manner are in perfect condition, such as the specimen of Hastigerina pelagica shown in Plate 1, Fig. 1, and they exhibit all the delicate structures first shown by Murray (1897).

Underwater observations in tropical and subtropical regions are limited to six spinose species of planktonic foraminifera (Hastigerina pelagica, Globigerinella aequilateralis, Globigerinoides ruber, G. sacculifer, G. conglobatus and Orbulina universa (Plate 1, Figs. 1-6), (Bé et al., 1977). These species can be discriminated from each other in the water, especially if they are of adult size. Hastigerina pelagica is more readily recognized than the other five species, because of its generally larger size, longer spines, orange-reddish protoplasm and bubble capsule. The 2 mm-diameter bubble capsule is a flotation device that is unique to this species. The total diameter of H. pelagica including spines may reach about 20 mm. Orbulina universa's spherical shell as well as its juvenile, spiral stage can also be distinguished in the water. Globigerinoides sacculifer and G. conglobatus tend to have an extremely dense concentration of spines which sometimes are iridescent. Juveniles of G. sacculifer frequently possess white specks which are visible in the water and which appear to be spindle-shaped algae under the microscope. Globigerinella aequilateralis has longer, but fewer spines and its rhizopodial network, often harboring Trichodesmium filaments, is more extended than the other species. Globigerinoides ruber can be recognized in-situ by its pink pigmentation and even the white variety of this species can be distinguished from G. sacculifer by the relative density of its spines and the size of its symbiont-bearing rhizopodial halo. The rhizopodial networks of G. ruber, G. sacculifer, G. conglobatus, G. aequilateralis and O. universa form spherical cytoplasmic capsules around their shells which may be homologous to the bubble capsule of H. pelagica and aid their buoyancy.

---

Plate 1. Spinose planktonic foraminifera collected by scuba-diving (from Bé et al., 1977).

Fig. 1. Hastigerina pelagica (d'Orbigny) with numerous dinoflagellates in its bubble capsule (x 32);

Fig. 2. Globigerinoides ruber (d'Orbigny) with hundreds of zooxanthellae sequestered in its rhizopodial network (x 74);

Fig. 3. Globigerinoides sacculifer (Brady) (x 44);

Fig. 4. Globigerinoides conglobatus (Brady) with large dinoflagellates of Pyrocystis fusiformis Murray (x 78);

Fig. 5. Orbulina universa d'Orbigny (x 32).

Fig. 6. Globigerinella aequilateralis (Brady) (x 32).

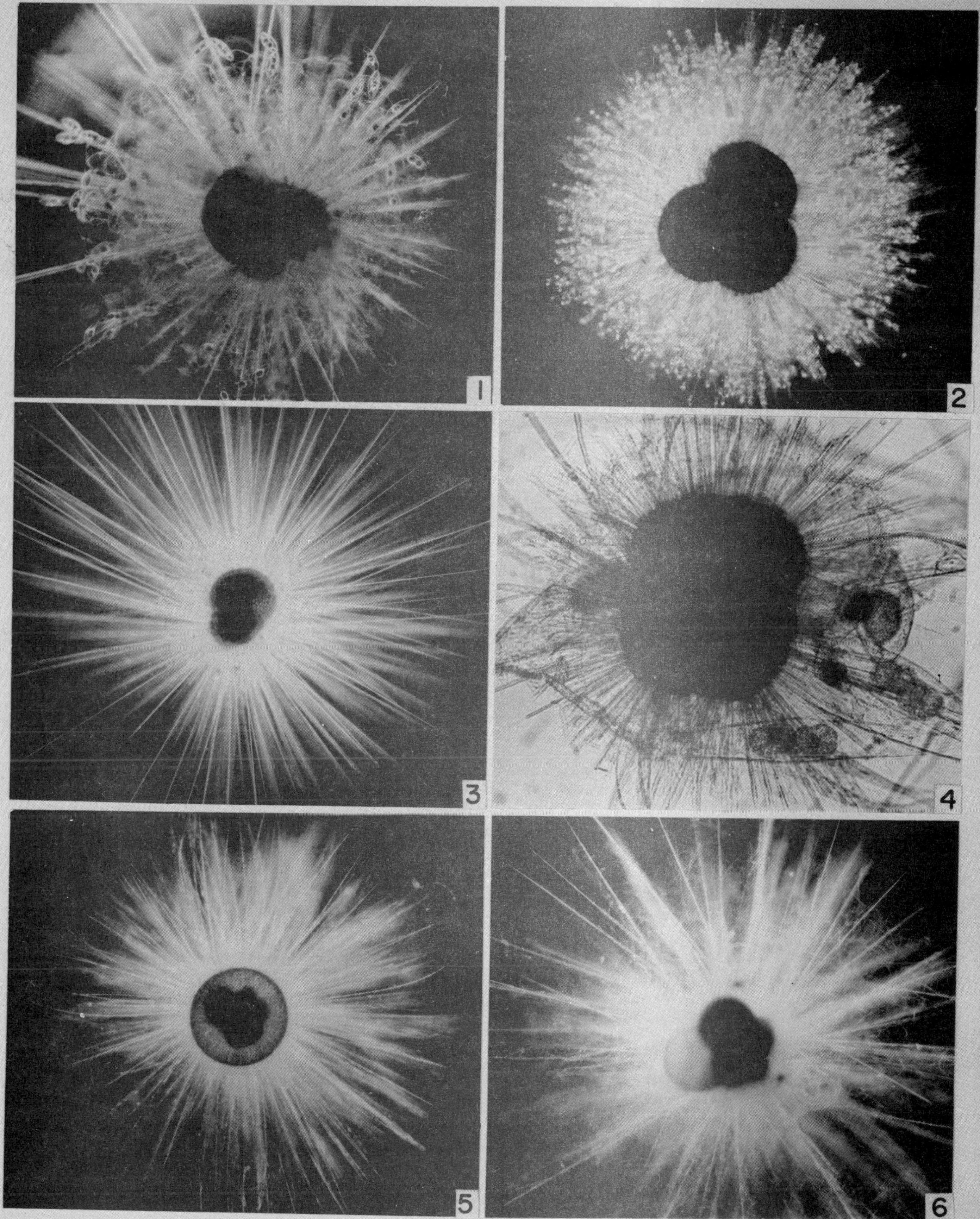


Plate 1

Plate 2

## NUTRITION

It is an ecological axiom that the availability of food largely determines the distribution and abundance of organisms. Species that subsist on a wide variety of prey have an ecological advantage over those with a more restricted diet. The fact that planktonic foraminifera are ubiquitous and prolific in epipelagic waters in all oceans indicates that they must be successful food gatherers and play an important role in the marine food web.

Mode of feeding.-- Planktonic foraminifera are opportunistic feeders, preying on whatever food organisms they encounter.
As passively drifting organisms, they have developed a unique method of trapping food organisms by means of a dense network of sticky rhizopodia, which in some species are supported by radiating spines. The ratio of spines and rhizopodial network diameter to shell diameter of spinose species is commonly 10:1 or higher (Plate 1, fig. 3). This greatly enlarges the effective radius for capturing prey than would be the case if the rhizopodia were limited to the shell periphery. Thus, *Hastigerina pelagica* and *Globigerinoides sacculifer* with adult shell diameters of about 1.4 and 0.5 mm can better be considered as relatively large microzooplankters with overall diameters of 15 mm and 5 mm, respectively. Planktonic foraminifera are therefore of approximately the same size range as many species of marine copepods, which constitute one of the most important groups in the pelagic ecosystem.

An adhesive substance secreted by the foraminifer's rhizopodia is used to snare food particles (animal, plant or organic detritus) which come into contact with the foraminifer. Somewhat analogous to a spiderweb, the rhizopodial network seldom allows snared organisms to escape. Actively motile prey immediately become entangled in the rhizopodial network and are inexorably drawn by the rhizopodia (Plate 2, fig. 2) until the prey is against the shell (Plate 2, fig. 1). Large prey normally lies at or near the shell's aperture before being reduced to smaller pieces, which are carried into the shell. Tintinnids, diatoms and other smaller prey are ingested. Copepods, which constitute the major food source of spinose planktonic foraminifera, are ruptured by the myriads of rhizopodial strands which invade their soft tissue and body cavity. This is followed by engulfment of muscle and other copepod tissue fragments within food vacuoles. Considerable amount of digestion of prey takes place outside the shell (Plate 2, fig. 3), but the bulk of food particles are transported through the shell

---

Plate 2. Carnivorous diet of planktonic foraminifera (Bé et al., 1977).

Fig. 1. *Hastigerina pelagica* (d'Orbigny) with empty carapace of a copepod which was drawn into and digested within the bubble capsule. x 80.

Fig. 2. Two snared *Artemia* nauplii are being drawn into the bubble capsule of *Hastigerina pelagica* (d'Orbigny) by a bundle of rhizopodial strands. x48.

Fig. 3. A piece of crustacian muscle tissue (M) surrounded by rhizopodia near the aperture of *Globigerinella aequilateralis* (Brady). x 8400.

Fig. 4. Muscle tissue is enclosed in a digestive vacuole within the intrashell cytoplasm of *Hastigerina pelagica* (d'Orbigny). x 7200.

aperture to the intrashell cytoplasm where further digestion takes place in digestive vacuoles (Plate 2, fig. 4) (Anderson & Bé, 1976a; Bé et al. 1977).

Although most if not all species of planktonic foraminifera are omnivorous (Table 1), individual species are probably either facultative herbivores or carnivores. In general, the spinose species appear to be predominantly carnivorous, and the nonspinose species predominantly herbivorous. This explains in part why there is a greater incidence of nonspinose species in eutrophic waters, where high phytoplankton productivity occurs in upwelling regions, boundary currents and periodically in central watermasses. In the latter, eutrophication of the mixed zone occurs on a seasonal basis. For example, *Globoquadrina dutertrei* is abundant in boundary currents and upwelling regions and *Globorotalia inflata* in the diatom-rich subpolar and cold-temperate waters. *Globorotalia hirsuta* and *G. truncatulinoides* occur prolifically during winter and early spring in the surface waters of the northern Sargasso Sea when phytoplankton blooms develop (Bé & Tolderlund, 1971; Tolderlund & Bé, 1971).

The spinose species by virtue of their long spines and denser, more extended rhizopodial network are better able to capture active, motile zooplankters than the nonspinose species. Since secondary producers such as copepods predominate in the oligotrophic central watermasses (Grice & Hart, 1962), this may explain why the carnivorous spinose species *Globigerinoides ruber*, *G. sacculifer*, *G. conglobatus*, *Orbulina universa*, *G. aequilateralis*, and *Hastigerina pelagica* usually constitute the bulk of the foraminiferal population in the mixed zone of the Sargasso Sea and other nutrient-poor oceanic regions. Moreover, the spinose species possess symbiotic or commensal algae which manufacture additional nutrients to sustain their hosts in an oligotrophic environment.

*Diversity of food organisms*.-- The types of prey that constitute the diet of the spinose species are quite diverse. Copepods are by far the greatest component of their food in terms of number as well as biomass. Tintinnids are also frequent prey, but their biomass is minute. Amphipods, ostracodes, tunicates, siphonophores, chaetognaths, polychaetes, acantharia and unidentified eggs have been observed trapped in the rhizopodia and are presumed to be food organisms. Diatoms, other phytoplankters and, perhaps, organic detritus also form part of the diet of foraminifera, but they are generally difficult to recognize under a light microscope and they can be identified with certainty only when they are found within the digestive vacuoles examined by transmission electron microscopy (Table 1).

Further investigations are necessary to determine the degree of preference for phytoplankton and zooplankton prey for each species of planktonic foraminifera. These studies should include laboratory experiments in which the ratio of phytoplankton protein to zooplankton protein consumed by the planktonic foraminifera is assessed.

Table 1.--Diet of planktonic foraminifera, identified in ultrathin sections by transmission electron microscopy (from Anderson et al., 1979). A = adipose animal tissue, C = connective tissue, D = Diatoms, Dn = dinoflagellates, E = eukaryotic pigmented algae, Ep = epithelial tissue, M = muscle tissue, N = nerve tissue, O = unidentifiable animal tissue, T = thecate algae. The number of specimens examined by electron microscopy is in parentheses.

| Spinose species | Algal prey | Animal prey |
|---|---|---|
| Hastigerina pelagica (d'Orbigny) (17) | - | A, M, O |
| Globigerinella aequilateralis (Brady) (7) | D, E | A, M |
| Globigerinoides conglobatus (Brady) (2) | E | O |
| G. ruber (d'Orbigny) (15) | D, Dn | A, M, N |
| G. sacculifer (Brady) (4) | - | Ep, O |
| Globigerina falconensis Blow (4) | D, E | C, M |
| Orbulina universa d'Orbigny (11) | Dn, E, T | A, O |
| Nonspinose species | | |
| Globoquadrina dutertrei (d'Orbigny) (4) | D, E, T | O |
| Globorotalia hirsuta (d'Orbigny) (6) | D, Dn, E | M, O |
| G. inflata (d'Orbigny) (2) | D, Dn | M, O |
| G. menardii (d'Orbigny) (3) | D, E | M, O |
| G. truncatulinoides (d'Orbigny) (8) | D, E, T | M, O |
| Pulleniatina obliquiloculata (Parker & Jones) (5) | D, Dn | M, O |
| Globigerinita glutinata (Egger)(3) | Dn | A, Ep |

## FEEDING FREQUENCY AND LIGHT INTENSITY

Laboratory experiments on the effects of varying feeding frequency and light intensity on life processes of Globigerinoides sacculifer were carried out by Bé et al. (1981) and Caron et al. (1982). G. sacculifer (Plate 3) was chosen because it is robust, amenable to culture and its shell morphology is particularly suitable for biometric studies. It is one of the most successful and widely distributed species of planktonic foraminifera and is especially abundant in the Caribbean Sea off Barbados, where the investigations took place. G. sacculifer may well become the Escherichia coli of foraminiferal biologists.

A matrix of experiments was established employing 882 specimens of G. sacculifer in three light regimes and four to six feeding regimes (Table 2). The organisms in the three light regimes were maintained in: (a) a relatively high light intensity of 400-500 μEinsteins/ $m^2$/sec (corresponding to a mid-day illumination level of 5-10 m water depth off Barbados) on a 12h light: 12h dark cycle (= HLI regime); (b) a relatively low light intensity of 20-50 μEinsteins/$m^2$/sec (= LLI regime), which corresponds to a mid-day illumination level of 30-50 m water depth, and similar light-dark cycle; and (c) continuous darkness (= CD regime), corresponding to light conditions below the photic zone.

Table 2. Experimental framework used by Bé et al. (1981) and Caron et al. (1982) for evaluating the effects of light intensity, feeding frequency, and algal symbiosis on the life processes of <u>Globigerinoides sacculifer</u>, based on specimens over a wide spectrum of ontogenetic stages in 3 light regimes, 4-6 feeding regimes, 2 temperature regimes and 3 photosynthesis-inhibited regimes.

| Experiment | Light Intensity | Feeding Frequency | Temperature (°C) | DCMU Herbicide | N 150≤200 µm | N 201-250 µm | N 251-300 µm | N 301-350 µm | N 351-400 µm | N 401-450 µm | N 451≥500 µm | Total N <150->500 µm |
|---|---|---|---|---|---|---|---|---|---|---|---|---|
| Feeding frequency and light intensity | HLI (Bright) 400-500 µE/m²/sec 12L:12D | daily | 28° | None | 9 | 9 | 9 | 9 | 9 | 9 | 9 | 63 |
| | | every 3d | | | 9 | 9 | 9 | 9 | 9 | 9 | 9 | 63 |
| | | every 7d | | | 9 | 9 | 9 | 9 | 9 | 9 | 9 | 63 |
| | | not fed | | | 9 | 9 | 9 | 9 | 9 | 9 | 9 | 63 |
| | LLI (Dim) 20-50 µE/m²/sec 12L:12D | daily | 28° | None | 9 | 9 | 9 | 9 | 9 | 9 | 9 | 63 |
| | | every 2d | | | 9 | 9 | 9 | 9 | 9 | 9 | 9 | 63 |
| | | every 3d | | | 9 | 9 | 9 | 9 | 9 | 9 | 9 | 63 |
| | | every 4d | | | 9 | 9 | 9 | 9 | 9 | 9 | 9 | 63 |
| | | every 7d | | | 9 | 9 | 9 | 9 | 9 | 9 | 9 | 63 |
| | | not fed | | | 9 | 9 | 9 | 9 | 9 | 9 | 9 | 63 |
| | Continuous dark | daily | 28° | None | 9 | 9 | 9 | 9 | 9 | 9 | 9 | 63 |
| | | every 3d | | | 9 | 9 | 9 | 9 | 9 | 9 | 9 | 63 |
| | | every 7d | | | 9 | 9 | 9 | 9 | 9 | 9 | 9 | 63 |
| | | not fed | | | 9 | 9 | 9 | 9 | 9 | 9 | 9 | 63 |
| Symbiont elimination & reinfection | HLI (Bright) 400-500 µE/m²/sec 12L:12D | every 2d | 28° | DCMU cont. | 1 | 7 | 1 | | | | | 9 |
| | | every 2d | | 72h DCMU/sw | 5 | 12 | 19 | | | | | 36 |
| | | every 2d | | 72h DCMU/sw+symb | 12 | 12 | 12 | | | | | 36 |
| | | every 2d | | SW + dark cont. | 5 | 14 | 17 | | | | | 36 |
| | | every 2d | | SW cont. | 11 | 9 | 16 | | | | | 36 |
| Temperature | HLI (Bright) 400-500 µE/m²/sec 12L:12D | daily | 20° | None | 9 | 7 | 7 | | | | | 23 |
| | | every 3d | | | 7 | 7 | 8 | | | | | 22 |

Table 3. Data on shell size, survival time, life span, shell growth rate, frequency of sac-like chamber and gametogenesis resulting from feeding frequency and light intensity experiments listed in Table 1. Each feeding regime is based on 63 specimens, ranging from 150-500 µm in shell size. Values are means ± 1 standard deviation. [Modified from Bé et al. (1981) and Caron et al. (1982)].

| Light Intensity | Feeding Frequency | No. of specimens | Mean initial shell size (µm) | Mean final shell size (µm) | Mean survival time (days) | Mean life span (days) | Mean shell growth rate (d/ch) | Frequency of sac chamber (%) | Total chambers formed | Frequency of Gametogenesis (%) |
|---|---|---|---|---|---|---|---|---|---|---|
| HLI (High Light Intensity) 400-500 µE/m$^2$/sec 12L:12D | Fed daily | 63 | 323 ±102 | 791 ±76 | 10 ± 3 | 16 | 3.3 ± 14.5 | 73 | 188 | 94 |
| | Fed every 3d | 63 | 326 ±102 | 705 90 | 14 ± 6 | 28 | 4.8 ± 12.0 | 78 | 161 | 92 |
| | Fed every 7d | 63 | 324 ±100 | 584 ±96 | 23 ± 15 | 54 | 11.1 ± 17.2 | 32 | 107 | 84 |
| | Starved | 63 | 323 ±103 | 289 ±176 | 31 ± 18 | 46 | 76.9 ± 32.2 | 2 | 102 | 32 |
| LLI (Low Light Intensity) 20-50 µE/m$^2$/sec 12L:12D | Fed daily | 63 | 324 ±102 | 707 ±131 | 7 ± 3 | 14 | 2.7 ± 7.9 | 76 | 157 | 95 |
| | Fed every 2d | 63 | 324 ±102 | 656 ±102 | 8 ± 4 | 18 | 3.1 ± 10.6 | 71 | 152 | 92 |
| | Fed every 3d | 63 | 324 ±104 | 654 ±88 | 9 ± 5 | 21 | 3.6 ± 12.2 | 70 | 146 | 94 |
| | Fed every 4d | 63 | 325 ±99 | 593 ±97 | 10 ± 6 | 25 | 4.1 ± 9.7 | 49 | 132 | 91 |
| | Fed every 7d | 63 | 324 ±103 | 577 ± 74 | 11 ± 8 | 29 | 4.8 ± 9.3 | 29 | 115 | 87 |
| | Starved | | 324 ±100 | 376 ±173 | 19 ±12 | 50 | 14.5 ± 9.3 | 10 | 36 (-31) | 44 |
| CD (Continuous Dark) | Fed daily | 63 | 325 ±103 | 474 ±85 | 4 ±2 | 7 | 2.9 ± 5.4 | 41 | 86 -1 | 92 |
| | Fed every 3d | 63 | 324 ±101 | 426 ±114 | 4 ±3 | 10 | 3.9 ± 5.5 | 22 | 57 | 79 |
| | Fed every 7d | 63 | 323 ±101 | 400 ±118 | 4 ±3 | 7 | 5.6 ± 5.9 | 19 | 38 | 83 |
| | Starved | 63 | 325 ±100 | 346 ±154 | 5 ±4 | 14 | 12.2 ± 6.9 | 5 | 16 | 75 |

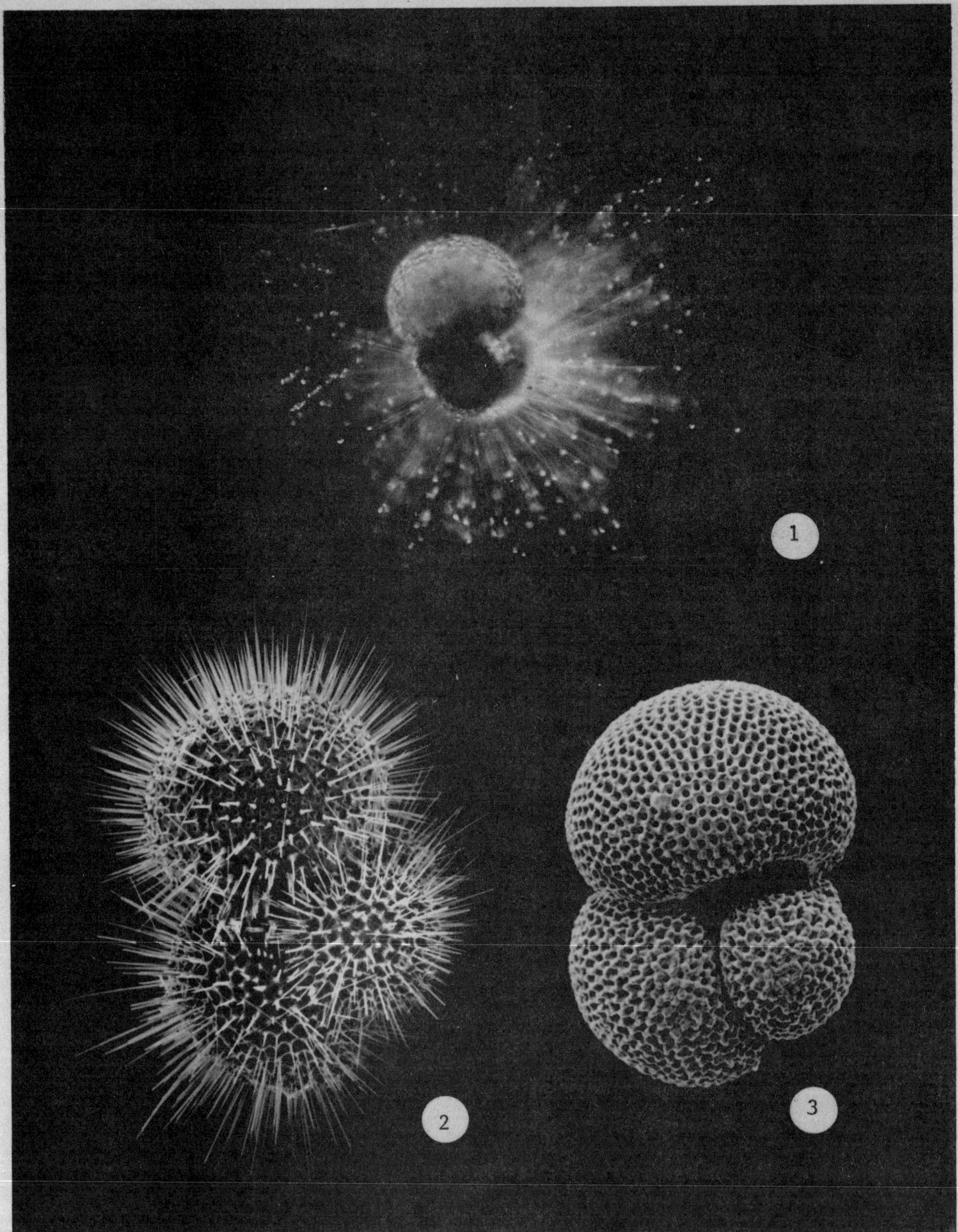


Plate 3. Globigerinoides sacculifer (Brady).

Fig. 1, living individual with numerous symbiotic dinoflagellates (x 70). Fig. 2, specimen that died in culture leaving a relatively thin shell with long spines (x 185); Fig. 3, specimen that underwent gametogenesis, showing encrusted shell due to gametogenic calcification (x 126).

The organisms were divided into four feeding groups in the HLI and CD regimes and six feeding groups in the LLI regime. Each feeding group (63 specimens) was made up of seven size classes at 50 µm intervals (150-200 µm, 201-250 µm, etc., to 451-500 µm in initial shell length), with nine organisms in each size class. Thus, the HLI and CD regimes consisted of a total of 252 specimens and the LLI regime of 378 specimens. Three feeding groups in each of the HLI, LLI and CD regimes were fed 1-day-old Artemia (brine shrimp) nauplii at the rate of 1 nauplius every, 1, 3 or 7 days; one feeding group in each light regime received no Artemia; and two additional groups in the LLI regime were fed every 2 and 4 days -- yielding a total of fourteen experimental groups (Table 2).

Data on shell size, spine length, rhizopod condition, amount of protoplasm in the shell, and the presence and distribution of symbionts in the rhizopodial network were recorded on the date of collection and daily thereafter as described by Bé et al. (1981). These observations provided information on survival time, life span, chamber formation and shell growth rates, and reproductive capacity (Table 3).

Effect of feeding frequency on survival time and life span.-- One measure of G. sacculifer's response to the experimental conditions is the longevity of each individual in culture from the day of collection to the day it either undergoes gametogenesis or dies. This duration is termed "survival time". Scatter plots of the relationship between survival time and initial shell length for each feeding group in the three light regimes are shown in Fig. 1. Regression lines were also plotted for each group of 63 individuals, except for the starved group in the HLI regime whose data points appear to be randomly distributed. The linear regression lines for the other nine groups in the HLI and LLI regimes show a consistent trend of increasing slopes with decreasing feeding frequency (Fig. 2D-F). Each experimental group clearly demonstrates that the survival time of an individual decreases the more frequently it receives food, and, conversely, that its survival time increases the less frequently it is fed, regardless of light condition. The explanation of this inverse correlation is that the more frequently an organism is fed, the sooner it will mature and undergo gametogenesis and become an empty shell, and thus the shorter its survival time.

Individuals cultured in continuous darkness showed a less distinct trend, because under this light condition most of the G. sacculifer organisms underwent gametogenesis within 5 days, resulting in uniformly low slopes for all feeding groups in the CD regime (Figs. 1K-N, 2D).

The mean survival times for the fourteen experimental groups are plotted in Fig. 4. When individuals with a mean initial shell length of 325 µm were fed daily, their mean survival times were minimal and ranged from 4 days in the CD regime to 10 days in the HLI regime. The mean survival times were progressively longer (regardless of light condition) when feeding frequency was decreased. The highest survival times were observed when the individuals were not fed, ranging from a mean of 5 days in the CD regime to 31 days in the HLI regime.

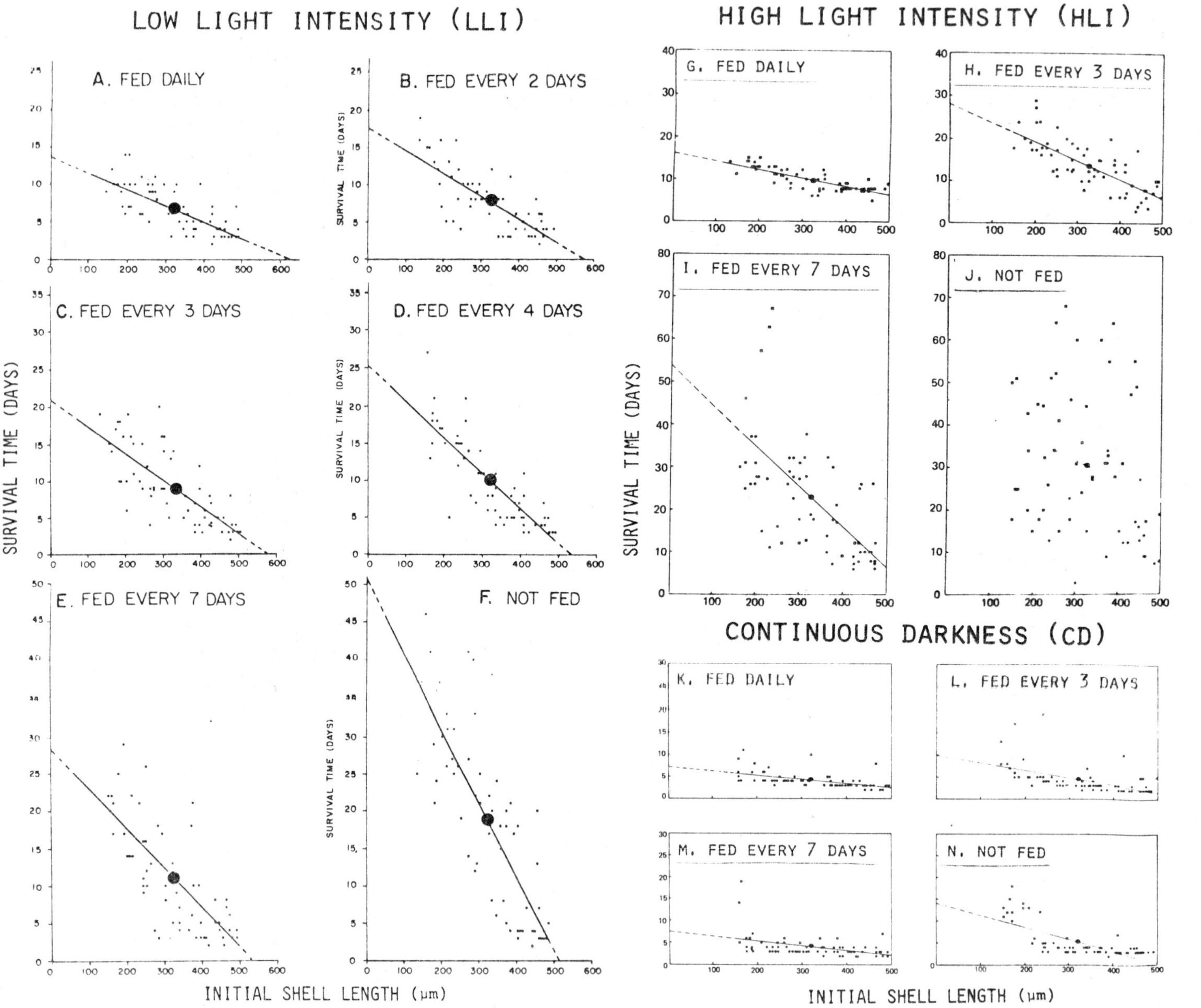


Fig. 1A-N. Scatter plots of relationship between survival time and initial shell length of G. sacculifer in 3 light regimes: A-F, 6 feeding regimes maintained in low light intensity (LLI); G-J, 4 feeding regimes in high light intensity (HLI); K-N, 4 feeding regimes in continuous darkness (CD). Black circles on regression lines are mean values. Each feeding regime consists of of 63 specimens (see Tables 2 & 3). [From Bé et al., 1981 and Caron et al., 1982]

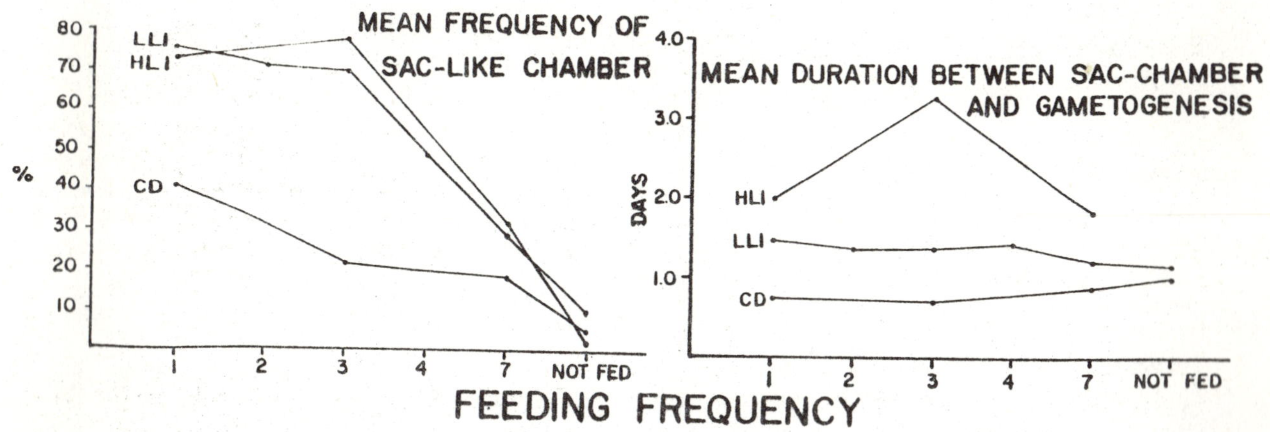


Fig. 7 (left). Effects of feeding frequency and light intensity on mean frequency of sac-like chamber formation. Fig. 8 (right). Effects of feeding frequency and light intensity on mean duration between sac-chamber formation and gametogenesis. See Figs. 4-5 legend.

Effect of feeding frequency and light intensity on mean duration between sac-chamber and gametogenesis.-- There is a clear trend showing that, after the formation of a sac-like chamber, the onset of gametogenesis was progressively delayed with increasing light intensity (Fig. 8). This delay was unrelated to feeding frequency. Thus, the mean time interval between sac-like chamber formation and gamete release was 18 h in CD, 36 h in LLI and 48-72 h in HLI.

## SHELL GROWTH

Prior to routine maintenance of planktonic foraminifera in the laboratory, the process of chamber formation and shell growth rates of planktonic foraminifera were matters of conjecture. In recent years periodic additions of chambers were directly observed and basic organic structures associated with calcite secretion were identified (Bé et al., 1979). Furthermore, shell growth rates of planktonic foraminifera can now be measured under varying experimental conditions in the laboratory (Bé et al., 1981; Caron et al., 1982).

Chamber formation.-- To date the process of chamber formation and its fine structure have been studied in nonspinose species, while those in spinose species are still in progress. It takes about 5 to 6 hours for nonspinose *Globorotalia hirsuta* or *G. truncatulinoides* to grow a new, thin-walled chamber (Plate 4; Bé et al., 1979). Chamber formation in either species began with the emergence of a cytoplasmic bulge from the aperture. After 1-2 hours the translucent cytoplasmic bulge expanded to the full shape of a new chamber. Specimens fixed at this moment for transmission electron microscopy showed the development of three basic organic structures, of which the "primary organic membrane" (POM) is associated with the actual secretion of calcite. After 2-3 hours, calcareous plaques were evident on both sides of the POM. Subsequently, the jigsaw-puzzled calcareous plaques increased in number and size until they coalesced to form a bilamellar wall. Further lamellar additions, mainly on the distal side of the POM resulted in thickening of the chamber wall.



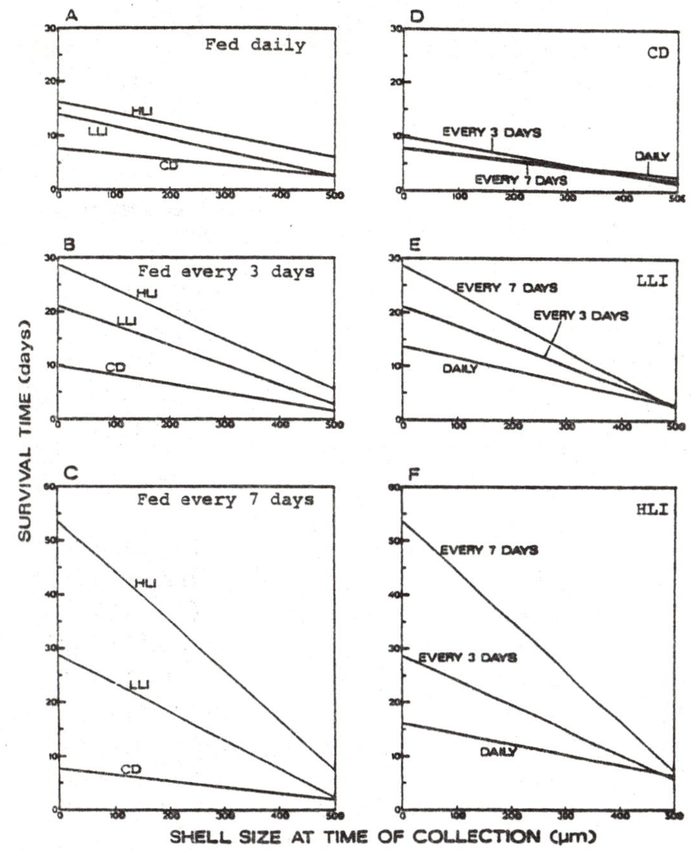


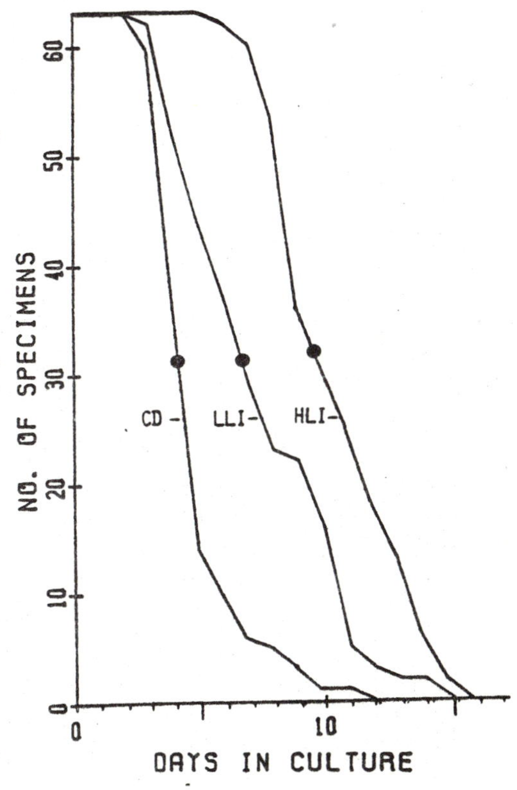


Fig. 2 (left) Regression lines from Fig. 1 of *G. sacculifer* cultured in 3 light regimes and 3 feeding regimes. A-C show variations in the regression lines due to different light conditions with constant feeding schedules. D-F show variations in the regression lines due to different feeding regimes under constant light conditions. In this and following figures, HLI = high light intensity regime, LLI = low light intensity regime, and CD = continuous darkness regime (see Tables 2 & 3). [From Caron et al., 1982].

Fig. 3 (right) Survivorship curves of daily-fed *G. sacculifer* in 3 light regimes (N = 63 specimens).

The mean life span of each experimental group was estimated by determining the intercepts of the y-axes with the extrapolated regression lines (Fig. 1).The difference between life span and survival time is the duration that the average organism (with shell length of 325 µm) lived in the ocean prior to being collected. The mean life spans for the fourteen groups obtained from these extrapolations are plotted in Fig. 5. Since life spans closely parallel survival times, there is a concomitant increase in life span with a decrease in feeding frequency. Life spans of individuals fed daily ranged from 7 to 16 days, whereas those not fed ranged from 14 to 50 days.

Effect of light intensity on survival time and life span.-- Survival time is directly correlated with light intensity. At equivalent feeding frequencies, organisms in HLI lived longer than those in LLI, which in turn lived longer than those in continuous darkness (Fig. 2A-C;Fig. 3). Organisms in CD survived on the average for only 4 to 5 days, those in LLI for 7 to 19 days, and those in HLI for 10 to 31 days (Fig. 4).

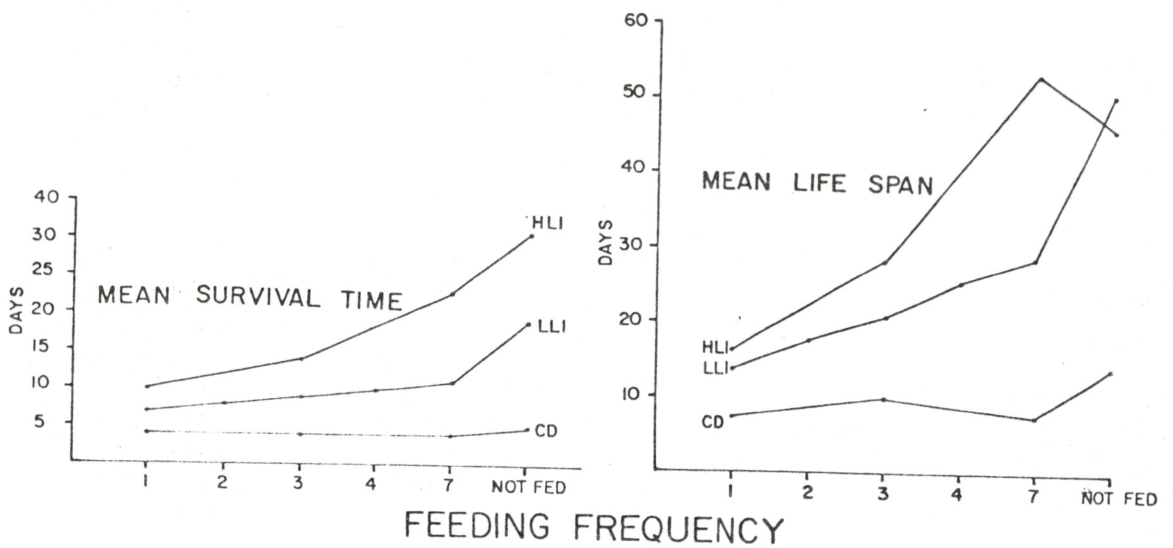


Fig. 4 (left). Effects of feeding frequency and light intensity on mean survival time of G. sacculifer. Fig. 5 (right). Effects of feeding frequency and light intensity on mean life span of G. sacculifer. Feeding frequency on x-axis decreases from left to right, i.e. 1 = fed daily, 2 = fed every 2 days, etc. Figs. 4-8, 10, 11 & 16 are based on experimental groups of G. sacculifer in 4-6 feeding regimes and 3 light regimes listed in Tables 2 and 3.

The mean life span of G. sacculifer is also directly correlated with light intensity (Fig. 5). The mean life span of fed individuals ranged from 7 to 10 days in CD, 14 to 29 days in LLI and 16 to 54 days in HLI. Long life spans of planktonic foraminifera are possible if growth and maturation are retarded and gametogenesis is delayed. This can be attained if the organism ingests a minimum amount of particulate food, as in the starved group. It is conjectured that the long survival time under starvation conditions is aided by the symbiotic dinoflagellates, which provide some nutrition to their hosts. Although symbiont-derived nutrition alone is not sufficient for foraminiferal growth under any of the 3 light regimes, it may be important for survival of individuals at times of extreme oligotrophy. Young specimens (<300 μm in shell length) which were starved, exhibited little or no shell growth. Thus, the estimated mean life spans for unfed individuals of 50 days in LLI and 46 days in HLI should be regarded as upper limits and an index of their survival capacity. It is questionable, however, whether complete starvation conditions occur in the photic zone of the ocean.

Effect of feeding frequency and light intensity on gametogenesis frequency.-- Gametogenesis frequency, i.e. the percentage of individuals which undergo gametogenesis vs. the percentage that die (due to disease, etc.), is a measure of the reproductive capacity of planktonic foraminifera. In laboratory culture, "death" can be distinguished morphologically from "gametogenesis" by the fact that the former results in a relatively thin shell with elongate spines intact (Plate 3, fig. 2), whereas the latter is accompanied by calcitic encrustment of the shell (Plate 3, fig. 3; Bé, 1980) as discussed in more detail on page 34.

There was a direct correlation between feeding frequency and mea gametogenesis frequency, while light intensity had no apparent effec on gametogenesis frequency (Table 3; fig. 6). The values were relativel high, ranging from 79% to 95% for the 11 groups fed Artemia under different light conditions. Mean gametogenesis frequency decrease markedly to 32% in HLI and 44% in LLI when the organisms were not fed at all (Fig. 6). Increasing light intensity caused a delay in the onset of gametogenesis, while continuous darkness accelerated the onset of gametogenesis. Gamete release, among pooled specimens of 3 equivalent feeding regimes, occurred on the average in 4.2 days in CD, 9.1 days in LLI , and 15.6 days in HLI. Because of the immediate effect of inducing gametogenesis, continuous darkness overrode any influence due to feeding frequency. This explains the relatively high gametogenesis frequencies for all feeding regimes in CD.

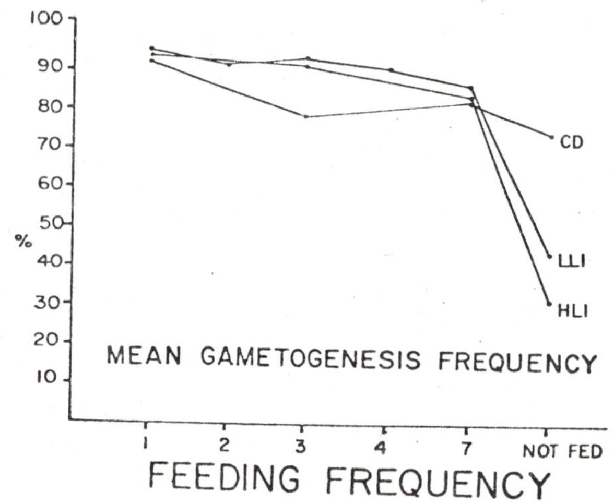


Fig. 6. Effects of feeding frequency and light intensity on mean gametogenesis frequency. See Figs. 4-5 legend.

Effect of feeding frequency and light intensity on frequency of sac-like chamber.-- When G. sacculifer approached reproductive maturity, a sac-like terminal chamber was frequently formed, signaling impending gametogenesis (Plate 6, fig. 1) (Bé, 1980). The frequency of sac-like chamber formation correlated directly with feeding frequency in all three light regimes (Fig. 7). There was a progressive increase in the mean frequency of sac-like chambers from a minimum of 2% in the starved group to a maximum of 76% in the daily-fed and 78% in the fed-every-3 days groups, regardless of light condition. Since the frequency of terminal, sac-like chambers is contingent upon the availability and consumption of food by G. sacculifer, this parameter holds great promise as an indicator of regional plankton standing stock and productivity.

The sac-like chamber frequency was considerably lower for the CD group than for the LLI and HLI groups. For some unknown reason, CD appeared to inhibit the formation of sac-like chambers, as only 5% of starved and 41% of daily-fed individuals formed sac-like chambers in the dark. The implication is that such terminal chambers are rarely formed below the euphotic zone.

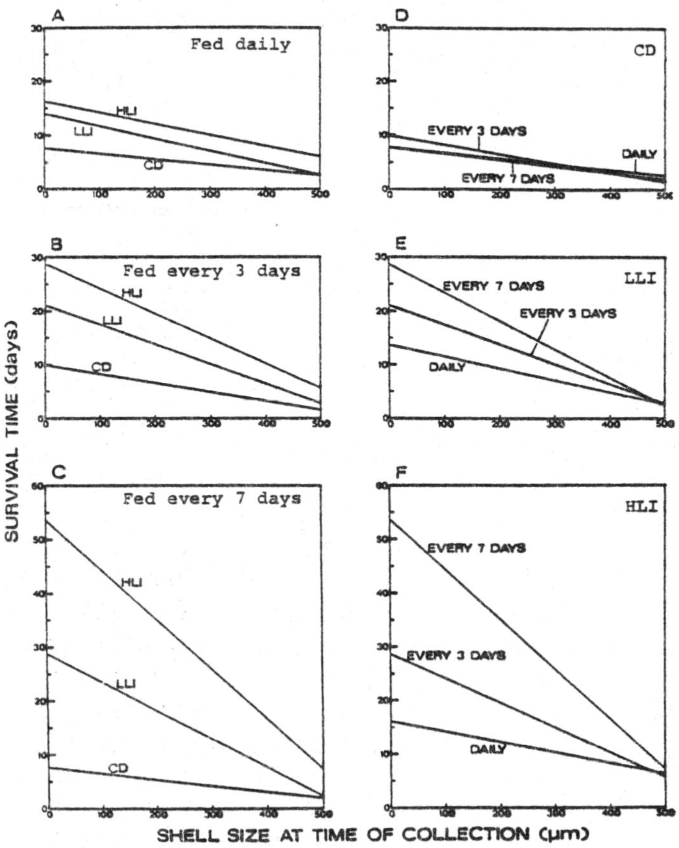


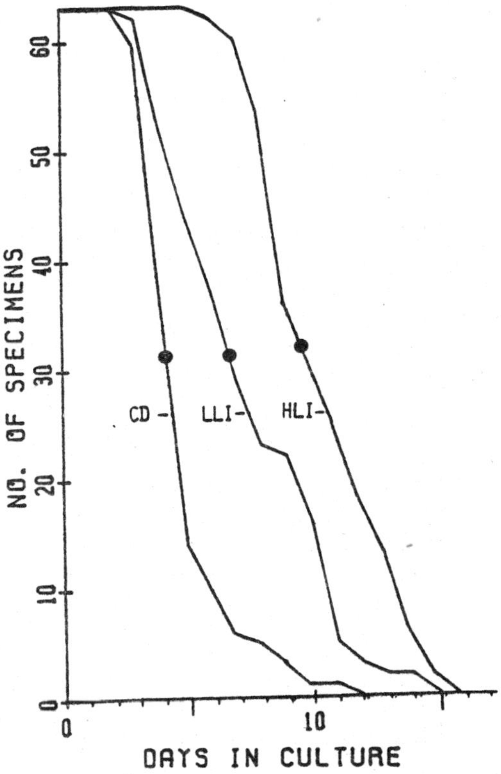


Fig. 2 (left) Regression lines from Fig. 1 of G. sacculifer cultured in 3 light regimes and 3 feeding regimes. A-C show variations in the regression lines due to different light conditions with constant feeding schedules. D-F show variations in the regression lines due to different feeding regimes under constant light conditions. In this and following figures, HLI = high light intensity regime, LLI = low light intensity regime, and CD = continuous darkness regime (see Tables 2 & 3). [From Caron et al., 1982].

Fig. 3 (right) Survivorship curves of daily-fed G. sacculifer in 3 light regimes (N = 63 specimens).

The mean life span of each experimental group was estimated by determining the intercepts of the y-axes with the extrapolated regression lines (Fig. 1). The difference between life span and survival time is the duration that the average organism (with shell length of 325 µm) lived in the ocean prior to being collected. The mean life spans for the fourteen groups obtained from these extrapolations are plotted in Fig. 5. Since life spans closely parallel survival times, there is a concomitant increase in life span with a decrease in feeding frequency. Life spans of individuals fed daily ranged from 7 to 16 days, whereas those not fed ranged from 14 to 50 days.

Effect of light intensity on survival time and life span.-- Survival time is directly correlated with light intensity. At equivalent feeding frequencies, organisms in HLI lived longer than those in LLI, which in turn lived longer than those in continuous darkness (Fig. 2A-C; Fig. 3). Organisms in CD survived on the average for only 4 to 5 days, those in LLI for 7 to 19 days, and those in HLI for 10 to 31 days (Fig. 4).

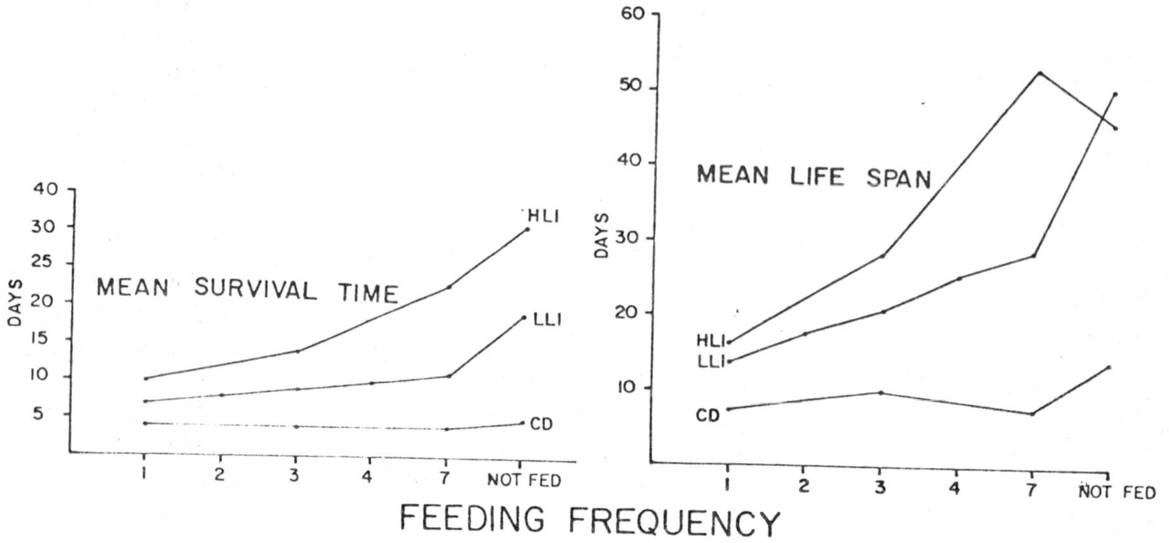


Fig. 4 (left). Effects of feeding frequency and light intensity on mean survival time of G. sacculifer. Fig. 5 (right). Effects of feeding frequency and light intensity on mean life span of G. sacculifer. Feeding frequency on x-axis decreases from left to right, i.e. 1 = fed daily, 2 = fed every 2 days, etc. Figs. 4-8, 10, 11 & 16 are based on experimental groups of G. sacculifer in 4-6 feeding regimes and 3 light regimes listed in Tables 2 and 3.

The mean life span of G. sacculifer is also directly correlated with light intensity (Fig. 5). The mean life span of fed individuals ranged from 7 to 10 days in CD, 14 to 29 days in LLI and 16 to 54 days in HLI. Long life spans of planktonic foraminifera are possible if growth and maturation are retarded and gametogenesis is delayed. This can be attained if the organism ingests a minimum amount of particulate food, as in the starved group. It is conjectured that the long survival time under starvation conditions is aided by the symbiotic dinoflagellates, which provide some nutrition to their hosts. Although symbiont-derived nutrition alone is not sufficient for foraminiferal growth under any of the 3 light regimes, it may be important for survival of individuals at times of extreme oligotrophy. Young specimens (<300 μm in shell length) which were starved, exhibited little or no shell growth. Thus, the estimated mean life spans for unfed individuals of 50 days in LLI and 46 days in HLI should be regarded as upper limits and an index of their survival capacity. It is questionable, however, whether complete starvation conditions occur in the photic zone of the ocean.

Effect of feeding frequency and light intensity on gametogenesis frequency.-- Gametogenesis frequency, i.e. the percentage of individuals which undergo gametogenesis vs. the percentage that die (due to disease, etc.), is a measure of the reproductive capacity of planktonic foraminifera. In laboratory culture, "death" can be distinguished morphologically from "gametogenesis" by the fact that the former results in a relatively thin shell with elongate spines intact (Plate 3, fig. 2), whereas the latter is accompanied by calcitic encrustment of the shell (Plate 3, fig. 3; Bé, 1980) as discussed in more detail on page 34.

There was a direct correlation between feeding frequency and mean gametogenesis frequency, while light intensity had no apparent effect on gametogenesis frequency (Table 3; fig. 6). The values were relatively high, ranging from 79% to 95% for the 11 groups fed Artemia under 3 different light conditions. Mean gametogenesis frequency decreased markedly to 32% in HLI and 44% in LLI when the organisms were not fed at all (Fig. 6). Increasing light intensity caused a delay in the onset of gametogenesis, while continuous darkness accelerated the onset of gametogenesis. Gamete release, among pooled specimens of 3 equivalent feeding regimes, occurred on the average in 4.2 days in CD, 9.1 days in LLI , and 15.6 days in HLI. Because of the immediate effect of inducing gametogenesis, continuous darkness overrode any influence due to feeding frequency. This explains the relatively high gametogenesis frequencies for all feeding regimes in CD.

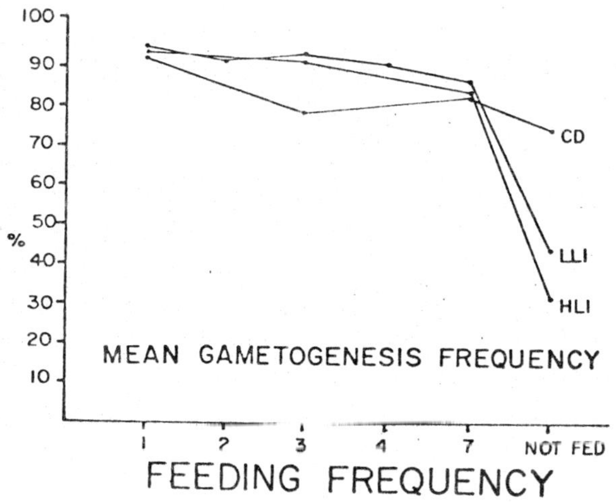


Fig. 6. Effects of feeding frequency and light intensity on mean gametogenesis frequency. See Figs. 4-5 legend.

Effect of feeding frequency and light intensity on frequency of sac-like chamber.-- When G. sacculifer approached reproductive maturity, a sac-like terminal chamber was frequently formed, signaling impending gametogenesis (Plate 6, fig. 1) (Bé, 1980). The frequency of sac-like chamber formation correlated directly with feeding frequency in all three light regimes (Fig. 7). There was a progressive increase in the mean frequency of sac-like chambers from a minimum of 2% in the starved group to a maximum of 76% in the daily-fed and 78% in the fed-every-3 days groups, regardless of light condition. Since the frequency of terminal, sac-like chambers is contingent upon the availability and consumption of food by G. sacculifer, this parameter holds great promise as an indicator of regional plankton standing stock and productivity.

The sac-like chamber frequency was considerably lower for the CD group than for the LLI and HLI groups. For some unknown reason, CD appeared to inhibit the formation of sac-like chambers, as only 5% of starved and 41% of daily-fed individuals formed sac-like chambers in the dark. The implication is that such terminal chambers are rarely formed below the euphotic zone.

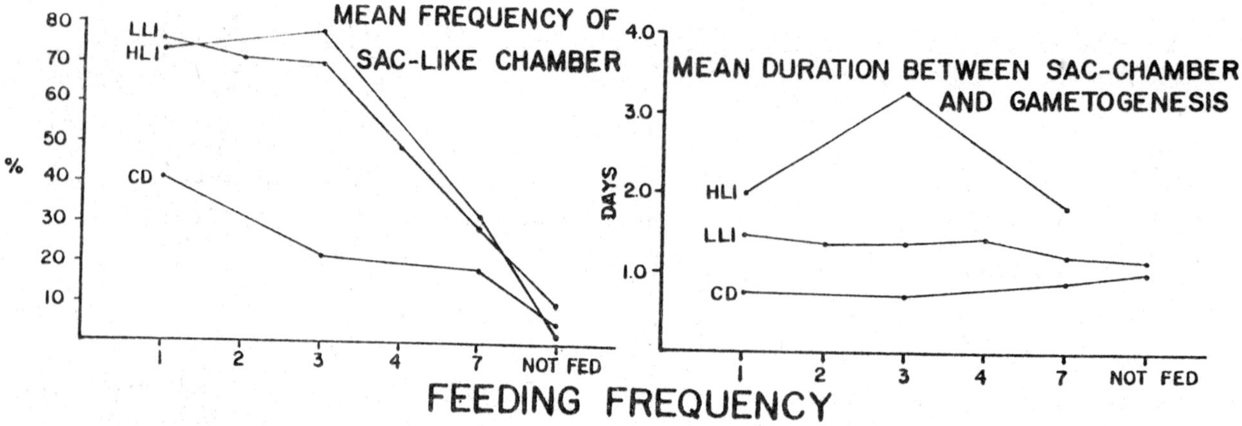


Fig. 7 (left). Effects of feeding frequency and light intensity on mean frequency of sac-like chamber formation. Fig. 8 (right). Effects of feeding frequency and light intensity on mean duration between sac-chamber formation and gametogenesis. See Figs. 4-5 legend.

Effect of feeding frequency and light intensity on mean duration between sac-chamber and gametogenesis.-- There is a clear trend showing that, after the formation of a sac-like chamber, the onset of gametogenesis was progressively delayed with increasing light intensity (Fig. 8). This delay was unrelated to feeding frequency. Thus, the mean time interval between sac-like chamber formation and gamete release was 18 h in CD, 36 h in LLI and 48-72 h in HLI.

## SHELL GROWTH

Prior to routine maintenance of planktonic foraminifera in the laboratory, the process of chamber formation and shell growth rates of planktonic foraminifera were matters of conjecture. In recent years periodic additions of chambers were directly observed and basic organic structures associated with calcite secretion were identified (Bé et al., 1979). Furthermore, shell growth rates of planktonic foraminifera can now be measured under varying experimental conditions in the laboratory (Bé et al., 1981; Caron et al., 1982).

Chamber formation.-- To date the process of chamber formation and its fine structure have been studied in nonspinose species, while those in spinose species are still in progress. It takes about 5 to 6 hours for nonspinose *Globorotalia hirsuta* or *G. truncatulinoides* to grow a new, thin-walled chamber (Plate 4; Bé et al., 1979). Chamber formation in either species began with the emergence of a cytoplasmic bulge from the aperture. After 1-2 hours the translucent cytoplasmic bulge expanded to the full shape of a new chamber. Specimens fixed at this moment for transmission electron microscopy showed the development of three basic organic structures, of which the "primary organic membrane" (POM) is associated with the actual secretion of calcite. After 2-3 hours, calcareous plaques were evident on both sides of the POM. Subsequently, the jigsaw-puzzled calcareous plaques increased in number and size until they coalesced to form a bilamellar wall. Further lamellar additions, mainly on the distal side of the POM resulted in thickening of the chamber wall.

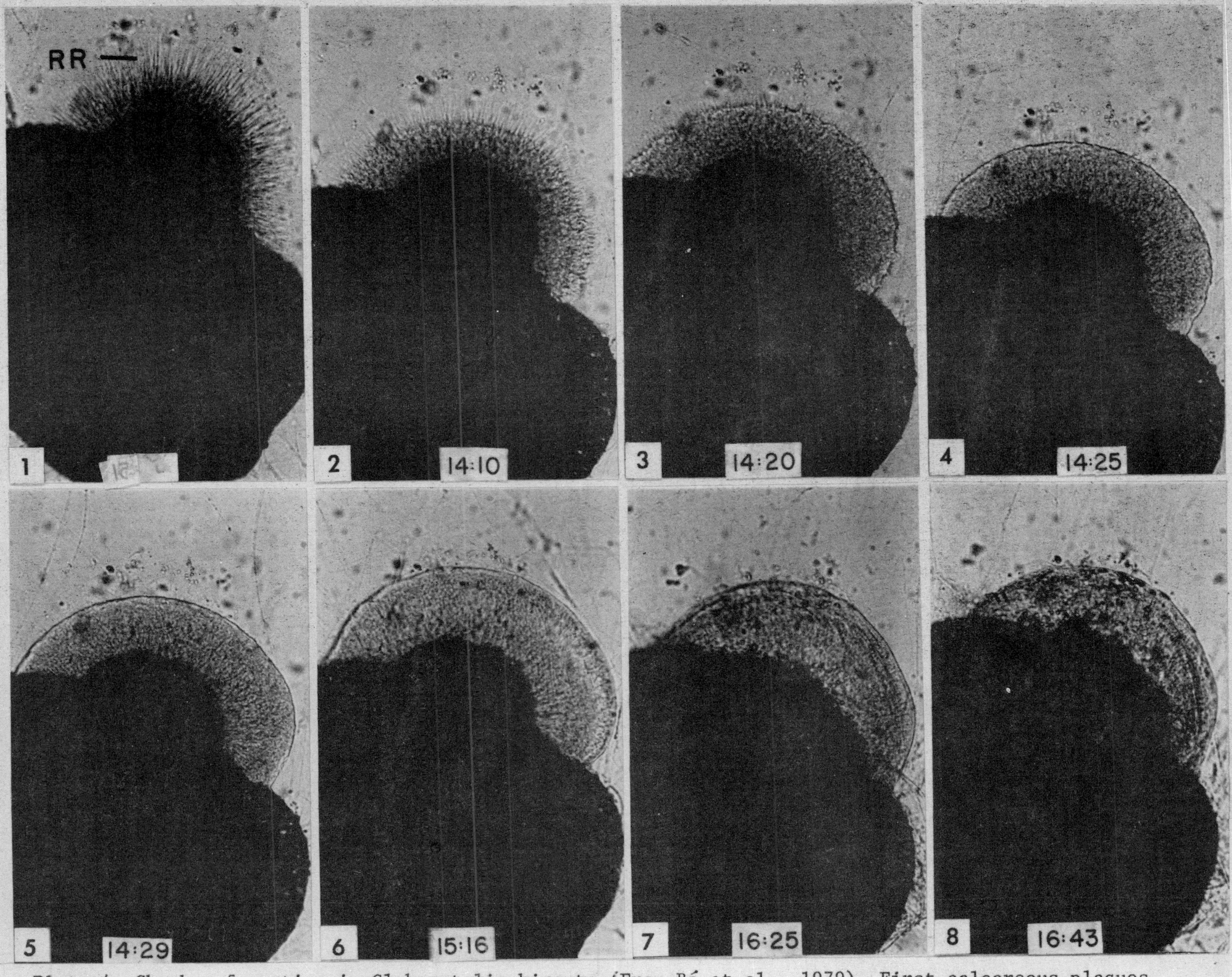


Plate 4. Chamber formation in Globorotalia hirsuta (From Bé et al., 1979). First calcareous plaques appear at 16:25, or 2 hours after first appearance of cytoplasmic bulge. RR=Radiating rhizopodia. x 166.

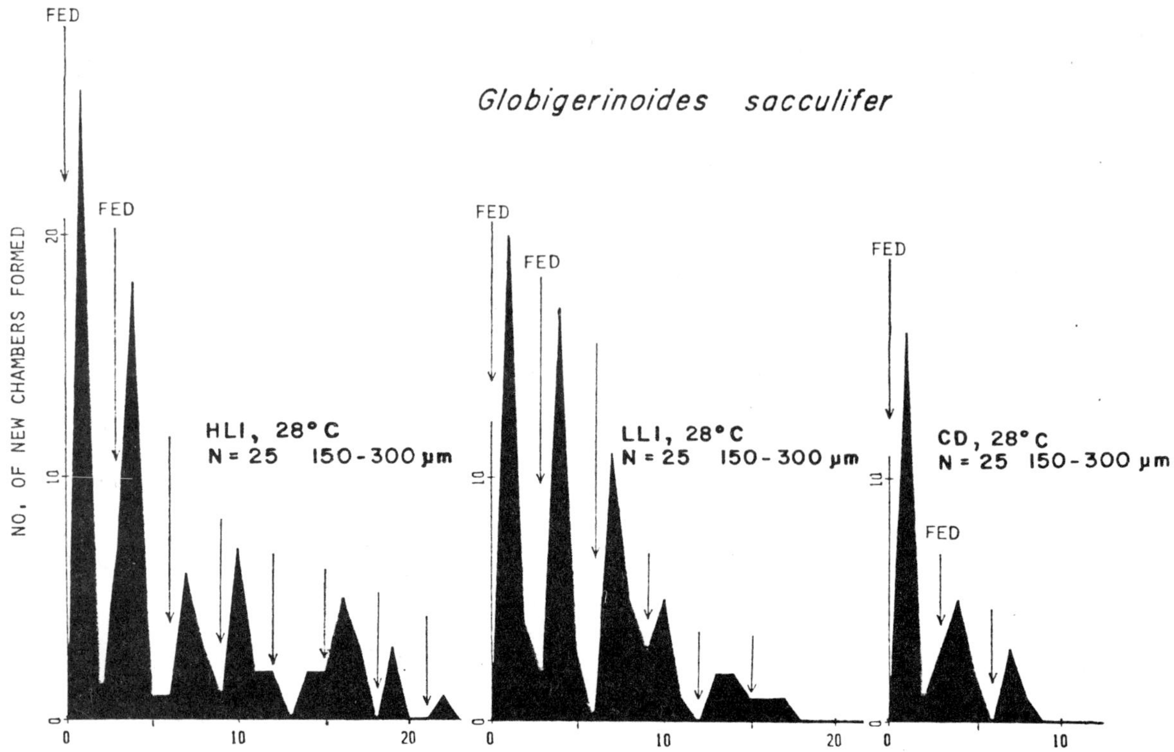


Fig. 9. A progressive decrease in chamber formation with decreasing light intensity as well as with ontogeny is evident in these three experimental groups of G. sacculifer which were fed every three days. Feeding (arrows) is followed by high frequencies of chamber formation on the following day and greatly reduced numbers on the second and third days.

Feeding time vs. chamber formation.-- The close link between time of feeding and chamber formation is strikingly apparent when the organisms are fed every 3 days (Fig. 9). Feeding of young individuals of G. sacculifer (150-300 μm in shell length) is followed by high percentages of new chamber formation on the following day, but greatly reduced percentages on the second and third days after feeding. Each of the three experimental groups showed a progressive decrease in the total number of new chambers formed. Hence, the three graphs in Fig. 9 show 3-day-cycle peaks which decrease exponentially. It is noteworthy that the total number of chambers formed also decreased with decreasing light intensity, i.e. a total of 88 chambers were formed in HLI, 78 chambers in LLI, and 31 chambers in CD by 27 individuals in each light regime.

Shell growth rates.-- Shell growth rates are directly proportional to feeding frequency. The decrease in mean growth rate with decreasing feeding frequency in each of the three light conditions is evident in Fig. 10. The approximately parallel slopes indicate that the growth rates decreased uniformly within and between each light regime. And although the organisms in CD formed fewer total number of chambers than those in LLI or HLI, their shorter survival times resulted in an intermediate shell growth rate.

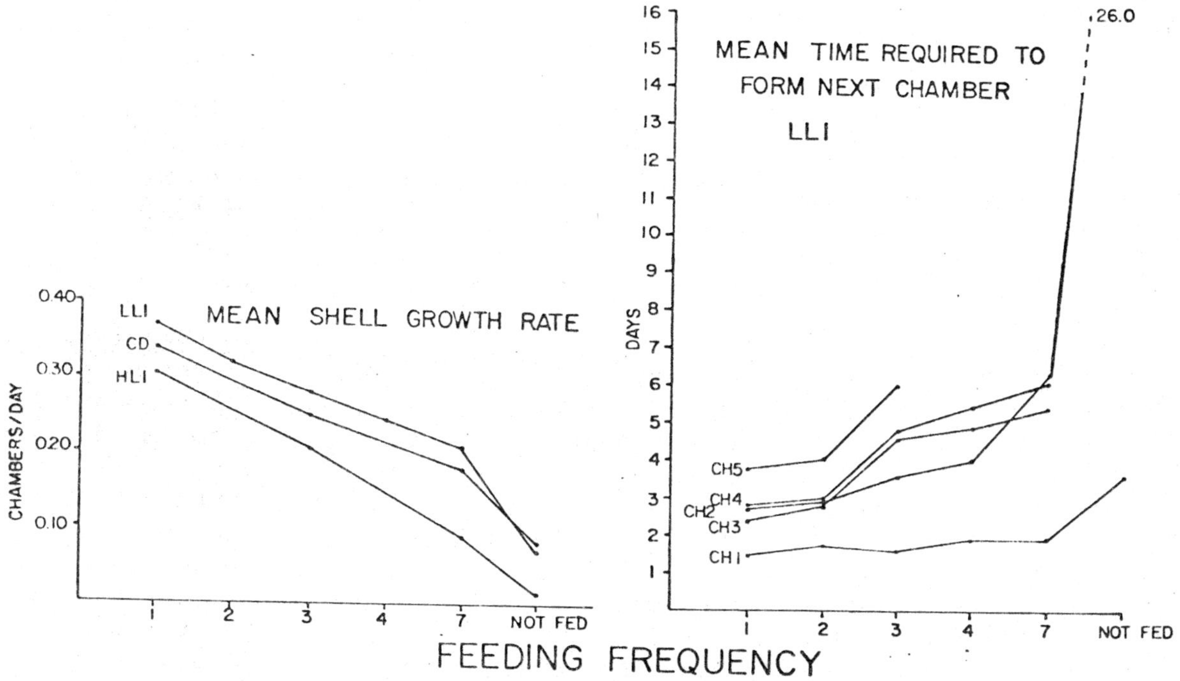


Fig. 10 (left): Effects of feeding frequency and light intensity on mean shell growth rate of G. sacculifer. Fig. 11 (right): Effect of feeding frequency on mean time required to form next chamber in G. sacculifer cultured in low-light intensity (LLI) regime. See Figs. 4-5 legend.

Variation of shell growth rate during ontogeny.-- The duration between successive chamber formations became progressively longer with ontogenetic development (Fig. 11). An organism with a mean shell length of 325 μm required only about 1.6 days on the average to form the first chamber in culture, if it were fed under any of the 5 feeding frequencies (3.6 days when starved). The durations between the second, third and fourth chambers were significantly longer, varying between 2.3 and 6.2 days for organisms that received food. The longest duration was between the formations of the fourth and fifth chamber, which varied between 3.9 and 6.0 days. Thus, Figs. 9-11 show that as the organisms matured, the rate of chamber formation decreased.

Daily measurements of shell size of the experimental organisms also enabled us to determine variations in shell growth rate during their ontogeny (Figs. 12-14). These graphs are mean growth curves based on day-to-day averagings of the shell sizes of all living individuals in each group whether or not they formed chambers during each 24h period. Each experimental group started with 23 to 27 specimens with initial shell lengths ranging between 150 and 300 μm and a mean of 220 μm. It is evident from each graph that G. sacculifer's most rapid growth was in early ontogeny and that growth slowed down in late ontogeny. The "negative" growth at the ends of some curves is due to one or more survivors of smaller shell size than their predecessors.

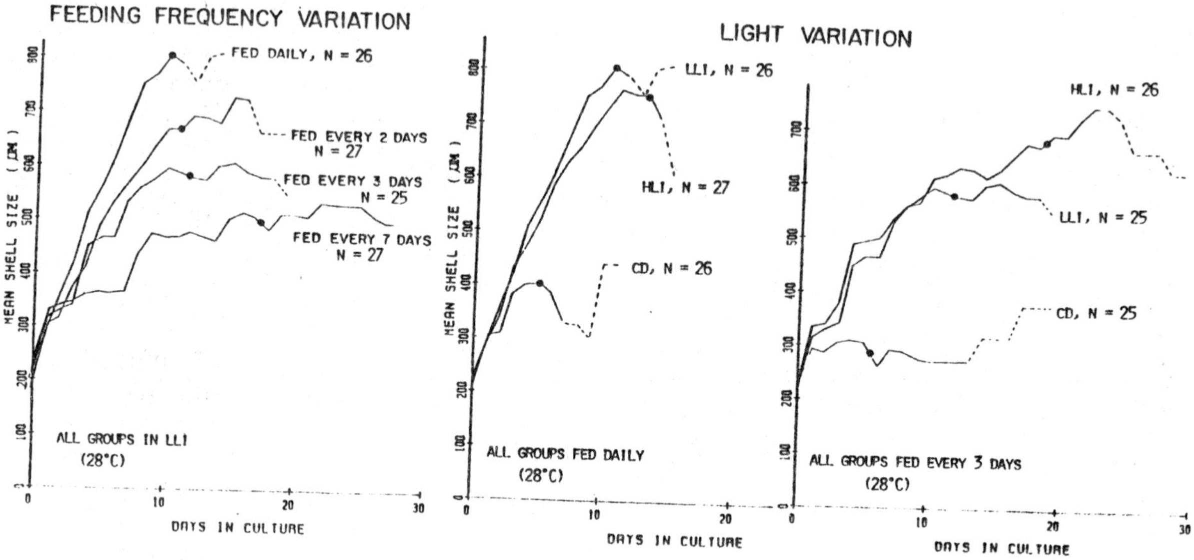


Fig. 12 (left). Shell growth curves of G. sacculifer as a function of feeding frequency. All 4 groups were cultured at LLI and 28°C.
Fig. 13 (center) and Fig. 14 (right). Shell growth curves of G. sacculifer as a function of light intensity.
Each growth curve in Figs. 12-14 is based on young specimens (150-300 µm initial shell length) listed in Table 2. Black circles are median values and dotted parts of curves are based on 2 or 3 individuals.

Figure 12 shows clearly that feeding frequency was the dominant factor in shell growth of G. sacculifer, whereas variation in light intensity (LLI vs. HLI) yielded approximately similar growth curves under equivalent feeding frequencies (Figs. 13, 14). On the other hand, shell growth in CD was drastically reduced, because the inhibition of algal photosynthesis apparently resulted in the death of the symbionts and premature gametogenesis of their hosts.

Our preliminary data indicate that the shell growth rates of spinose species of Globigerinoides are greater than those of nonspinose species belonging to Globorotalia or Globoquadrina, but further detailed investigations are necessary to determine the variability in shell growth rates of each planktonic foraminiferal species.

Mean final shell size.-- There is a direct correlation between final shell size, feeding frequency, and light intensity (Table 3). Histograms of frequency distributions of final shell lengths of 252 specimens in four pooled feeding regimes and three light regimes indicate that in general those cultured in HLI attained larger final shell sizes than those in LLI, which in turn were larger than those in CD

(Fig. 15A-C). The prolongation of the growth period in HLI and the early onset of gametogenesis in CD resulted in a difference of 259 μm in the mean final shell size for individuals cultured under these two light regimes. The data in Fig. 15 also indicate that a minimum size of about 250 μm was necessary before G. sacculifer attained reproductive maturity (gametogenesis).

The progressive increase in mean final shell size with increase in feeding frequency and light intensity is evident in Fig. 16. The mean shell size for all fourteen experimental groups ranged from a maximum of 791 μm for the daily-fed group in HLI to a minimum of 289μm for the starved group in HLI.

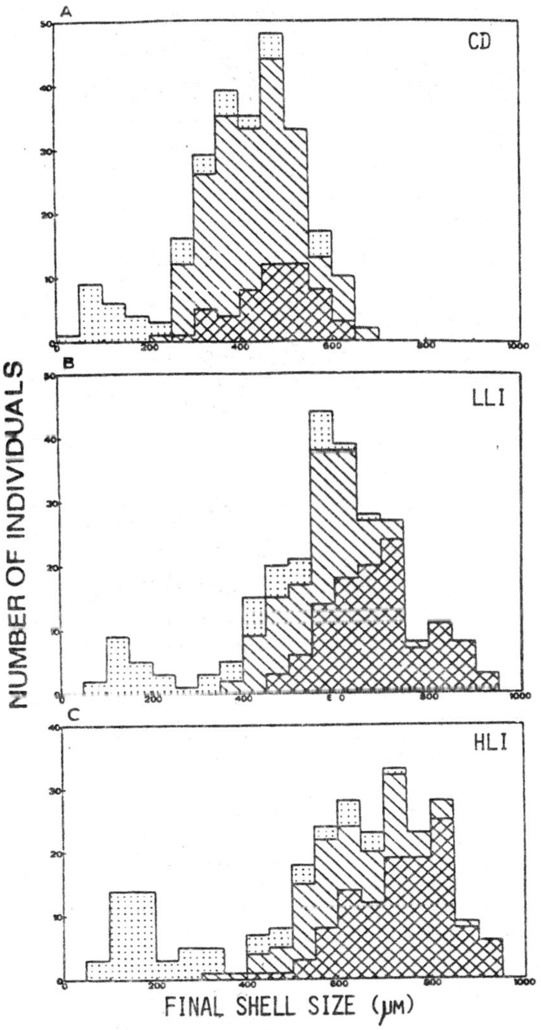


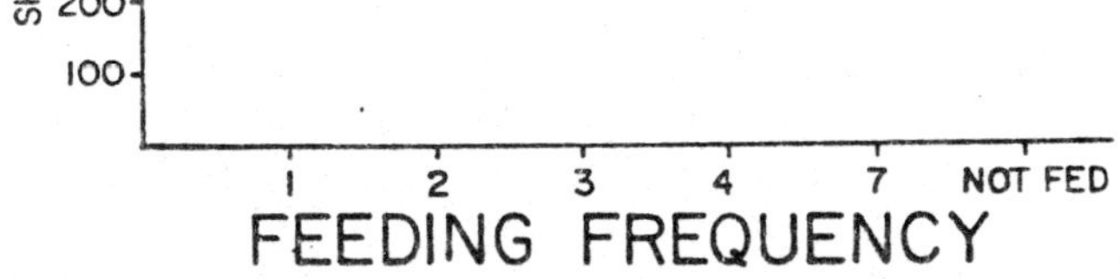


Fig. 15 (left). Histograms of frequency distributions of final shell lengths of 4 pooled feeding regimes (252 specimens) in 3 different light regimes. Non-gametogenic specimens that died are stippled. Gametogenic specimens without sac-like chamber are oblique-ruled. Gametogenic specimens with sac-like chamber are crossed (from Caron et al., 1982).

Fig. 16 (right). Effects of feeding frequency and light intensity on mean final shell size of G. sacculifer. See Figs. 4-5 legend.

Temperature effect on shell growth.-- Our previous discussions were on the results of G. sacculifer cultured at 28°C -- the surface-water temperature over a large part of its geographic distribution. In order to evaluate the effect of low temperature on its shell growth, we maintained two groups of G. sacculifer at 20°C (Table 2), which approximates the surface-water temperature at its distributional limit in the North Atlantic. This temperature also corresponds to a water depth of 150 ± 25 m off Barbados that is below the photosynthesis compensation depth.

The two experimental groups were cultured in HLI and fed daily and every 3 days -- thus enabling a comparison with two corresponding experimental groups cultured at 28°C. The results are preliminary and still inconclusive. For the daily-fed groups, G. sacculifer's shell growth rate at 20°C was significantly lower and its mean final shell size considerably smaller than those cultured at 28°C (Fig. 17). However, for the organisms fed every 3 days, the shell growth rate at 20°C was closely similar to that at 28°C, except that the latter group survived 10 days longer (Fig. 18). The results of both experimental sets agree that G. sacculifer maintained at 20°C have generally shorter survival times and smaller final shell sizes than those at 28°C.

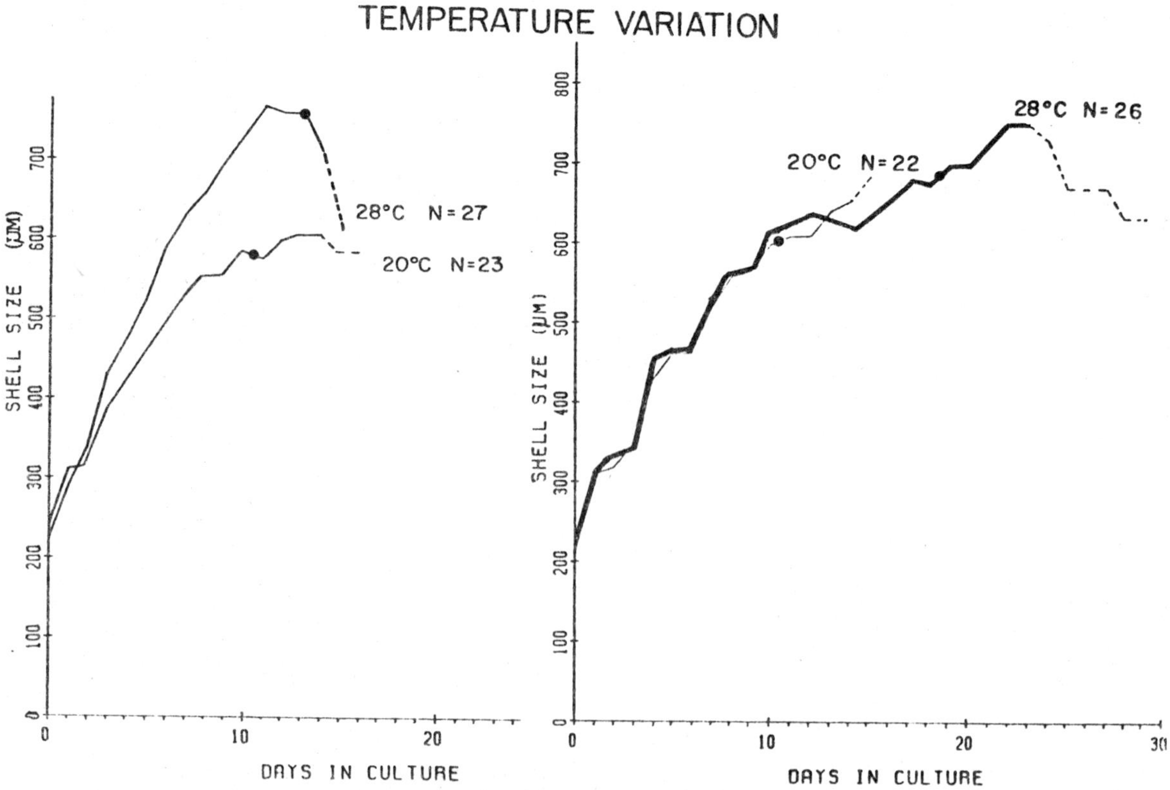


Fig. 17 (left) and Fig. 18 (right). Shell growth curves of young individuals of G. sacculifer (150-300 µm initial shell length) as a function of water temperature. Both groups in Fig. 17 were fed daily and those in Fig. 18 were fed every 3 days. All groups were maintained in HLI.

## SHELL STRUCTURE, POROSITY AND SURFACE TEXTURE

Investigations of the wall structure of planktonic foraminifera prior to 1971 were limited to the mineralized constituents of the shell (e.g. Reiss, 1957, 1958, 1963; Bé and Ericson, 1963; Lipps, 1966; Pessagno and Miyano, 1968; Hansen et al., 1969; Hemleben, 1969). An important finding was that of Reiss (1957, 1958, 1963) who demonstrated the bilamellar nature of the chamber walls of globigerinid and globorotaliid species. Towe (1971) showed that the delicate organic structure of pore plates can be examined with the transmission electron microscope after embedding, decalcifying and ultrathin-sectioning shells of freshly collected living specimens. After further refinement in analytical methods, Hemleben et al. (1977) were able to demonstrate the intimate structural relationship between organic and calcite layers in Globorotalia menardii. Bé et al. (1980) examined the nature of organic pore structures and their close connection with the organic layers embedded between the calcite layers.

There are some basic similarities and differences in the organic layers and pore structures of spinose and nonspinose species (Fig. 19). Besides the possession of elongated spines, species belonging to Globigerinoides and Globigerinella have an inner organic lining (IOL) that is relatively thin and in contact with the perforated pore plates, which may allow exchange of soluble or colloidal material between the intrashell cytoplasm and the environment. Nonspinose species of Globorotalia, Globoquadrina and Pulleniatina have a thick imperforate IOL which is separated from the pore plates, thus preventing any cytoplasmic communication through the pores.

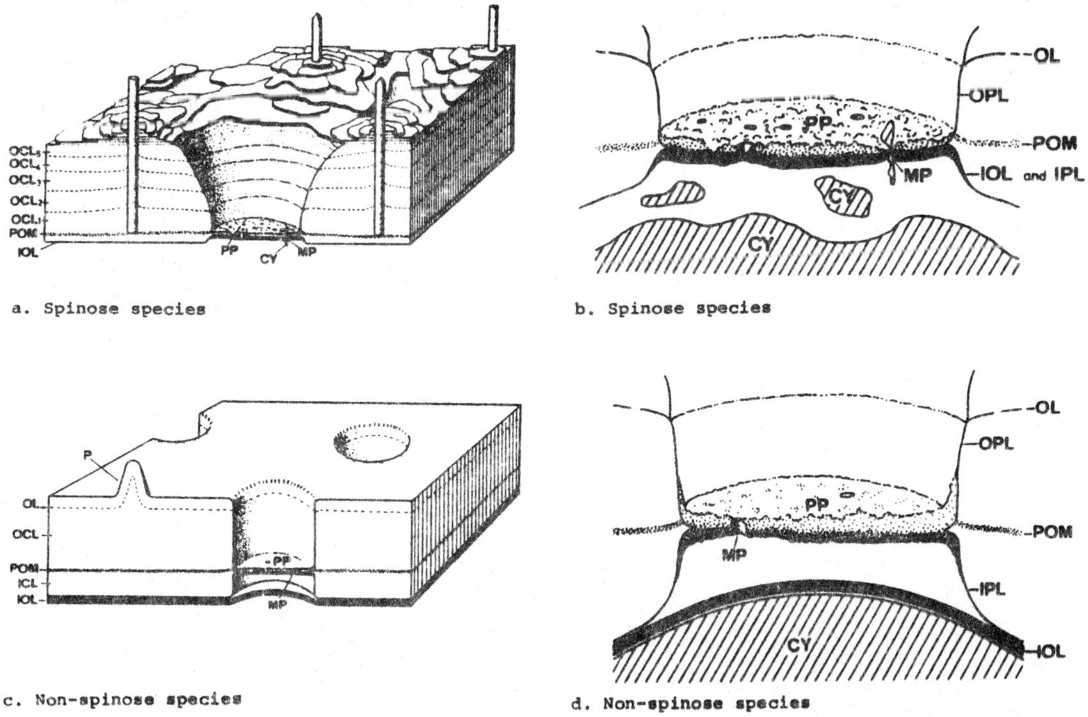


Fig. 19. Schematic wall and pore structures of planktonic foraminifera. The pore diameters of spinose and nonspinose species are 4-7 and 2-4 μm, respectively. CY = cytoplasm; ICL = inner calcite layer; IOL = inner organic lining; IPL = inner pore lining; MP = micropore; OCL = outer calcite layer; OL = outer organic layer; OPL = outer pore lining; POM = primary organic membrane; PP = pore plate (from Bé et al., 1980).

Variations in pore diameter and pore concentrations appear to be species-specific among the 22 species examined by Bé (1968). However, the shell porosity, i.e. percentage of open pore area per unit of shell area, is relatively uniform for species co-occurring in the same latitudinal belts. The shell porosity of tropical-subtropical species is over 10%, that of temperate species 5 to 10%, and that of subpolar-polar species less than 5%. Hence, shell porosities of planktonic foraminifera can be used as general latitudinal indicators.

Surface textures of Recent and Pleistocene planktonic foraminifera have been examined with the scanning electron microscope by Saito et al. (1981), but the original features of most of their specimens from deep-sea sediments show dissolution alterations. Cifelli's (1982) scanning electron micrographs of living specimens show the original surface textures of planktonic foraminifera.

## ROLE OF SYMBIOTIC ALGAE

Symbiotic algae have been observed in planktonic foraminifera by Murray (1897), Rhumbler (1911) and later investigators. The endosymbionts appear to be dinoflagellates, whereas diatoms and other algae are often associated with planktonic foraminifera as external commensals. *Hastigerina pelagica* harbors many large dinoflagellates, *Dissodinium* spp. and *Pyrocystis* spp., which occur within its bubble capsule, but not inside its shell. They are believed to be commensals deriving nutrition from remains of their host's carnivorous diet of copepods.

In general, the spinose planktonic foraminifera in tropical and subtropical regions possess endosymbiotic dinoflagellates. They include *Globigerinoides ruber*, *G. sacculifer*, *G. conglobatus*, *Orbulina universa*, and *Globigerina cristata* (Bé et al., 1977). A notable exception is *Globigerinella aequilateralis* which may be associated with small diatoms and *Trichodesmium*, but not with endosymbiotic dinoflagellates. *Globigerina bulloides*, a cold-water species, also does not appear to have endosymbionts.

Nonspinose planktonic foraminifera, such as *Globorotalia menardii* and *P. obliquiloculata*, may contain small eucaryotic algae within their cytoplasm. The algae are probably freshly ingested food organisms, but we have not ruled out the possibility that they are symbionts. The fact that they are not nearly as numerous as the symbionts of some spinose species and are not dispersed outside the shell lends support to the view that nonspinose species lack symbionts. These observations explain why symbiont-bearing, spinose planktonic foraminifera usually live in shallow depths of the euphotic zone, whereas nonspinose species (*Globorotalia* and *Globoquadrina* species) which presumably do not possess symbionts occur over a greater depth range and frequently below the euphotic zone.

A typical individual of *Globigerinoides sacculifer* possesses hundreds

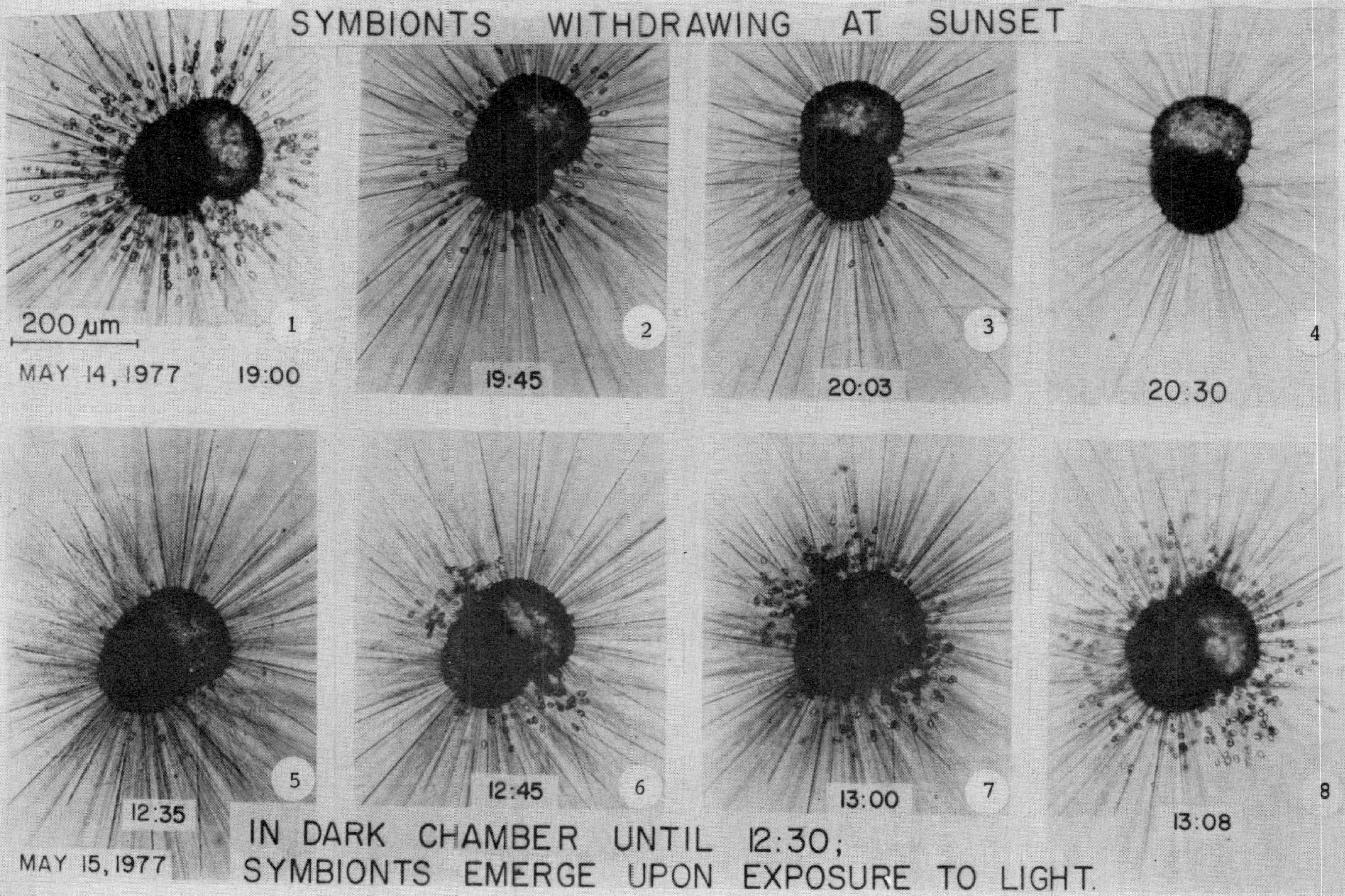


Plate 5. Figs. 1-4: Sequence of ingress of symbionts into the shell of Globigerinoides ruber at sunset, following the natural light cycle. Figs. 5-8: An exogenous control of symbiont movement is demonstrated by the location of symbionts inside the shell when kept in a dark chamber during the day and their immediate emergence upon exposure to artificial light (All magnified x 115).

of endosymbiotic dinoflagellates which are sequestered within vacuoles in its cytoplasm or dispersed in the rhizopodial network between its spines (Plate 3, fig. 1). The fine structure of these symbionts or zooxanthellae, ovoid in shape and measuring approximately 5x8 µm, was described by Anderson and Bé (1976b). They exhibit an interesting pattern of circadian migration. At sunrise they move distally along the spines until fully dispersed in the rhizopodial network outside the shell. At sunset they enter the host's cytoplasm (Plate 5, figs. 1-4). It is conjectured that the dinoflagellates are distributed outside the shell during the day to enhance photosynthesis, while they are sequestered within the foraminiferal cytoplasm at night to enable more efficient transfer of organic substances from the algal cells to the host.

We conducted an experiment to determine whether the circadian migration of symbionts follows an endogenous rhythm mediated by the host according to a biological clock or an exogenous rhythm controlled by an environmental source such as light. Individuals of G. sacculifer were maintained in four groups: (a) in a natural light-dark cycle of 12h L from 06:00 to 18:00 and 12h D from 18:00 to 06:00; (b) in a reversed light-dark cycle of 12h L from 18:00 to 06:00 and 12h D from 06:00 to 18:00, (c) in continuous light and (d) in continuous darkness. The results shown in Fig. 20 indicate that the symbionts responded primarily to the presence or absence of light regardless of the time of day. When the foraminifera and symbionts in the reversed light cycle group were exposed to artificial light during the night, the majority of foraminifera had their symbionts distributed outside the shell and when kept in total darkness during the day most of the foraminifera had their symbionts sequestered inside the shell. The foraminifera exposed to light continuously showed a minority of individuals with symbionts outside the shell for the first week and a gradual increase in number in the second week. In continuous darkness, foraminifera with symbionts outside the shell were very rarely observed. The latter two unnatural light conditions apparently caused completely erratic distributions of the symbionts. In a similar experiment, a symbiont-bearing foraminifer kept in a dark chamber during the day retained its symbionts inside its shell and the symbionts emerged immediately upon exposure to artificial light (Plate 5, figs. 5-8). Thus, the fact that the patterns of migration of the symbionts in natural and reversed light cycles are largely controlled by ambient light conditions supports the hypothesis of an exogenous rhythm of symbiont activity.

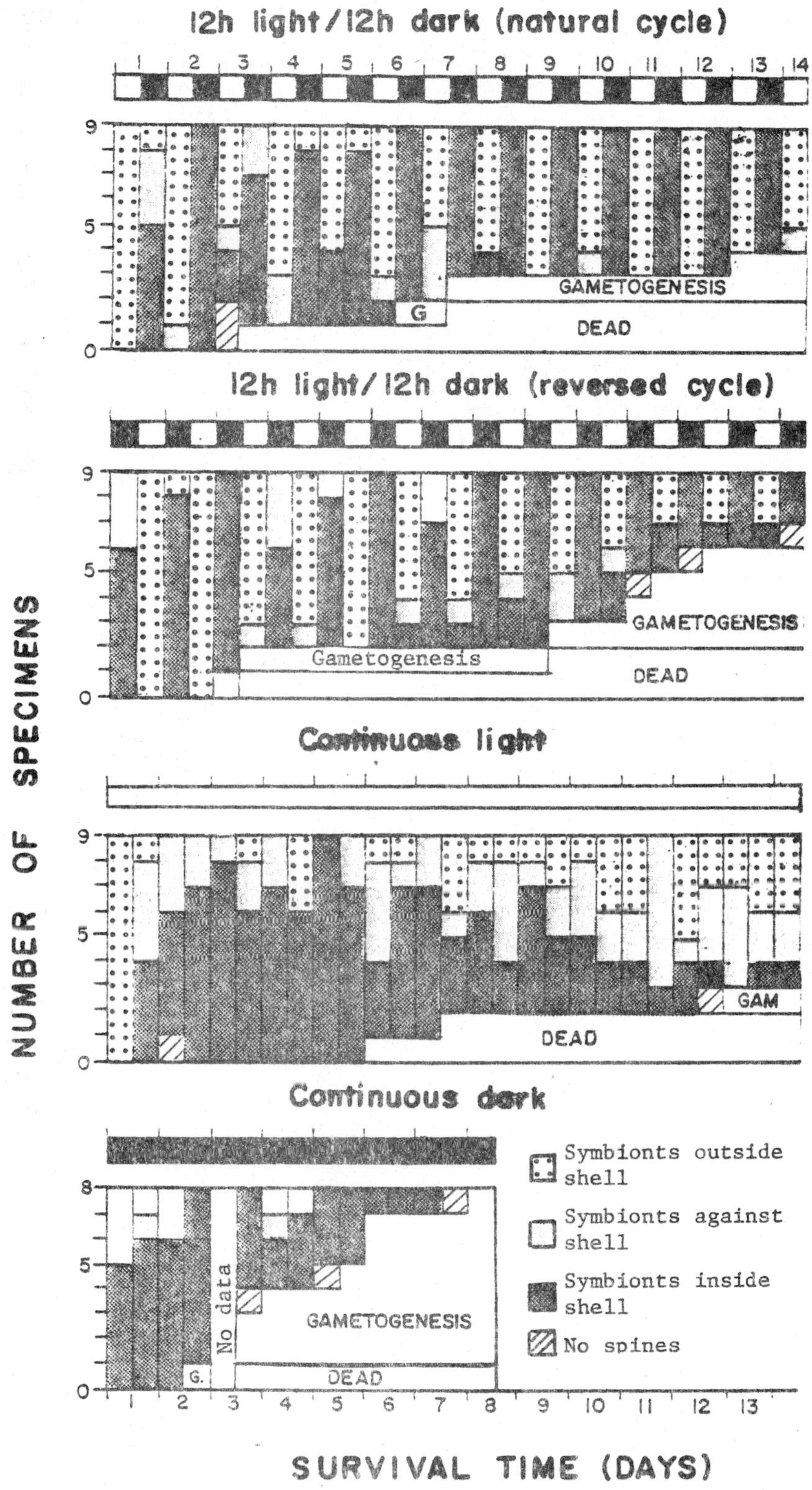


Fig. 20. The exogenous rhythm of circadian migration of symbionts is also evident from this series of experiments, in which symbionts emerged from the shell at sunrise or when exposed to artificial light, and entered the shell at sunset or when kept in a dark chamber. White bar = lights on; black bar = lights off.

Effects of symbiont elimination and reinfection.-- In order to evaluate what benefit foraminifera derive from their symbiotic algae, Bé, Spero and Anderson (1982) posed the following questions: (1) What is the effect of symbiont elimination upon the survival time, shell growth and reproduction of their foraminiferal host? (2) Can an aposymbiotic (= symbiont-free) G. sacculifer be experimentally reinfected in the laboratory with symbionts from a donor of the same species? If so, what are the comparative effects of elimination and reinfection of symbionts on the longevity, shell growth and reproduction of such perturbed foraminifera vs. normal foraminifera maintained in a 12h light: 12h dark cycle and in continuous darkness?

A series of experiments were conducted by maintaining five groups of G. sacculifer under the following conditions (Table 2):

Group 1: in DCMU-sea water continuously.
Group 2: in DCMU-sea water for 72 h, then transferred to freshly filtered sea water.
Group 3: in DCMU-sea water for 72 h, then transferred to freshly filtered sea water and offered zooxanthellae from donor G. sacculifer 24 h after sea water transfer.
Group 4: in sea water and total darkness continuously.
Group 5: in sea water continuously.

The foraminifera in Groups 1, 2 and 3 were treated with $10^{-5}$M 3-(3, 4 dichlorophenyl)-1, 1-dimethylurea (DCMU) in order to totally eliminate their symbionts. This concentration and 72h treatment in DCMU were not deleterious to most of the foraminifera, because 75% or more of the organisms continued shell growth and underwent gametogenesis after transfer to fresh sea water. However, continuous maintenance of G. sacculifer in DCMU-sea water (Group 1) drastically reduced gametogenesis frequency and new chamber formation (Figs. 21, 22).

Globigerinoides sacculifer, rendered aposymbiotic after 72 h in DCMU-sea water and then placed in fresh sea water (Group 2) had a mean survival time of 8.8 days, final shell size of 467 µm, and shell growth rate of 4.3 days/chamber. The equivalent values for the organisms of Group 3, also rendered aposymbiotic after 72 h in DCMU-sea water, then transferred to fresh sea water, but which were reinfected with symbionts from freshly crushed donor individuals of G. sacculifer, were 13.9 days, 697 µm, and 4.1 days/chamber, respectively. These latter values closely approached those of symbiont-bearing organisms maintained continuously in sea water (Group 5).

Figs. 21 and 23 indicate that the mean survival times of aposymbiotic individuals of Groups 1 and 2 were significantly shorter than those of the symbiont-bearing Groups 3 and 5. In addition the elimination of symbionts resulted in a reduced shell growth rate, fewer total number of chambers formed and smaller final shell size (Figs. 22 , 23). Individuals of Group 4, which were not treated with DCMU, but were kept in continuous darkness, also exhibited short survival times and small shell sizes. These observations show that G. sac-

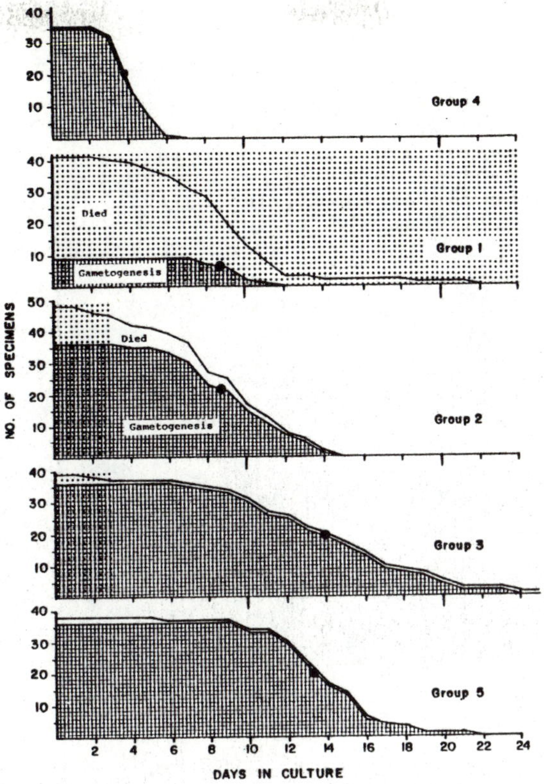


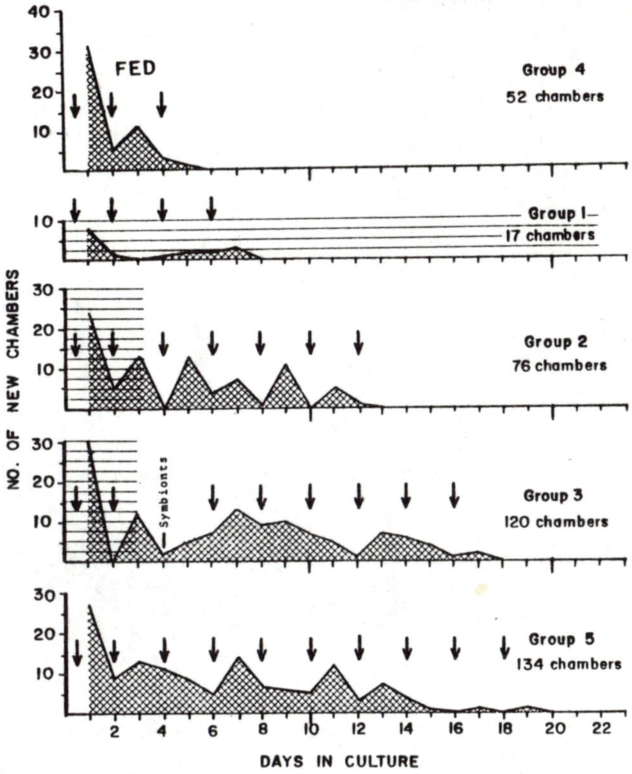


Fig. 21 (left). Survivorship curves of non-gametogenic and gametogenic individuals in 5 groups of G. sacculifer used in evaluating the role of symbionts. Black circles are mean values. Stippling indicates duration of DCMU treatment (from Bé et al., 1982).

Fig. 22 (right). Daily and total production of new chambers in the 5 groups of G. sacculifer, based on a total of 36 gametogenic individuals in each group (Groups 2 to 5) and 9 gametogenic individuals in Group 1. Chamber formation increased in Group 3 after reinfection with symbionts. Horizontal lines indicate duration of DCMU treatment (from Bé et al., 1982).

culifer grows very poorly without symbionts, whereas reintroduction of symbionts to formerly symbiont-free organisms enabled them to attain a shell growth rate, final shell size, and sac-chamber frequency nearly equivalent to those of normal, unperturbed individuals (Figs. 23, F, G; Plate 6).

Two methods of depicting shell growth curves are presented in each graph of Fig. 23. The step-like graph is a "periodic" growth curve of a given population, based on their mean shell lengths and mean durations between successively formed chambers in culture. The second method is based on day-to-day calculations of the average shell sizes of all individuals and yields a relatively smooth curve of daily increments of shell growth. The steplike growth curve is useful for showing the time of chamber formation (vertical lines) and the number of chambers (horizontal lines) formed in culture. For example, Fig. 23B indicates that Group 2 grew an average of 3 chambers and did so after 1.6, 6.2 and 8.9 days in culture, whereas Group 3 grew an average of 5 chambers after 1.3, 5.9, 9.8, 13.4 and 18.0 days (Fig. 23C).

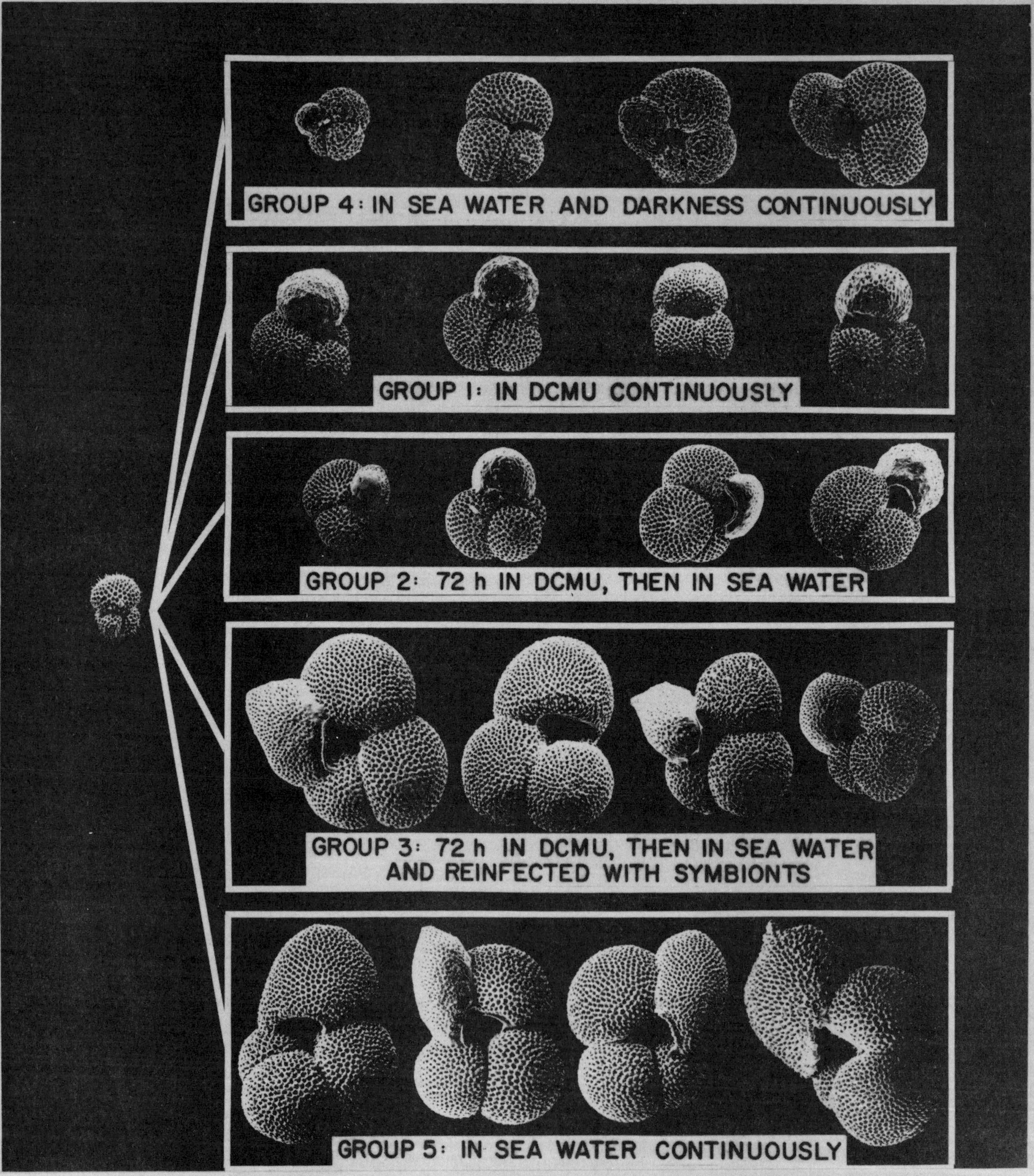


Plate 6. Four specimens of G. sacculifer from each of 5 groups were randomly selected to show the morphological variations and shell sizes obtained under different experimental conditions. The left-hand organism represents the mean initial size (230 µm) of each group of organisms (from Bé et al., 1982).

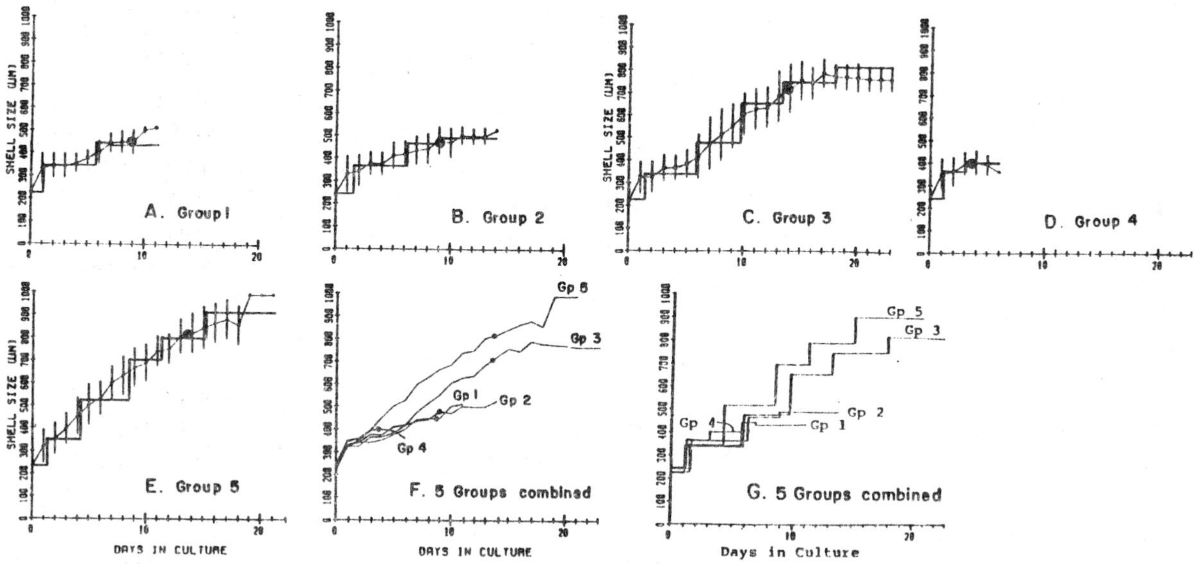


Fig. 23, A-G. Daily and periodic shell growth curves of 5 groups of G. sacculifer in symbiont elimination and reinfection experiments. The daily growth curves are based on the means (± 1 SD = vertical bars) of the shell sizes of all individuals alive each day. The step-graphs are based on the mean durations between successively formed chambers (horizontal sections) and the mean shell size increases at successive growth stages (vertical sections). Black circles are the median values for 36 gametogenic individuals in Groups 2 to 5, and 9 in Group 1 (from Bé et al., 1982).

It is evident from these experiments that the elimination of symbionts or inhibition of symbiont photosynthesis induced early onset of gametogenesis in G. sacculifer. This result is strikingly similar to that when individuals of G. sacculifer are maintained in continuous darkness. Such an unfavorable situation would be encountered in the ocean when the symbiont-bearing foraminifera stay voluntarily or involuntarily below the euphotic zone for extended duration.

It is also clear that the symbionts not only play an essential role in shell growth and calcification, but in addition serve a photoreceptive function integrating host physiology with environmental changes in amount of available light. This finding opens new opportunities for research on the effects of variations in incident radiation and light intensity at different depth habitats on such vital processes as chamber formation and cellular morphogenesis.

## DEPTH HABITATS

Planktonic foraminifera live predominantly in the euphotic zone, where their phytoplankton and zooplankton food occur in highest concentration. Yet, differences in ecological and physiological requirements have resulted in a segregation of species by depth habitat. Symbiont-bearing species are limited to the sunlit, near-surface waters for the optimum maintenance of their photosynthetic algae. Symbiont-free species, on the other hand, do not have such a restriction and tend to live in deeper waters. The range in vertical distribution of individual species is frequently so wide and varies both regionally and seasonally that it is only valid to characterize "depth stratification" of various species in a statistical sense.

An example of such "stratified" depth habitats is shown in Fig. 24, which presents the vertical distribution of planktonic foraminifera in the western Gulf of Mexico. The data were obtained from a series of 8 opening-and-closing plankton tows (MOCNESS with 333 μm mesh aperture) from a depth of 200 m to the surface. On the basis of the vertical distribution patterns in Fig. 24, it is possible to group the 10 species of planktonic foraminifera in the following 3 depth habitats:

"Shallow" depth habitat (maximum abundance in 0-65 m):
*Globigerinoides ruber* (pink and white variety)
*Hastigerina pelagica*
*Globigerinoides sacculifer*
*Globoquadrina dutertrei*

"Intermediate" depth habitat (maximum abundance in 65-85 m):
*Globigerinella aequilateralis*
*Orbulina universa*

"Deep" depth habitat (maximum abundance in 105-200 m):
*Globorotalia truncatulinoides*
*Globigerinella calida*
*Globorotalia crassaformis*
*Pulleniatina obliquiloculata*

It should be noted that 5 out of 6 "shallow"- and "intermediate-water" species are spinose and harbor algae and that 3 of them (*G. ruber*, *G. sacculifer* and *O. universa*) contain large numbers of endosymbionts. Conversely, 3 out of 4 of the "deep-water" species are nonspinose and all probably lack symbiotic algae. These variations in depth habitats among planktonic foraminiferal species are now more understandable in view of our laboratory observations of the presence of algal symbiosis in some species and its absence in other species.

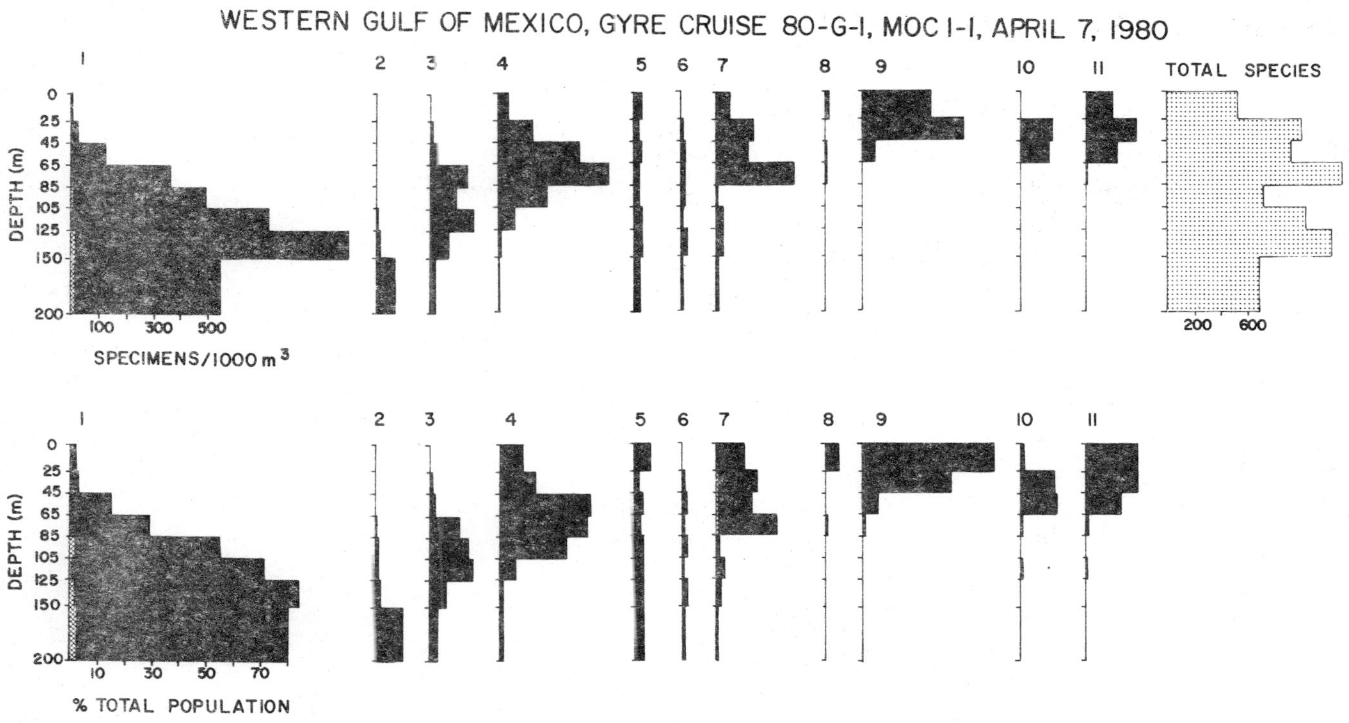


1. G. truncatulinoides (dark = right-coiling)
(crossed = left-coiling)
2. G. crassaformis
3. G. calida
4. G. aequilateralis
5. G. dutertrei
6. P. obliquiloculata
7. O. universa (dark = spherical stage)
(crossed = spiral stage)
8. G. sacculifer
9. G. ruber (white)
10. G. ruber (pink)
11. H. pelagica

Fig. 24. Vertical distribution of planktonic foraminifera in the western Gulf of Mexico, based on serial opening-and-closing plankton tows (MOCNESS 1-1, April 7, 1980) on Gyre Cruise 80-G-1. Location of station is 26°5.0'N, 94°35.5'W.

## REPRODUCTION

Reproduction in planktonic foraminifera takes place by gametogenesis. A definite and predictable sequence of morphological changes and cellular differentiation stages prior to the release of gametes were observed in the spinose species by Bé and Anderson (1976), Anderson and Bé (1978), and Spindler et al. (1978). Gametogenesis in the nonspinose species is rarely observed in the laboratory and, hence, their reproductive process is not adequately understood.

Gametogenesis in *Globigerinoides sacculifer* has recently been described by Bé et al. (in prep.) in more detail than an earlier paper by Bé and Anderson (1976). As with most organisms, there is some variability in the time of onset and duration of reproductive events. The following account is typical of the majority of specimens observed in laboratory culture. Reproduction in this species is normally preceded by the formation of a sac-like chamber (Plate 7, fig. 1) and starts with the progressive resorption and shortening of the spines and digestion of its symbionts at mid-day (Plate 7, figs. 2, 3, 4). The spines, which are normally 2 mm in length, are broken up into short segments (10-30 m), which form a ring of debris around the shell. By midnight all its spines are shed, most of the symbionts are digested, and nuclear division is underway (Plate 7, fig. 7). The digestion of the symbionts and expulsion of their remains from the shell is accompanied with a color change of the foraminiferal cytoplasm from an olive-brown during non-reproductive growth to a milky-white in advanced gametogenic stages. Digesting its symbionts is believed to provide the foraminifer with a source of extra energy during nuclear division. Development and expansion of large vacuoles and lacunae and continued nuclear division cause the cytoplasm to fill the sac-like chamber. (Plate 7, figs. 4 to 7). It is possible that the function of this terminal chamber is as a "brood chamber" protecting the expanding multinucleated cytoplasm during the development of the delicate gametes.

Cytoplasmic bulges emerge from the shell's apertures in the morning, from which myriads of free-swimming flagellated gametes are released at mid-day (Plate 7, fig. 8). An empty shell remains after all the gametes are released several hours later. Bé and Anderson (1976) estimated that about 300,000 gametes are produced from a single shell of *G. sacculifer*. The synchronous developmental sequence of gametogenesis in this and other spinose species may be of adaptive significance for greater potential fertilization of gametes in the natural environment.

*Gametogenic calcification*.-- Comparative studies of shells of *G. sacculifer* before and after gametogenesis have shown that a considerable amount of calcite is secreted over the last-formed chambers, during a period of about 16 hours prior to gamete release (Bé, 1980). The process, termed "gametogenic calcification", differs from "ontogenetic calcification", which occurs during non-reproductive growth. Ontogenetic calcification (Plate 8, figs. 1 and 2) is directed to overall shell growth by chamber-by-chamber formation, lamellar thickening over previous chambers and spine growth. In an actively growing *G. sacculifer*, it can be recognized by a relatively thin-walled shell,

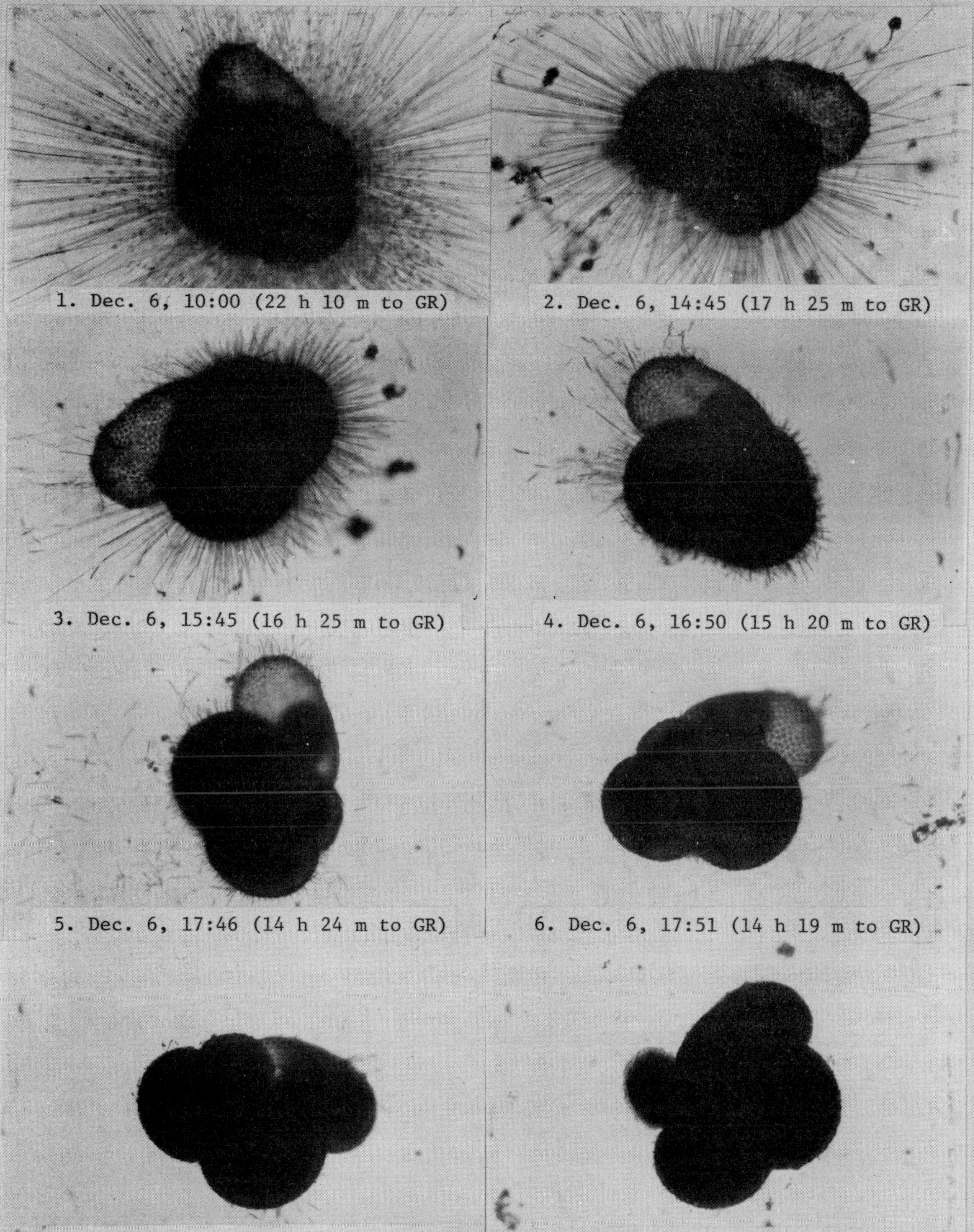

1. Dec. 6, 10:00 (22 h 10 m to GR)
2. Dec. 6, 14:45 (17 h 25 m to GR)
3. Dec. 6, 15:45 (16 h 25 m to GR)
4. Dec. 6, 16:50 (15 h 20 m to GR)
5. Dec. 6, 17:46 (14 h 24 m to GR)
6. Dec. 6, 17:51 (14 h 19 m to GR)
7. Dec. 6, 22:00 (10 h 10 m to GR)
8. Dec. 7, 08:10 (GR)

Plate 7. Successive observations of the same individual of G. sacculifer undergoing gametogenesis in laboratory culture. Time before gamete release (GR) in brackets. x64.

knife-edged ridges forming polygonal enclosures around the pores, elongate spines located at the corners of the polygons, and deep bowl-shaped pore basins with terraced walls.

The precise point when gametogenic calcification starts is difficult to determine, but it may coincide with the gradual shortening of the spines (∿24 h before gamete release). Whereas prior to gametogenesis the rhizopodial network is widely dispersed distally along the elongate spines, the rhizopodia are wrapped closely around the shell surface once the spines are shed. It is this rhizopodial envelope that apparently continues to secrete calcite during gametogenesis.

Gametogenic calcification in G. sacculifer (Plate 8, figs. 3 and 4) is characterized by (a) complete resorption of spines, leaving only "spine holes" as traces of the spines, (b) pore basins which have been filled to varying degrees, so that the pores are more constricted and cylindrical in cross-section, (c) broad and rounded inter-pore ridges, and (d) euhedral or irregularly lumpy calcareous deposits which are frequently covered by a smooth veneer and (e) generally thicker walls than pre-gametogenic shells. The "spine holes" are crater-like depressions (∿0.5 μm in diameter) produced by continued calcitic accretion of the spine bases. Their presence or absence is one of the best criteria for distinguishing between gametogenic and pre-gametogenic shells of spinose planktonic foraminifera. Bé (1980) estimated that gametogenic calcification adds on the average about 28% by weight to the pre-gametogenic shell.

This secondary process of calcification is believed to occur well below the euphotic zone, as the organism is sinking through the water column as a result of loss of buoyancy due to spine shedding and nuclear division (Bé and Anderson, 1976). We estimate that an adult shell (about 500 μm in length and 24 μg in weight) of G. sacculifer which normally lives in the euphotic zone will sink and continue to calcify to a depth of about 800 m in 24 hours.

Duplessy et al. (1981) presented corroborative evidence for gametogenic calcification by their finding that empty shells of G. sacculifer on the sea-bed are significantly enriched in $^{18}O$ as compared with the shells of their living counterparts in surface waters. The enrichment occurs as the organism sinks below the euphotic zone and extracts calcium carbonate from deeper and colder waters.

Gametogenic calcification explains why thick-walled shells of G. sacculifer and other species are seldom collected in surface waters, whereas they occur predominantly on the sea-bed. Such encrusted shells are occasionally collected below the euphotic zone, but their residence time in the water column is relatively short because the loss of their buoyancy mechanism and increase in shell weight cause them to sink rapidly. Further investigations are necessary to determine whether and to what degree gametogenic calcification occurs in other species of planktonic foraminifera. There is a good likelihood that it could explain the origin of the "calcite crust" that is so common in globorotaliid as well as globigerinid species (Bé and Ericson, 1963; Bé and Lott, 1964; Orr, 1967).

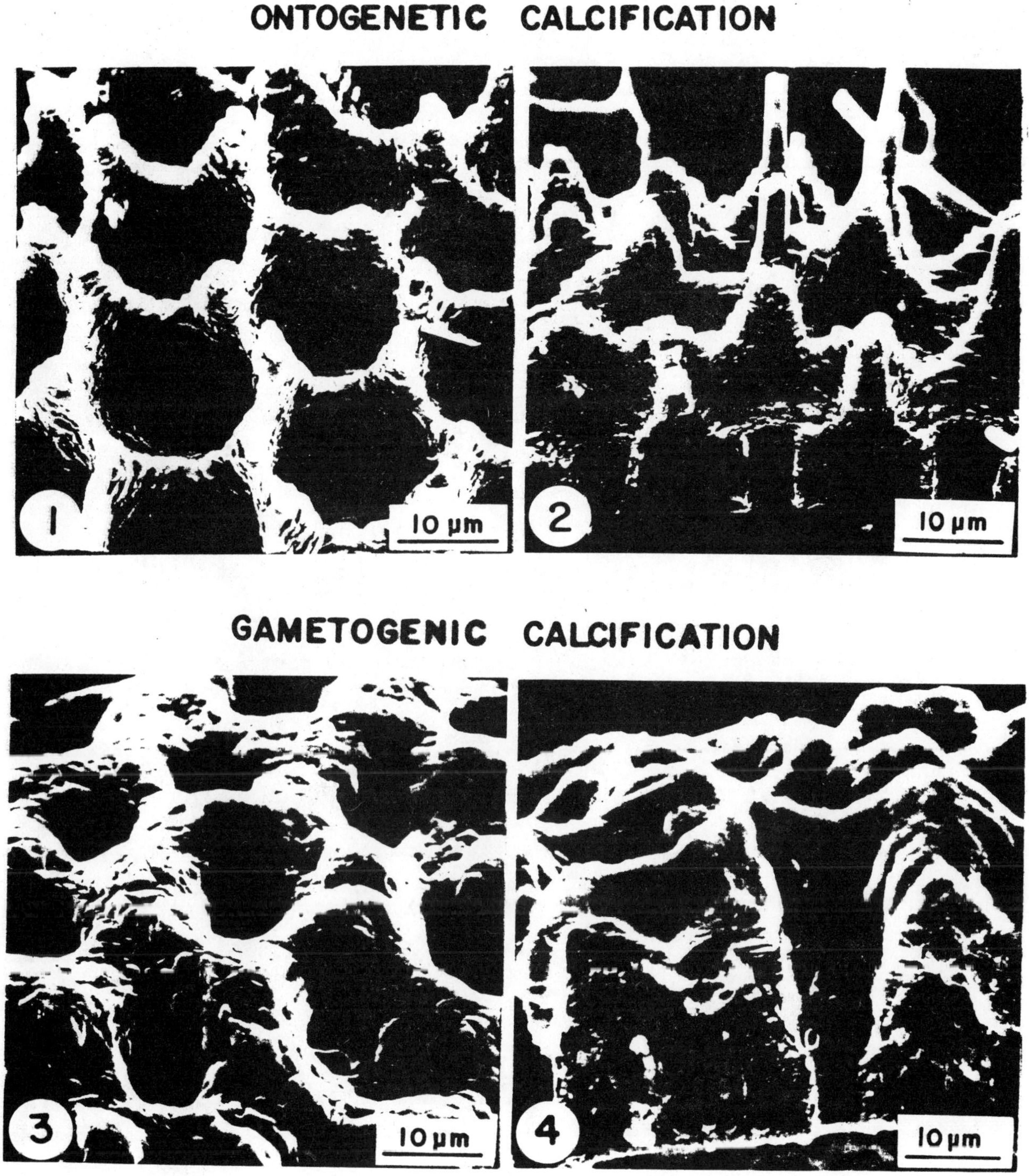


Plate 8. Comparison of shell ultrastructure of antepenultimate (= F-2) chamber of _G. sacculifer_. Figs. 1 & 2 show ontogenetic calcification of a pre-gametogenic specimen. Figs. 3 and 4 show gametogenic calcification in a post-gametogenic specimen.

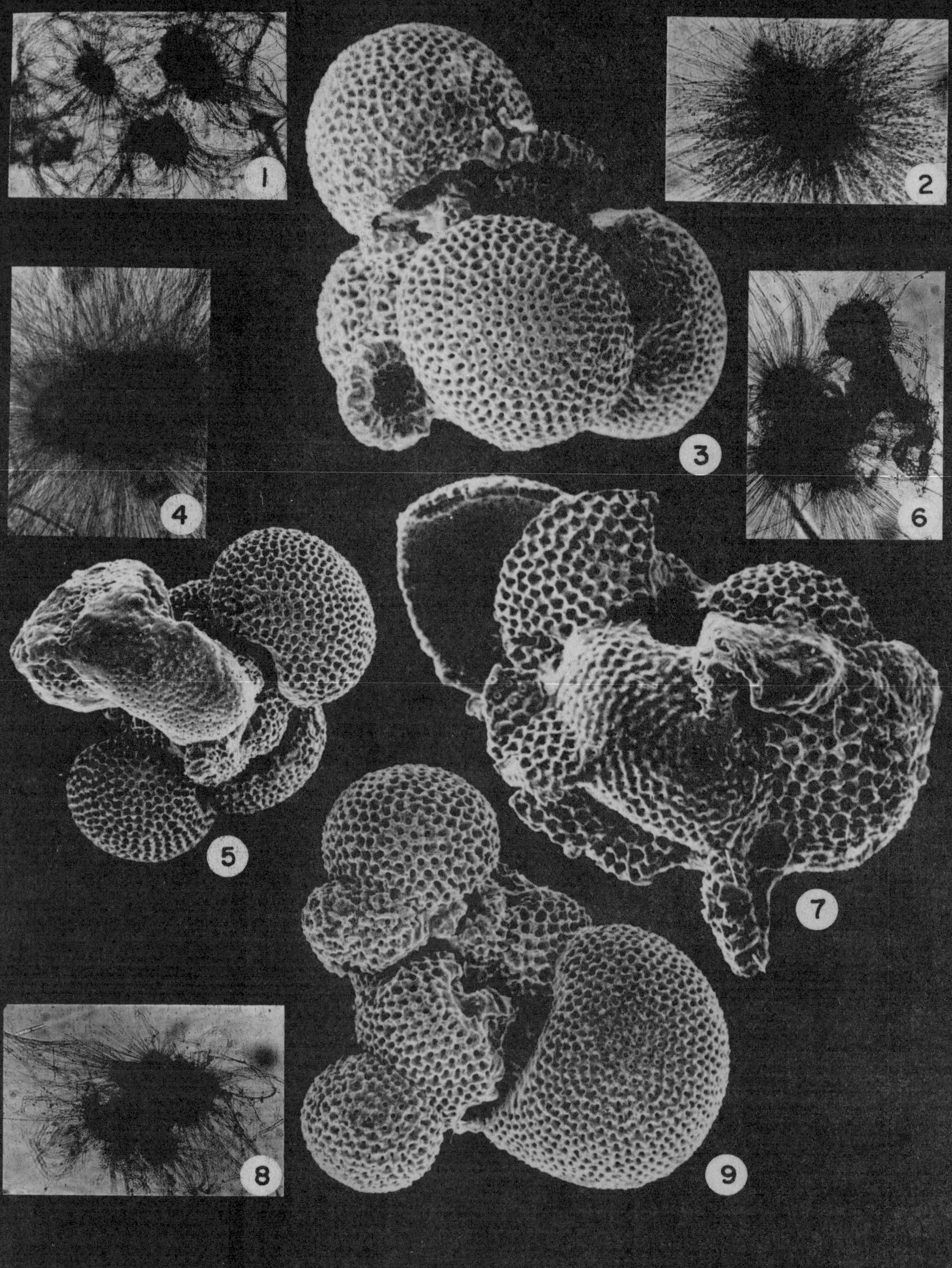
1
2
3
4
5
6
7
8
9

Shell regeneration experiments.-- A series of experiments was conducted by Bé and Spero (1981) in which living specimens of G. sacculifer were subjected to varying degrees of damage. In the first experimental group, one or more chambers of their shells were crushed, whereas in the second group successive chambers were amputated and completely removed from the shells. The results showed that G. sacculifer has a remarkable ability to survive after physical injury (Plate 9). Although one or more chambers were damaged, the majority of crushed individuals recovered their basic biological functions of chamber formation, ingestion of crustacean prey, symbiosis with zooxanthellae, and gametogenesis.

Amputation and removal of the cytoplasm and shell fragments of the last 3 chambers also did not have an adverse effect on most specimens which were able to regenerate their shells and undergo gametogenesis. The successive amputations indicate that the nucleus is normally located in an early chamber (prior to the antepenultimate chamber in individuals of G. sacculifer ∿600 μm in shell length). The cementation of shell fragments and addition of new chambers in crushed specimens generally result in bizarre shell shapes. By contrast, the organisms whose chambers were entirely amputated frequently regrew chambers of normal shape.

Planktonic foraminifera in the ocean are probably preyed upon by fish, squids, crustaceans and other carnivores. It is likely that there is a certain ratio of foraminifera which are ingested to those which are only partly eaten. If physical damage among the latter group is not too severe and results in partially crushed shells, a large number of these crushed organisms will recover and repair their shells. Consequently, the frequency of "orthopedic" shells showing evidence of healing after mechanical damage may be considered an index of the relative predation pressure on natural populations of planktonic foraminifera.

## ACKNOWLEDGMENTS

The assistance of Wailes Gray, Marc Sverdlove and Mei Bé in the preparation of this review and financial support from National Science Foundation Grant OCE 81-17715 are gratefully acknowledged. Special thanks are due to John Wormuth of Texas A & M for providing the MOCNESS plankton samples and to the Journal of the Marine Biological Association of the United Kingdom, Marine Biology and Journal of Foraminiferal Research for permission to republish some illustrations. This publication is Lamont-Doherty Geological Observatory Contribution no. 3346.

---

Plate 9. The robustness of planktonic foraminifera is demonstrated by these four individuals of G. sacculifer, which after they were crushed repaired their shell, formed new chambers and underwent gametogenesis. The last 2 chambers of specimen 1 (Figs. 1-3), specimen 2 (Figs. 4, 5) and specimen 3 (Figs. 6, 7) were crushed, but each formed 2 or more new chambers. All chambers of specimen 4 were crushed, but it regenerated 3 new chambers. All light micrographs x 40. Fig. 3, x 153; Fig. 5, x 100; Fig. 7, x 171; Fig. 9, x 126 (From Bé and Spero, 1981).

# Biogeography of Modern Benthic Foraminifera

Martin A. Buzas
Department of Paleobiology
Smithsonian Institution
Washington, D. C. 20560

Stephen J. Culver
Department of Geophysical Sciences
Old Dominion University
Norfolk, Va. 23508

## INTRODUCTION

Biogeography is concerned with recording and explaining the distribution of organisms in space and time. In their search for patterns of distribution, biogeographers have classified the surface of the earth into a hierarchical scheme of ecological units (e.g. Kauffman and Scott, 1976). The basic biogeographic units are realms and provinces. A faunal realm is a large area about the size of a continent while provinces are subdivisions within a realm. An appreciation of the scale involved is gained by considering a current classification of the terrestial world (Udvardy, 1975) which contains 8 realms and 172 provinces.

Marine macrofaunal biogeographers such as Milne-Edwards (1838), Dana (1853a, b), Woodward (1856), Ekman (1953), Hedgpeth (1957) and Briggs (1974) classified the shallow seas of the world into realms and provinces but, unfortunately, little agreement seems to exist on the definition of a faunal province. Woodward (1856) believed a province should contain at least 50% endemic species, Kauffman and Scott (1976) required from 25 to 50% endemism (as low as 20% in practice), while Briggs (1974) settled for 10% endemism. Another view expressed by Menzies et al. (1973) defined a province on the basis of the rate of change of genera within it. Valentine (1968, p.257) defined provinces as "...collections of communities associated in space and time." If a verbal definition is required, we would choose the latter because it frees the investigator from any preconceived and arbitrary limits on endemism.

Today, numerous mathematical techniques exist for the

quantification of distributions (Buzas, 1970, 1979). The definition of any ecological unit is contained in the mathematics of the technique and the interpretation of the results by the investigator. For example, Hazel (1971) used Q-mode cluster analysis on the ostracode fauna north and south of Cape Hatteras. The dendrogram showed two large clusters and several smaller ones. Hazel called the large clusters provinces and the smaller ones biofacies. In our view this is a perfectly reasonable procedure. Based on similarity and the scale involved the investigator must interpret the analyses and assign ecological units. Because so many different techniques are now available, no simple definition of subdivisions of the biota is possible. This may be discomforting to those who seek agreement on definitions by all workers, but it is, nevertheless, the reality of the situation. While some differences in distributional pattern may appear depending upon the technique used, experience suggests that different techniques produce similar patterns (Buzas, 1966; Mello and Buzas, 1968; Hayward and Buzas, 1979).

The original designations of marine faunal provinces by macrofaunal workers were made on the basis of shallow water species. Some researchers extend the provincial name far offshore and describe major faunal changes with depth as facies or zones within the province. Menzies et al. (1973), however, place the photic and aphotic zones of the ocean in different realms. Thus, off the Carolinas Menzies et al. (1973) recognize the Intertidal Province, Shelf Province, Archibenthal Zone of Transition, and the Abyssal Province. This is a much more reasonable approach because the world of the deep is so different from that of shallower waters. We recognize major faunal changes with depth as provinces, and smaller ones as zones or biofacies within them.

Many foraminiferal workers have not concerned themselves with giving formal names to the areal distribution of foraminifera (Phleger, 1960; Murray, 1973). Rather, distributions are simply shown on a map, called faunal areas or zones, and related to some abiotic variable(s). This lack of formality has not hampered distributional studies of foraminifera and the literature on foraminiferal distribution is vast. In the following review of what is known of modern benthic foraminiferal biogeography we will cite relatively few references, but ones which contain extensive bibliographies for those interested in pursuing the subject further.

## THE WORLD OCEAN

In the first edition of his famous textbook "Foraminifera, Their Classification and Economic Use", Cushman (1928b) discussed the geographic distribution of foraminifera. He presented two maps, one for cold water faunas and one for warm water faunas. The cold water faunas he subdivided into Arctic and Antarctic pointing out that each of these extends part way up the coast of the continents (Fig. 1a). Cushman depicted the entire West coast of the Americas as a cold water fauna. According to Cushman the warm water faunas are more easily divided and he recognized East African, Indo-Pacific, Mediterranean, and West Indian faunas (Fig. 1b). Cushman did not indicate how he recognized or defined faunas and probably relied on his considerable experience from having looked at foraminiferal faunas from all over the world. In any case, Cushman's chapter on geographic distribution still makes interesting reading.

Boltovskoy (1965b) and more recently Boltovskoy and Wright (1976) prepared a world map showing three major faunas: warm, temperate and cold. Each of these was broken down into the provincial scheme shown on Text-Figure 2. Boltovskoy and Wright (1976) gave a list of references they found most useful for distributional data, but did not indicate in detail by what criteria they distinguished their faunal provinces other than that the various provinces have faunas differing in some respect from each other. Water temperature, however, was stated to be the major ecologic factor controlling subdivision of the global foraminiferal biota into separate faunas. Like the earliest marine biogeographers' schemes, the subdivisions proposed by Cushman and Boltovskoy and Wright are for shelf species. A great deal of information is available on depth zonations of foraminifera, and their distribution in the deep ocean basins. Recently, Douglas and Woodruff (1981) thoroughly reviewed the literature on deep-sea benthic foraminifera. Assemblages of benthic foraminifera have been identified over vast areas of ocean basins. Some common species reported from the deep sea fauna are cosmopolitan in their distribution; the same dominant species are reported from the Indian, Atlantic, Pacific and Antarctic oceans (see Douglas and Woodruff, 1981, for references). We must point out, however, that the amount of available information is small compared to those available for the continental margins.

Analyzing foraminiferal distribution on a world wide basis is most difficult because, at the same time, the literature is vast and yet not nearly comprehensive enough. While the use of large computers can accomodate the vast literature, problems of taxonomy, and the need for more distributional studies still plague any detailed worldwide analysis.

Progress is being made on a more modest scale, and we will now review biogeographic studies from Western Europe, the South American continental shelves, the Arctic Ocean, the North American Pacific inner shelf, the North American Atlantic continental margin, the Gulf of Mexico, and the Caribbean.

## WESTERN EUROPE

Numerous papers have dealt with the distribution of modern foraminifera off Western Europe but Murray's brief review (1971b, 1973) is the only one summarizing distributions at the provincial scale. Murray (1971b) recognized three groups of inner shelf species indicating that Britain "..spans a diffuse faunal boundary between northern and southern influences" (Murray, 1973). His map, redrawn here (Fig. 3), shows the southern limit of a northern group of species and the northern limit of a southern group and indicates that their distribution can be related to water masses. A third group of species is widely distributed over the area covered by the map.

We suggest that this large-scale distribution pattern can be interpreted as representing two inner shelf foraminiferal provinces, a northern and a southern, with the transition between them at the British Isles. The species restricted to these two provinces are listed in Murray (1971b).

## SOUTH AMERICAN ATLANTIC AND PACIFIC CONTINENTAL SHELVES

Boltovskoy and Wright (1976) discussed the foraminiferal biogeography of the continental shelves of South America. Four provinces extend around South America. The West Indian Province (Fig. 4) was divided into Venezuelan, North Brazilian, and South Brazilian Subprovinces. The Argentine province consists of the North Pategonian and South Pategonian Subprovinces with Malvin Subprovince occurring further offshore at about 80 to 100m depth. The Chilean-Peruvian Province is also subdivided into three subprovinces, the South Chilean, the North Chilean and the

Peruvian. The Panamanian Province extends from Baja California to the north-western corner of South America and passes into the Peruvian Subprovince just south of the equator.

Although they listed species characteristic of each province, Boltovskoy and Wright (1976) did not fully document their methodology in provincial and subprovincial subdivision. In 1980 Boltovskoy et al. gave more detail on the Argentine Province. In all about 650 species were recorded from the area. Maps were drawn indicating the limits of favorable environmental conditions for many of the shelf-dwelling species. Again, some characteristic species for each province were listed and Boltovskoy et al. (1980) stated that the foraminiferal zoogeography of South America is controlled by water mass distributions.

## ARCTIC OCEAN

Lagoe (1979) studied the distribution of modern benthic foraminifera in the Arctic Ocean by analyzing data from several publications. Cluster analysis defined three biofacies, two of which were subdivided into subfacies. These biofacies, succeeding each other with depth (17-350m, 350-900m, >900m), were not mapped by Lagoe but they cover such a large area of the Amerasian Basin that we suggest they are provinces rather than biofacies. The subfacies described by Lagoe may be interpreted as depth (or other zones) within major faunal provinces that Lagoe related to major water masses.

## NORTH AMERICAN PACIFIC CONTINENTAL MARGIN

The distribution of benthic foraminifera in the nearshore turbulent zone (0 to 40m depth) off the North American Pacific coast was reported by Lankford and Phleger (1973). They recognized four major assemblages based on living populations. A cosmopolitan group of species occurred from the Strait of Juan de Fuca, Washington to Cabo San Lucas, Baja California. This stretch of inner shelf was then subdivided on the basis of three major assemblages defined by the species limited to them and by wider ranging species having characteristic frequencies within the areal limits of those assemblages.

An Oregonian Province stretches from the northern limit of their data south to Point Conception (Fig. 5). There is a relatively sharp boundary here between the Oregonian

Province and the Californian Province which continues south to Baja California. Between Bahia Magdalena and Cabo San Lucas there is a more gradational boundary between the Californian Province and the Panamanian Province (Fig. 5). Lankford and Phleger related their faunal provinces to differences in coastal water masses and showed that three depth assemblages occurred within each province.

## NORTH AMERICAN ATLANTIC CONTINENTAL MARGIN

Sen Gupta (1972) summarized the distribution of benthic foraminifera on the Atlantic Continental Shelf of North America. He discussed distributional patterns in terms of the provincial scheme proposed by some macrofaunal workers based mainly on molluscs. From Nova Scotia to Florida this classification has four provinces: Nova Scotian, Virginian, Carolinian, Caribbean. The boundaries of the provinces are at Cape Cod, Cape Hatteras, and Cape Canaveral. While noting that Cushman (1918) and Parker (1948) recognized a faunal change at Cape Cod, Sen Gupta pointed out a strong similarity between the Nova Scotian and Virginian provinces. Sen Gupta (1972) stressed the major faunal change was at Cape Hatteras. Three shelf depth zones were recognized on the basis of the more common species.

The large-scale distribution of modern benthic foraminifera on the Atlantic continental shelf of North America was also summarized by Murray (1973), who compared faunas listed in several papers. Six groups were recognized. Murray discussed these distribution patterns in terms of temperature and salinity and concluded that "...general correlation exists between the overall environmental conditions and the geographic distribution of the shelf foraminiferids" (Murray, 1973, p.135).

The comparison of faunal lists from many papers is a time consuming and difficult task. If an attempt is made to use all the literature for all the species recorded, it becomes impossible without the aid of a large computer. Realizing this, Culver and Buzas in 1978 embarked upon a plan to computerize all existing distributional data of foraminiferal species around the continental margins of North America. Around 700 papers containing distributional data have been published over the past 130 years (Culver, 1980). Computer files were constructed containing the name of the author and year of publication, generic and specific name, geographic location, and latitude-longitude. Because the object of the files was to record large scale patterns and many of the authors were studying distributions on relatively small scales with many sampling stations, in many instances a single locality was recorded for all stations. For the Atlantic continental margins 1303 species names were used and 542 localities were recorded . Through use of the literature and examination of specimens at the National Museum of Natural History, and the British Museum (Natural History), the initial file was reduced to 876 species names. The distributions of the 149 most commonly recorded species were plotted on computer-generated maps (Culver and Buzas, 1980).

Q-mode cluster analysis discriminated 7 provinces from the available species-level information (Fig. 6). The diagnostic and characteristic species of each province were presented in Culver and Buzas (1981a). The change in the foraminiferal fauna noted at Cape Hatteras by earlier workers is clearly discriminated even offshore in slope waters. The lack of a coastal province in the southern area may be due to insufficient data. In general, the distribution coincides with the major water masses of the western north Atlantic (Buzas & Culver, 1979).

## GULF OF MEXICO

Poag (1981) summarized the distribution of genera in the Gulf of Mexico. His analysis included data from about 400 new stations off Campeche and the Yucatan. The distribution of the genera, shown on a large map, indicated that many of the shelf and slope genera have a circum-Gulf distribution. A miliolid facies occurs off southwestern Florida and the Yucatan. The abyssal depths were divided into eastern and western facies. Following macrofaunal researchers (e.g. Briggs, 1974), Poag (1981) placed the northern Gulf in the Carolinian Province and the southern Gulf in the Caribbean Province.

Using the same format as for the Atlantic margin, Culver and Buzas (1981b) published all available species level distributional data for the Gulf of Mexico. The initial computer file contained 1219 species names and 426 sample localities. Synonymization reduced the number of species to 848. Maps for the 295 most commonly recorded species were presented in Culver and Buzas (1981b). A cluster analysis of the most commonly occurring species provided the provincial scheme shown in Fig. 7 (Culver and Buzas, 1981c). The large non-stippled areas are places where no information was available. As for the Atlantic continental margin, there is a relationship between foraminiferal provinces and Gulf of Mexico water masses (Culver and Buzas, 1981c).

## CARIBBEAN

The foraminiferal fauna of the Caribbean region has also been investigated by Culver and Buzas. The names of 1868 taxa were recorded from 338 localities (Culver and Buzas, 1982a). Synonymization reduced the number to 1189 species. Cluster analysis of the 130 most commonly recorded species showed no meaningful clusters. However, in the Caribbean region few sample localities are in deep water. Consequently, further research is needed to shed light on the make-up of the slope, rise and basin foraminifera and their relationship to the Atlantic and Gulf of Mexico species. At present the Caribbean data can best be considered as representing a Caribbean Shelf Province.

## DISCUSSION

On the North Atlantic Continental margin of North America and in the Gulf of Mexico some differences exist among the provincial schemes proposed by various researchers. On the Atlantic continental shelf, early macro-faunal workers (e.g. Dana, 1853a,b) proposed boundaries for faunal provinces at Cape Cod, Cape Hatteras, and Cape Canaveral. More recent workers (Coomans, 1962; Franz and Merrill, 1980) suggested that the boundary at Cape Cod does not mark a significant enough faunal change to warrant a separate Nova Scotian province. Among foraminiferal workers, Cushman (1918) and Parker (1948) recognized a boundary at Cape Cod, but Sen Gupta (1972), while discussing foraminiferal distribution in terms of molluscan provinces, noted little difference in faunal lists north and south of Cape Cod. Based on cluster analysis of all available data Buzas and Culver (1979) and Culver and Buzas (1981a) recognized only Cape Hatteras and Cape Canaveral as boundaries for faunal provinces. Thus, the most recent molluscan and foraminiferal provincial schemes for the North American Atlantic shelf are in agreement.

In their world-wide provincial scheme for continental shelves. Boltovskoy and Wright (1976) recognized Cape Hatteras as a faunal boundary, but not Cape Cod or Cape Canaveral (Fig. 2). They extended their West Indian Province (Caribbean of this paper) to Cape Hatteras. Culver and Buzas (unpublished data) found that of the 156 commonly recorded species of foraminifera (occurrence of greater than 4%) occurring in the Caribbean and the Southern Shelf Province off the east coast of North America (Text-fig. 6) only 22% are in common. Consequently, the area from Cape Hatteras to Cape Canaveral warrants a separate provincial designation.

Some macrofaunal researchers (Hedgpeth, 1957; Briggs 1974) assigned the northern Gulf of Mexico to the Carolinian Province and the Southern Gulf (South of Cabo Rojo and Cape Romano) to the Caribbean Province. Poag (1981) followed this provincial scheme for foraminifera while Boltovskoy and Wright (1976) assigned the northwestern Gulf to the North American Atlantic Province and the southern and northeastern Gulf to the West Indian province. Culver and Buzas (1981c and in press), however, recognized a circum-Gulf Inner Shelf Province. If we compare the Southern Shelf Province of the Atlantic to the Inner Shelf Province of the Gulf of Mexico, among the 189 commonly recorded species only 22% are in common (Culver and Buzas, 1982b). Comparing the Caribbean with the Inner Shelf of the Gulf of Mexico 24% of the 243 commonly recorded species are in both areas. Thus, the Gulf of Mexico Inner Shelf Province is distinct from the east coast Southern Shelf Province (Carolinian) and the Caribbean Province. In his recent molluscan compilation Ode (1974) also suggested that the Gulf of Mexico is distinct from the Carolinian province. The analysis by Culver and Buzas (1981c and in press) contained relatively few localities for the southern Gulf. Poag (1981) utilized unpublished information from many stations off Yucatan in preparing his map of the distribution of foraminiferal genera. He indicated the presence of a miliolid facies on the Yucatan Shelf. It seems likely that the Caribbean Province extends to the southern tip of Florida and the Yucatan. The exact extent of the Caribbean influence into the Gulf remains to be determined.

Pflum and Frerichs (1976) and Poag (1981) recognized an east-west division of abyssal facies in the Gulf of Mexico while Culver and Buzas (1981c) did not. This is probably the case because Culver and Buzas (1981c) used presence or absence data and did not consider species or generic abundance. But Poag (1981) recognized only Carolinian (northern) and Caribbean (southern) provinces in the Gulf. Culver and Buzas (1981c and in press) chose to name separate provinces for the deep-dwelling species. We believe, along with Menzies et al. (1973), that differences in environment and species composition between shelf and deeper faunas warrant provincial status.

The detailed studies on the eastern margin of North America and in the Gulf of Mexico indicate that the worldwide provincial scheme presented by Boltovskoy and Wright (1976) is too simple (Fig. 2). For example, the West Indian Province (Caribbean) extends only to the tip of

Florida and onto the Yucatan shelf of the southern Gulf. Further, the Southern Shelf Province of the western North Atlantic, the Caribbean Shelf Province, and the Gulf of Mexico Inner Shelf Province are all distinct entities. As more information becomes available, more provinces will be recognized. In the marine world there may be as many provinces as there are major water masses.

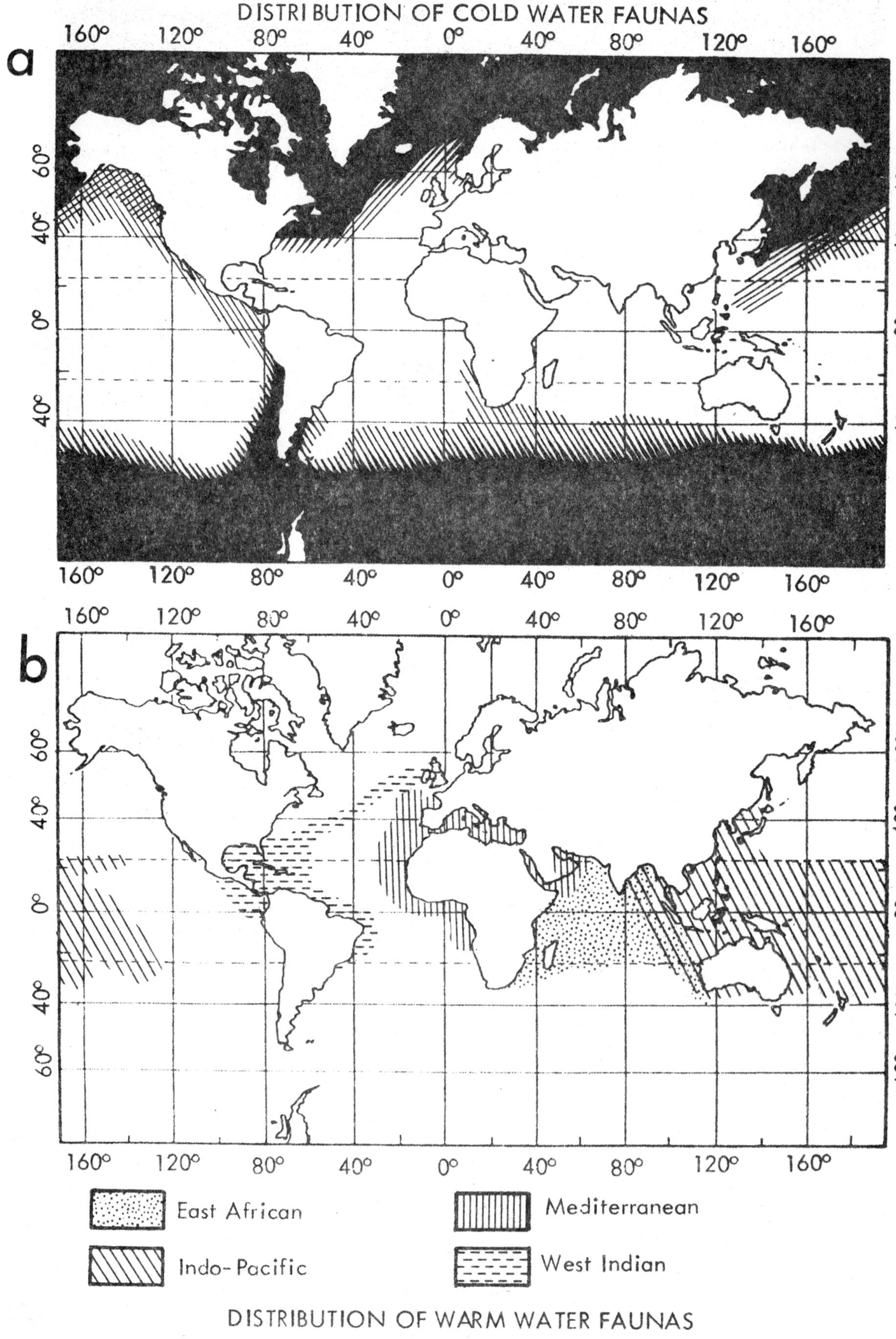


FIG. 1--After Cushman, 1928b.

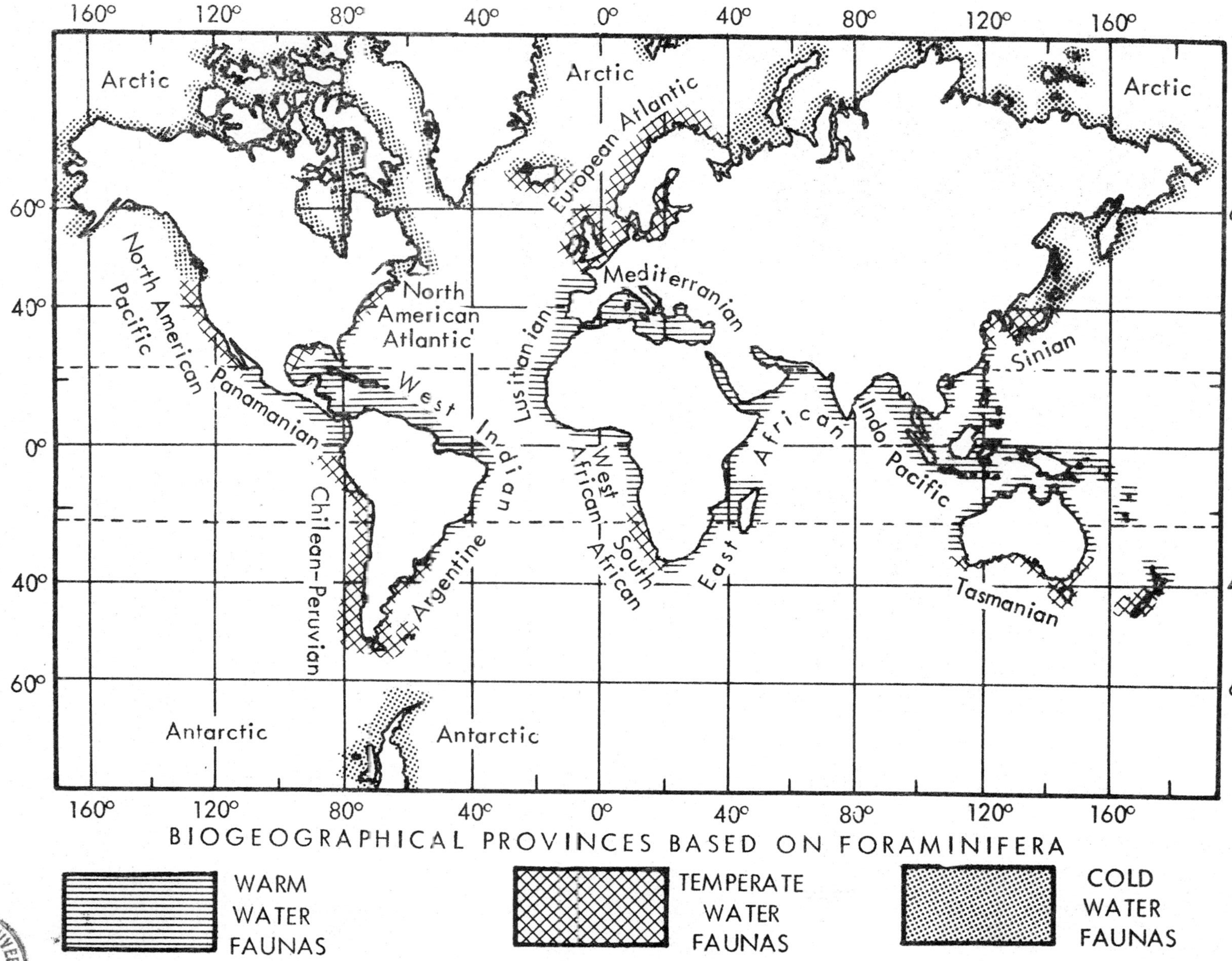




FIG. 2-- Benthic foraminiferal zoogeography of continental shelves (after Boltovskoy and Wright, 1976).

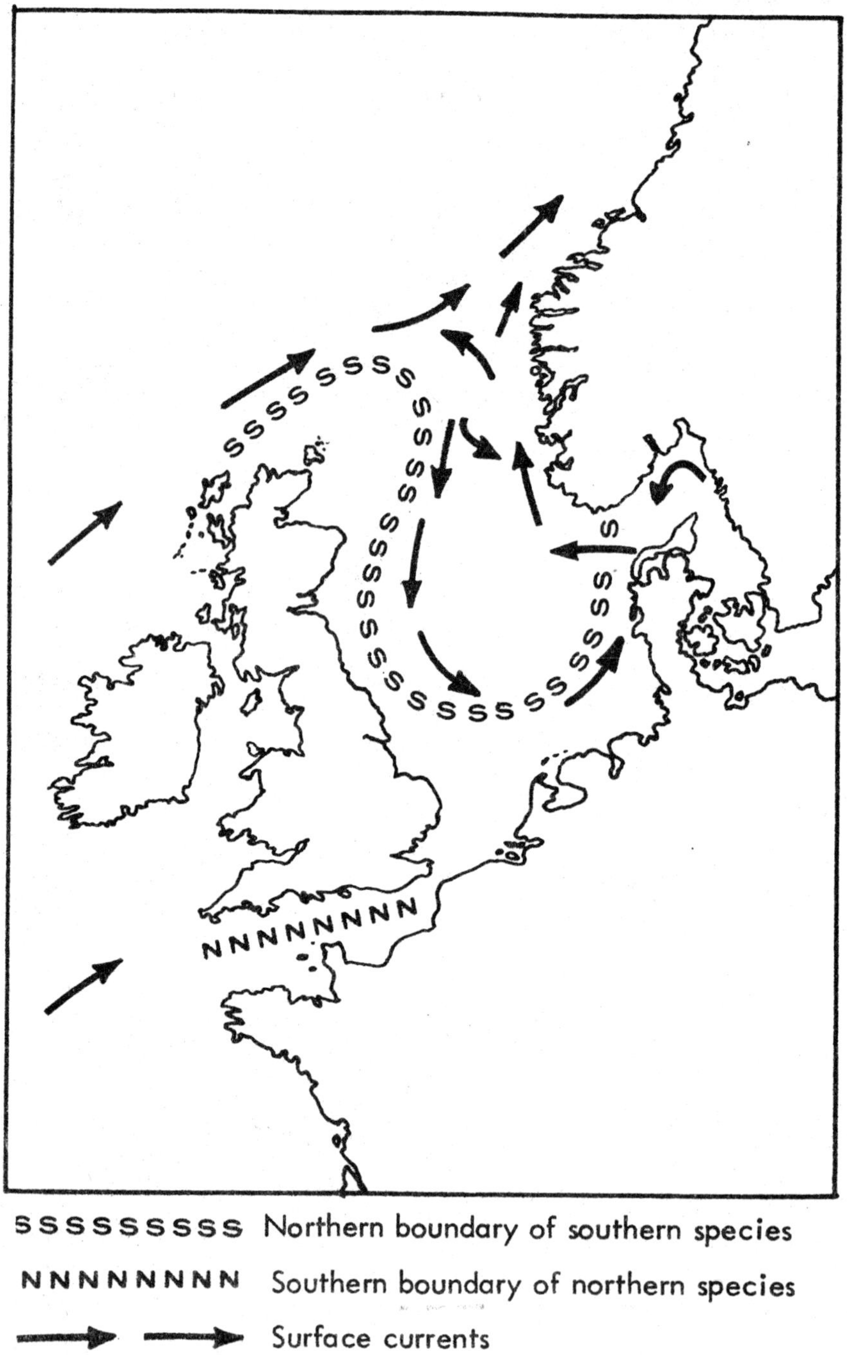


FIG. 3--Foraminiferal faunal boundaries along the European seaboard (after Murray, 1971b).

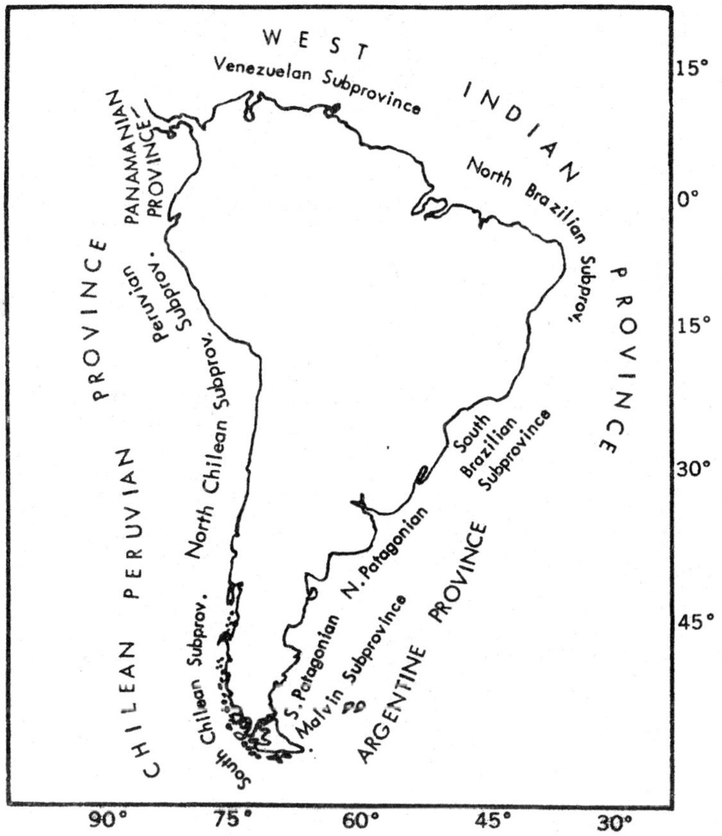


FIG. 4--Foraminiferal zoogeography of the South American continental shelves (after Boltovskoy and Wright, 1976).

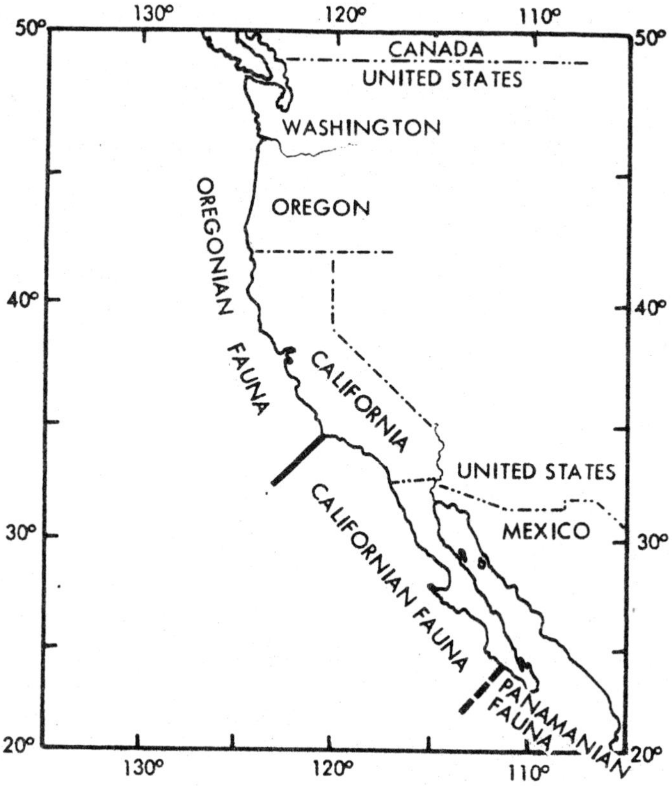


FIG. 5--Foraminiferal zoogeography of the Pacific inner shelf of North America (after Lankford and Phleger, 1973).

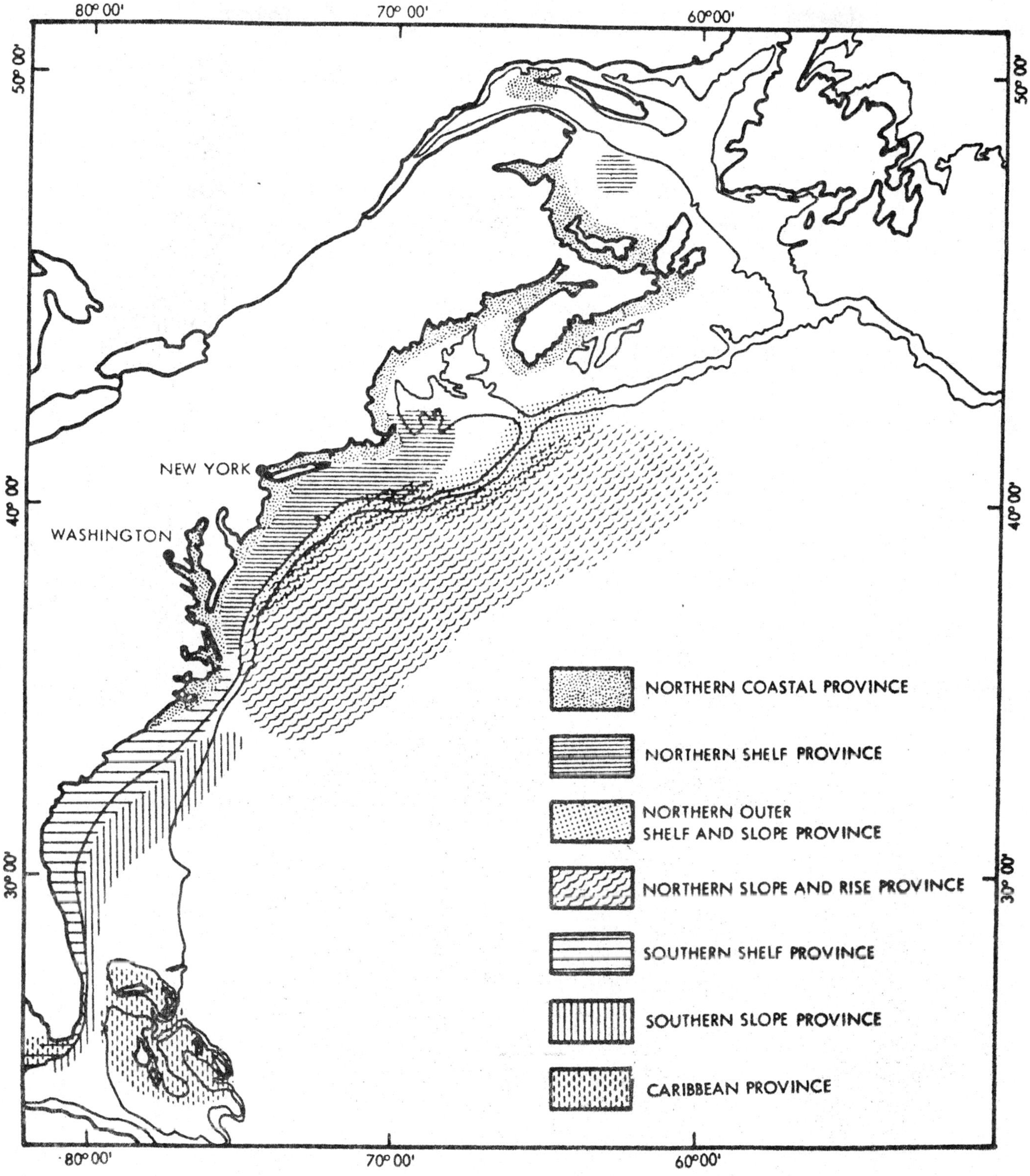


FIG. 6--Foraminiferal provinces of the North American Atlantic continental margin (after Culver and Buzas, 1981a).

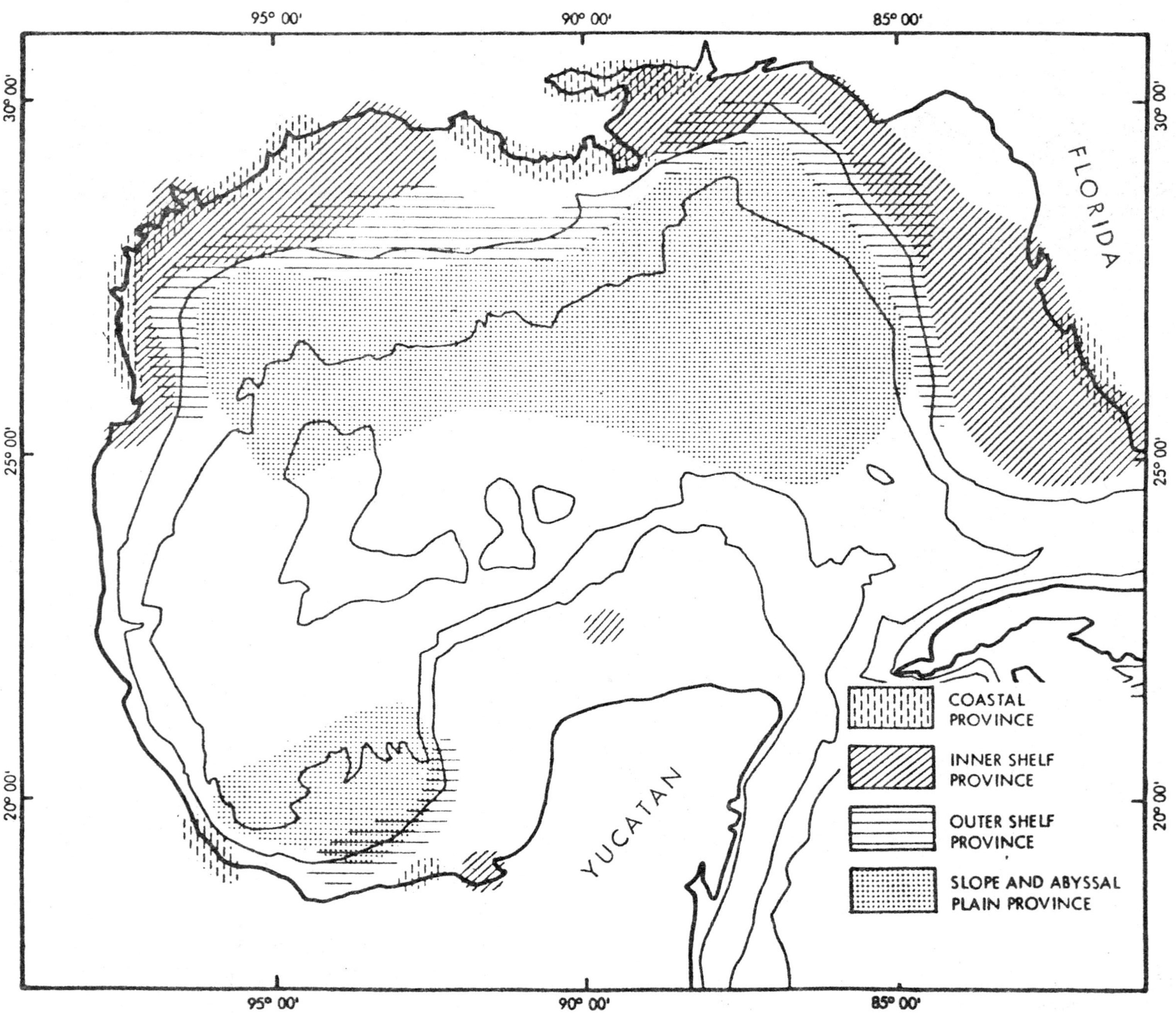


FIG. 7--Foraminiferal provinces of the Gulf of Mexico (after Culver and Buzas, 1981c).

# Cenozoic Benthic Foraminifera Case Histories of Paleoceanographic and Sea-Level Changes

Kenneth G. Miller
Woods Hole Oceanographic Institution/Massachusetts Institute of Technology Joint Program in Oceanography, Woods Hole, MA 02543

"The reports of my death have been greatly exaggerated." (Twain, 1897)

## INTRODUCTION

Since the inception of their use in commercial micropaleontology, benthic foraminifera have proven to be eminently useful in the solution of geological problems. The utilitarian credentials of benthic foraminifera in estimating paleodepths from marsh through neritic environments with a reasonable degree of accuracy and to indicate approximate ages (viz. subdivision of series/epochs) have been established in both commercial and academic applications. Benthic foraminifera are generally more resistant to dissolution than planktonic foraminifera, and have wide distributions; many taxa have restricted stratigraphic ranges, making them suitable for correlation and paleoenvironmental studies. Yet, three problems have tended to limit the utility of benthic foraminifera: 1) there is a lack of uniformity in taxonomy (Boltovskoy, 1980; Douglas & Woodruff, 1982); 2) attempts to erect zonal schemes using benthic foraminifera have resulted in boundaries which are later proven to be diachronous relative to planktonic zonations (e.g. the California provincial stages, Poore, 1976); and 3) attempts to interpret paleodepths from deep-sea benthic foraminifera have produced widely-varying results. One could perhaps conclude, as Boltovskoy (1965a) did over a decade ago, that these problems indicate "...the near future of this science is rather bleak."

Such obstacles can be overcome by: 1) revising taxonomic and biostratigraphic assignments; and 2) evaluating paleonvironmental interpretations of benthic foraminifera in view of other paleoenvironmental indicators (isotopic, seismic, lithologic, independent estimates of paleodepth, etc.), for interpretations of faunal changes are often equivocal without such corroborative data. Two examples illustrate the success of this procedure. The paleobathymetric interpretations of Cretaceous-Early Tertiary flysch-type agglutinated benthic foraminiferal faunas have ranged from paralic to abyssal (Gradstein & Berggren, 1981), and their biostratigraphic and paleoecologic utility was doubted.

Although largely depth-independent, flysch-type foraminifera have been shown to be predominantly deep-sea forms (bathyal-abyssal; Gradstein & Berggren, 1981; Gradstein et al., 1981). Despite diachrony of certain taxa, flysch-type foraminifera have been correlated to planktonic biostratigraphic zonations, and have proven useful in zonations in the Canadian margin, North Sea, and deep sea. Comparisons of distribution of this group with other paleoenvironmental indicators (see below) have shown that they can be useful in indicating restricted circulation (Gradstein & Berggren, 1981; Miller et al., 1982).

The paleobathymetric and stratigraphic utility of deep-sea (bathyal-abyssal) calcareous benthic foraminifera similarly have been doubted. Boltovskoy (1980) found that deep-sea "...benthonic foraminifera are poor stratigraphic index fossils..." In addition, it has been shown that the distributions of some Recent and Pleistocene deep-sea foraminifera are not consistently depth stratified. The abundances of deep-sea benthic foraminifera are correlated with water masses (T, S, $O_2$) and properties that covary with water masses independent of depth (see discussion below). Nevertheless, due to the regional depth stratification of water masses, deep-sea benthic foraminiferal biofacies are often regionally depth consistent (e.g. Streeter & Lavery, 1982; Miller & Lohmann, 1982). Ranges of deep-sea benthic foraminifera can be calibrated to planktonic biostratigraphic zonations and independent estimates of paleodepth obtained from subsidence models of oceanic crust. Once so calibrated, deep-sea benthic foraminifera can be useful in age and paleobathymetric control (Tjalsma & Lohmann, 1982).

I will outline several examples where Cenozoic benthic foraminifera have figured in the interpretation of paleoceanographic and sea-level changes. Both shallow- (neritic) and deep-water (bathyal-abyssal) applications will be presented, although I will stress the relationships among deep-sea benthic foraminifera, other indicators of paleoenvironment, and abyssal circulation changes. Ecologic, paleoecologic, and biogeographical controls on benthic foraminiferal distribution are dealt with elsewhere in this volume. Historical background, sampling methods, preservation, and studies of larger foraminifera are omitted; for these the reader is referred to other reviews (Boltovskoy & Wright, 1976; Douglas, 1979; Douglas & Woodruff, 1982; Adams, 1967).

## MIDWAY FAUNA AND THE SUBSIDENCE HISTORY OF ROCKALL PLATEAU

Berggren and Aubert (1975) used the term "Midway fauna" for Paleocene middle-outer shelf ( ~ 50-200m water depth) smaller benthic foraminifera, originally described from Gulf Coast outcrops (Plummer, 1927; Cushman, 1951; Kellough, 1959, 1965). They differentiated this fauna from a shallow water ( $<$ 50m water depth), inner-middle shelf assemblage dominated by <u>Cibicides</u> spp., nonionids, discorbids, <u>Elphidium</u> spp., and larger benthic foraminifera (= Tethyan Carbonate Fauna; further subdivided and used by Berggren, 1974a)(Fig. 1). The Midway fauna is a rich, diverse fauna characterized by nodosariids, lenticulinids, dentalinids, marginulinids, vaginulinids, polymorphinids, <u>Gavelinella</u>,

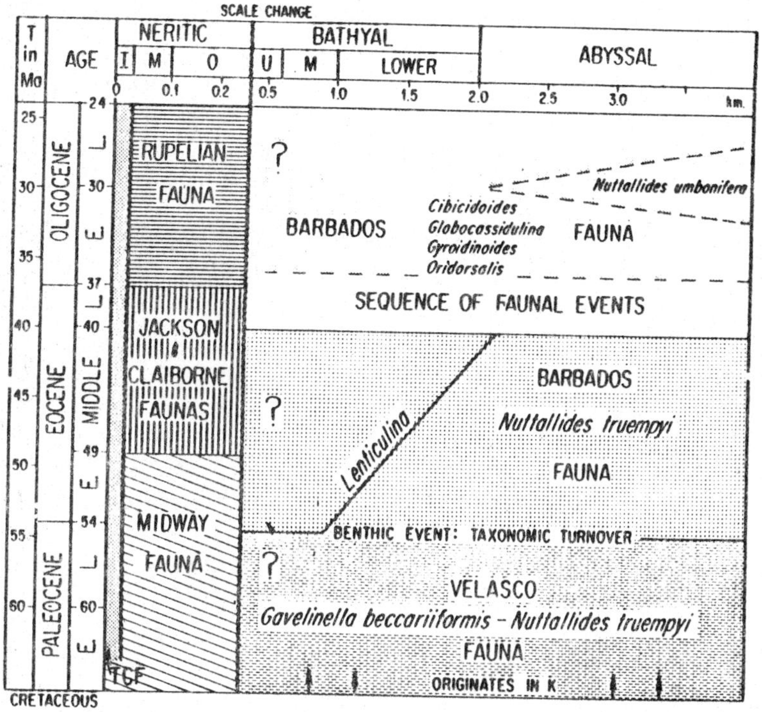


Fig. 1. Simplified scheme of North Atlantic Paleogene benthic foraminiferal assemblages. TCF = Tethyan Carbonate Fauna. (Modified after Berggren & Schnitker, in press). Data from Tjalsma & Lohmann (1982), Miller, (in press), and Miller et al. (in press). The upper depth limits of the deep-sea faunas are poorly constrained; for example, the Barbados Oligocene fauna may be strictly abyssal.

and Cibicides, found primarily in clastic sediments. These faunas are distributed widely in the Gulf Coast (nominate Midway Fm., Kincaid Fm., Salt Mountain Fm., Wills Point Fm.), Atlantic Coastal Plain, and other Circum-Atlantic and Tethyan localities including Nugssuaq, West Greenland (60° paleolatitude, reconstruction of Sclater et al. (1977) (Fig. 2). In addition, they have been recorded from California, Argentina, New Zealand, Australia, Siberia, Turkmenia, Crimea, Caucasus, Ukraine, and Saudi Arabia (Fig. 2 in Berggren & Aubert, 1975).

Paleobathymetric data derived from the Midway fauna have proven useful in subsidence-history interpretations of Rockall Plateau. This foundered microcontinent lies to the west of the Irish continental shelf (Fig. 2). Berggren (1974b) and Berggren and Schnitker (in press) noted the occurrence of Midway faunas in the early Eocene of the Rockall Plateau (Sites 117, 403, 404; also see Murray, 1979). These neritic faunas are succeeded by deep-water assemblages: 1) a middle Eocene Nuttallides truempyi assemblage (Site 403); 2) an Oligocene Cibicidoides-Planulina assemblage (Site 117); and 3) an Oligocene Bulimina marginata-Eggerella bradyi assemblage (Site 404). The punctuated nature of the stratigraphic record recovered in the Rockall region (Miller & Tucholke, in press), prevents firm dating of the timing of these

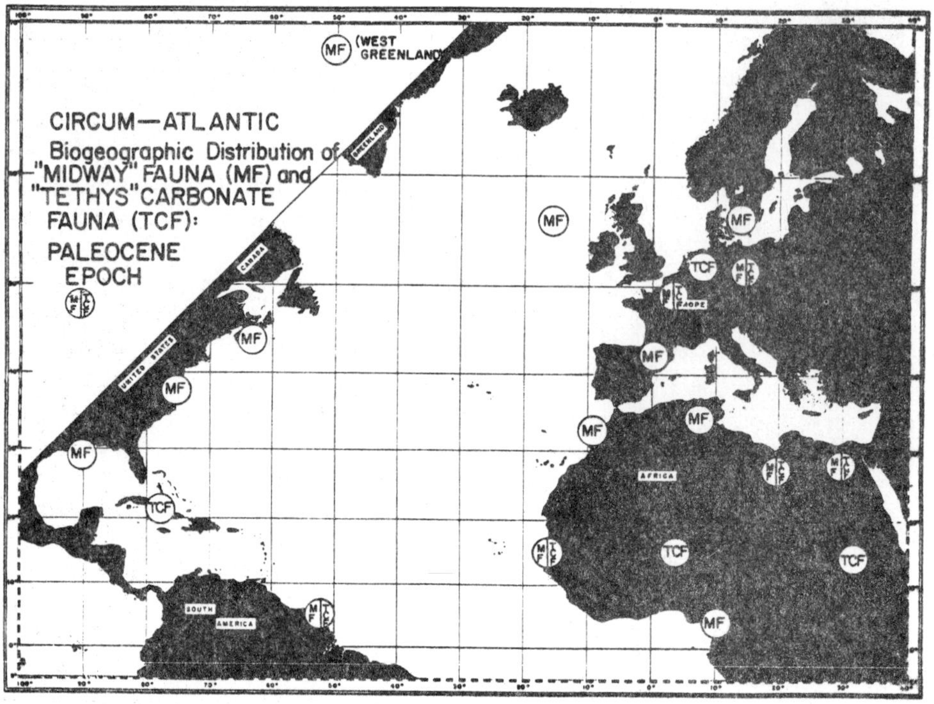


Fig. 2. Circum-Atlantic biogeographic distribution of the Midway and Tethyan carbonate fauna. (After Berggren, 1974a).

successions. However, these data show that the subsidence of Rockall Plateau began in the late Paleocene-early Eocene and continued throughout the Eocene into the Oligocene (Berggren, 1974b; Berggren and Aubert, in press; Murray, 1979). There are significant differences in the paleobathymetric interpretations of various studies (cf. Murray, 1979 and Roberts et al., 1979 with Berggren, 1974b, Berggren & Aubert, 1976, and Berggren & Schnitker, in press). Nevertheless, these studies are in agreement that Rockall Plateau underwent differential subsidence as indicated by: 1) early Eocene deep-water benthic foraminifera noted in Sites 405 and 406 (S transform margin) suggest that this portion of the plateau subsided earlier (Murray, 1979); 2) a comparison of Sites 403 and 404 (SW margin) with Site 117 (Hatton-Rockall Basin) shows that although these sites contain the Midway fauna in the early Eocene, they are bathymetrically offset by over 1.5km today. Thus, benthic foraminifera indicated the importance of post-rifting (Eocene-Oligocene) differential subsidence in the morphologic development of Rockall Plateau (Roberts et al., 1979).

## VELASCO FAUNA, CRETACEOUS/TERTIARY BOUNDARY, AND PALEOCENE BENTHIC CRISIS

Berggren and Aubert (1975) applied the term "Velasco-type" fauna for deep-water Paleocene benthic foraminifera originally described from the Velasco Shale (Cushman, 1926) and the Lizard Springs Formation (Cushman & Renz, 1946)(Fig. 1). Schnitker (1979a) noted the Velasco assemblage in the deep Bay of Biscay (Site 400A and 401), while Tjalsma and Lohmann

(1982) made a detailed taxonomic comparison of Paleocene deep-water benthic foraminifera from Atlantic/Caribbean DSDP sites and the Velasco and Lizard Springs Formations. However, due to lack of material from the upper bathyal realm, the upper depth limit of the Velasco fauna (and the corresponding lower depth limit of the Midway fauna) remains poorly constrained (Fig. 1). Tjalsma and Lohmann (1982) updated the taxonomy, and reported detailed distribution data (both quantitative and qualitative) of deep-water Paleocene benthic foraminifera. Despite terming all deep-water Paleocene benthic foraminifera "Velasco-fauna," it should be noted that the Velasco-Lizard Springs samples are richer in agglutinated benthic foraminifera than the DSDP material examined by Tjalsma and Lohmann (1982). Still, both the Velasco-Lizard Springs and deep-water DSDP samples are dominated by Gavelinella (mostly G. beccariiformis) and Nuttallides (mostly N. truempyi). This G. beccariiformis-N. truempyi assemblage is typical of Velasco-type deep-water Paleocene assemblages (Fig.1).

Comparison of the Velasco Paleocene benthic foraminiferal faunas with Cretaceous benthic faunas shows that, unlike planktonic organisms, there was no crisis in benthic foraminifera at the end of the Cretaceous (= K/T boundary). In Trinidad, most Cretaceous species (~ 82 o/o) range into the Paleocene Lizard Springs Formation (Beckmann, 1960). Cushman and Renz (1946) also noted the similarity of the lower Lizard Springs fauna with the underlying Cretaceous fauna, although they did not recognize the Tertiary age of the former. Douglas and Woodruff (1982) observed that Maestrichtian and early Paleocene deep-water assemblages are nearly identical on the generic level, although some changes did occur at the species level. All of the Paleocene deep-water species described by Tjalsma and Lohmann (1982) were recorded from Cretaceous strata. The flysch-type agglutinated benthic foraminiferal assemblages found in deep-sea sediments show little change between the Cretaceous and the Tertiary (see Miller et al., 1982 for full references). However, Cushman and Renz (1946) and Webb (1973) suggested that only ~ 50 o/o of deep-sea benthic foraminiferal species survived the Cretaceous. Although some debate may exist as to the extent of the changes associated with the K/T boundary, catastrophic mass extinctions did not occcur in deep-sea benthic foraminifera. As a result, Paleocene deep-water benthic foraminifera are predominantly relict Cretaceous taxa (Tjalsma & Lohmann, 1982).

A similar lack of numerous extinctions apparently occurs in shallow-water benthic foraminifera. Although Plummer (1927) noted that the Midway Formation fauna was distinct from the underlying Navarro Formation (Maestrichtian) fauna on the specific level, the assemblages are similar on the generic level. Olsson (1960) observed the Cretaceous benthic foraminifera assemblages in New Jersey to be similar to early Paleocene fauna on the familial and generic levels. He reported that many Cretaceous species range into the Paleocene (Cibicides marylandicus, Pulsiphonina prima, Parrella expansa, Bulimina quadrata, Tappanina selmensis, among others), although many other species are restricted to the Cretaceous or Paleocene sections. It is not clear, therefore, exactly what impact the K/T boundary event had on shallow-water benthic foraminifera. It is clear that the extinctions associated with this group were much less severe than the widespread extinctions suffered by planktonic organisms. A detailed comparison of Maestrichtian and early Paleocene assemblages is needed to document the extent to which shallow-water benthic foraminifera were affected by the K/T event.

In general, stability characterizes Paleocene deep-sea assemblages studied so far. Most of the faunal variation in the Paleocene is attributable to the gradual bathymetric restriction of the shallower G. beccariiformis assemblage and the bathymetric expansion of the N. truempyi assemblage (Fig. 3). Such depth-migrations, both expansions and restrictions, are prominent among the faunal changes noted in deep-sea benthic foraminifera studies to date, not only for the Paleocene, but for the entire Cenozoic. There is some tendency for Paleocene assemblages to be latitudinally differentiated. Tjalsma and Lohmann (1982) noted that a Nuttallides crassaformis association is more important in tropical sites. However, throughout the Cenozoic, deep-sea benthic foraminifera remain generally cosmopolitan in distribution (Douglas & Woodruff, 1982).

A major benthic faunal crisis occurs in the latest Paleocene (Zone P6a) (Fig. 3). Following the gradual depth-restriction of the G. beccariiformis assemblage, rapid massive extinctions occur at the generic and specific levels (Tjalsma & Lohmann, 1982). Most of the extinctions occur in the shallower G. beccariiformis assemblage, which contained predominately Cretaceous relict species (Tjalsma & Lohmann, 1982). Schnitker (1979a) also suggested that the extinction event occurred primarily at shallower sites such as Site 401 ( ~ 2.0-2.5km paleodepth), and that the deepest (N. truempyi) assemblage was characterized more by appearances than extinctions.

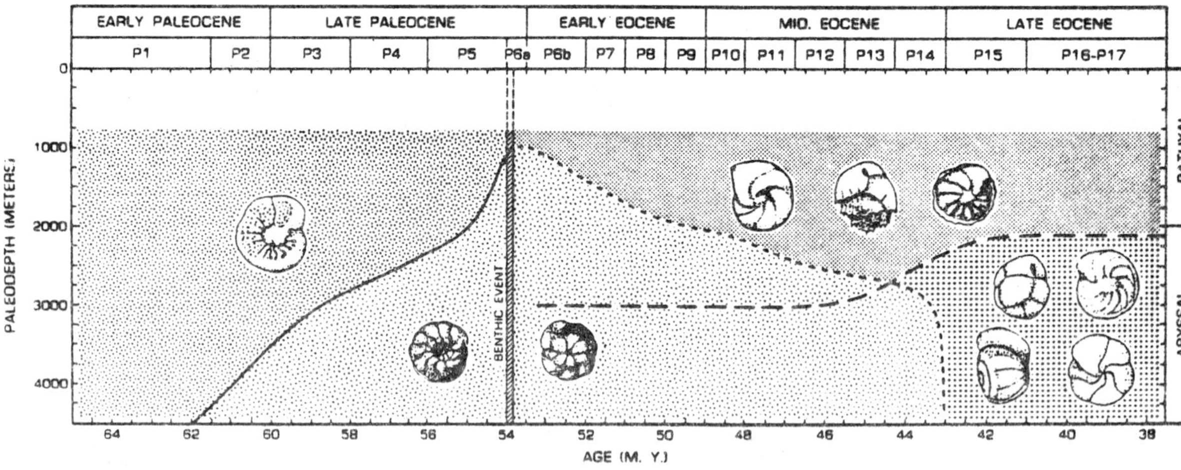


Fig. 3. Distribution of Paleocene-Eocene deep-water benthic foraminifera (after Tjalsma & Lohmann, 1982). The age-paleodepth diagram was obtained by "backtracking" DSDP sites along an empirical age-versus-subsidence curve (Sclater et al., 1971; Parsons & Sclater, 1977) in the manner of Berger and Winterer (1974) and Sclater et al. (1977); this allows an independent estimate of paleobathymetry with geophysical constraints. For other examples of "backtracking" benthic foraminifera see Tjalsma and Lohmann (1982), Douglas and Woodruff (1982), Miller (in press), Miller et al. (in press).

The Paleocene benthic faunal event was catastrophic, and it resulted in low diversity early Eocene assemblages (Tjalsma & Lohmann, 1982). This event constitutes the most dramatic of all Cenozoic deep-sea benthic foraminiferal changes noted to date, yet remains the greatest enigma in Cenozoic deep-sea benthic foraminiferal studies. Although major paleoceanographic changes occurred at the K/T and Eocene/Oligocene boundaries, and various scenarios have been erected to explain these changes (see summaries Haq, 1981; Alvarez et al., 1982; Miller et al., in press), no major paleoceanographic changes have been documented in association with the Paleocene faunal event. Without such corroborative data, the meaning of the event will remain obscure.

## EUSTATIC SEA-LEVEL EFFECTS: AN EARLY EOCENE EXAMPLE

Estimates of Cenozoic eustatic changes in sea level based upon analyses of seismic stratigraphic sequences have been made by P.R. Vail and collegues at Exxon Production Research (Vail et al., 1977; Vail & Hardenbol, 1979; Vail & Mitchum, 1980; Hardenbol et al., 1981). They noted several major lowstands of sea level in the Cenozoic. However, the relationship of eustatic sea level and the sense of relative sea-level change obtained from benthic foraminifera may differ. Depending upon the rates of subsidence and sediment input, biofacies may be deepening, shallowing, or compensating during a eustatic sea-level rise. A regression indicated by benthic foraminifera (such as the basinward migration of shallow-water biofacies), in fact, may occur during rising sea level (Vail et al., 1977). Still, paleobathymetric changes often follow the sense of sea-level fall, and paleodepths derived from benthic foraminifera can be used in evaluating eustatic sea level changes if properly corrected for sediment compaction, loading, and tectonic subsidence (e.g. Van Hinte, 1978; Watts & Steckler, 1979; Hardenbol et al., 1981).

Vail et al. (1977) noted a major early Eocene (49.5 Ma) sea-level lowstand. Benthic foraminiferal biofacies show 3 major regressions that correlate with this lowstand (Fig. 4). 1) In the California Coast Ranges, Berggren and Aubert (in press) noted that a marked shallowing occurred in the early Eocene (late Zone P8-early Zone P9). They based this upon a change from a deep-sea benthic foraminiferal assemblage (including *Nuttallides truempyi*; estimated paleodepth ~ 600m) to a neritic (< 200m) *Florilus-Uvigerina-Eponides-Cyclammina-Trochammina* assemblage. 2) In Libya, a change from an upper bathyal *Nuttallides truempyi-Bulimina trinitatensis-B. bradburyi* assemblage to an outer neritic *Bulimina-Uvigerina* (*B. jacksonensis*, *B. gardnerae*, *B. cocoaensis*) assemblage and to an inner neritic *Nummulites* assemblage occurs within Zone P9. Barr and Berggren (1982) interpreted this as reflecting a rapid (< 1 m.y.) regression of ~ 400m. 3) In the New Jersey Coastal Plain, Charletta (1980) observed that a marked shallowing occured near the early/middle Eocene boundary. This shallowing was indicated by local terminations of deep-water benthic foraminifera (including *N. truempyi*), their replacement by shallow water benthic foraminifera, and decreases in abundance of planktonic foraminifera. These three studies vary as to the exact timing of the change in sea level, but agree that a regression occurred in each area near the end of the early Eocene (within Zone P9).

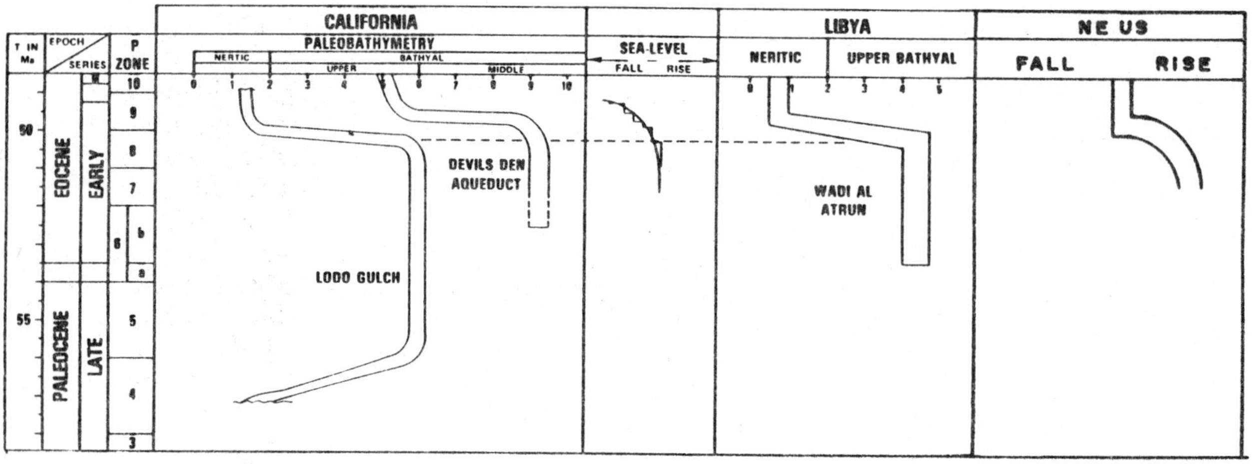


Fig. 4. Comparison of benthic foraminiferal estimates of early Eocene water-depth changes. (Modified after Vail & Hardenbol, 1979).

There are differences as to the magnitude of the regression (Fig. 4) (California and Libya ~ 400m; New Jersey, updip wells, ~ 150m; New Jersey downdip wells 300-400m). Vail et al. (1977) estimated that the early Eocene eustatic sea-level drop was ~ 150-200m (from ~ 250m to 100m above present sea level), although such large estimates are controversial (Watts & Steckler, 1979). The benthic foraminiferal data come from tectonically-varied regimes, and are not corrected for sediment input, crustal response to loading and unloading, and local tectonic changes. It is remarkable, given all of these uncertainties, that the foraminiferal data indicate 3 rapid (Zone P9) regressions that approximately correlate with this assumed eustatic lowering.

## FLYSCH-TYPE BENTHIC FORAMINIFERA AND BASIN EVOLUTION

Agglutinated (arenaceous) benthic foraminiferal assemblages (= flysch-type assemblages of Gradstein and Bergrren, 1981; = Rhabdammina faunas of Brouwer, 1965) have been noted in various flysch basins. These assemblages consist of predominantely (or exclusively) agglutinated benthic foraminifera. Gradstein and Berggren (1981) described taxonomically similar assemblages from Maestrichian-Eocene mudstones of the Labrador Shelf/northeast Newfoundland Shelf and North Sea Paleocene-Eocene strata. Miller et al. (1982) differentiated between two types of predominantely agglutinated benthic foraminifera found in DSDP sites. Type A assemblages resemble flysch-type faunas, occur in bathyal-abyssal Late Cretaceous to Early Tertiary sediments, are composed of larger ( > 150 μ m), coarser-grained tests, and are dominated by "simple" forms. Type-B assemblages are taxonomically distinct from flysch-type faunas, occur in Cretacous zeolitic and variegated abyssal clays, are composed of smaller, finer-grained tests, and are dominated by Labrospira, Plectorecurvoides, and Haplophragmium (see Krashenninikov, 1973, 1974 for illustrations of type-B faunas).

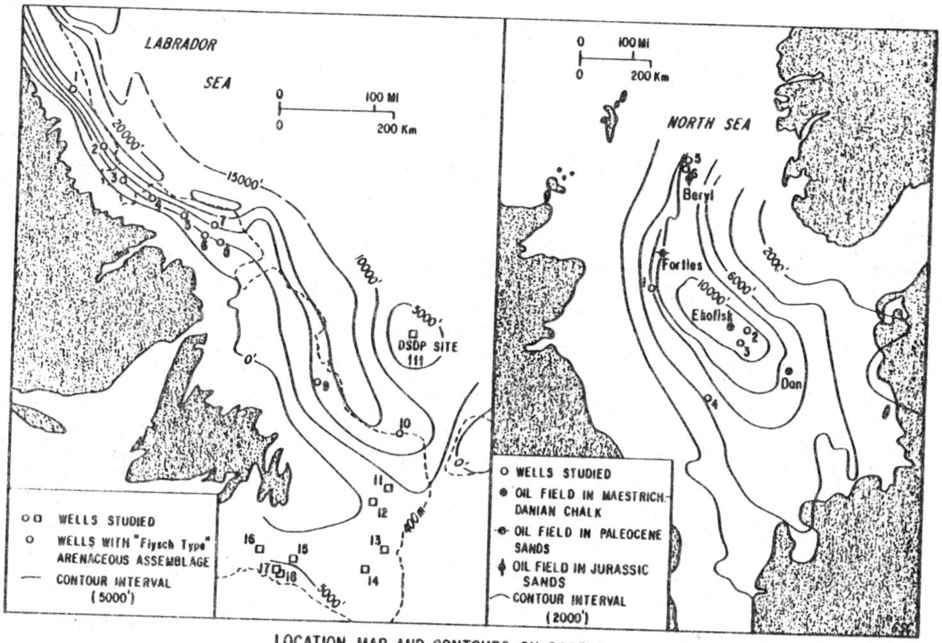

LOCATION MAP AND CONTOURS ON BASE OF TERTIARY

Fig. 5. Location of wells containing flysch-type assemblages from the Canadian margin and North Sea. Structural contours drawn on the base of the Tertiary. (After Gradstein & Berggren, 1981).

Flysch-type agglutinated foraminifera in the Canadian margin and North Sea are associated with high sedimentation rates; highly-diverse flysch-type assemblages are found in organic-rich shales in wells at the depositional center of the basins (Fig. 5) These wells have calcareous benthic foraminifera indicative of deeper water whereas updip wells containing a less diverse assemblage (not flysch-type), are dominated by Bathysiphon and Trochammina. Paleoenvironmental factors controlling the distribution of agglutinated assemblages in the Canadian margin and North Sea were related to rapid deposition of organic-rich, fine-grained clastics (see discussion in Gradstein & Berggren, 1981). Here, differential subsidence resulted in restricted basins which coincided with deltaic outbuilding into the deep basin and the development of agglutinated foraminiferal assemblages. Near the end of the Eocene, infilling of the basins resulted in the demise of the agglutinated assemblages.

Predominantly agglutinated foraminiferal assemblages have been accounted for by water depth, carbonate availability, substrate characteristics, and certain hydrographic properties. The extensive paleobathymetric range of type-A (= flysch-type) assemblages suggests that they are depth-independent. Miller et al. (1982) noted that in the deep southern Labrador Sea (Site 112), type-A assemblages are found in carbonate sediments. Here, the agglutinated foraminiferal assemblages are replaced by calcareous benthic foraminifera near the end of the Eocene. Lithologic characteristics remain constant across this change, but the change correlates with a change in abyssal circulation (inferred from the seismic stratigraphic record; Miller and Tucholke, in press) and hydrographic properties (inferred from the isotopic record, Miller and Curry, 1982) which occurred near the Eocene/Oligocene boundary in this region. Thus, although low carbonate content and high organic content may be associated with type-A assemblages, these properties are not necessary for development of flysch-type faunas. Rather, certain hydrographic properties (viz. low oxygen, low pH, high $CO_2$, and more corrosive waters) associated with restricted circulation are critical to the development of the predominantly agglutinated assemblages. Similar conditions may develop in reducing substrates associated with high organic matter and poor circulation. Assemblages of predominantly agglutinated foraminifera are probably favored in such substrates and/or in areas with the given hydrographic properties, and therefore may be used as generally indicative of restricted circulation.

## LATE PALEOGENE BENTHIC FORAMINIFERA AND THE DEVELOPMENT OF ABYSSAL CIRCULATION

Various lines of evidence (seismic, lithostratigraphic, isotopic) indicate that a major change in abyssal circulation occurred near the end of the Eocene. Early studies (Savin et al., 1975; Shackleton & Kennett, 1975; Kennett & Shackleton, 1976) showed that a major $\delta^{18}O$ increase in planktonic and benthic foraminifera occurred near the Eocene/Oligocene boundary in the Southern and Pacific Oceans. This was interpreted to represent a cooling of surface waters and the initial formation of cold bottom water of southern (Antarctic) origin. Some of this enrichment may represent the buildup of the Antarctic ice cap (Matthews & Poore, 1980). More recent studies have shown that the bulk of the $^{18}O$ enrichment in benthic foraminifera occurred in a $< 0.5$ m.y. interval just above the Eocene/Oligocene boundary as defined by the extinction of Hantkenina and Globorotalia cerroazulensis lineage in the Southern, Pacific, and North Atlantic Oceans (Keigwin, 1980; Miller & Curry, 1982; Miller et al., in press). Evidence of widespread unconformities in the Southern and Pacific Ocean (Moore et al., 1978) supports the idea that a major change in abyssal circulation occurred at this time. Based upon studies of seismic stratigraphy and sediment distribution, Miller and Tucholke (in press) noted that major abyssal circulation changes, from a sluggish Eocene ocean to a more vigorously ciculating Oligocene ocean, occurred in the North Atlantic. They suggested that the North Atlantic had a northern bottom water source from the Arctic and/or Norwegian-Greenland Sea. Thus, climatic cooling in the late Eocene to early Oligocene (ref. in Miller & Tucholke, in press) resulted in the formation of cold bottom water of northern origin and southern origin.

As some distributions of modern benthic foraminifera correlate with modern water-mass distributions (see below), benthic foraminifera may be expected to have responded to hydrographic changes associated with the $\delta^{18}O$ increase. Early workers noted that Eocene deep-sea benthic foraminiferal assemblages were distinct from Oligocene assemblages (Douglas, 1973; Berggren, 1972a). Because the late Eocene to early Oligocene interval is associated with numerous hiatuses (Moore et al., 1978), the timing of this faunal change remained debatable. Kennett and Shackleton (1975) and Schnitker (1979a) suggested that a crisis in benthic

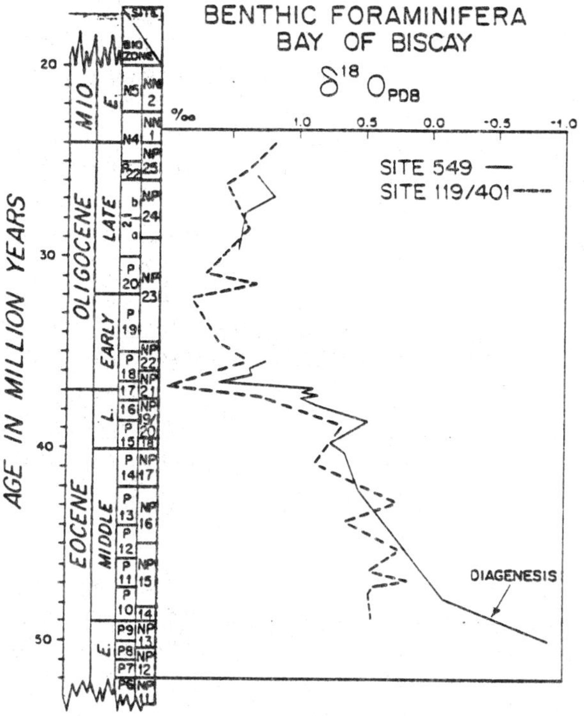


Fig. 6. Benthic foraminiferal isotopic record. (After Miller et al., in press). Site 549 was drilled on the Goban Spur in 2533m water depth; Site 119 was drilled on Cantabria seamount in 4447m water depth; Site 401 was drilled on the Armorican margin in 2495m water depth. For details on the Site 119/401 record see Miller & Curry, 1982).

foraminifera occurred near the Eocene/Oligocene boundary. However, Corliss (1979a) observed no major benthic foraminiferal changes associated with the boundary in Site 277 in the Southern Ocean; Corliss (1981) evaluated previously published results and concluded that a series of gradual changes in benthic foraminifera occurred between the middle Eocene and the early Oligocene in response to gradual climatic cooling. Tjalsma (1982 and personal communication) and Tjalsma and Lohmann (1982) also noted no change associated with the Eocene/Oligocene boundary; they reported that a series of sequential changes in benthic foraminifera occurred during the middle Eocene to early Oligocene. The most dramatic change seen in their Caribbean/South Atlantic sites is the decrease in abundance of Nuttallides truempyi near the middle/late Eocene boundary (Fig. 3).

The evolution of the late Eocene-Oligocene fauna occurred as a series of events over several m.y. Miller et al. (in press) noted that the decrease in abundance of N. truempyi was apparently a synchronous event in the Atlantic Ocean that occurred just above the middle/late Eocene boundary (Fig. 6)( ~ 38.5-40 Ma). This pre-dates (~2 m.y.) the major $^{18}O$ enrichment and change in abyssal circulation regime inferred from the seismic stratigraphic record (Fig. 5). In Site 549 in the Bay of Biscay (Fig. 6), the Bulimina spp. assemblage that succeeds dominant N. truempyi is itself replaced by an assemblage dominated by stratigraphically long-ranging and bathymetrically wide-ranging taxa (e.g. Globocassidulina subglobosa, Gyroidinoides spp., Cibicidoides ungerianus, Oridorsalis umbonatus) just below the Eocene/Oligocene boundary at approximately 37.5 Ma (Fig. 6). The record at the deepest sites (Sites 119 and 400A; both paleodepths > 3km) shows the greatest changes, for here N. truempyi is associated with many endemic deep-water taxa (Abyssammina, Clinapertina, Aragonia, Alabamina dissonata, among

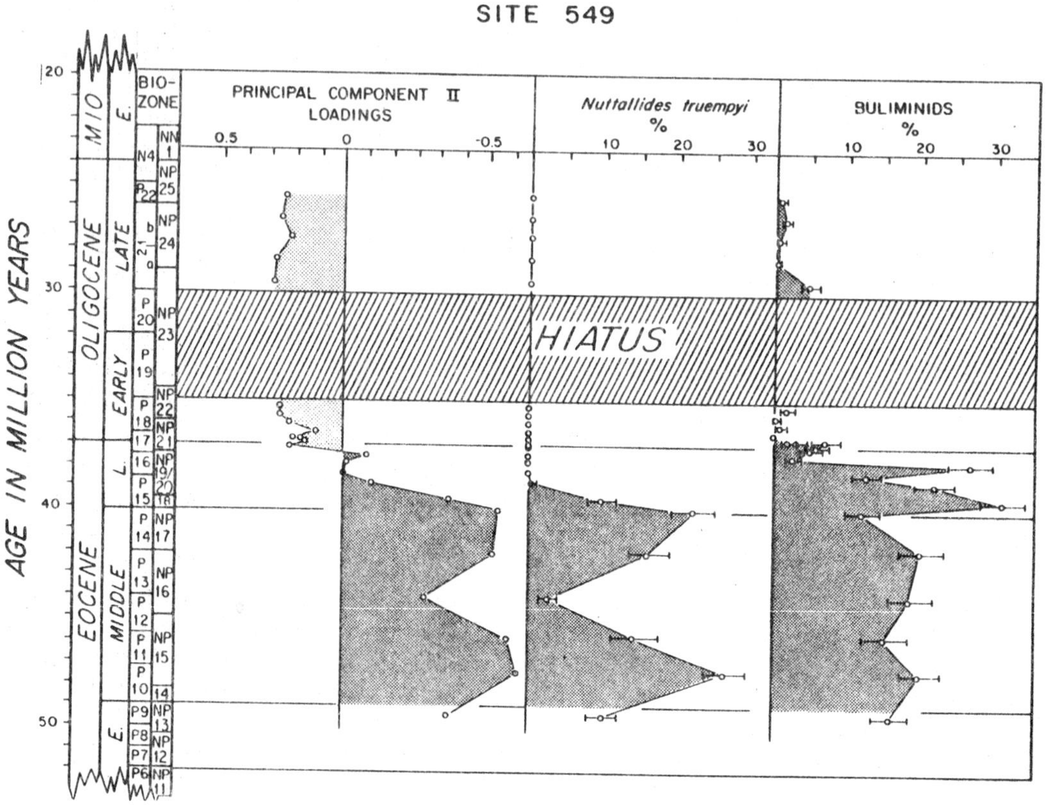


Fig. 7. Benthic foraminifera assemblage changes during the Late Paleogene, Site 549, Goban Spur. Interval estimates reflect 80 o/oo confidence limits. Principal component II represents the dichotomy between Nuttallides truempyi-buliminid assemblage (negative loadings) and Globocassidulina subglobosa-Cibicidoides ungerianus-Gyroidinoides assemblage (positive loadings) (After Miller et al., in press).

others) that decrease in abundance and become extinct prior to the Oligocene. Unfortunately, the timing of the abundance changes and extinctions in these deepest sites cannot be established due to poor recovery of upper Eocene sections there (Miller, in press). However, in the intermediate depths (Site 549; ~ 2-2.5km paleodepth), a series of first and last appearances occurred in the late Eocene to earliest Oligocene (Miller et al., in press). Thus, like the K/T boundary, the Eocene/Oligocene boundary was not catastrophic for benthic foraminifera.

The Oligocene abyssal fauna is dominated by stratigraphically long-ranging and bathymetrically wide-ranging taxa (Miller, in press). These taxa, which are interpreted to be tolerant of paleoenvironmental changes, range through Late Paleogene oceanographic changes. Many species that appeared in the Eocene and Oligocene range into the Recent (Boltovskoy, 1980; Douglas & Woodruff, 1982). Oligocene deep-sea faunas were generically similar to modern faunas, and several species important in the modern deep ocean made their appearance in the late Eocene to early Oligocene (e.g Nuttallides umbonifera, Epistominella exigua, Eggerrella bradyi, among others). Oligocene deep-sea faunas mark a change from Paleocene (Cretaceous relict) and Eocene taxa (e.g. N. truempyi, Alabamina dissonata, Aragonia spp.) toward the development of modern assemblages. This led Boltovskoy (1980) to claim that modern bathyal assemblages developed no later than the Oligocene. However, many important modern taxa did not make their appearance until the Miocene (see below).

One obstacle to determining Oligocene faunal and paleoceanographic changes is that this series is poorly represented in the deep-sea stratigraphic record (Moore et al., 1978; Miller & Tucholke, in press), The records that have been recovered, such as the benthic foraminiferal $\delta^{18}O$ record, show few changes (Shackleton & Kennett, 1975; Miller & Curry, 1982). However, Miller and Curry noted that $\delta^{13}C$ values reach a distinct minimum in the middle Oligocene of the Bay of Biscay, and that this minimum correlates with the acme of Nuttallides umbonifera (Fig. 7). Today, the abundance of N. umbonifera is positively correlated with increased corrosiveness of bottom water (Bremer & Lohmann, in press), while lower $\delta^{13}C$ values are often associated with older water masses that are more corrosive to carbonate (Kroopnick et al., 1972; Kroopnick, 1974, 1980). Large numbers of N. umbonifera and light $\delta^{13}C$ values in the middle Oligocene were interpreted as reflecting older, more corrosive bottom water (Miller & Curry, 1982; Miller, in press). This inferred increase in bottom water age correlates with a reduction in the intensity of bottom water flow in the North Atlantic inferred from the seismic stratigraphic record of the Rockall region (Miller & Tucholke, in press).

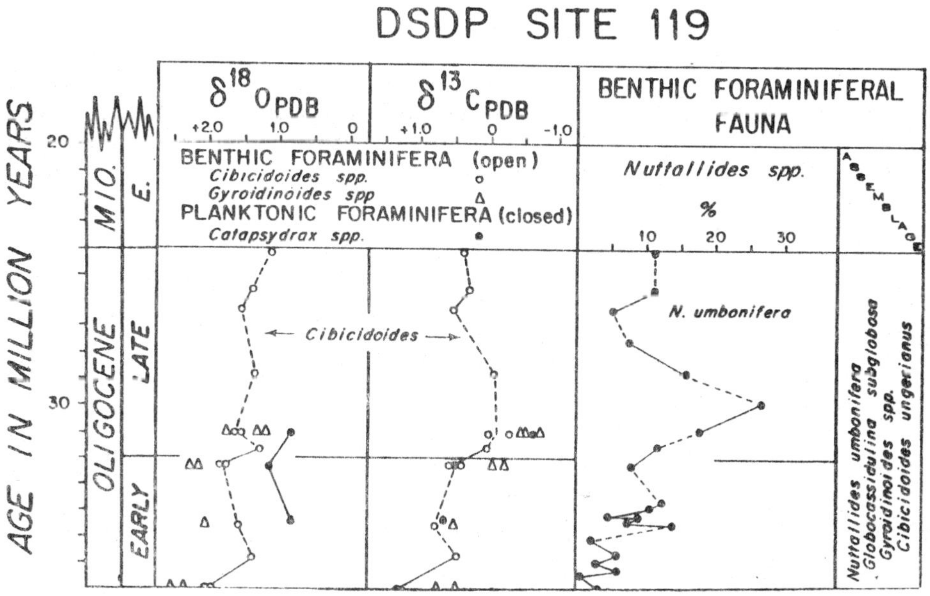


Fig. 8. Oligocene isotopic and faunal record in the abyssal Bay of Biscay, DSDP Site 119 (Modified after Miller & Curry, 1982; Miller, in press).

## NEOGENE DEEP-SEA BENTHIC FORAMINIFERA: DEVELOPMENT OF MODERN ASSEMBLAGES

Considerable debate exists as to the timing and interpretation of the development of the modern deep-sea assemblage. Berggren (1972) suggested that the modern deep-sea fauna (including Planulina wuellerstorfi, Uvigerina peregrina, Pyrgo murrhyna) appeared in the middle Miocene ~ 15 Ma, while Schnitker (1979a) thought that the modern assemblage became established near the end of the middle Miocene. At the other end of the spectrum, Boltovskoy (1980) maintained that the modern fauna originated in the pre-Miocene. These discrepancies may result, in part, from the material studied (as a function of recovery, preservation, geographic location, paleobathymetry) and approach in analysis (viz. size-fraction examined, studying ranges versus abundances, R-mode versus Q-mode principal component, factor, or cluster analyses, plotting against time versus depth). The timing of the faunal change may be resolved, for Woodruff and Douglas (1981) and Douglas and Woodruff (1982) noted that major faunal changes occurred not only in the early middle Miocene (Zones N9-N11) but also in the late Miocene (Zones N16-N17). They stated that the middle Miocene change is due primarily to appearances and extinctions of species, while the late Miocene development is chiefly the result of changes in abundance. Thus, the latter event resulted in assemblages on the Ontong-Java Plateau that resembled Recent assemblages in composition and abundance, while the former resulted in extinctions of deep-water taxa that appeared in the Late Paleogene (Bulimina jarvisi, Buliminella grata, Hanzawaia cushmani, and Cassidulina havanensis).

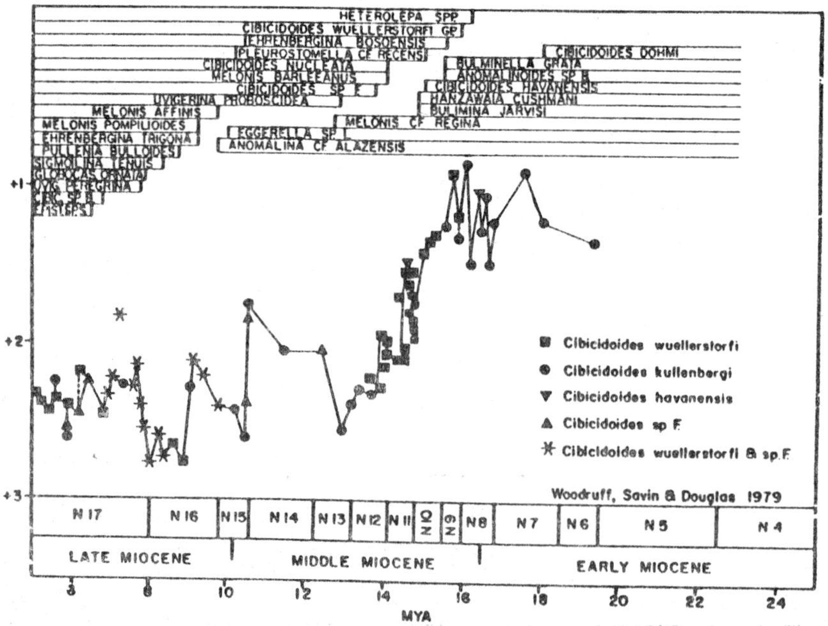


Fig. 9. Stratigraphic distribution of important species compared to bottom water $\delta^{18}O$ changes, Site 289. $^{18}O$ given in $^{o}/_{oo}$ PDB. (After Douglas & Woodruff, 1982).

A major enrichment of $^{18}O$ in benthic foraminifera occurs in the middle Miocene. This has generally been attributed to global changes in the $^{18}O$ composition of seawater due to the buildup of continental ice on Antractica (Savin et al., 1975; Shackleton & Kennett, 1975; Woodruff et al., 1981). (Ice is depleted in $^{18}O$ relative to seawater, thus ice buildup results in seawater enriched in $^{18}O$.) However, Matthews and Poore (1980) argued that, as in the Pleistocene, ice buildup should be reflected in enrichments of $^{18}O$ in both benthic and tropical planktonic foraminifera. This was not the case with the middle Miocene, for tropical planktonic foraminifera became depleted in $^{18}O$. Savin et al. (1975) and Savin (1977) interpreted this as a tropical sea-surface warming. In any case, these authors agree that the middle Miocene $^{18}O$ enrichment (Fig. 8) represents some drop in bottom-water temperature. Douglas and Woodruff (1982) and Woodruff et al. (1981) suggested that this enrichment occurred in Zones N9-N11 (13-16.5 Ma), and that it approximately coincided with the major Miocene benthic foraminiferal changes.

Disagreements exist as to the interpretation of the middle Miocene faunal and isotopic events and the timing of the first influences of cold bottom water formed in the subpolar-polar basins of the North Atlantic and adjacent Norwegian-Greenland Sea or Arctic ocean. Schnitker (1979a, 1980a,b) and Blanc et al. (1980) ascribed the middle Miocene event in the North Atlantic to the first input of bottom water from such northern sources, while Schnitker (1979a, 1980a,b) related Eocene/Oligocene events in the North Atlantic to the initial formation of bottom water derived from the Southern Ocean. In contrast, the Eocene/Oligocene faunal and isotopic events in the North Atlantic have been attributed to the input of northern sources of bottom water (Miller et al., 1982; Miller & Curry, 1982; Miller, in press; Miller et al., in press).

It is clear from the data reviewed here that there are two major post-Paleocene events in the development of deep-sea foraminiferal assemblages and isotopic composition. Some of the ambiguity in interpretation of these events results because changes in deep-sea benthic foraminifera alone do not indicate bottom water sources. Although benthic foraminiferal isotopic data potentially can be used to indicate bottom water sources (especially $\delta^{13}C$ data, Curry & Lohmann, in press), insufficient isotopic data exist to resolve these ambiguities. The seismic stratigraphic record provides less ambiguous evidence for changes in abyssal circulation. Miller and Tucholke (in press) used seismic stratigraphic evidence to indicate that northern sources of vigorously circulating bottom water began in the latest Eocene to earliest Oligocene in the North Atlantic. This suggests that the middle Miocene faunal and isotopic events did not represent the first influx of northern sources of cold bottom water into the North Atlantic. Woodruff and Douglas (1981) attributed the middle Miocene faunal event to increased production of southern bottom water. It is reasonable to speculate that the middle Miocene event represents variations in production of both southern and northern sources of bottom water in response to climatic changes. However, there is not a simple relationship between bottom-water temperature change and faunal changes. The Eocene/Oligocene faunal changes begin prior to a temperature drop (Figs. 6,7), while the middle Miocene changes lag a temperature drop (Fig. 9). The benthic faunas may have responded to other paleoenvironmental changes associated with changing bottom waters (corrosiveness, $O_2$ content, etc.).

OLIGOCENE-PLIOCENE SHALLOW WATER BENTHIC FORAMINIFERAL AND SEA-LEVEL

Due to the paucity of Oligocene sediments and to difficulties in Oligocene biostratigraphy, the chronostratigraphic distribution of Oligocene shallow-water strata and thus the precise record of sea-level changes is equivocal. One or more major sea-level lowstands occurred in the late Eocene to middle Oligocene interval, but the timing of the event(s) is unclear. Vail et al. (1977) placed the lowest sea-level lowstand of the Cenozoic in the "middle" Oligocene (base of Zone P21A; 29 Ma). Based upon a widespread upper Eocene to lower Oligocene disconformity in the Middle Atlantic Coastal Plain of the U.S. (and a similar record in the Canadian margin, Gradstein & Srivastava, 1981), Olsson et al. (1980) suggested that this sea-level drop occurred prior to deposition of Zone P20, probably near the Eocene/Oligocene boundary. The sea-level curve of Vail et al. (1977), in general, corresponds well with the $\delta^{18}O$ record (Vail & Hardenbol, 1981), due to the linkage of sea-level, climate (lower sea level results in greater continentality and cooler climates), and ice volume (increasing ice volume lowers sea level). However, there is a striking lack of correspondence between their late Eocene to Oligocene sea-level curve and the $\delta^{18}O$ record (Miller, in press). No $^{18}O$ enrichment is associated with their middle Oligocene sea-level drop (Fig. 6; also see Shackleton & Kennett, 1975; Savin et al., 1975). The major late Eocene to early Oligocene $^{18}O$ enrichment (Fig. 6) correlates only with a minor sea-level drop (Vail & Mitchum, 1980). In fact, the most recent estimates of sea-level from the Exxon group (Baum et al., 1982) suggest that the late Eocene to early Oligocene was actually a highstand and that the numerous disconformities associated with this boundary are attributable to basin starvation. This serves to accentuate the differences between the $\delta^{18}O$ and sea-level records.

Examples of Oligocene inner, middle, and outer neritic faunas have been described from the Gulf Coast, Atlantic Coastal Plain, and Europe. Poag (1966) noted an inner neritic assemblage of *Hanzawaia*-*Discorbis*-*Elphidium* in the Gulf Coast (upper Oligocene Paynes Hammock Fm). A middle neritic late Oligocene assemblage was described from the U.S. middle Atlantic Coastal Plain: *Buliminella elongata*, *B. elgantissima*, *Cassidulina crassa*, *Cibicides ornatus*, *C. lobatulus*, *Epistominella pontoni*, *Florilus pizzarensis*, *Hanzawaia concentrica*, *Melonis affine*, *Trifarina bradyi*, and *Uvigerina juncea* (Olsson et al., 1980). A similar fauna has been noted in the late Oligocene of SW France (Butt, 1966) and the Stampian stage of the Paris Basin (Cushman, 1928a). In the NW European Rupelian stage (early Oligocene), Batjes (1958) and Ulleberg (1974) described faunas, indicative of outer neritic-upper bathyal depths (Berggren, 1977), with *Turrilina alsatica*, *Angulogerina tenuistriata*, *Melonis affine*, *Gyroidina girardana*, and *Sphaeroidina bulloides* (= Rupelian fauna, Fig. 1).

Neogene shallow-water benthic foraminifera are poorly represented in the U.S. Atlantic Coastal Plain, Gulf Coast outcrops, and Western Europe (Berggren, 1977). Many of the middle neritic fauna found in the late Oligocene of the U.S. Atlantic Coastal Plain are also found in the Miocene there (Cushman & Cahill, 1933; McLean, 1956). Gibson (1967) and

Schnitker (1970) discussed neritic benthic foraminifera from Miocene strata in North Carolina (Pungo River and Yorktown fms.). These assemblages are similar to the late Oligocene assemblages in the middle Atlantic Coastal Plain, with dominant *Florilus pizzarensis*, *B. elegantissima*, *Cassidulina norcrossi*, *C. crassa*, *Uvigerina*, *Nonionella auris*, and *Cibicides lobatulus*. Although Oligocene North American and European shallow-water faunas have close affinities, similarities between European and North American faunas decrease in the middle Miocene. Berggren and Phillips (1971) noted that circum- Mediterranean benthic foraminifera exhibited a marked drop in similarity with North American forms at this time. They attributed this to isolation of the Mediterranean due to the rotation of Iberia towards Africa.

Poag (in press) discussed the Neogene stratigraphy of the submergent U.S. Atlantic Coastal Plain. He noted a correspondence of the major middle Miocene (Zones N9-N14) highstand of sea level (Vail et al., 1977) with widespread deposition on the margin. Early Miocene strata (mostly Zones N4-N5) are less well represented, and late Miocene to Pliocene strata are very poorly represented; these coincide with lowerstands of sea level. In the southeast Georgia Embayment, he noted a neritic *Buliminella elongata-Florilus-Epistominella* assemblage similar to that discussed above, and attributed it to increased upwelling. An upper bathyal *Brizalina-Siphogenerina-Uvigerina* assemblage was also observed in upper Miocene strata in this region.

Perhaps the most dramatic Late Neogene oceanographic event is the desiccation of the Mediterranean Sea (Hsu et al., 1972, 1973; Van Couvering et al., 1976). Benthic foraminiferal evidence from the

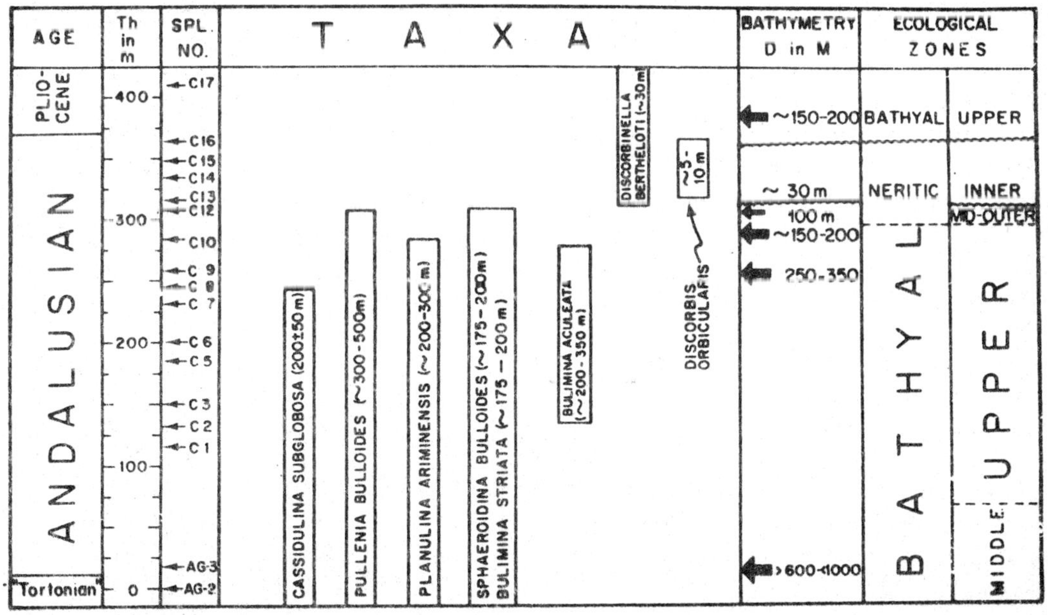


Fig. 10. Estimated paleobathymetry of the upper Miocene-lower Plicoene sequence at Carmona-Dos Hermanas, SW Spain. Modern upper depth limits and stratigraphic range in the section are shown for the benthic foraminifera (after Berggren & Haq, 1976).

Andalusian (late Miocene) stage shows evidence of shallowing (Fig. 9) that may represent the desiccation event. Berggren and Haq (1976) and Berggren et al. (1976) showed that in a 4-5 m.y. interval of the late Miocene, a lower bathyal assemblage (Cibicidoides mediocris-Melonis soldanii-Planulina wuellerstorfi) was succeeded by an upper bathyal assemblage that was, in turn, succeeded by an inner neritic assemblage (Fig. 9). In the latest Miocene ( ~ 5.5 Ma), a sharp decrease in water depth occured ( ~ 50-70m, Fig. 9). This sharp drop was interpreted to represent a eustatic sea-level drop (see Vail et al., 1977) that served to isolate fully the already partially-isolated Mediterranean, resulting in its desiccation. Berggren and Haq believed that the absence of rapid variations of sea level in the uppermost Andalusian section argued against periodic refilling of the Mediterranean as suggested by Hsu et al. (1972). Wright (1979) reported that following the early Pliocene reconnection and refilling of the Mediterranean, North Atlantic benthic foraminifera quickly ( ~0.5 m.y.) repopulated the former evaporite basin.

## QUATERNARY DEEP-SEA BENTHIC FORAMINIFERA, WATER MASSES, GLACIAL HISTORY

Deep-sea benthic foraminiferal bathymetric distribution patterns are not maintained interregionally, and downcore analyses show that changes in bathymetric distribution of greater than 1km occurred in the Pleistocene (Lohmann, 1978a). Thus, the distribution of deep-sea taxa cannot be controlled by depth alone. The abundances of some deep-sea benthic foraminifera are correlated with water massses (T, S, $O_2$) and properties that covary with water masses independent of depth (Streeter, 1973; Schnitker, 1974, 1979b; Lohmann, 1978b; Corliss, 1979b; Bremer & Lohmann, in press). These water-mass benthic foraminiferal relationships have been extended into the Pleistocene record in order to determine the response of the deep ocean to glacial-interglacial changes

Presently, deep and bottom waters formed in the Norwegian-Greenland Sea overflow the Greenland-Scotland Ridge (Worthington, 1976) and influence hydrography as far away as the deep Pacific (Reid & Lynn, 1971). However, during the last glacial period, the Norwegian-Greenland Sea apparently was covered with ice year-round (Kellogg, 1975), and convection neccessary for the formation of the Norwegian-Greenland Sea overflow (Worthington, 1976) probably could not have occurred. Thus, bottom water in the North Atlantic may have become warmer and "older" (lower $O_2$, higher nutrients, more corrosive) during glacial times. Such a scenario has been supported by $\delta^{18}O$ analyses of benthic foraminifera that indicated warmer bottom-water temperatures during glacials (Streeter & Shackleton, 1979), although this interpretation has been recently disputed (Duplessy et al., 1980). Glacial benthic foraminiferal assemblages in the North Atlantic show evidence of older bottom water. Uvigerina peregrina expanded in abundance and bathymetric distribution during the last glaciation. Since this species is often associated with lowered oxygen values (Lohmann, 1978a; Schnitker, 1979b), this was interpreted as reflecting older bottom water (Streeter & Shackleton, 1979; Schnitker, 1979b). Such an interpretation may be valid, but needs to be evaluated in view of the fact that abundant U.

*peregrina* is not consistantly associated with low oxygen values in the modern ocean (Miller & Lohmann, 1982).

Lohmann (1978a) found very little correlation between glacial-interglacial cycles and benthic foraminiferal response in the South Atlantic. Here, the major faunal change, from a Circumpolar Deep Water (CPDW) fauna, dominated by *Uvigerina peregrina* and *Globocassidulina subglobosa*, to an Antarctic Bottom Water (AABW) fauna, dominated by *Nuttallides umbonifera*, occurred between 700 and 900 kyr ago. This change correlates with general climatic deterioration (Peterson and Lohmann, 1982). In the southeast Indian Ocean, Corliss (1979b) also found no simple relationship of benthic foraminiferal variations and inferred water masses with glacial-interglacial cycles. However, based upon abundant *Uvigerina* and *Melonis* in the glacial southeast Indian Ocean, Corliss (in press) suggested that there was reduced influence of NADW on CPDW during glacial periods. Thus, although benthic foraminifera (and by inference, bottom water masses) show coherent responses to oceanographic changes in the Pleistocene, their response to climatic changes has been more irregular than the response of the surface ocean (Lohmann, 1978b).

It should be noted that the Pleistocene-Recent studies outlined here were conducted on different size fractions (Schnitker, 1974; 1979b: > 123 μm; Lohmann, 1978a,b, Peterson and Lohmann, 1981: > 250 μm; all others, >149 μm). Differences in size fraction can yield significantly different results. This is illustrated by differences between biofacies mapped in the modern western North Atlantic by Streeter (1973) and Schnitker (1974). Similar differences arise in comparing fine-fraction studies of Paleogene benthic foraminifera with studies typically performed on >150 μm size fraction; such biases may be present in the quantitative data reviewed here (Poag, personal communication).

## USES AND LIMITS IN BIOSTRATIGRAPHY, PALEOBATHYMETRY, AND PALEOCEANOGRAPHY

The foregoing discussion of Cenozoic benthic foraminifera is far from complete in detail. The examples discussed here illustrate that benthic foraminifera can yield valuable supportive data to solving geological problems. However, potential pitfalls in using benthic foraminiferal data lie in failure to recognize several factors. 1) Benthic foraminiferal zonal schemes, although satisfactory for local zonations and use in the absence of planktonic age criteria, are liable to be diachronous due to control by local changes in paleoenvironment (substrate, water depth, water masses). 2) Paleobathymetric assignments made using benthic foraminifera are estimates, and probably can be used to indicate Cenozoic depth zonations no finer than marginal marine (and various subdivisions, see Scott & Medioli, 1978), neritic (inner, middle, and outer), bathyal (upper, middle and lower), and abyssal zones (Fig. 1) (although Tjalsma & Lohmann (1982) and Miller (in press) have shown that the abyssal zone may be further subdivided). The uncertainties inherent in making these assignments must be understood before these data can be used in basin analysis (e.g. Van Hinte, 1978; Watts & Steckler, 1979).

3) Although distributions of deep-sea benthic foraminifera correlate with water masses or properties that covary with water masses, changes in assemblages are not diagnostic of the nature of deep-sea circulation changes. Supplementary geological and geophysical data (e.g. isotopic, lithologic, seismic) are needed to help interpret the significance of deep-sea benthic foraminiferal changes.

The benthic foraminiferal case histories reviewed here show a sharp change from earlier studies, which were more of a taxonomic nature. While establishing a firm taxonomic base is crucial to further interpretations of benthic foraminifera, modern foraminiferal studies need to integrate a wide range of data. Isotopic stratigraphy, planktonic biostratigraphy, lithostratigraphy, seismic stratigraphy, magnetostratigraphy, physical and chemical oceanography, geophysically constrained subsidence models ("backtracking") and well-histories ("backstripping"), and plate tectonic reconstructions and history are critical to interpreting benthic foraminiferal changes, while benthic foraminiferal studies form integral parts of such geologic and geophyscial studies. The procedure in the past has been to evaluate fossil benthic foraminiferal changes in view of the distribution of modern taxa, and then to use benthic foraminifera to interpret changes in paleoenvironment, often largely independent of other constraints. While it is valid to base paleoenvironmental changes upon Recent analogues, the case histories illustrate that benthic foraminiferal changes alone cannot be used to unequivocally interpret abyssal circulation and sea-level changes. The need for supportive data will continue to draw paleontologists and paleontologist-to-be out of isolation. With the continued integration of benthic foraminiferal and other geological and geophysical data, "...the near future of this science is rather..." optimistic.

I thank W.A. Berggren, K. Chapneys, W.B. Curry, A.E. Edwards, G.P. Lohmann, C.W. Poag, and R.C. Tjalsma for reviews and B. Sen Gupta and M. Buzas for inviting me to participate in the short course on foraminifera. Permission to reproduce or modify the following figures is gratefully acknowledged: Figs. 2,3, Micropaleontology Press; Fig. 4, Oceanus Magazine; Figs. 5,10, Figs. 6,7 Elsevier Scientific Publishing Co.; Fig. 8, Nature Magazine; Fig. 9, John Wiley & Sons, Inc. This is Woods Hole Oceanographic Institution Contribution number 5194.

# Cenozoic Planktonic Foraminifera: a Paleobiogeographic Summary

Richard K. Olsson
Department of Geological Sciences
Rutgers University, New Brunswick, New Jersey 08903

## INTRODUCTION

The Study of Cenozoic planktonic foraminifera has benefited in the last 1-1/2 decades by Deep Sea Drilling retrieval of many cored sections from the major ocean basins of the world. The stratigraphy of planktonic foraminiferal species in these ocean basins has provided important data on evolutionary lineages of species and biostratigraphic utility of species. The geographic coverage of the oceans has given a clearer picture of the biogeography of species and of past oceanographic conditions. It has become clear that the evolution of planktonic foraminifera was controlled by oceanographic and climatic events related to the redistribution of land and sea. The purpose of this presentation is to summarize the major patterns of the paleobiogeography of planktonic foraminifera during the Cenozoic.

## ORIGIN OF CENOZOIC PLANKTONIC FORAMINIFERA

Evolution of Cenozoic planktonic foraminifera began with the mass extinctions at the end of Cretaceous time. The iridium concentrations at the K-T boundary first reported by Alvarez and others (1980) has focused attention on an extraterrestrial body impact with the Earth as the principal factor that triggered the extinctions in the sea and on the land. Current arguments center on the size and composition of the extraterrestrial body, the physical and geochemical implications of impact and the suddenness with which various groups died off.

The pattern of the extinction of species of Cretaceous planktonic foraminifera is not clearly understood. The recognition of the "*Globigerina*" *eugubina* Zone in the Gubbio section of Italy (Luterbacher and Premoli-Silva, 1964) as the lowest zone of the Cenozoic indicated that all Cretaceous species of the *Abathomphalus mayaroensis* Zone became extinct at the end of the Mesozoic and that none survived morphologically intact into the "*G.*" *eugubina* Zone. Cretaceous species such as *Hedbergella monmouthensis* and *Guembelitria cretacea* have been linked as ancestors of Cenozoic species (Berggren, 1962; Olsson, 1970) but evolutionary transition of these Cretaceous species into their earliest Cenozoic descendents has not yet been described from boundary sections. A newly discovered interval between the *A. mayaroensis* and "*G.*" *eugubina* Zones in Spain (Smit, 1977) has introduced yet another intriguing element to the Cretaceous boundary event. This so-called intermediate zone, some 10 cm in thickness, is characterized by the absence of globotruncanids of the A. mayaroensis Zone and the absence of the globigerinids of the "G." eugubina Zone. Species thought to be indigenous to this zone include Hedbergella monmouthensis (Olsson), Archeoglobigerina blowi Pessagno, Guembelitria cretacea Cushman, and possibly some heterohelicid species. The reporting of A. blowi in this interval is suspect because this species is not known elsewhere to range above the Campanian. Nevertheless, if the Barranco Del Gredero section of Spain represents an expanded interval (albeit very thin) containing data on species that survived the massive extinctions of Cretaceous planktonics, it provides a valuable insight into the origins of the Cenozoic planktonic foraminifera. The data suggest that H. monmouthensis and G. cretacea are definitely linked to the origin of the first planktonic species of the Cenozoic. Whether other Cretaceous species are related to the first Cenozoic species is uncertain; there seems to be no general agreement among workers on the details of phylogeny. The data also indicate that the "G." eugubina assemblage appears abruptly in the stratigraphic record, a fact which must be taken into account in the evolution of this foraminiferal oceanic assemblage.

This author favors a two species model in which H. monmouthensis and G. cretacea are the only links with the Cretaceous. The first true globigerinid of the Cenozoic, Globigerina fringa, is derived from H. monmouthensis by modification of wall structure, principally the buildup of laminae around the pores (i.e. primitive pore pits and cancellate wall structure), a characteristic feature of Cenozoic globigerinids. Woodringina, a genus with initial triserial chambers, is derived from Guembelitria by later biserial chamber development and suppression of the initial biserial morphology. These genera appear further linked by similar wall structure which is hispid and is penetrated with very small pores. Chiloguembelina would appear linked, as well, to Woodringina and Guembelitria by similarity in wall structure and by the asymmetrical aperture. Even though the biserial arrangement of chambers might suggest a linkage with Heterohelix the twisted test and the offset initial chamber (proloculus) of Chiloguembelina can be interpreted as a suppressed initial triserial chamber arrangement. In this case Woodringina can be interpreted as a first step in the evolution of a biserial form from a triserial one.

"Globigerina" eugubina presents a dilemma because it does not possess a clear morphologic similarity with any known Cretaceous species. The wall structure appears closer to Guembelitria but the trochospiral chamber arrangement would suggest linkage with a trochospiral ancestor. Although "G." eugubina has an aperture which is positioned near the axial periphery it is not an extraumbilical opening but rather an areal aperture. The species does not have an open umbilicus like Cretaceous trochospiral species. Furthermore the initial chambers of "G." eugubina are irregularly arranged, unlike the initial coil of planktonic trochospiral forms. All of these morphologic features suggest that "G." eugubina may have been derived from a triserial ancestor, such as G. cretacea, by ontogenetic arrangement of later formed chambers in a coil around an initial triserial portion which would cause the aperture to shift from a triserial position to one near the periphery of the coil.

The abruptness with which the "G." eugubina assemblage appears and the abundance of individuals indicates that evolution occurred rapidly and that the new species were able to quickly populate the oceanic realm in great numbers. In the later case, "G." eugubina, may be regarded as an opportunistic species. "Globigerina" eugubina survived for a brief interval of geologic time before its extinction. Chiloguembelina survived as a conservative stock, never having more than a few species at any given time, before extinction in the Oligocene. Out of Globigerina fringa with its primitive cancellate wall structure and interiomarginal aperture began the radiation of the important groups of Cenozoic globigerinid stocks.

## OCEANOGRAPHIC AND CLIMATIC SETTING OF PALEOGENE

The early Cenozoic ocean originated during the Cretaceous where temperature of ocean waters was moderate and no strong longitudinal gradients existed (Haq, 1981). As the continents drifted further apart oceanographic conditions changed as the Earth's climatic regime adjusted to the changing distribution of land and water masses (Berggren and Hollister, 1977). This, in turn, had a profound effect on the evolution and distribution of species of planktonic foraminifera during the Cenozoic. Many lineages of planktonic foraminifera responded to these environmental pressures by speciation or extinction, thus providing a stratigraphic record that could be finely divided by biostratigraphers into a high resolution zonal scheme.

Several paleoceanographic events appear to have had a strong effect on the evolution of planktonic foraminifera (Text-fig. 1). During the Paleocene and Eocene the Tethys provided a major westward flowing circum-global tropical current (Haq, 1981). Antarctica and Australia separated in the late Paleocene and Australia began a northwards drift. Nevertheless these continents remained attached in an oceanographic sense in that the Drake Passage connection with South America and the South Tasman Rise between Australia and Antarctica prevented development of a circum-polar current (Kennett, et al., 1975). Connection of the North Atlantic with the Arctic Ocean via the Norwegian-Greenland Sea and the Labrador Sea developed during the Paleocene and Eocene (Gradstein and Srivastava, 1980; Talwani and Eldholm, 1977) but Arctic Basin flow of cool waters southward was prevented by shallow submarine topography (e.g. Iceland-Faeroe sill).

The oxygen isotope record (Shackleton and Kennett, 1975; Douglas and Savin, 1975; Savin, 1977) for the Cenozoic indicates that ocean climate during the Paleocene and Eocene was relatively warm and that mild temperatures endured in the polar regions. The record also shows a number of step-like transitions to cooler and warmer conditions. The effect of this Paleocene-Eocene climatic setting was a world ocean dominated by mild climates and a weak temperature contrast between high and low latitudes. Planktonic foraminiferal provinces were broad and less sharply delineated than later in the Neogene. Species diversity in high latitudes was fairly high (Table 1).

## PALEOCEANOGRAPHY OF PALEOCENE AND EOCENE SPECIES

The depletion of species in the ocean following the Cretaceous terminal event set the stage for the radiation of Paleogene planktonic foraminifera. Table 1 shows that speciation was highest in the low latitudes and moderately high in the mid latitudes. Origination of genera was also highest in the low latitudes. Important genera of the Paleocene include *Subbotina*, *Planorotalites*, *Acarinina*, *Morozovella*, *Chiloguembelina*, and *Zeauvigerina*. All of these genera, except *Morozovella*, were cosmopolitan in distribution (Tables 2,3). The keeled *Morozovella* was limited to mid and low latitudes. The number of species of *Subbotina* and *Acarinina* was greater in the warm low latitudes whereas speciation in *Zeauvigerina* appears to have been favored in the high latitudes.

Speciation led to increased diversity in the Early Eocene (Table 1). The majority of genera were cosmopolitan in distribution including *Morozovella* which appeared in the high latitudes for the first time (Table 4). *Zeauvigerina* and *Truncorotaloides* were limited in occurrence to low and mid latitudes. Speciation of *Subbotina*, *Acarinina*, and *Morozovella* was greatest in the low latitudes.

Strong faunal differentiation developed in the Mid Eocene. The greatest number of species and genera appeared in the Mid Eocene than in any other time in the Paleogene (Tables 1,5). A large number of genera had restricted geographic distributions and the warm low latitudes were characterized by a peak number of species. Of importance are species of *Subbotina*, *Acarinina*, *Chiloguembelina*, *Globigerinatheka*, *Hantkenina*, *Orbulinoides*, and *Clavigerinella*. The increase in diversity in the low and mid latitudes did not occur in the high latitudes where it fell somewhat and *Morozovella* disappeared.

Diversity in the Upper Eocene dropped significantly in the low latitudes (Table 1) where a large number of genera disappeared (Table 6). Faunal contrast remained fairly strong between low, mid, and high latitudes (Table 7). *Turborotalia* and *Globigerina* were important genera in speciation in mid and low latitudes. About 1/3 of the Upper Eocene genera were cosmopolitan in distribution.

Haq and others (1977) analyzed the migratory patterns of calcareous nannofossils and planktonic foraminifera during the early Cenozoic. They showed (Text-fig. 2) that migrations towards higher latitudes, caused by climatic warming, ocurred in the early Paleocene (63-60 Ma) and in the late Paleocene-early Eocene (53-49 Ma). Important migrations towards the low latitudes, indicating climatic cooling, took place during the middle Paleocene (60-57 Ma), the middle Eocene (46-44 Ma), and near the Eocene/Oligocene boundary (38-36 Ma). These paleobiogeographic trends agree fairly well with the oxygen-isotopic paleoclimatic data (Haq, 1981).

The data presented here are too generalized to compare with the migratory patterns detected by Haq and others (1977) but the data correlate with the oxygen-isotopic curve (Text-fig. 3). The warming trend that occurred from the late Paleocene into the early part of middle Eocene time compares favorably with the trend (Table 1) of increasing diversity in species and genera during this interval of time. Genera are mostly cosmopolitan in distribution during the late Paleocene and early Eocene. The cooling trend that began in the mid-Eocene and lasted into the late Eocene correlates with the drop in diversity of species and genera that occurred in the low latitudes (Tables 1,6). Also during the mid and upper Eocene a sharper contrast is evident between low, mid, and high latitude assemblages which suggests stronger temperature gradients. Thus it appears that oceanographic events as discussed earlier influenced climatic events which in turn strongly controlled the development of planktonic forminiferal assemblages in the low, mid, and high latitudes. The low latitudes in particular experienced the greatest impact of these events in the origination and extinction of species and genera (Table 6).

## PALEOCEANOGRAPHIC SETTING FOR THE NEOGENE: PALEOGEOGRAPHY OF NEOGENE SPECIES

Paleoceanographic events in the late Eocene and Oligocene had a profound effect on the evolution of and the geographic organization of planktonic foraminifera. The subsidence of the South Tasman Rise in the late Eocene provided a surface connection between the Indian and Pacific Oceans, a connection which led to the development of the circum-Antarctic Current (Kennett et al., 1975). With the increasing isolation of Antarctica in the south polar region the first large scale freezing of surface water led to the development of Antarctic Bottom Water (AABW) near the beginning of Oligocene time. This event is evident in the oxygen isotope record of benthic foraminifers (Text-fig. 3), indicating the formation of the psychrosphere (Benson, 1975; Kennett and Shackleton, 1976).

The opening of the Drake Passage in the Oligocene led to the total thermal isolation of Antarctica (Barker and Burrell, 1977; Kennett, 1977). In the North Atlantic further separation of the Greenland Sea and the Labrador Passage allowed colder Arctic water to flow southward. The Tethys Sea closed. As a consequence global circulation patterns took on the major features of present day (Haq, 1981).

The development of Oligocene planktonic foraminiferal assemblages was strongly controlled by these paleoceanographic events. Diversity of species and genera in low to high latitudes dropped significantly in the lower Oligocene (Table 1). Assemblages were dominated by species of Globigerina and Turborotalia (Table 8). Low latitude assemblages were similar to mid latitude assemblages and about 1/2 of the genera were cosmopolitan in distribution. Most of the typical genera of the Paleocene and Eocene were extinct. By the end of the Oligocene all had become extinct.

Diversity remained low in the upper Oligocene (Tables 1,9). Most of the genera had a cosmopolitan distribution and species of Globigerina and Turborotalia were most abundant. Near the end of Oligocene time conditions had developed for the radiation of the Neogene planktonic foraminifera and the beginning of some important lineages within Globigerinoides, Globigerina, and Globorotalia had begun. Deep Sea Drilling sites in the tropical and southern Pacific Ocean and in the Southern Ocean have provided valuable information on biogeography of planktonic foraminifera in the Neogene

(Kennett, 1978, 1980; Jenkins, 1975; Srinivasan and Kennett, 1981a, b). It seems clear that the evolutionary radiation of Neogene planktonic foraminifera took place in an ocean with sharp temperature contrasts between polar and tropical regions. Distinct tropical to polar foraminiferal assemblages can be identified (Text-figs. 4-8).

In the early Miocene important radiations took place within Globorotalia, Globigerina, Globigerinoides and Globoquadrina (Text-fig. 4). Planktonic foraminifera began to take on a modern looking appearance and a steep diversity gradient developed between high and low latitudes (Table 10).

Radiation of Globorotalia expanded in the mid Miocene and the first keeled species developed in the G. fohsi and G. menardii lineages in tropical waters and in the G. miozea lineage in temperate waters (Text-fig. 5). Orbulina added to the increasing modern aspect of the foraminiferal assemblages. This trend continued in the late Miocene (Text-fig. 6) with the appearance of Pulleniatina, Neogloboquadrina, and Candeina.

Continued radiation occurred in Globorotalia in the Pliocene (Text-figs. 7,8) and important evolutionary developments took place in Pulleniatina and in Neogloboquadrina. Sphaeroidinella dehiscens evolved at the beginning of the Pliocene and N. pachyderma first appeared in polar waters. By the beginning of the Pleistocene almost all of the extant species of today had appeared, all of the extant genera were present, and Globoquadrina and Sphaeroidinellopsis were extinct.

## NEOGENE BIOSTRATIGRAPHY AND PALEOCEANOGRAPHY

Prior to Deep Sea Drilling planktonic foraminiferal zonation was based on stratigraphic sequences from land sections. Planktonic zonation was pioneered by Bolli (1957b, c) who first published a scheme for the Paleocene to Miocene of Trinidad. Since then numerous modifications and other new zonations (notably Blow, 1969) have been proposed. These first zonations which were based on tropical sequences could not be applied to temperate regions where in New Zealand marine sequences Jenkins (1966, 1967, 1971) established an entirely different biostratigraphic scheme. With DSDP drilling in different ocean basins it became obvious that it was not possible to uniformly apply any one zonation because of past oceanographic conditions and problems of dissolution of less resistant species (Jenkins and Orr, 1972). Srinivasan and Kennett (1981a) have summarized the history of development of biostratigraphy of planktonic foraminifera. They pointed out that the results of Deep Sea Drilling show that different ocean basins are marked by differences in biogeographic provinciality and paleoceanography and that foraminiferal assemblages reflect these differences.

Kennett (1973) developed the first Neogene zonal scheme for areas transitional between temperate and tropical regions. Srinivasan and Kennett (1981a) noted in an analysis of Neogene biostratigraphy in the equatorial and South Pacific that very few species appeared restricted to the transitional areas. They concluded that almost all the evolution of Neogene planktonic foraminifera occurred either in the true temperate or tropical areas. They summarized the major lineages of the Neogene (Text-fig. 9). Major lineages that developed in tropical regions include the Globigerina praebulloides phylogeny, the G. druryi phylogeny, the Globigerinoides primordius phylogeny, the G. quadrilobatus phylogeny, and the Orbulina phylogeny. Other important phylogenies include Neogloboquadrina, Sphaeroidinella, Pulleniatina, Globorotalia fohsi, G. menardii, G. tumida and G. truncatulinoides.

Five major temperate lineages were identified; Globigerina woodi, Globigerinoides trilobus, Globorotalia zealandica, G. miozea and Neogloboquadrina pachyderma. One middle latitude lineage was identified, that of Globorotalia kugleri.

The lineage concept is the basis for most of the planktonic foraminiferal zonation used in the Cenozoic. This concept has been applied to the taxonomy of several groups of foraminifera, in particular Globorotalia (Bandy, 1972, 1975; Fleisher, 1974; Srinivasan and Kennett, 1981a). Seven subgenera are erected to represent the different major globorotaloid lineages of the Neogene. These include Fohsella, representing the G. fohsi lineage, Menardella - G. menardii, Globorotalia-G. tumida, Hirsutella-G. hirsuta, Tenuitella - G. gemma, Truncorotalia-G. truncatulinoides, and Globoconella - G. miozea.

In addition to the lineage concept as a basis of biostratigraphy the datum concept has also been widely employed (Hornibrook, 1966; Jenkins, 1966). This concept based on the abrupt evolutionary appearances and the extinction levels of planktonic species has been used to identify correlation planes believed to be isochronous. First appearances (F.A.) and last appearances (L.A.) have gained a preeminent position in biostratigraphy. However, as Srinivasan and Kennett have shown (Text-fig. 9), a F.A. may be a migration event if the appearance occurs outside of the area of the ancestral species. A L.A. may also be a migration event. The distances over which datums are useful will vary considerably depending on the oceanographic adaptation of the species involved.

Nevertheless datums have been widely applied with success and they have been integrated along with zones into a chronological framework which has been intercalibrated with radiometric data and paleomagnetic stratigraphy (Berggren, 1969, 1971, 1972b; Berggren and Van Couvering, 1974, 1978; Saito, 1977).

## EVOLUTION

Much study has been directed at understanding the major evolutionary lineages of planktonic foraminifera in order to better enhance biostratigraphy and correlation. As a result of such analyses over the last 2-1/2 decades most of the major evolutionary lineages of the Neogene are well known (Srinivasan and Kennett, 1981a) and many in the Paleogene are well understood (Stainforth, et al., 1975). Nevertheless, relatively few studies have been conducted on the tempo and mode of speciation. Interest in such studies has been heightened by the introduction of an alternate model to gradualism or the slow steady change of a species population. The alternate model, known as punctuated equilibria (Eldredge and Gould, 1972; Gould and Eldredge, 1977), proposes that speciation occurs rapidly in small peripherally isolated populations after which the new species spreads rapidly to other areas.

The punctuated equilibrium model has appeal in planktonic foraminiferal biostratigraphy because rapid origination and geographic spread of a species is ideal for accuracy in correlation. The fact that a great many species seemingly have abrupt appearances in the stratigraphic record might be best explained by a punctuated mechanism but quantitative testing is needed to support this mode. Deep Sea Drilling with the recovery of nearly continuous stratigraphic sequences has provided an excellent opportunity to study speciation in planktonic foraminifera. Furthermore the climatic history of the oceans is fairly well known through oxygen and carbon isotope studies, thus offering clues to environmental pressures that may have acted upon a population.

Malmgren and Kennett (1981) and Scott (1982) conducted important quantitative studies testing which evolutionary model best explains speciation in the *Globorotalia conoidea* - *G. inflata* lineage. This lineage which begins in the Late Miocene is characterized by the loss of a keel and the rounding or inflation of the periphery in the evolution of *G. puncticulata puncticulata* from *G. puncticulata sphericomiozea* in earliest Pliocene time. In latest Miocene time highly conical keeled tests had speciated in *G. conomiozea* from the ancestral *G. conoidea*. In the late Pliocene further rounding of the periphery and reduction of the number of chambers in the final whorl to 4 occurred in *G. inflata*.

Of the six parameters measured by Malmgren and Kennett gradual change occurred in five. The loss of a keel was a rapid event, however, but this was interpreted as a consequence of the increased conicalness and subsequent peripheral rounding of chambers (i.e. the conical angle had increased to the point where rounding of the periphery resulted with subsequent loss of keel). A reduction in mean size of specimens also accompanied these changes. Malmgren and Kennett correlated the evolution of the *G. conoidea* - *G. inflata* lineage with major paleoceanographic changes of the oceans that occurred in Late Miocene-Early Pliocene time, specifically a change in the turnover of oceans and vertical water structure coincident with northward expansion of Antarctic waters, and possibly the isolation of the Mediterranean with the resulting salinity crisis.

In contrast to Malmgren and Kennett, Scott (1982) interpreted *G. puncticulata sphericomiozea* as a sequence of populations composing the speciation event for *G. puncticulata puncticulata*. He emphasized that the rapid morphologic changes involved in the inflation of chambers and elimination of the keel in *G. puncticulata sphericomiozea* was a major structural modification. He pointed out that the morphologic changes in the speciation of *G. puncticulata puncticulata* probably occurred in less than 0.3 m.y. The ancestral *G. conoidea* with its variant *G. conomiozea* was extant for 6-7 m.y. and the descendant *G. puncticulata puncticulata* for

2.6 m.y. Thus in Scott's view speciation in this lineage simulates the punctuational model except that the tempo of the speciation event is not necessarily as rapid nor is the stratigraphic trace of the event as restricted as is predicted by the model. Also the reduction in number of chambers in ancestral and descendant populations, as also noted by Malmgren and Kennett, may have been gradualistic.

Scott noted that the transforming populations of G. puncticulata sphericomiozea may have been limited to the SW Pacific south of the tropical convergence and that all records of the appearance of G. puncticulata puncticulata north of the convergence possibly reflect migration. Even so, he pointed out that the paleogeographic distribution of G. puncticulata sphericomiozea was still much greater than the punctuational model envisions.

These studies indicate great potential for future studies on the speciation of planktonic foraminifera. It would appear that fruitful lines of research should focus on stratigraphically short-ranging species (the speciating populations) and the paleogeographic distribution of the speciating populations. Did peripheral isolation in the sense of the punctuational model operate in the evolution of planktonic foraminifera? Finally, as alluded to by Scott (1982), it is necessary to understand the mode of speciation when a first appearance (FAD) is used as a biochronological datum.

## ACKNOWLEDGEMENTS

I thank B.U. Haq, J.P. Kennett, S.M. Savin, and M.S. Srinivasan for permission to use their illustrations. Appreciation is given to M.G. Hoffman, A.J. Melillo, and F.B. Szablewski for their assistance in preparation of the manuscript.

TABLE 1 - Diversity during Paleocene, Eocene, and Oligocene

| Age | HIGH-LATITUDE Genera | HIGH-LATITUDE Species | HIGH-LATITUDE Zones | MID-LATITUDE Genera | MID-LATITUDE Species | MID-LATITUDE Zones | LOW-LATITUDE Genera | LOW-LATITUDE Species | LOW-LATITUDE Zones |
|---|---|---|---|---|---|---|---|---|---|
| Late Oligocene | 8 | 19 | 1 | 7 | 24 | 1 | 9 | 22 | 2 |
| Early Oligocene | 5 | 12 | 1.5 | 9 | 23 | 1.5 | 9 | 26 | 2 |
| Late Eocene | 8 | 14 | 2.5 | 14 | 29 | 2.5 | 11 | 35 | 3 |
| Mid Eocene | 9 | 13 | 2.5 | 15 | 32 | 2.5 | 15 | 54 | 5 |
| Early Eocene | 8 | 16 | 1.5 | 9 | 26 | 1.5 | 6 | 43 | 4 |
| Late Paleocene | 5 | 11 | 1 | 7 | 27 | 1 | 6 | 31 | 2 |
| Mid Paleocene | ? | ? | ? | ? | ? | ? | 6 | 25 | 3 |
| Early Paleocene | ? | ? | ? | 4 | 7 | 1 | 6 | 15 | 3 |

TABLE 2 - Species diversity in the LOWER PALEOCENE

| Genera | HIGH LATITUDE | MID LATITUDE | LOW LATITUDE |
|---|---|---|---|
| "Globigerina" | | | 1 |
| Globoconusa | ? | 1 | 1 |
| Subbotina | | 3 | 8 |
| Planorotalites | ? | 1 | 1 |
| Chiloguembelina | | 2 | 2 |
| Woodringina | ? | | 2 |

TABLE 3 - Species diversity in the UPPER PALEOCENE

| Genera | HIGH LATITUDE | MID LATITUDE | LOW LATITUDE |
|---|---|---|---|
| Subbotina | 1 | 2 | 4 |
| Planorotalites | 2 | 4 | 2 |
| Acarinina | 3 | 5 | 9 |
| Morozovella | | 10 | 9 |
| Chiloguembelina | 2 | 4 | 6 |
| Zeauvigerina | 3 | 1 | 1 |
| Globorotaloides | | 1 | |

TABLE 4 - Species diversity in the LOWER EOCENE

| Genera | HIGH LATITUDE | MID LATITUDE | LOW LATITUDE |
|---|---|---|---|
| Subbotina | 1 | 3 | 10 |
| Planorotalites | 1 | 2 | 2 |
| Acarinina | 3 | 6 | 16 |
| Morozovella | 3 | 7 | 11 |
| Chiloguembelina | 1 | 1 | 3 |
| Pseudohastigerina | 2 | 3 | 1 |
| Zeauvigerina | 2 | 1 | |
| Truncorotaloides | 3 | 2 | |
| Globorotaloides | | 1 | |

TABLE 5 - Species diversity in the MIDDLE EOCENE

| Genera | HIGH LATITUDE | MID LATITUDE | LOW LATITUDE |
|---|---|---|---|
| Subbotina | 1 | 4 | 10 |
| Planorotalites | 1 | 3 | 1 |
| Acarinina | 2 | 5 | 10 |
| Morozovella | | 4 | 4 |
| Chiloguembelina | 1 | | 5 |
| Pseudohastigerina | 1 | 1 | 2 |
| Zeauvigerina | 2 | 2 | |
| Truncorotaloides | 3 | 2 | 2 |
| Globigerinatheka | 1 | 2 | 5 |
| Globorotaloides | 1 | 1 | 1 |
| Turborotalia | | 3 | 3 |
| Hantkenina | | 1 | 5 |
| Catapsydrax | | 1 | 1 |
| "Candeina" | | 1 | |
| Testacarinata | | 1 | |
| Globigerina | | 1 | |
| Orbulinoides | | | 1 |
| Clavigerinella | | | 3 |
| Guembelitria | | | 1 |

TABLE 6 - Number of appearances (+) and disappearances (-) of genera during Paleocene, Eocene, and Oligocene

| Age | HIGH-LATITUDE + | HIGH-LATITUDE - | MID-LATITUDE + | MID-LATITUDE - | LOW-LATITUDE + | LOW-LATITUDE - |
|---|---|---|---|---|---|---|
| Late Oligocene | 4 | 1 | 1 | 3 | 1 | 1 |
| Early Oligocene | 1 | 5 | 1 | 6 | 1 | 3 |
| Late Eocene | 1 | 1 | 0 | 2 | 3 | 7 |
| Mid Eocene | 2 | 1 | 7 | 1 | 9 | 0 |
| Early Eocene | 3 | 0 | 2 | 0 | 1 | 1 |
| Late Paleocene | 5? | ? | 4? | 1 | 1 | 1 |
| Mid Paleocene | ? | ? | ? | ? | 2 | 2 |
| Early Paleocene | ? | ? | 4 | 0 | 6 | 0 |

TABLE 7 - Species diversity in the UPPER EOCENE

| Genera | HIGH LATITUDE | MID LATITUDE | LOW LATITUDE |
|---|---|---|---|
| Subbotina | 2 | 4 | 7 |
| Pseudohastigerina | 1 | 1 | 1 |
| Globigerinatheka | 2 | 3 | 5 |
| Globorotaloides | 2 | 2 | 1 |
| Turborotalia | 2 | 6 | 6 |
| Chiloguembelina | 1 | 1 | 3 |
| Zeauvigerina | 2 | 1 | |
| Truncorotaloides | 1 | 1 | |
| Acarinina | 1 | | |
| Catapsydrax | | 2 | 3 |
| Hantkenina | | 2 | 2 |
| Globigerina | | 3 | 5 |
| "Candeina" | | 1 | |
| Testacarinata | | 1 | |
| Planorotalites | | 1 | |
| Cribrohantkenina | | | 1 |
| Globoquadrina | | | 1 |

TABLE 8 - Species diversity in the LOWER OLIGOCENE

| Genera | HIGH LATITUDE | MID LATITUDE | LOW LATITUDE |
|---|---|---|---|
| Subbotina | 1 | 2 | 5 |
| Chiloguembelina | 1 | 1 | 1 |
| Globorotaloides | 2 | 2 | 1 |
| Turborotalia | 3 | 5 | 4 |
| Globigerina | 5 | 8 | 9 |
| Pseudohastigerina | | 1 | 2 |
| Catapsydrax | | 2 | 2 |
| Globoquadrina | | 1 | 1 |
| "Candeina" | | 1 | |
| Cassigerinella | | | 1 |

TABLE 9 - Species diversity in the UPPER OLIGOCENE

| Genera | HIGH LATITUDE | MID LATITUDE | LOW LATITUDE |
|---|---|---|---|
| Chiloquembelina | 1 | 1 | 1 |
| Globorotaloides | 1 | 3 | 1 |
| Turborotalia | 4 | 6 | 4 |
| Cassigerinella | 1 | 1 | 1 |
| Globigerina | 8 | 11 | 9 |
| Catapsydrax | 2 | 1 | 2 |
| Globoquadrina | 1 | 1 | 2 |
| Guembelitria | 1 | | |
| Subbotina | | | 1 |
| Globigerinoides | | | 1 |

TABLE 10 - Species diversity in the MIOCENE and PLIOCENE

| Age | TROPICAL | SUBTROPICAL | TEMPERATE | SUBPOLAR | POLAR |
|---|---|---|---|---|---|
| Late Pliocene | 29 | 21 | 20 | 10 | 2 |
| Early Pliocene | 27 | 27 | 19 | 7 | 1 |
| Late Miocene | 40 | 32 | 25 | 10 | 4 |
| Mid Miocene | 31 | 29 | 26 | 8 | 3 |
| Early Miocene | 30 | 29 | 27 | 10 | 5 |

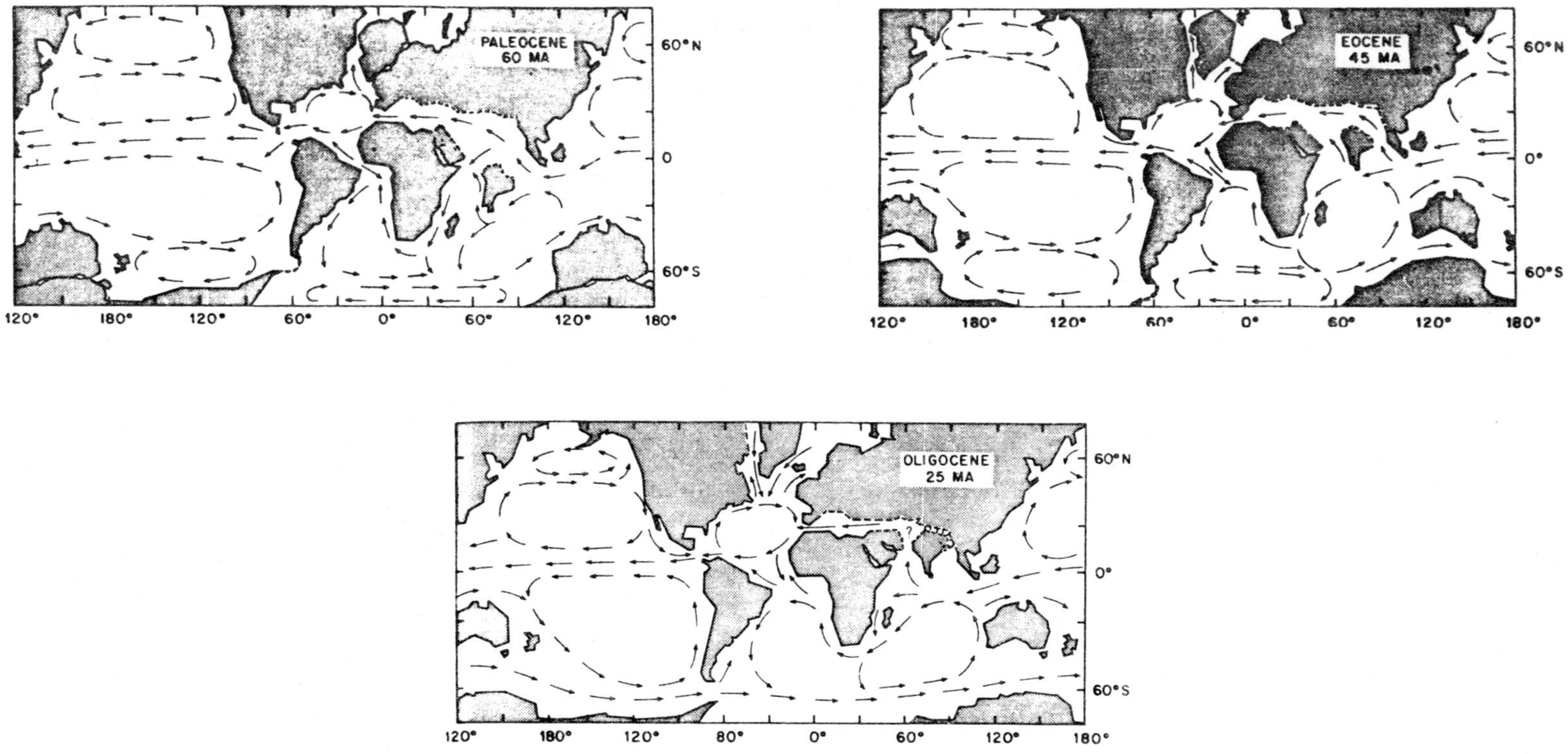


TEXT - FIG. 1 - Inferred global circulation patterns during the early Paleocene, middle Eocene, and late Oligocene (after Haq, 1981).

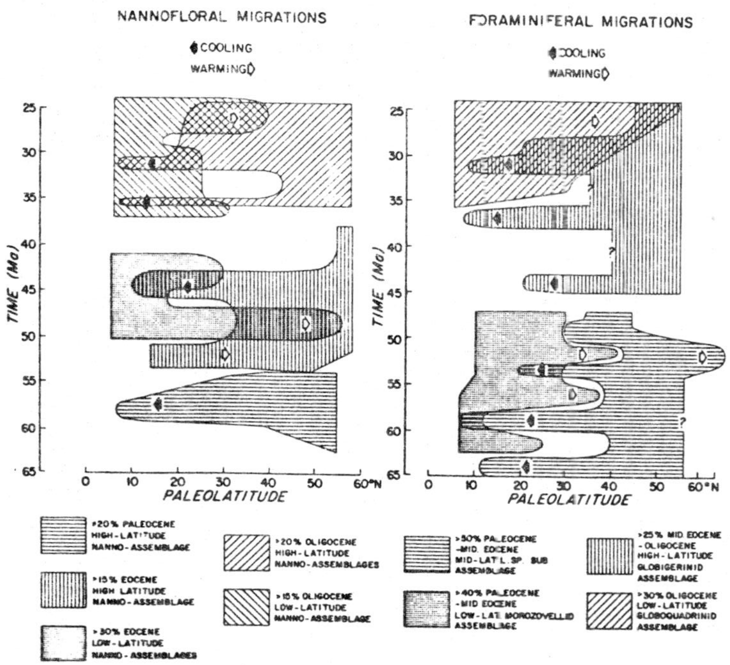


TEXT - FIG. 2 - Faunal migrations through latitudes during the Paleogene in the North Atlantic Ocean (after Haq et al., 1977).

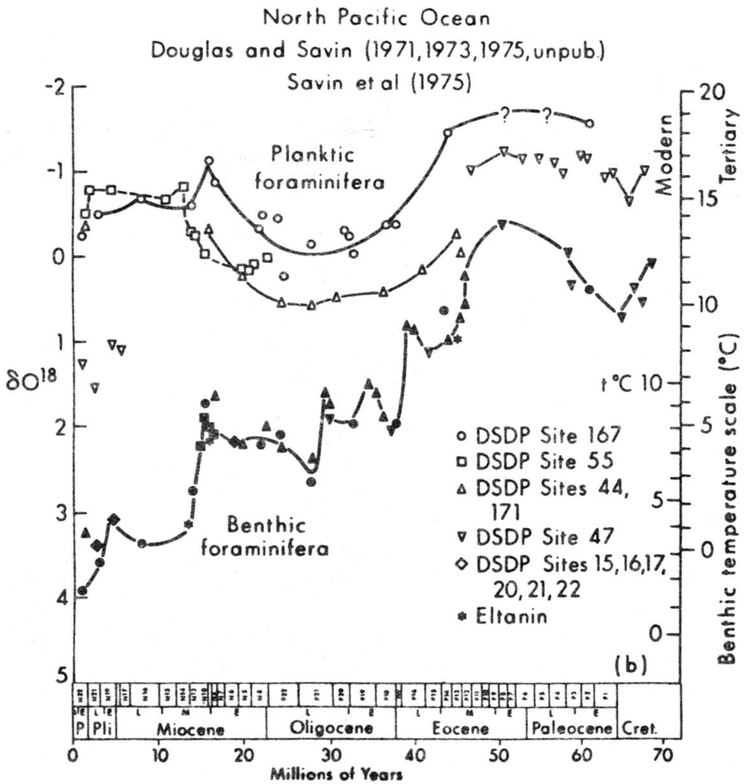


TEXT - FIG. 3 - Cenozoic planktonic and benthic foraminiferal oxygen-isotopic data (after Savin, 1977).

TEXT-FIG. 4

Paleobiogeographic distribution of species

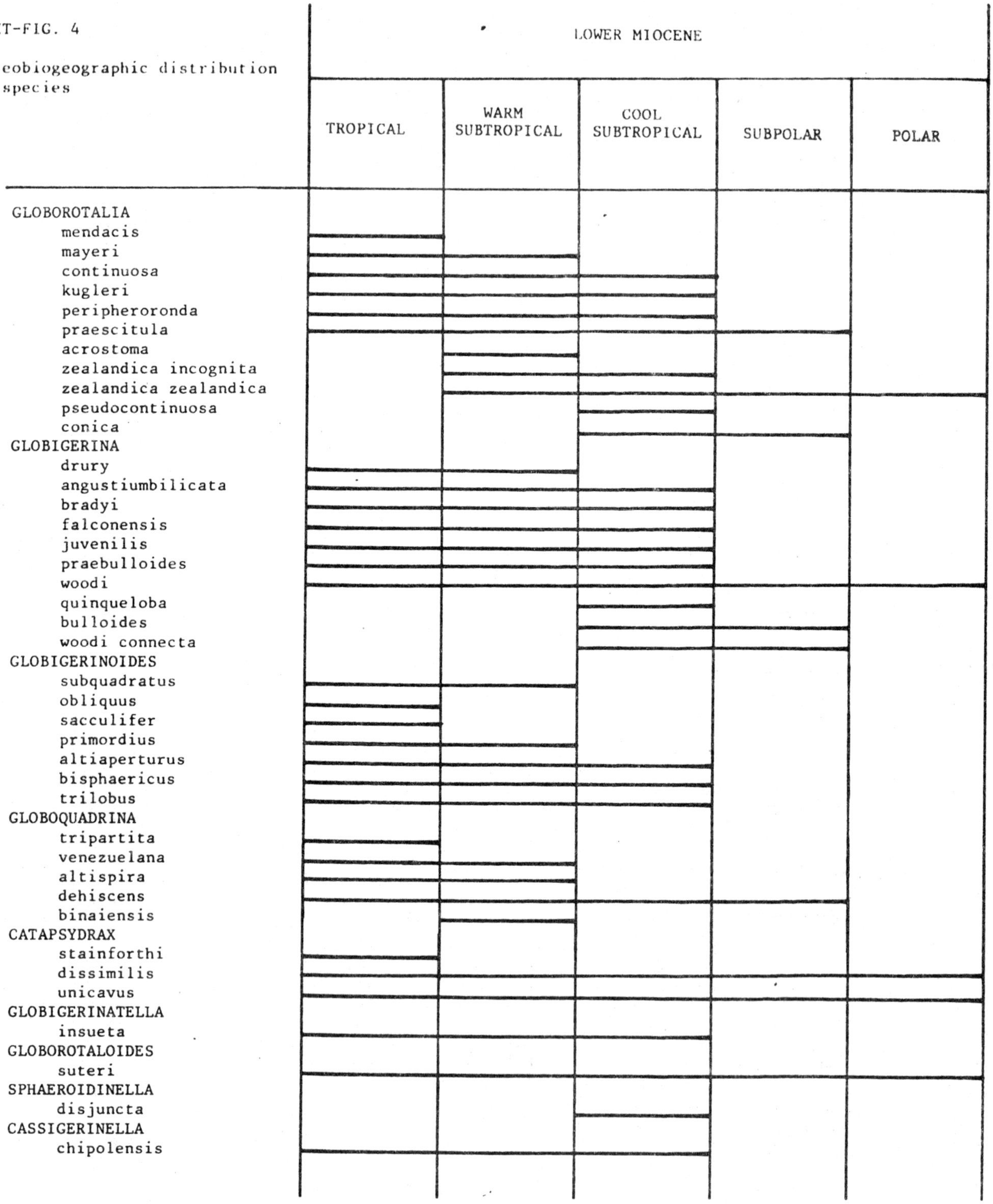

TEXT-FIG. 5

Paleobiogeographic distribution of species

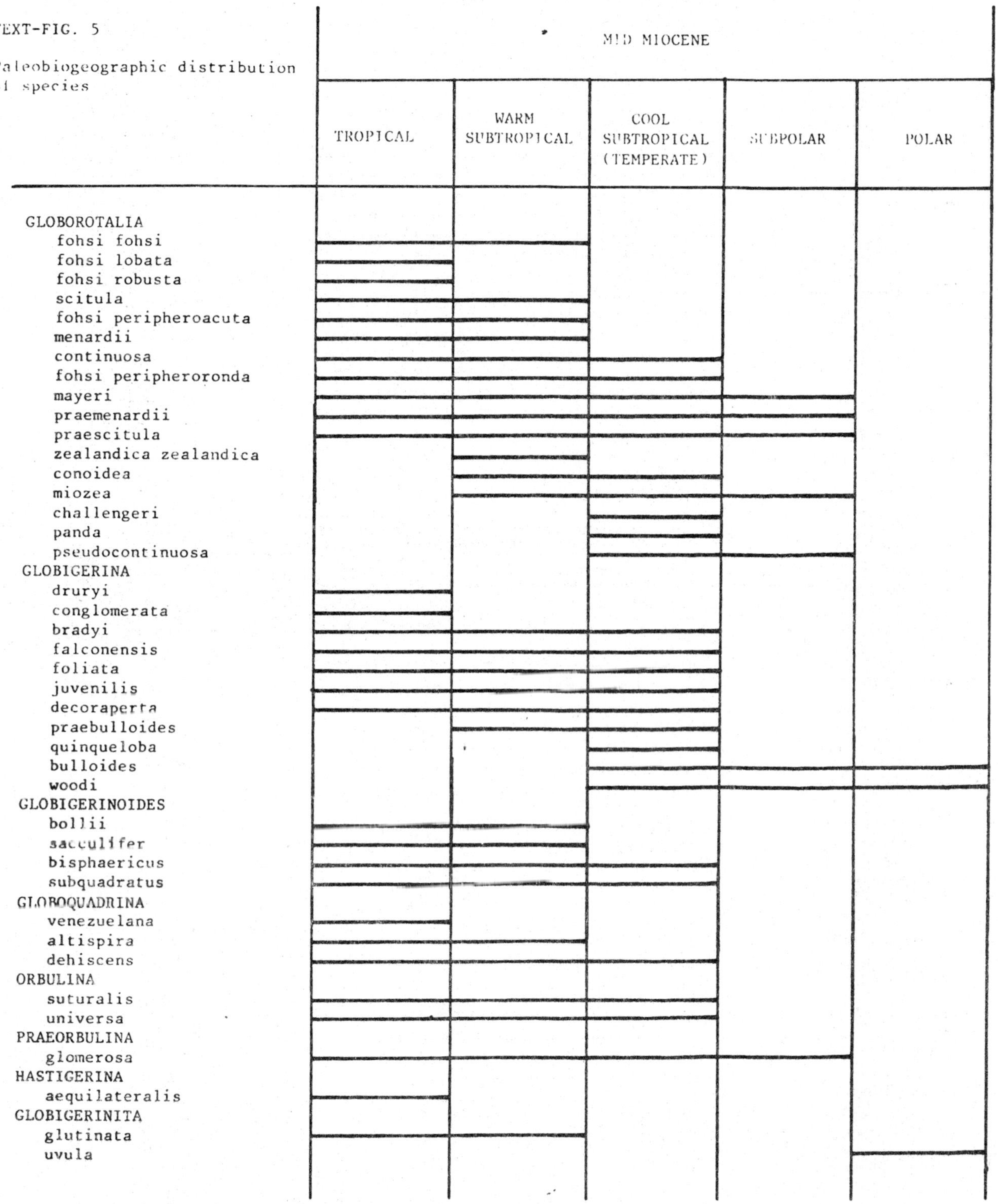

TEXT-FIG. 6

Paleobiogeographic distribution of species

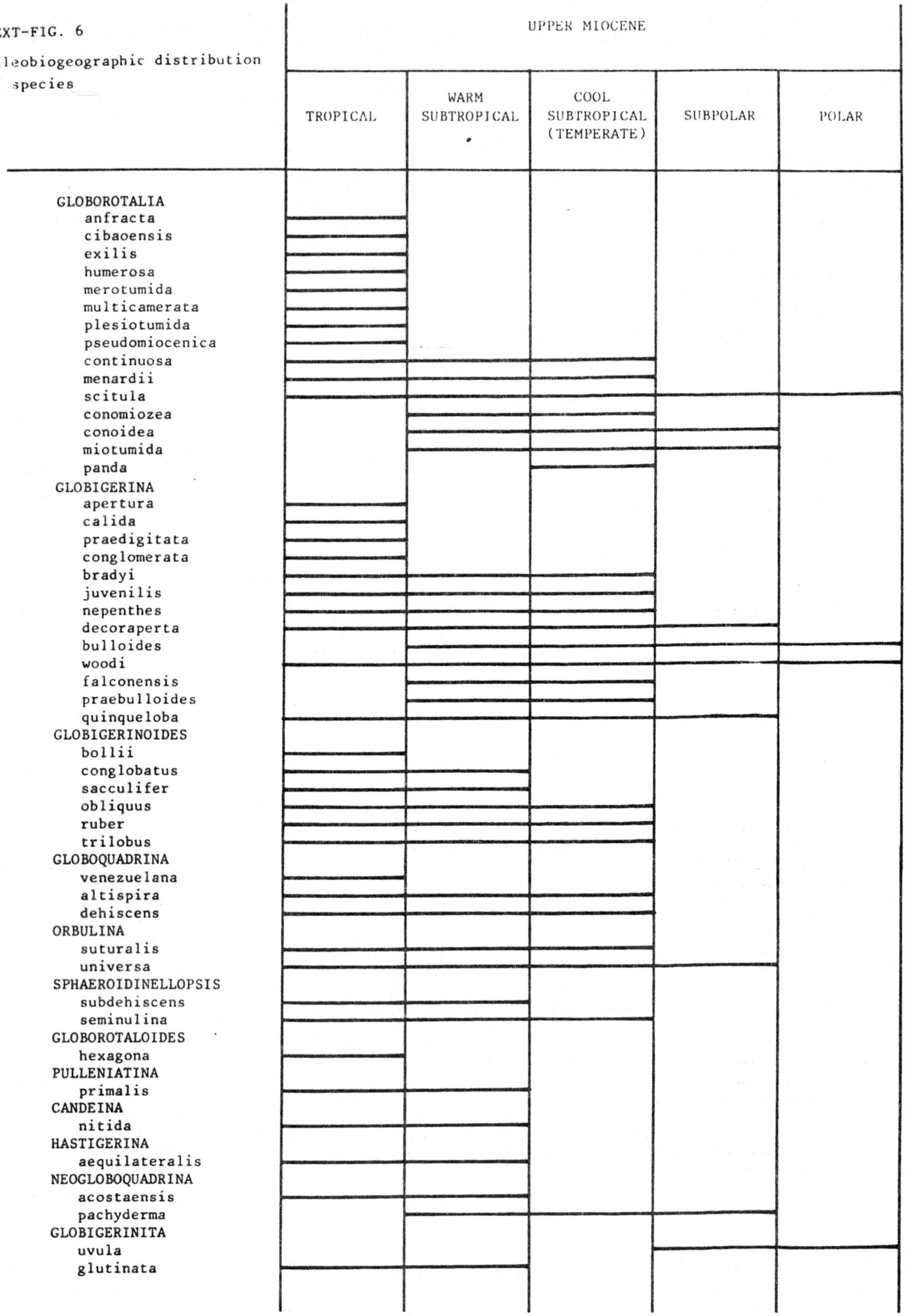

TEXT-FIG. 7

Paleobiogeographic distribution of species

| | LOWER PLIOCENE | | | | |
|---|---|---|---|---|---|
| | TROPICAL | WARM SUBTROPICAL | COOL SUBTROPICAL (TEMPERATE) | SUBPOLAR | POLAR |
| GLOBOROTALIA | | | | | |
| exilis | X | | | | |
| menardii | X | | | | |
| tumida flexuosa | X | | | | |
| margaritae | X | X | | | |
| multicamerata | X | X | | | |
| pseudomiocenica | X | X | | | |
| scitula | X | X | | | |
| crassaformis | | X | X | | |
| conoidea | | X | X | | |
| conomiozea | | X | X | | |
| puncticulata | | X | X | X | |
| GLOBIGERINA | | | | | |
| calida | X | | | | |
| nepenthes | X | X | | | |
| apertura | X | X | X | | |
| decoraperta | X | X | X | | |
| falconensis | | X | X | | |
| bulloides | | X | X | X | |
| quinqueloba | | X | X | X | |
| woodi | | | X | | |
| antarctica | | | | X | |
| GLOBIGERINOIDES | | | | | |
| conglobatus | X | X | | | |
| sacculifer | X | X | | | |
| obliquus | X | X | X | | |
| ruber | X | X | X | | |
| trilobus | X | X | X | | |
| NEOGLOBOQUADRINA | | | | | |
| humerosa | X | | | | |
| acostaensis | X | X | | | |
| pachyderma | | | X | X | X |
| SPHAEROIDINELLA | | | | | |
| dehiscens | X | X | | | |
| CANDEINA | | | | | |
| nitida | X | | | | |
| PULLENIATINA | | | | | |
| primalis | X | X | | | |
| HASTIGERINA | | | | | |
| aequilateralis | X | X | | | |
| SPHAEROIDINELLOPSIS | | | | | |
| seminulina | X | X | X | | |
| subdehiscens | X | X | X | | |
| GLOBOQUADRINA | | | | | |
| altispira | X | X | X | | |
| GLOBIGERINITA | | | | | |
| glutinata | X | X | X | X | |
| uvula | | | X | X | |
| GLOBOROTALOIDES | | | | | |
| hexagonus | X | | | | |

TEXT-FIG. 8

Paleobiogeographic distribution of species

| | UPPER PLIOCENE | | | | |
|---|---|---|---|---|---|
| | TROPICAL | WARM SUBTROPICAL | COOL SUBTROPICAL (TEMPERATE) | SUBPOLAR | POLAR |
| **GLOBOROTALIA** | | | | | |
| exilis | — | | | | |
| miocenica | — | | | | |
| menardii | — | | | | |
| tumida flexuosa | — | | | | |
| multicamerata | — | — | | | |
| pseudomiocenica | — | — | | | |
| tumida | — | — | | | |
| scitula | — | — | — | | |
| tosaensis | — | — | — | — | |
| crassaformis | — | — | — | — | |
| inflata | | | — | — | |
| puncticulata | | | — | — | |
| **GLOBIGERINA** | | | | | |
| calida | — | | | | |
| apertura | — | — | — | | |
| decoraperta | — | — | — | | |
| falconensis | | — | — | | |
| woodi | | | — | | |
| quinqueloba | | | — | — | |
| bulloides | | | — | — | — |
| **GLOBIGERINOIDES** | | | | | |
| fistulosus | — | | | | |
| conglobatus | — | — | | | |
| obliquius | — | — | | | |
| sacculifer | — | — | | | |
| ruber | — | — | — | | |
| trilobus | — | — | — | | |
| **NEOGLOBOQUADRINA** | | | | | |
| acostaensis | — | | | | |
| dutertrei | — | — | — | | |
| dutertrei subcretacea | | — | — | | |
| pachyderma | | | — | — | — |
| atlantica | | | | — | |
| **PULLENIATINA** | | | | | |
| obliquiloculata | — | | | | |
| **SPHAEROIDINELLA** | | | | | |
| dehiscens | — | | | | |
| **GLOBOQUADRINA** | | | | | |
| altispira | — | — | | | |
| **CANDEINA** | | | | | |
| nitida | — | — | | | |
| **ORBULINA** | | | | | |
| universa | — | — | — | | |
| **HASTIGERINA** | | | | | |
| aequilateralis | — | — | — | | |
| **GLOBIGERINITA** | | | | | |
| glutinata | — | — | — | — | |
| uvula | | | — | — | |
| **GLOBOROTALOIDES** | | | | | |
| hexagonus | — | | | | |

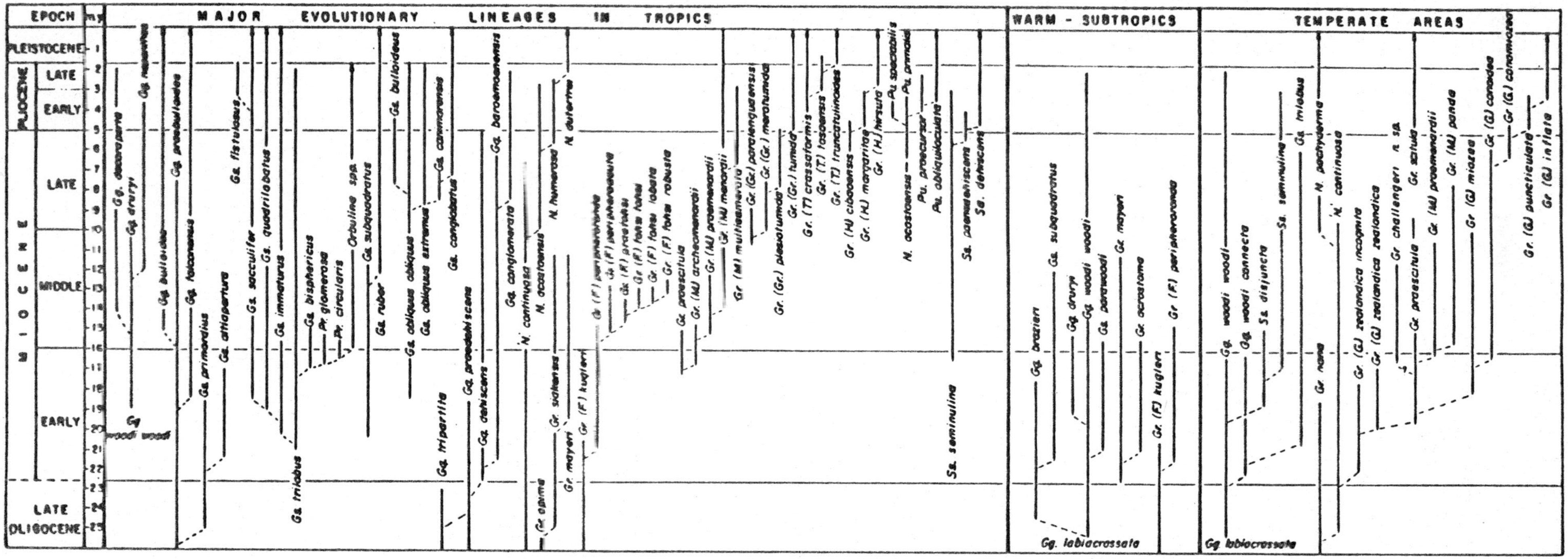


TEXT - FIG. 9 - Major evolutionary lineages for tropical to subantarctic areas of the South Pacific (after Srinivasan and Kennett, 1981).

# Mesozoic Planktic Foraminifera

William E. Frerichs
Department of Geology and Geophysics
University of Wyoming, Laramie, Wyoming 82071

## INTRODUCTION

The evolutionary development of Mesozoic planktic foraminifera is almost exclusively restricted to the Cretaceous Period. Diversity is relatively low until the Late Cretaceous (Turonian to Maastrichtian) when numerous architectural forms are found in the record. Just as remarkable, and more so to this author, is the nearly complete extinction of the group at the end of the Maastrichtian. My purpose will be to attempt to briefly categorize the major areas of study and to describe the numerous problems yet to be solved by workers in the field.

## PALEOECOLOGY

Introduction: Studies of the distribution and ecology of Recent planktic foraminifera are the basis for Mesozoic paleoecological interpretations. Distributional information, both horizontal and vertical, has most often been the basis for Mesozoic studies, all of which are confined to the Cretaceous. At present it is assumed that Cretaceous planktic foraminifera had distributional patterns similar to Recent faunas in that forms with small unornamented tests and globular chambers were found in cold temperate waters, with ornamented tests and a greater diversity found in subtropical and tropical waters. The vertical distribution in subtropical and tropical waters had forms with small unornamented tests and globular chambers living at shallower depths and the larger ornamented forms living at deeper depths. These data are interpreted oxygen isotope studies (Douglas and Savin, 1978) and numerous publications from diverse areas and sedimentological settings.

If the western interior of North America is used as an example, the horizontal and vertical distribution patterns just described can be seen from the Cenomanian to Maastrichtian. In the southern part of the western interior subtropical and tropical species are abundant when a connection to the south was extant (Frerichs and Dring, 1981), but during these same intervals subtropical and tropical species are missing from the northern part of the western interior. The vertical distributional stratification can be found in the marginal marine sediments of western Wyoming and eastern Utah. In these areas major transgressive pulses are represented by sediments which at first contain only benthic foraminifera, then small unornamented planktic species with globular chambers followed by unornamented species with keeled forms. In the regressive phase of the pulse the sequence is reversed. The planktic sequence is substantiated by the benthic sequence of faunas in the same pulse. Therefore, the distributional patterns in today's oceans are similar to the same patterns in an epicontinental sea.

There are, however, many difficulties or potential problems when Recent models are applied to Cretaceous faunas. These include:

1) Continental drift: North America has not rotated or drifted across a significant number of degrees of latitude, but other continents have. Continents must be placed in their Cretaceous positions before the Recent models are applied.

2) Paleoceanography: The Cretaceous and especially the Late Cretaceous was the time of one of the great transgressions of the Phanerozoic. Entirely different current patterns, a much milder temperature gradient, and the oceanic anoxic events (Arthur and Schlanger, 1979) all characterize the Cretaceous and make it distinctly different from the Recent oceans. One result of this phenomenon (coupled with point 3) is that the faunas were not as cosmopolitan even though the subtropical and tropical faunas extended farther north and south. Another is the effect of the anoxic events. At least one author (Wonders, 1980) would explain the evolutionary pattern of Cretaceous planktic foraminifera by the loss of ecological niches to expanded oxygen deficient bottom waters or much thickened oxygen minimum zones at this time. In North America, however, these events do not correlate so well with evolutionary development. In the western interior anoxic events have been documented for the late Cenomanian and early Turonian (Eicher and Worstell, 1970) during which time numerous pulses are documented and in the early Santonian to late Campanian - early Maastrichtian (Frerichs and Dring, 1981) where the event is more continuous. The latter continuous event occurs during the greatest diversification of planktic foraminifera. Therefore, the effect of anoxic events is not yet understood and more study of this phenomenon is needed.

3) Evolutionary Development: Planktic foraminifera first appear in the Mesozoic and it is not until the late Cretaceous that the Recent model can be applied with satisfactory results. Even during the Late Cretaceous there are intervals of time (early Albian and early Turonian) when only the small unornamented forms with globular chambers are present and the model works imperfectly.

# PALEOECOLOGICAL APPLICATIONS

There is space to discuss only a few paleoecological studies. Therefore, I will restrict myself to the western interior and related areas of North America.

Paleoceanography. Eicher and Worstell (1970) recognized three major assemblage zones in Cenomanian and Turonian strata of the western interior. They called these the "lower planktonic zone", the "benthonic zone" and the "upper planktonic zone". At that time they attributed this alternation of small or nonexistent benthic assemblages with rich benthic assemblages to pulses of anoxic conditions. Later Frush and Eicher (1975) found this same sequence of faunas in the late Cenomanian and early Turonian in the Big Bend region of Texas and Mexico. They again interpreted this sequence of faunas as pulses of anoxic events and indicated a preference for water depths in excess of 200 meters during deposition. The Cenomanian-Turonian boundary interval coincides with one of the major anoxic events (Arthur and Schlanger, 1979). It is interesting however that in the Big Bend area of Texas and Mexico Rotalipora becomes extinct in the "lower planktonic zone", but does not become extinct until the lower part of the "benthonic zone" in the western interior. On the basis of the above evidence it is difficult to illustrate a relationship between the anoxic event and planktic foraminiferal evolution.

A similar, but more extensive anoxic event has been recorded in the highly calcareous Smoky Hill Shale of west central Kansas (Frerichs and Dring, 1981). Much of the Santonian and Campanian in this area is characterized by P/B ratios of infinity and benthic specimens are either absent or very rare. This event seems of a much greater duration than those of the Cenomanian and Turonian, but their direct effect upon planktic evolution is not evident or unrecognized at present.

That relatively well defined water masses were extant in the western interior has been illustrated by test porosity (pore space per unit area). Frerichs (1978) attempted to apply work on this topic in the western interior. Following the work of Frerichs et al. (1971) and Bé and others (1976) on Recent faunas he was able to illustrate probable surface water temperature correlation to test porosity and pore density in the western interior, but as a secondary result of his study a different water mass may be defined overlying present day Kansas, Nebraska and part of South Dakota during the late Cenomanian anoxic event.

Faunal Migrations: One of the major areas of paleoecological research in all parts of the geologic column has dealt with changes in the biogeography of species. Most of the Pleistocene work has been involved with studying changes in the geographic distribution of species and the few studies in the Cretaceous have been based on the same change. One of the limiting factors, however has been the lack of detailed biostratigraphic study. Bandy (1967) first called attention to a "keeled line" which indicated the presence of tropical planktic foraminifera farther north than Recent tropical species in sediments of Europe and North America. Again using the western interior as an example, study of Cretaceous planktic foraminifera indicated that the central western interior seaway was a cold temperate province as only six species of planktics had been noted during the Santonian and Campanian (Douglas, 1971). This interpretation was on the information available at that time. Since its publication the Coniacian, Santonian and Campanian sediments of the western interior have been shown to have more than 50 planktic species, many of which are best developed in subtropical and tropical areas (Frerichs and others, 1975, 1977, 1978, 1981, 1982; and Frerichs, 1979, 1980a). Therefore, subtropical conditions are probable for the southern part of the western interior at this time.

Additionally, the occurrence of keeled species in marginal marine areas of western Wyoming and eastern Utah normally seem to mark the peaks of transgressive pulses and mark the deeper waters of the transgressions (Frerichs and Adams, 1972).

Eicher (1967) also has noticed what he considers to be a preference for keeled distributions in deeper water Late Cenomanian sediments of the western interior. Rotalipora is found as far north as central South Dakota and would indicate warmer waters in the western interior during deposition of late Cenomanian sediments. This genus is mainly confined to the central water mass covering Kansas, Nebraska and part of South Dakota, but is not restricted to it.

There are other examples of planktic migrations in the Cretaceous, but as examples it seems best to restrict ourselves to one area. The work of Douglas and Savin (1979) certainly belongs in any discussion of paleoecology of planktic foraminifera, but it has earlier been used to establish the basic parameters upon which so many studies are based.

## ZONATION

Introduction: Any zonation scheme should optimally be worldwide in application. This one criterion although nowhere achievable is particularly difficult to approach. If only North America and Europe are examined certain inherent problems, as with all zonation attempts, are probably most pronounced.

1) Environmental restraints restrict the most elaborate schemes to the subtropical and tropical areas of the Cretaceous and to the moderately deep and deep water parts of these areas. In temperate areas most subtropical and tropical species have environmentally truncated ranges and appearances and extinctions may not be synchronous with subtropical ranges.

2) Due to the recent evolutionary development some species appeared appreciably sooner on one continent. An excellent example of this is the first occurrence of Globotruncana elevata (Brotzen) on the two continents. Van Hinte (1976) places its first appearance in the Santonian of Europe, but Pessagno (1967) first records it in the Campanian.

3) Some genera and species, especially those belonging to the Heterohelicidae have not had sufficient biostratigraphic work completed to establish their ranges. Pseudotextularia for example ranges as old as the early Coniacian, but most authors indicate a Campanian and younger range for the genus. Other examples are commonly found within the Mesozoic planktics and these create many interpretive problems.

Correlation with the stratotypes of Europe: European stage names are traditionally used throughout by Mesozoic planktic workers. They are neither ideal nor characterized by abundant planktic faunas. Numerous gaps occur between stages, and significant faunal breaks at other localities are often used to represent boundaries between adjoining ages. Veerbeck (1977) gives a short review of most of the Cretaceous stages.

The correlation chart and time scale presently used by most workers is that of Van Hinte (1976) although Wonders (1980) has attempted to update and revise his work. Only time will tell if his effort is successful and accepted. At present, I will follow the scheme of Van Hinte realizing that it is imperfect and a best efforts system. His ages and his correlation chart are illustrated in Text-Fig. 1-3. I have added some "superzones" and the temperate zonation of Frerichs and others (1975; 1977; 1978; 1981; 1982) and Frerichs (1979, 1980b).

Evolutionary lineages: In the Late Jurassic-Early Cretaceous Mesozoic planktic assemblages are normally sparse and poorly diversified. The more abundant faunas of the Aptian are characterized by numerous species of Hedbergella and zonation schemes in this part of the section are normally local in nature.

It is only with the beginning of the Turonian that evolutionary lineages of the keeled forms become excellent correlation aids. Most, but not all species of Marginotruncana in the Turonian and early and middle Coniacian have depressed ventral sutures with portici. Middle Coniacian to late Santonian forms have raised ventral sutures and portici, and late Santonian to Maastrichtian species have raised ventral sutures without portici.

One particularly helpful lineage in Coniacian and Santonian sediments is the Marginotruncana concavata primitiva (Dalbiez) - M. concavata concavata (Brotzen) - M. concavata carinata (Dalbiez). M. concavata primitiva is characterized by a nearly flat dorsal side, extraumbilical umbilical aperture, a narrow double keel and a slightly inflated ventral side. It ranges from the early Coniacian to late

Coniacian. M. concavata concavata ranges from the late Coniacian to middle Santonian and differs from M. concavata primitiva in having concave dorsal chambers and greatly inflated ventral chambers. M. concavata carinata differs from M. concavata concavata in having more greatly inflated ventral chambers and in possessing periumbilical ridges. It ranges from the early Santonian to late Santonian.

Another particularly useful lineage during this interval is the Marginotruncana canaliculata (Reuss) - M. pseudolinneana Pessagno - Globotruncana linneana (d'Orbigny). M. canaliculata is characterized by depressed ventral sutures, an extraumbilical-umbilical aperture and a wide double keel. It ranges from middle Turonian to middle Santonian. M. pseudolinneana is similar to M. canaliculata, but has raised sutures and ranges from the middle Turonian to middle Santonian. G. linneana ranges from the middle Santonian to early Maastrichtian and differs from M. pseudolinneana by possessing a true tegillum and umbilical aperture.

The third major lineage amongst the keeled forms is the Marginotruncana sinuosa Porthault-Globotruncana fornicata (Plummer) - G. contusa (Cushman). M. sinuosa has a medium wide double keel, crescentic chambers, raised ventral sutures and an extraumbilical-umbilical aperture. It ranges from the late Turonian to late Santonian. G. fornicata differs from M. sinuosa in having fewer chambers, a more biconvex test, plicated dorsal chambers and an umbilical aperture. Morphotypes similar to G. fornicata first appear in the early Santonian, but the form does not become common until the late Santonian. It ranges to the middle Maastrichtian where it gives rise to G. contusa. This species differs from G. fornicata in being tremendously high dorsally and ranges from the middle Maastrichtian to the end of the Cretaceous.

Other important faunal horizons in the Cretaceous include the:

1) First appearance of Hastigerinoides at or near the Coniacian-Santonian boundary.

2) First appearance of Rugoglobigerina at the base of the Campanian.

Other horizons could be mentioned, but they are more controversial and are in the realm of the specialist.

Correlation with the magnetic reversal scale: Although work is proceeding on the magnetostratigraphy of the Late Jurassic and Cretaceous the results are not yet entirely satisfactory. The interval is mainly normal and reversals are few with mixed results in some parts of the section. The Cretaceous Long Normal ends near the end of the Santonian according to most authors (Vandenberg and Wonders, 1979; Wonders, 1980), but Shive and Frerichs (1974) could not find a reversal at this horizon in the western interior of North America. Reversals in the Campanian and Maastrichtian are also noted and small events have been reported from the Cenomanian and Albian within the "Cretaceous Long Normal Zone". At present the future is promising for successful correlation between the planktic foraminiferal zone and the magnetostratigraphy, but more study is still needed.

## TAXONOMY

The classification of Mesozoic planktic foraminifera is somewhat primitive compared to that of the Cenozoic planktic foraminifera. There is considerable disagreement as to the use of generic names and only a small amount of test structure work has been attempted. This means that higher taxonomic categories are based, for the most part, on modes of coiling, aperture position, and apertural structures.

For this Short Course I will list the genera I recognize and discuss the genera which are often cited, but which I have placed in synonomy with the genera in Table 1.

1) Ventilabrella Cushman is placed in synonomy with Planoglobulina. The morphotypes assigned to this genus are found sporadically in Mesozoic sediments as old as the middle Coniacian (Frerichs and Dring, 1981), but they do not become a consistent element until the early Campanian. Therefore, I assign all consistent stratigraphic occurrences to Planoglobulina and prefer to assign the earlier sporadically occurring forms to their normally easily recognizable counterparts in the biserial genus Heterohelix.

2) Thalminella Sigal is placed in synonomy with Rotalipora. The placement of secondary sutural apertures is often used as a specific character, rather than a generic character, and I prefer to follow this usage at the present time.

3) Dicarinata Porthault is placed in synonomy with Marginotruncana Hofker. The lack of imperforate calcite sutures is not considered to be of generic importance at this time. This eliminates biostratigraphic problems related with this genus, especially in North America.

I have, however, retained at least two genera not always recognized by Mesozoic specialists. These are Clavihebergella and Hastigerinoides. Each is characterized by an elongation of the final one to three chambers, and they are recognized because of their importance in temperate faunas of North America.

A phylogenetic scheme for the genera retained is presented in Text-Fig. 1. Such a scheme allows for numerous polyphyletic genera, especially among the Hedbergella-Whiteinella-Archeoglobigerina-Rugoglobigerina and Praeglobotruncana-Marginotruncana-Globotruncana lineages. These lineages can also be expressed in terms of apertural modification. The boundaries between these structural features are transitional, but they represent the basis for the phylogenetic scheme of trochospirally coiled genera presented in Text-Fig. 3.

One problem with this classification is that the apertural flaps, portici or tegilla are not often preserved. Therefore, except in especially abundant or well preserved material, generic assignments must be made on aperture position, a feature which may change during the ontogeny of an individual. Consequently, in small assemblages such as are found in marginal environments, identifications may be difficult.

TABLE 1

SYSTEMATIC PALEONTOLOGY

Phylum PROTOZOA
Subphylum SARCODINA
Class RHIZOPODEA
Subclass GRANULORETICULOSIA
Order FORAMINIFERIDA
Superfamily GLOBIGERINACEA

Remarks. The diagnosis of Loeblich and Tappan (1964, C652) as amended by Pessagno (1967, p. 256-258) is followed herein.

Family HETEROHELICIDAE

Type genus. - *Heterohelix* Ehrenberg, 1843.
Remarks. - The diagnosis of Loeblich and Tappan (1964, p. C652) is accepted herein.

Subfamily GUEMBELITRIINAE

Type genus. - *Guembelitria* Cushman, 1933.
Remarks. - The diagnosis of Loeblich and Tappan (1964, C652) is followed herein.

Genus GUEMBELITRIA Cushman, 1933

Type species. - *Guembelitria cretacea* Cushman, 1933.
Range. - Lower Cretaceous - Eocene.

Subfamily HETEROHELICINAE

Type genus. - *Heterohelix* Ehrenberg, 1843.
Remarks. - The diagnosis of Pessagno (1967, p. 258) is followed herein.

Genus HETEROHELIX Ehrenberg, 1843

Type species. - *Spiroplecta americana* Ehrenberg, 1843.
Remarks. - The diagnosis of Loeblich and Tappan (1964, pp. C652-C654) is followed herein.
Range. - Barremian to Maastrichtian.

Genus GUBLERINA Kikoine, 1948

Type species. - *Gublerina cuvilieri* (= *Ventilabrella missima* Cushman and Church, 1929).
Remarks. - The diagnosis presented by Loeblich and Tappan (1964, pp. C654-655) is followed herein.
Range. - Late Cretaceous.

Genus PSEUDOTEXTULARIA Rzehak, 1891

Type species. - *Cuneolina elegans* Rsehak, 1891.
Remarks. - The diagnoses presented by Gallitelli (1957, p. 138) and by Loeblich and Tappan (1964, p. C656) are followed herein.
Range. - Campanian to Maastrichtian.

Genus RACEMIGUEMBELINA Gallitelli, 1957

Type species. - *Guembelina fructicosa* Egger, 1900.
Remarks. - The diagnoses of Gallitelli (1957, p. 142) and of Loeblich and Tappan (1964, p. C656) are followed herein.
Range. - Late Maastrichtian

Genus PLANOGLOBULINA Cushman, 1927

Type species. - *Guembelina acervulinoides* Egger, 1899.
Remarks. - The diagnoses of Gallitelli (1957, pp. 141, 142) and Loeblich and Tappan (1964, p. C655-656) are followed herein.
Range. - Early Coniacian to Maastrichtian.

Family PLANOMALINIDAE Bolli, Loeblich, and Tappan, 1957

Type genus. - *Planomalina* Loeblich and Tappan, 1946.
Remarks. - The diagnosis of Loeblich and Tappan (1964, p. C656) is accepted herein.
Range. - Aptian to Maastrichtian.

Genus HASTIGERINOIDES Bronnimann, 1952

Type species. - *Hastigerinoides alexanderi* Cushman, 1931.
Remarks. - The diagnoses of Bolli, et al., (1957, pp. 24, 25) and Loeblich and Tappan (1964, p. C658) are followed herein.
Range. - Early Campanian.

Genus GLOBIGERINELLOIDES Cushman and Ten Dam, 1948

Type species. - Globigerinelloides algeriana Cushman and Ten Dam, 1948.

Remarks. - The diagnoses of Loeblich and Tappan (1964, p. C656) and Pessagno (1967, p. 274) are followed herein.

Range. - Aptian to Maastrichtian.

Family SCHACKOINIDAE Pokorny, 1958

Type genus. - Schackoina Thalmann, 1932.

Remarks. - The diagnosis for this family presented by Loeblich and Tappan (1964, p. C658) is followed herein.

Genus SCHACKOINA Thalmann, 1932

Type species. - Siderolina cenomana Schacko, 1897.

Remarks. - The diagnosis of Loeblich and Tappan (1964, p. C658) is followed herein.

Range. - Aptian to Maastrichtian

Family ROTALIPORIDAE Sigal, 1958 emended

Type genus. - Rotalipora Brotzen, 1942.

Remarks. - The diagnosis of Loeblich and Tappan (1964, p. C659) is followed herein.

Subfamily HEDBERGELLINAE Loeblich and Tappan, 1961

Type genus. - Hedbergella Bronnimann and Brown, 1958.

Remarks. - The diagnosis presented by Loeblich and Tappan (1961a, p. 273; 1964, p. C659) is followed herein.

Genus HEDBERGELLA Bronnimann and Brown, 1958

Type species. - Anomalina lorneiana d'Orbigny var. trochoidea Gandolfi, 1942.

Remarks. - The diagnosis presented by Loeblich and Tappan, 1964, p. C659) is followed herein.

Range. - Hautervian to Campanian; Maastrichtian (?).

Genus CLAVIHEDBERGELLA Banner and Blow, 1959

Type species. - Hastigerinella subcretacea Tappan, 1943.

Remarks. - The diagnosis of Loeblich and Tappan (1961a, pp. 278, 279; 1964, p. C659) is followed herein.

Range. - Aptian to Coniacian.

### Genus PRAEGLOBOTRUNCANA Bermudez, 1952

Type species. - *Globorotalia delrioensis* Plummer, 1931.
Remarks. - The diagnosis presented by Loeblich and Tappan (1961a, p. 280; 1964, p. C659) is followed herein.
Range. - Albian to Lower Turonian.

### Subfamily ROTALIPORINAE Sigal, 1958

Type genus. - *Rotalipora* Brotzen, 1942.
Remarks. - The diagnosis presented by Loeblich and Tappan (1961a, p. 290; 1964, p. C659) is followed herein.
Range. - Albian to Cenomanian.

### Genus ROTALIPORA Brotzen, 1942

Type species. - *Rotalipora turonica* Brotzen, 1942 (= *Rotalipora cushmani* (Morrow, 1934).
Remarks. - The diagnosis of Loeblich and Tappan (1961a, p. 296; 1964, pp. C659-661) is followed herein.
Range. - Cenomanian.

### Family MARGINOTRUNCANIDAE Pessagno

Type genus. - *Marginotruncana* Hofker (1956) (emended) by Pessagno (1967).
Remarks. - The diagnosis presented by Pessagno (1967, p. 298) is followed herein.

### Genus WHITEINELLA Pessagno, 1967

Type species. - *Whiteinella archaeocretacea* Pessagno.
Remarks. - The diagnosis of Pessagno (1967, p. 298) is followed herein.
Range. - Turonian to Maastrichtian.

### Genus MARGINOTRUNCANA Hofker, 1956

Type species. - *Rosalina marginata* Reuss, 1845.
Remarks. - The diagnosis of Pessagno (1967, p. 299) is followed herein.
Range. - Early Turonian to Late Santonian.

### Family GLOBOTRUNCANIDAE Brotzen, 1942

Type genus. - *Globotruncana* Cushman, 1927.
Remarks. - The diagnosis presented by Loeblich and Tappan (1964, p. 662) and the emendations of Pessagno (1967, p. 314) are followed herein.

Genus ARCHAEOGLOBIGERINA Pessagno, 1967

Type species. - Archaeoglobigerina blowi Pessagno, 1967.
Remarks. - The diagnosis of Pessagno (1967, p. 315) is followed herein.
Range. - Coniacian to Maastrichtian

Genus GLOBOTRUNCANA Cushman, 1927

Type species. - Pulvinulina arca Cushman, 1926.
Remarks. - The diagnosis of Pessagno (1967, p. 318) is followed herein.
Range. - Early Santonian to Maastrichtian.

Genus RUGOGLOBIGERINA Bronnimann, 1952

Type species. - Globigerina rugosa Plummer, 1927.
Remarks. - The diagnosis of Pessagno (1967, p. 364) is followed herein.
Range. - Campanian to Maastrichtian.

Genus RUGOTRUNCANA Bronnimann and Brown, 1956

Type species. - Rugotruncana tilevi Bronnimann and Brown, 1956 (= Rugotruncana subcircumnodifer (Gandolfi) 1955).
Remarks. - The diagnosis of Pessagno (1967, p. 368) is followed herein.
Range. - Late Campanian to Maastrichtian.

Family ABATHOMPHALIDAE Pessagno

Type genus. - Abathomphalus Bolli, Loeblich and Tappan, 1957.
Remarks. - The diagnosis of Pessagno (1967, p. 371) is followed herein.

Genus ABATHOMPHALUS Bolli, Loeblich and Tappan, 1957.

Type species. - Globotruncana mayaroensis Bolli, 1951.
Remarks. - The diagnosis of Pessagno (1967, p. 371) is followed herein.
Range. - Late Maastrichtian.

Genus GLOBOTRUNCANELLA Reiss, 1957

Type species. - Globotruncana citae Bolli, 1951 (= Globotruncana havanensis Voorwijk, 1937.
Remarks. - The diagnosis as emended by Pessagno (1967, p. 372-373) is followed herein.
Range. - Late Campanian to Maastrichtian.

| | ZONES | | BIOHORIZONS |
|---|---|---|---|
| ALBIAN | LC 19 | P. buxtorfi – R. apenninica | ← TICINELLA |
| | LC 18 | R. ticinensis – P. buxtorfi | ← ROTALIPORA |
| | LC 17 | T. (B.) breggiensis | |
| | LC 16 | T. praeticinensis | |
| | LC 15 | T. bejaouaensis – Gl. gyr. – T. primula | |
| | LC 14 | T. bejaouaensis – Gl. gyroidinaeformis | |
| APTIAN | LC 13 | Gl. ferreolensis–T. bejaouaensis | ← TICINELLA |
| | LC 12 | H. trocoidea – Gl. ferreolensis | |
| | LC 11 | Gl. algerianus | |
| | LC 10 | Schackoina cabri | |
| | LC 9 | Gl. blowi | ←GLOBIGERINELLOIDES |
| BARREMIAN | LC 8 | H. sigali | |
| | LC 7 | H. aff. H. simplex | |
| NEOCOMIAN | LC 6 | "H." hoterivica | ← HEDBERGELLA |

Text - Fig. 1 and 2(on next page). Cretaceous planktic foraminiferal zonation (after Van Hinte, 1976).

| | ZONES | | BIOHORIZONS |
|---|---|---|---|
| MAASTRICHTIAN | UC 17 | Ga. mavaroensis | |
| | UC 16 | G. contusa | |
| | UC 15 | G. stuarti | |
| | UC 14 | G. gansseri | |
| | UC 13 | G. scutilla | |
| CAMPANIAN | UC 12 | G. calcarata | |
| | UC 11 | G. subspinosa | |
| | UC 10 | G. stuartiformis | |
| CAMPANIAN / SANTONIAN | UC 9 | G. elevata | ← SIGALIA |
| SANTONIAN | UC 8 | G. concavata - G. elevata | |
| CONIACIAN | UC 7 | G. sigali - G. concavata | |
| | UC 6 | G. renzi - G. sigali | |
| TURONIAN | UC 5 | "G." helvetica | |
| | UC 4 | H. lehmanni ("zone a grandes globigerines") | ← PRAEGLOBOTRUNCANA |
| CENOMANIAN | UC 3 | R. cushmani | ← ROTALIPORA |
| | UC 2 | R. gandolfii - R. reicheli | |
| | UC 1 | R. gandolfii - R. greenhornensis | |

Text - Fig. 2

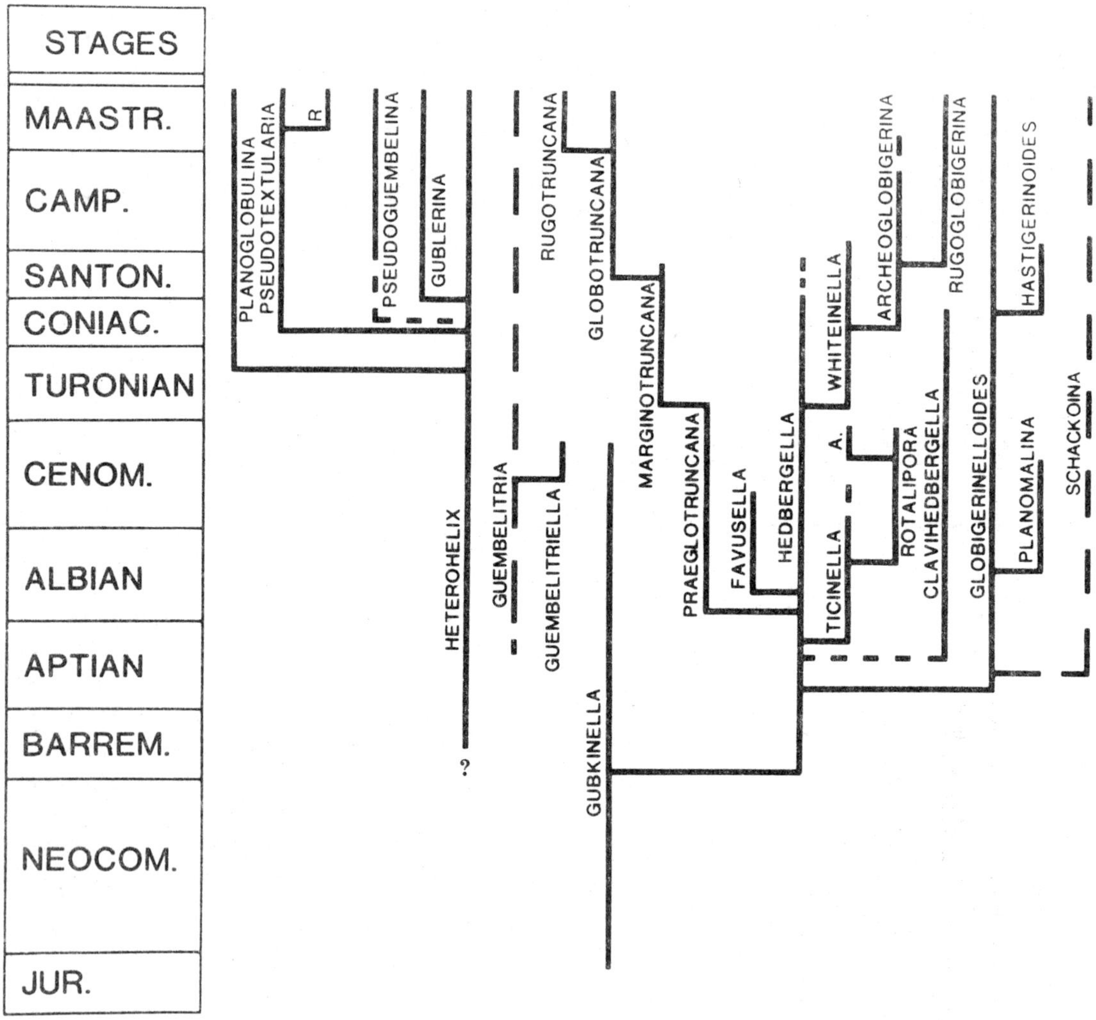


Text - Fig. 3. Cretaceous planktic foraminiferal phylogeny. R = Racemguembelina and A = Anaticinella.

# Paleozoic Foraminifera - Fusulinids

Charles A. Ross
Gulf Oil Exploration and Production Company
P. O. Box 36506, Houston, TX 77236

## INTRODUCTION

Fusulinacea are an extinct superfamily of small to large, to very large, calcareous Foraminiferida. They were entirely marine and made their first appearance near the end of Early Carboniferous time (Text-fig. 1). They thrived during Middle and Late Carboniferous and Permian time. During that 100 million year existence, they evolved more than 150 genera and more than 6000 species before starting a decline in the later part of the Permian and finally becoming extinct near the end of that period. Members of the Fusulinacea became giants, typically reaching 1 cm or more in length, and some became enormous giants, reaching 4 and 5 cm in length in the mid-Permian, and a few as much as 10 cm! Not bad for a single celled organism. Fusulinaceans are usually large enough that field geologists can spot them quickly in their stratigraphic units.

As a group, Fusulinacea are characterized by rapid evolutionary changes in distinctive morphological features and this has enabled paleontologists to establish detailed stratigraphic zonations and phylogenies for them. Most of these features are studied in oriented thin sections (Text-fig. 2). The most useful of these is an axial section oriented so that the section lies in the axis of coiling and thus passes through the proloculus. A sagittal section is perpendicular to the axial section and also passes through the proloculus. It shows the spiral shape of the shell and the outline of each chamber. A third type of oriented section, the tangential section, is cut so the plane of the section is parallel, or nearly so, to the axis of coiling but cuts a plane just above the outer surface of a previous volution. Tangential sections show the relationships at the toes of septal folds, particularly whether these toes have been resorbed to form little passageways called cuniculi.

## MORPHOLOGY

The general structure and appearance of each of the seven families is illustrated in Text-figure 2. The shell typically is constructed as a simple, planispiraled coil and in nearly all families and most subfamilies this growth form is retained throughout the life of the individual. The small, spherical proloculus is the first formed part of the shell and has a large pore that opens into the first chamber. The orientation of that prolocular pore lies at about 90° to what becomes the axis of coiling. As in other chambered Foraminiferida, growth of the protoplasm in fusulinaceans was accommodated by periodic additions of new shell in the form of chambers. These were added in a planispiral pattern in most fusulinaceans so that axial and sagittal sections show

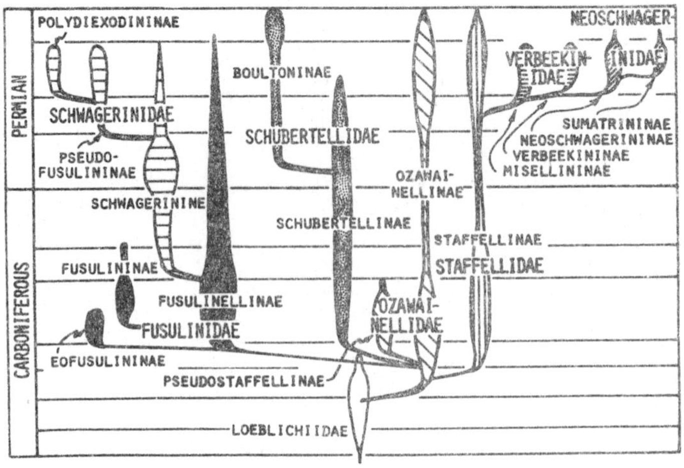


Text-figure 1. Distribution of fusulinacean families and subfamilies and their relative importance during Carboniferous and Permian time.

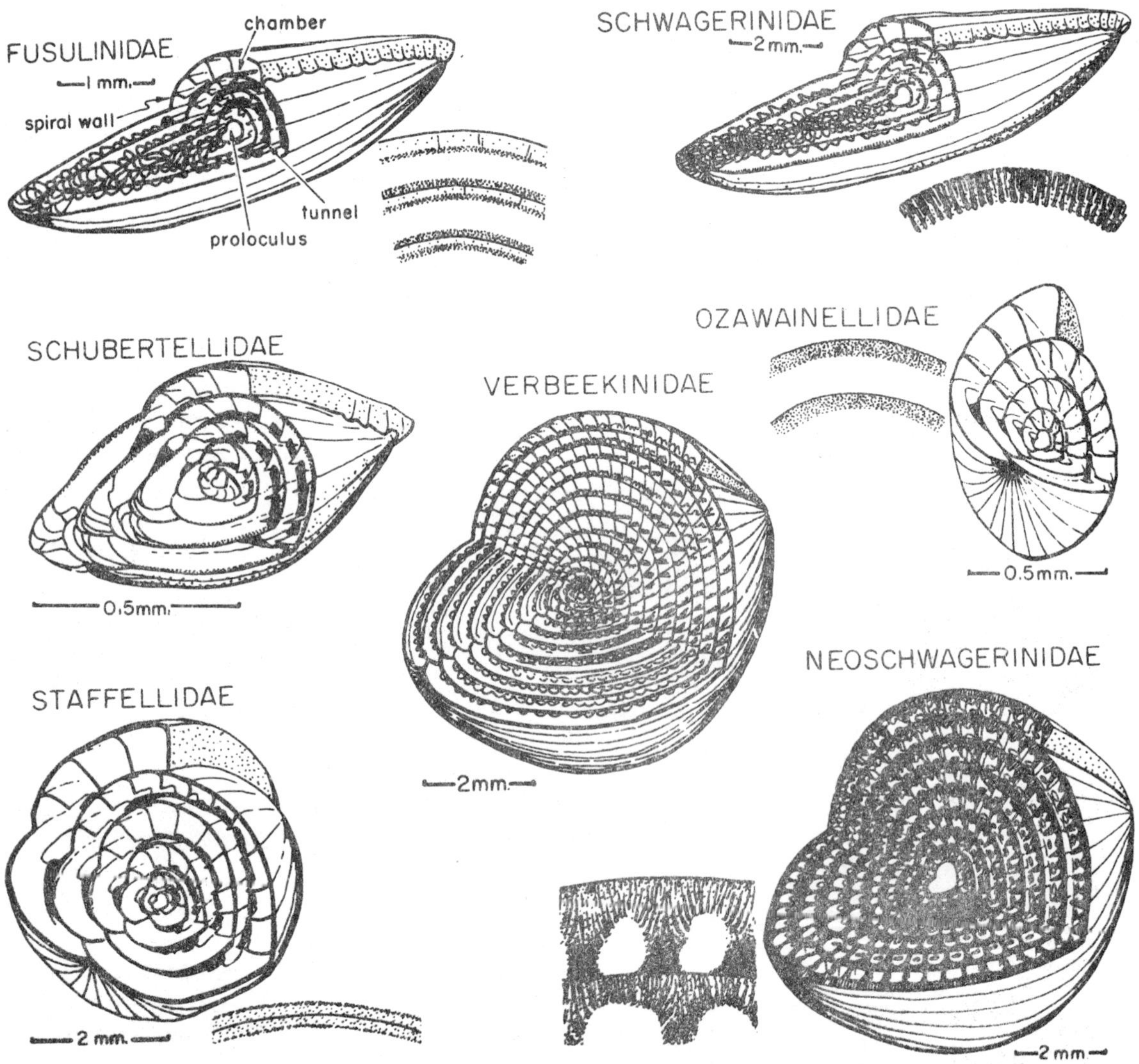


Text-figure 2. Representatives of seven families of fusulinaceans as seen in cutaway diagrams showing internal axial and sagittal features. The inserts show enlargement of walls to illustrate differences in wall structure and comparative thickness.

a great regularity. In most genera, succeeding chambers are slightly larger than previous ones and they overlap at the ends of the axis of coiling so the shell becomes elongate along the axis. After several volutions of growth they begin to look like grains of wheat or barley.

Some fusulinaceans, particularly some of the smaller genera in the middle and later parts of the Permian, have a tendency to lose their regular coiling patterns and become uncoiled. In the Upper Permian of Texas, for example, the boultonin Paradoxiella becomes uncoiled and forms a flattened, disk-shaped shell.

The spiral wall (spirotheca) of fusulinaceans contains many features which have proven useful in separating the families (Text-fig. 3). Initially, the shape and outline of the wall are formed by a thin, dark, organic-rich layer (tectum). Beneath this a thicker, calcareous-rich layer (diaphanotheca) is deposited which is translucent, at least to some degree. The tectum and diaphanotheca form the two basic layers of the spiral wall in fusulinaceans and it is to these layers that various modifications are added. The wall of each chamber is divided into two parts, the spirotheca, which forms the outer surface of the volution, and the septum, which is the front of the chamber and is covered by the succeeding chamber. The spirotheca is usually thicker and has many, small mural pores (about 2$\mu$m) that extend through it to the exterior. The septa are usually thinner, darker, and have larger, septal pores (about 10$\mu$m).

The septa may be nearly planar from axial pole to axial pole, particularly in some of the Middle Carboniferous lineages. In most genera, however, the septa are folded, usually more strongly folded near the poles than across the equatorial region. In some the septa are so strong folded and so closely spaced that the folds of one septa overlap those of adjacent septa along the base of the chambers. These places of septal overlap may be resorbed in some Permian genera of Schwagerinidae to form low, secondary passageways, call cuniculi, leading through the shell. The height and intensity of these septal folds are useful in determining stages of evolution and in determining species assignments.

The three earliest families, Ozawainellidae, Fusulinidae, and Staffellidae, have well developed tectum and diaphanotheca. In the early Ozawainellidae, these two layers are relatively thin. In the early Staffellidae, the diaphanotheca is relatively thicker, but it is also typically strongly recrystallized and appears coarsely granular. In some silicified specimens from Atokan rocks from northwestern United States, this diaphanothecal wall is exceptionally well preserved by siliceous replacement and shows closely spaced, fine tubules that extend through the diaphanotheca into the tectum to form mural pores. Early Fusulinidae show the gradual development of secondary calcareous deposits within the chambers. These deposits (tectorial deposits) are linings within the chambers. At first they coat only the floor of the chambers so that the wall of all but the last volution appears to have three layers (Text-fig. 3a). Members of the fusulinid lineage gradually evolve tectorial deposits that coat the sides of the septa and finally coat the tops of the chambers as well (Text-fig. 3b) which, except for the outermost volution, gives the wall a four-layered appearance. The three-layered wall is called a profusulinellid wall and the four-layered wall

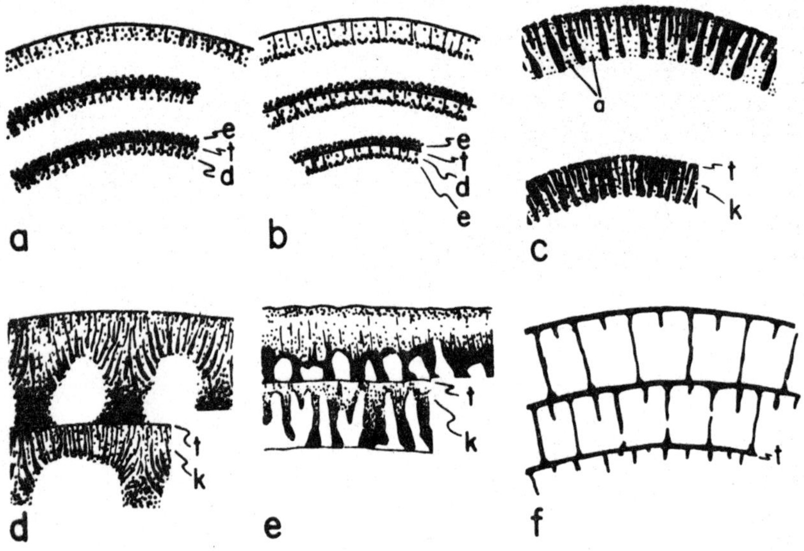


Text-figure 3. Wall structure of various families of fusulinaceans, see text for discussion. a, alveoli; d, diaphanotheca; e, tectorial deposits; k, keriotheca; t, tectum.

is called a fusulinellid wall. Because these secondary wall layers (tectorial deposits) form a continuous gradation in some lineages but not in others within the Fusulinidae, they are commonly hard to interpret.

A significantly different modification occurs in the lineage that forms the Schwagerinidae. In this family, the diaphanotheca thickens and becomes modified into a series of large tubes that are subdivided into smaller tubes near the tectum that then pass individually into small mural pores (Text-fig. 3c). These appear as miniature honeycombs that have even smaller honeycombs near their ends. The name keriotheca has been applied to this type of wall construction. Very similar construction is found in many members of the Verbeekinidae and Neoschwagerinidae (Text-fig. 3d), apparently independently derived, as they are not closely related to the Schwagerinidae. (This might be an example of genetic character exchange between protozoan lineages, but that is going to be very difficult to prove in extinct foraminifers.) In the Neoschwagerinidae additional modifications to the keriothecal wall are so extreme that without intermediate steps being preserved it would be difficult to work out the relationships. Only three are shown. In the Neoschwagerininae the keriotheca is very thick and forms wall partitions known as transverse and longitudinal septula (Text-fig. 3d). In the daughter lineage that includes Yabeina, the keriothecal structure is almost entirely lacking and closely spaced, densely infilled septula essentially take their place (Text-fig. 3e). In the Sumatrininae this process of wall reduction is taken another step and the tectum and septula are thin, organic-rich calcareous structures that partition the upper part of the chambers into small honeycomb-like structures (Text-fig. 3f). Finding the reason for the development of keriothecal walls in several dominant lineages of fusulinaceans is difficult to completely substantiate, however, I have (Ross, 1972) been struck by the similarities with structures in large miliolids in which photosynthetic symbionts are stored as if in little greenhouses.

In the Fusulinidae and some Schwagerinidae, other secondary deposits include the chomata, which are low, dark calcareous ridges that lie at the bottom of the chambers and border the tunnel, which is itself a secondary resorption feature. Chomata come in a tremendous variety of sizes and shapes. In thin section, they may appear low and broad, tapered toward the poles, or they may be narrow and tall or square and blocky. They may change dimensions from the early volutions into later volutions. They may become incomplete, occurring only as small basal thickenings where the septa are cut by the tunnel but not occurring between septa (pseudochomata). In many Permian Schwagerinidae chomata are present only in the first volution on the outside of the proloculus and in some chomata are completely lacking. Chomata in the Fusulinidae and Carboniferous Schwagerinidae are extremely useful in gauging the stage of evolution and combined with the amount of septal folding can be used to separate and identify many species that are externally about the same size and shape.

Another secondary feature in fusulinaceans, is the development of false walls, or phrenothecae. These are thin, dark calcareous, diaphragm-like layers that extend across chambers from one septal fold to the next. They are similar in appearance to the tectum but their significance is not understood. They are common only in the genus Pseudofusulina in the Schwagerinidae. This genus and

Rugosofusulina have another interesting feature. Both may have crenulated, or rugose, spirothecae in about the same part of the Lower Permian succession. The reasons for the development of rugose walls are not understood either.

Finally among the secondary deposits are a set simply known as "other secondary deposits." Nearly all of these are deposits that either coat or variously fill the folds of septa, or coat the septula in the Neoschwagerinidae. What is particularly helpful about these "other secondary deposits" is that when present they normally form a distinctive pattern which is species specific. In some species these deposits occur only along the axis of coiling, in others they are in the shoulders of the shell, in still others they are only in the early chambers, and in others they are only in the later chambers.

In contrast to secondary deposits, the tunnel is a secondary resorption feature. It forms by the resorption of the basal part of the septa near the midplane and results in a nearly planispiraled passageway that begins in the first volution and extends outward to, but not including, the last formed three or four chambers at the outer or growing edge of the shell (Text-fig. 2a). Most fusulinaceans have a single tunnel and it is likely that the tunnel served an important function in life, perhaps as a path for the cell nucleus to remain near the outer margin of the cell or as a means of passing embryonic fusulinaceans to the shell margins during reproduction. A few Permian genera of Schwagerinidae have multiple or auxiliary tunnels and these may have had several nuclei, perhaps vegetative and reproductive nuclei as in some of the recent Rotaliina.

The general shape of fusulinacean shells is another morphological feature of considerable taxonomic importance at the generic and specific levels. Relatively little has been published about the possible adaptive signficance of the different shapes found in fusulinaceans, although by just looking at the distribution of different fusulinacean shapes in contrasting sedimentary rock types, there is obviously much to learn.

Very elongate, subcylindrical fusulinacean species are normally associated with well sorted, laminated, shallow water, calcarenitic sandstones. When present these species are usually numerous and are well aligned in bedding planes or in the troughs or ripple marks. Fusulinaceans with the more common wheat grain outline occur in a wider range of carbonate rock types. Of these, the species with heavier secondary deposits tend to occur in better sorted, coarser grained, bioclastic limestones. Larger species show a similar trend.

One of the more puzzling morphological trends is the development of globose, subspherical, and inflated genera in Early Permian time. There are six or seven inflated genera in the Schwagerinidae, several in the Verbeekinidae, and a number in the Staffellidae. Geometrically the sphere holds the greatest volume per surface area and this suggests an adaptation to reduce shell weight. Inflation of the chambers reduces the number of chambers and number of volutions for the equivalent shell volume and this also seems to be an adaptation toward reduced shell weight. Pseudoschwagerina is a common inflated genus in North America and is associated with small bioherms and shelf edge carbonate banks. In the Southern Urals, Robustoschwagerina, Zellia, and some other inflated genera are widely distributed in several rock types, including a few in evaporitic successions that have few other fossils, suggesting those genera may have been pelagic when adults.

Most fusiform fusulinaceans are associated with shallow water, marine limestones. They are particularly common in post-Middle Carboniferous light colored limestones having algal and echinodermal fragments. They are less common in molluscan shell debris channels in calcareous mud flats. They are rare in "deeper water" dark, fine grained limestone unless they have been transported there from topographically higher habitats. Middle Carboniferous fusulinaceans perhaps lived in slightly deeper waters than their descendents. Most fusulinaceans lived at depths of less than about 25 meters and the great majority lived at depths less than 10 meters.

## DEVELOPMENTS IN FUSULINACEAN STUDIES

In the nearly twenty years since the Fusulinacea section of the Treatise on Invertebrate Paleontology (Thompson, 1964) was prepared, fusulinacean studies have taken many new directions. In general, prior to the 1960's most fusulinacean studies concentrated on taxonomy and biostratigraphic zonation. Since then, taxonomy and zonation have become the important tools used in solving still other types of problems.

### Taxonomic Studies

Taxonomic studies in the past twenty years have greatly increased and clarified our knowledge of the distribution of genera and their age relationships in many parts of the world, for example in Japan, Southeast Asia, China, New Zealand, Soviet Central Asia, Darvas, Northern Urals, Canadian Arctic, and the western Cordilleran of North America. Many more of these studies are still needed to fill in the taxonomy and geographical record of fusulinaceans.

### Phylogenetic Studies

A few major studies on fusulinacean phylogenetic relationships have appeared. Rozovskaya (1975) published an extensive review and revision of the group (Text-fig. 4). She considers the group to be an order (Fusulinida) and recognizes eight families and seventeen subfamilies. It is a particularly thorough study, however, there is still considerable uncertainty about some of the phylogenetic relationships that she illustrated, as well as the validity of a few of the generic names (because the rules of zoological nomenclature were disregarded in the selection of several of them).

A second review of fusulinacean phylogeny appeared in the following year (Kanmera, Ishii, and Toriyama, 1976) and examined the evolution and extinction of Permian Fusulinacea (Text-fig. 5). While there are many similarities between the phylogenies proposed by Rozovskaya (1975) and Kanmera and others (1976),

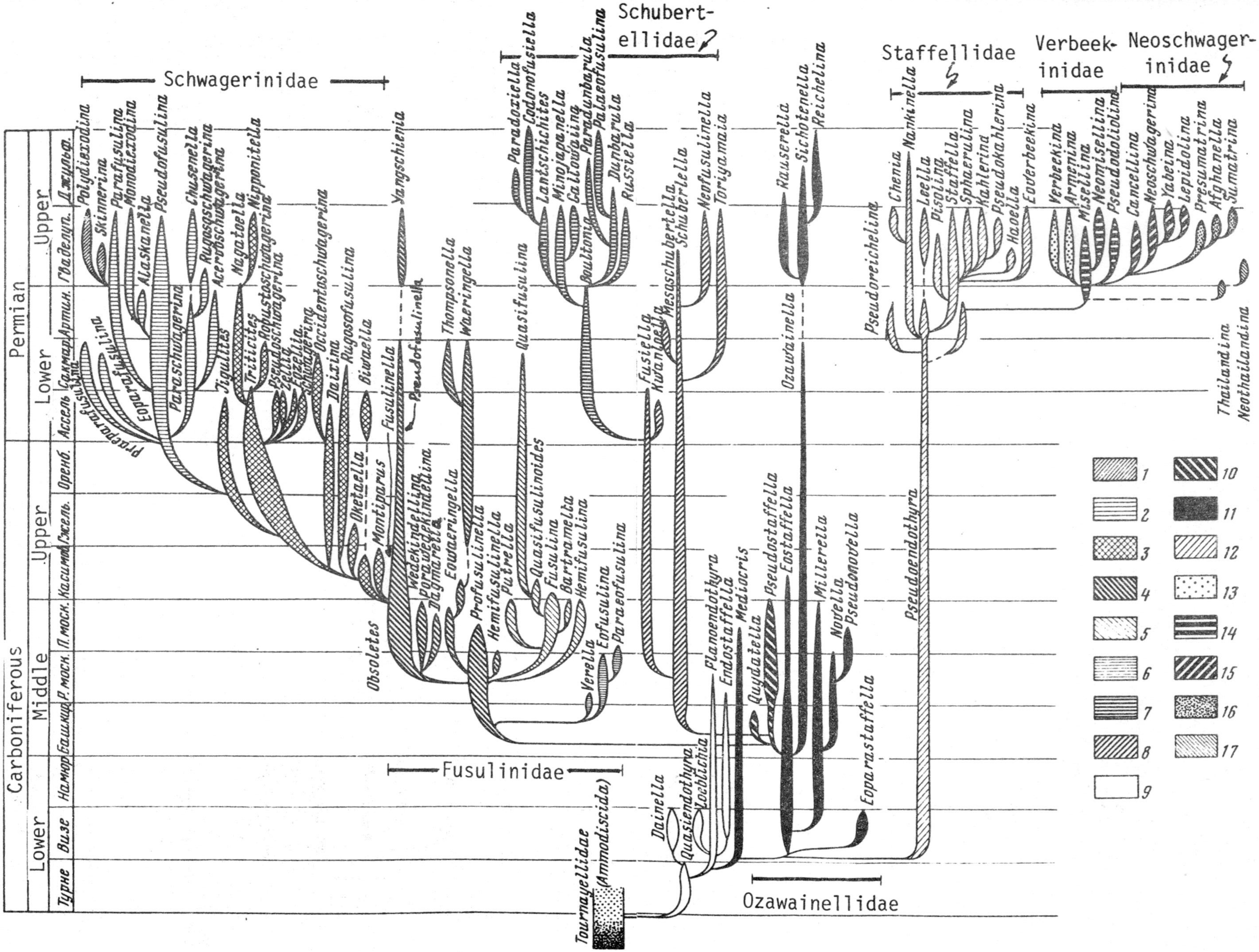


Text-figure 4. Phylogeny of fusulinaceans according to Rozovskaya (1975).

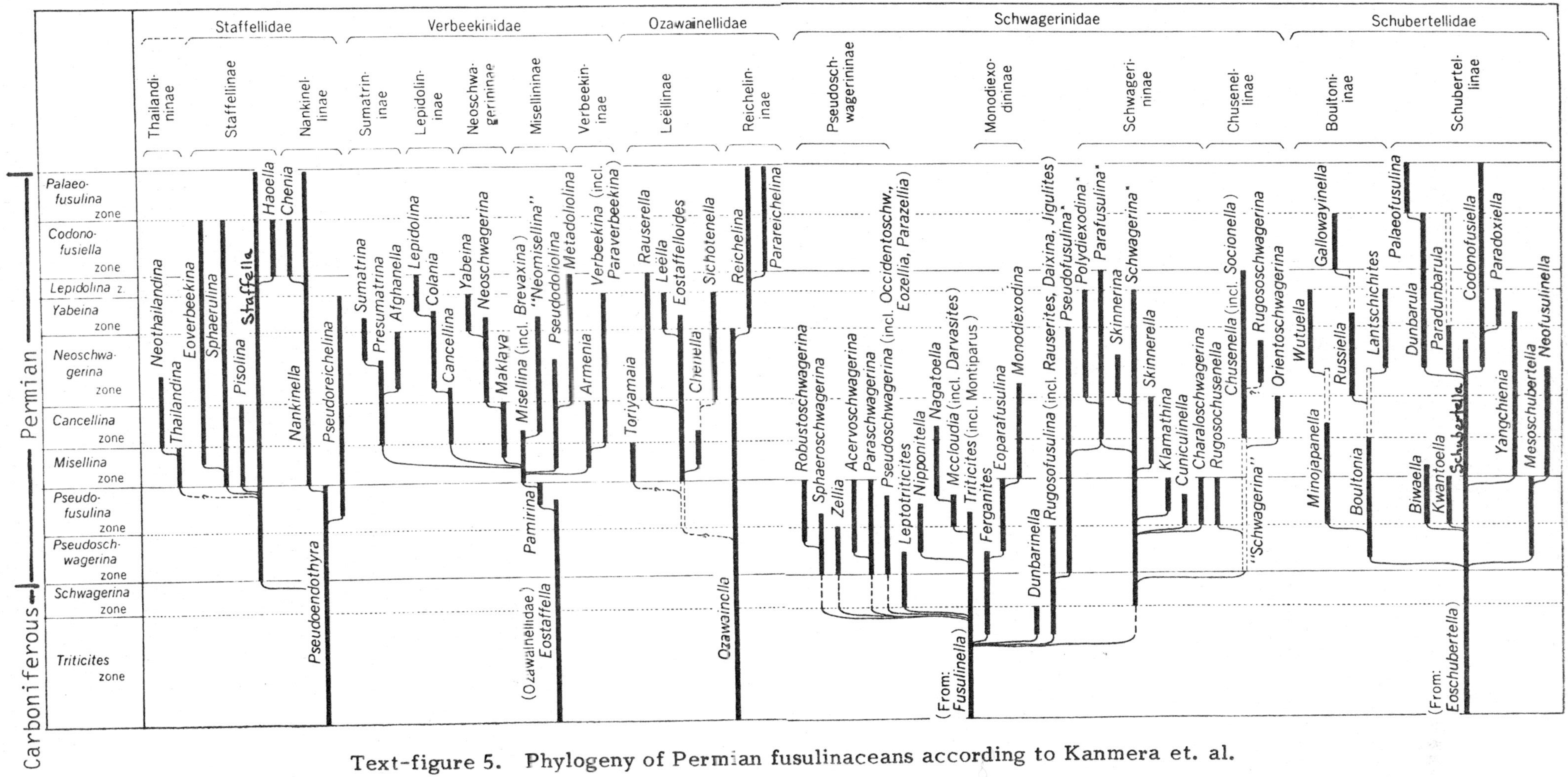


Text-figure 5. Phylogeny of Permian fusulinaceans according to Kanmera et. al. (1976).

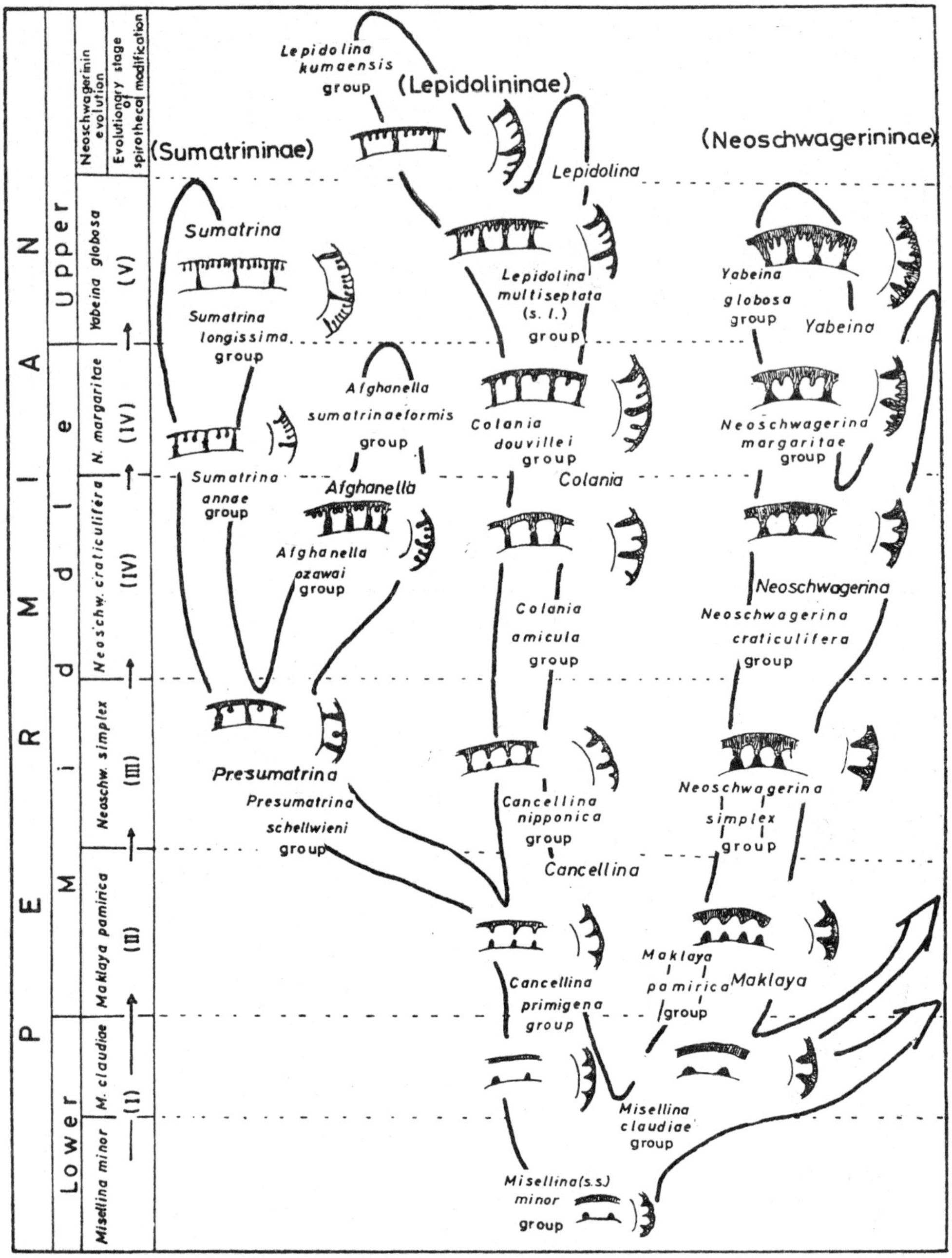


Text-figure 6. Evolution of wall structures in Neoschwagerinidae (modified from Ozawa, 1970).

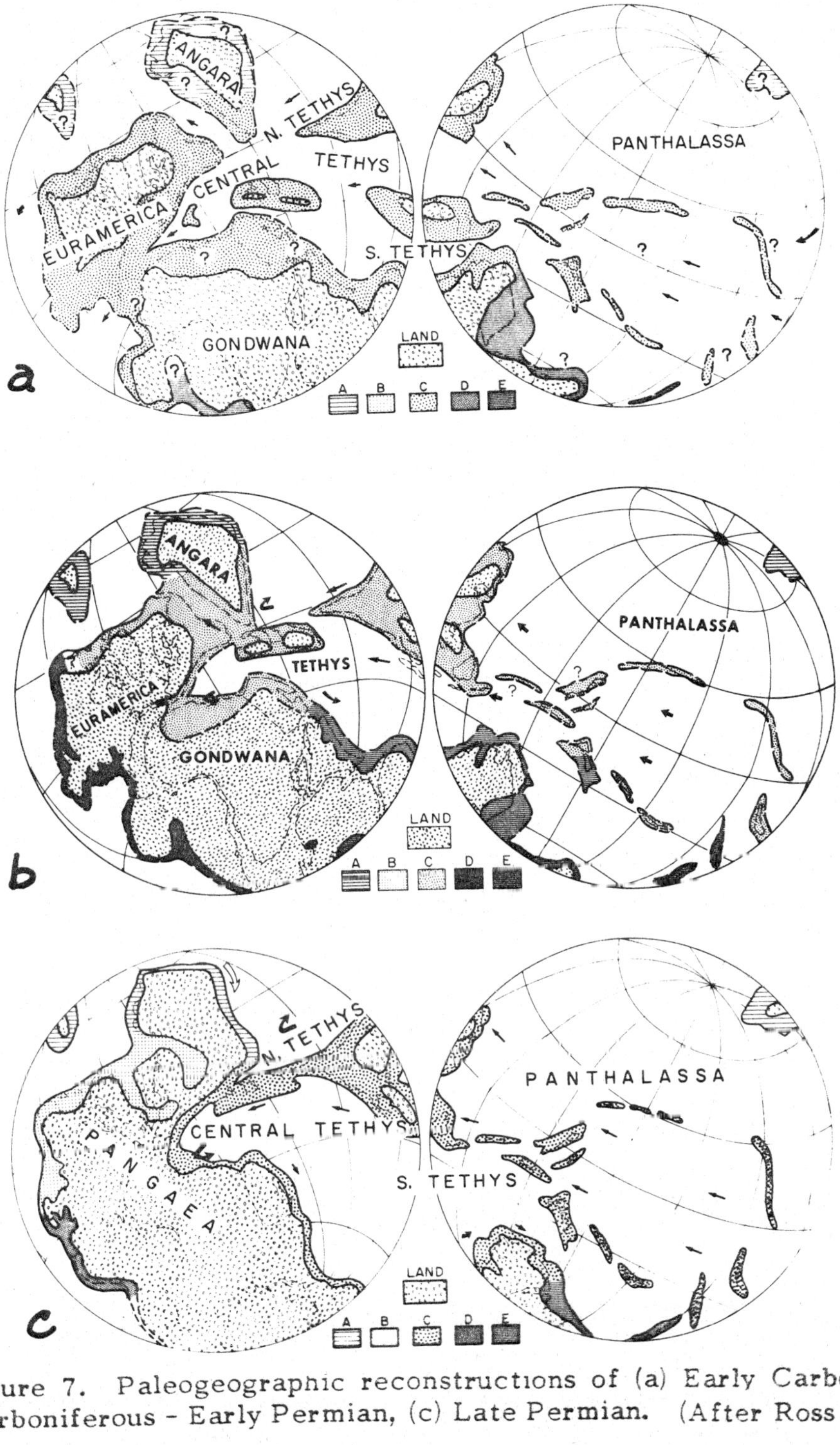


Text-figure 7. Paleogeographic reconstructions of (a) Early Carboniferous, (b) Late Carboniferous - Early Permian, (c) Late Permian. (After Ross and Ross, 1981)

there are also a remarkably large number of differences, particularly the inclusion of many additional genera by Kanmera and others (1976) and significant differences in several stratigraphic ranges.

One paper on the phylogeny and classification of the neoschwagerinids and verbeekinids by T. Ozawa (1970) has greatly clarified the structures and evolutionary relationships within that important and complex group. Ozawa was able to show that three lineages are distinguishable in the neoschwagerinids (Text-fig. 6) and two lineages in the verbeekinids and that these lineages are characterized by morphologic structures that change with time and hence have stratigraphic significance. Ozawa's results are particularly helpful because these two families form the dominant groups in the Permian Tethyan fusulinacean faunas. They nearly always occur in poorly bedded, reefy limestones in tectonically disturbed belts so that their stratigraphic succession has been difficult to work out in any detail.

## Paleoecology

Relative little information was included about the paleoecology of Fusulinacea in the Treatise (Thompson, 1964) based on the assumption that fusulinaceans were widespread on most continents (except Australia and Antarctica) and that they were common in nearly all limestones from the Middle Carboniferous to the Upper Permian, if only we would look hard enough for them.

Several developments have shown these assumptions to be oversimplifications. One major development has been the rapid growth in our understanding of recent and past carbonate depositional environments during the 1960's and 1970's. Most species of fusulinaceans appear to be marine, benthonic, and associated with relatively shallow, well aerated, carbonate depositional settings in which corals and algae are also common associates. Fusulinaceans normally do not occur in dark, fine-grained, "basinal" limestones unless they have been redeposited or washed into a basin from a nearby, shallow-water, shelf edge. Some species of fusulinaceans were almost the single inhabitant of certain environments, such as those in and around back-reef lagoons where each species is abundant only in limestones of a particular type.

Bioclastic limestones that show considerable wave sorting may have three or four common species of fusulinacean, but, by the very nature of the limestone, they are probably a death assemblage mixture from several closely adjacent microcommunities. A few inflated, subglose to globose Permian genera may have been pelagic. Some of these, such as Robustoschwagerina or Zeelia, are widely scattered in otherwise inhospitable deposits that may be associated with evaporites, as if the fusulinacean had been transported by surface currents and then settled to the bottom upon death.

Fusulinacea are not usually associated in the same lithology with brachiopod- or bryozoan-rich limestones, although they may occur higher or lower in the same succession. In some parts of the world, such as Australia, fusulinaceans are not known, but thick Permian limestones and rich and diverse brachiopods and bryozoans are common. Such distributions suggest that fusulinaceans were adapted to warmer, shallower water then the brachiopods and bryozoans.

Comparison with Recent large, shallow marine calcareous foraminifers, such as the soritids and alveolinids (Miliolida), also has been helpful in interpreting fusulinacean life histories and habitats (Ross, 1972). Large soritids and alveolinids have symbiotic zooxanthellae which apparently aid the foraminifers in calcium carbonate production and nutrition. Soritids mature after two or more years and it is likely that large fusulinaceans also had life cycles of two or more years. In some fusulinacean families, such as the Schwagerinidae and Neoschwagerinidae, the individual honeycombs of the keriothecal wall may have been an adaptation to serve as the resting place for similar symbionts and may account for their unusual success in shallow, warm marine conditions. In others, such as the Schubertellidae and Staffellidae, the wall was nearly transparent. The tectorial deposits in the Fusulinidae may have been an early mechanism to deposit excess calcite from such symbiotic relations. The Fusulinidae and Ozawainellidae probably ranged downward to water slightly deeper and, in some species, slightly cooler than the other families based on their common association with medium gray, calcareous grainstones.

## Paleogeography

The second major assumption that fusulinaceans were widespread on most continents also has proven to be a great simplification. Almost immediately after the Fusulinacea part of the Treatise was published, the theories of sea-floor spreading and plate tectonics revolutionized our concepts of the mobility of continents and the pieces that composed them. Reconstruction of late Paleozoic geographies showed that none of the ocean floor sediments of that age remained except as strongly deformed belts dominated by cherts, basalts, and turbidites, that had been crumpled to form accretionary margins of continents during Mesozoic and Cenozoic times. Further, these reconstructions showed the position of the continents lay in quite different positions to the latitudinal climate belts than they do at present.

Reconsideration of fusulinacean geographical patterns of distributions in light of these models of late Paleozoic geography, led to the recognition of increasing provincialism in these faunas during the Middle and Late Carboniferous and Permian (Ross, 1967). Based on these paleogeographic reconstructions, fusulinaceans were tropical to warm temperate in their distribution. As more and more continents combined with Gondwana to form Pangaea during the late Paleozoic (Text-fig. 7a, b, c), the Early Carboniferous Hercynian equatorial seaway was closed and uplifted during Middle and Late Carboniferous time to create two tropical shallow water provinces with a filter dispersal route around temperate northern Euramerica. This was followed in the middle of Early Permian time by the closing and uplift of the Ural seaway. The polar northern end of Angara was too cold for fusulinacean dispersal. Therefore, the only dispersal between the opposite shallow-water tropical sides of Pangaea was chance dispersal across the single, huge world ocean, Panthalassa. This situation gave rise to distinctive fusulinacean realms. In the Permian the eastern one is called the Tethyan realm and the western one the Midcontinent-Andean realm. Recognition of fusulinacean faunal realms has contributed to the restudy of geographic distribution in other late Paleozoic groups, especially corals, bryozoans, and brachiopods.

Fusulinacean provinciality has aided the unraveling of the structure and history of the western Cordillera of North America (Monger and Ross, 1971). There, different tectonic-sedimentary terrains (or assemblages) aligned in juxtaposed belts more or less parallel to the Paleozoic margin of North America, contain quite different fusulinacean faunas. These relations clearly indicate that the different tectonic-sedimentary terrains did not originate where they are today and have been accreted as fragments against western North America. It appears as if a number of island arcs with tropical reefs, small continental blocks with shelf carbonates, fore- and back-arc trench sediments, and ocean rise sequences have all been added during the Mesozoic and Cenozoic. While it is difficult at present to reconstruct exactly where each late Paleozoic fragment originated, the fusulinaceans suggest they almost certainly did not originate next to the fragments with which they presently are in contact. Northern warm temperate, tropical, and southern warm (and perhaps cool) temperate faunas seem to be present in these various belts, as if the whole of Panthalassa was swept clean, leaving only a scattering of its contents against the present Pacific margin.

### Correlations

Correlations using fusulinaceans have also improved in the past twenty years. On stable, shallow carbonate shelves where fusulinaceans are abundantly present, it is possible to establish correlations of stratigraphic intervals that have about 3 to 5 million years duration. This level of correlation needs to be extended into other sedimentary settings and should be a great aid in establishing more detailed sedimentary and tectonic histories for late Paleozoic events around the world.

## SUMMARY

Fusulinacean studies are progressing well, however, like other foraminiferal groups, much is still to be learned. Many fusulinacean faunas still need to be described, and a great deal more needs to be understood about their ecology, geographical distribution, dispersals, and the adaptive significance of different types of shell morphologies and shell structures. In such an extinct group, analysis of these facets will require considerable care and thought because it will not be possible to do more than make analogies and comparisons with living foraminifers.

# North American Paleozoic Agglutinate Foraminifera

James E. Conkin
Department of Geology
University of Louisville, Louisville, KY 40292
and
Barbara M. Conkin
Jefferson Community College, Louisville, KY 40202

## INTRODUCTION

Literature on foraminifera in general is quite accessible and useable. Cushman's (1948) textbook, combined with the treatise of Loeblich and Tappan (1964), suffices for most generic definitions. Loeblich and Tappan's (1964) work is, in addition, the most comprehensive treatment of the general biology, ecology, classification, and stratigraphic distribution of the foraminifera. On a specific level, the preeminent reference works on the foraminifera are Ellis and Messina's (1940-1982) Catalogue of the Foraminifera, Sherborn's (1888) Bibliography of the Foraminifera from 1765-1888 and his (1893-1896) Index to Genera and Species of the Foraminifera, and Cushman's (1950) Card Catalogue on microfilm. Important as well are Brady's Challenger report (1884), Cushman's (1910-1916) monograph on the North Pacific Ocean foraminifera and his (1918-1931) Foraminifera of the Atlantic Ocean, to mention a few of the works on the Recent foraminifera; a knowledge of Recent foraminifera is important, for a large number of agglutinate foraminiferal genera have stratigraphic ranges from Paleozoic to Recent. Brady's (1876) monograph of Carboniferous and Permian foraminifera is also important.

Summaries of knowledge on North American Paleozoic agglutinate foraminifera include papers on the Mississippian (Gutschick, 1960), pre-Pennsylvanian (Conkin and Conkin, 1973a), Late Devonian-Early Mississippian (Conkin and Conkin, 1967), Kinderhookian (Conkin and Conkin, 1970a), and Paleozoic (Conkin and Conkin, 1977).

Moreman (1930 and 1933), Ireland (1939), and Dunn (1942) set the pattern of methodology in North America, which rather limited the study of Paleozoic agglutinate foraminifera to acidization of limestones (exceptions being Cooper's 1947 work on the Chesterian Kinkaid Shale and Conkin's 1961 study of the Mississippian of the east-central United States). Conkin and Conkin's continuing works have utilized both acidization of calcareous rocks and washing of shales and siltstones. In Europe, thin section methodology has been in vogue, with notable success among Paleozoic calcareous forms with layered walls. European workers recently have begun to use acidization of limestone with resulting increase in the discovery of agglutinate foraminifera.

## ECOLOGY

Marked scarcity or abundance of agglutinate foraminifera within Paleozoic sequences is both real, due to the evolutionary process and to ecological control of the distribution of foraminifera, and apparent, due to unequal study of different sequences and use of differing extraction and study methods; however, foraminifera are no more ecologically controlled than most other animals and plants. The lithologies most likely to yield agglutinate foraminifera are silty shales and/or fine-grained shaly sandstones, fine-grained, impure (argillaceous) limestones, even sublithographic limestones if they bear fine-grained detrital material. Sparry calcarenites notably yield few or no agglutinate foraminifera nor do argillites or orthoquartzitic sandstones and siltstones. Dolomites bear few or no foraminifera whereas calcareous dolomites and dolomitic limestones may bear varying numbers. Sideritic concretions have yielded agglutinate foraminifera where surrounding siltstones were essentially barren. Some agglutinates were able to live (and even thrived) in conditions which were so unfavorable as to preclude many other life forms; an example of this is *Ammodiscus* in black shales in certain parts of the geologic column.

## STRATIGRAPHIC OCCURRENCE OF PALEOZOIC AGGLUTINATE FORAMINIFERA

### Cambrian

There is no substantiated record of agglutinate foraminifera in the Cambrian; most previously reported forms are referable to calcareous algae. Alexandrowicz (1969) recorded *Thuramminoides sphaeroidalis* (long considered an agglutinate) from the Middle Cambrian of Poland; however, *T. sphaeroidalis* Plummer, 1943, emend. Conkin, 1961, has been allocated to the spumellarine radiolarian genus *Entactinosphaera* Foreman, 1963, by Conkin, Conkin, Walton, and Neff (1981). Conkin, Conkin, and Canis (1968) reported *T. sphaeroidalis* at least from Ordovician into Permian.

We have examined the types of *Psammosphaera*? *greenlandensis* Howell and Dunn, 1942 from the Lower Cambrian of Ella Island, Greenland, (deposited in the Department of Geology, Princeton University), but are unable to ascertain their affinities because of extreme alteration; they do not appear to be beekite structures as we suspected before examining them.

### Ordovician

The validated North American Ordovician agglutinate foraminiferal fauna presently consists of at least 14 species belonging to 13 genera (Figure 1) distributed among three families (Astrorhizidae, Saccamminidae, and Ammodiscidae).

Moreman (1930) described *Hyperammina minuta* from the Early Ordovician Arbuckle Limestone of Oklahoma, *Bathysiphon exiguus*, *Raibosammina* mica, R. aspera, and Tholosina elongata from the Late Ordovician Viola Formation, and (1933) Kerionammina fava, Marsipella aggregata, Rhabdammina trifurcata, and Webbinella tholus from the Viola. Kjellesvig (1934) reported (without descriptions) 16 genera and 29 species from the Trenton of New York; one he identified as a poorly preserved Reophax, but the earliest known Reophax is from the Hamilton of southern Indiana (Conkin, Conkin, Walton, and Neff, 1981). Moore (1952) reported unidentified middle Trentonian foraminifera from Catawba Mountain, Virginia. Conkin and Conkin (1965) reported Blastammina sp. and Tolypammina sp. from the Richmondian of Oklahoma, and (1979) Amphitremoida sp. from Iowa. We herein report Psammosphaera cava Moreman, 1930 from the Maysville Mt. Auburn Formation of Carroll County, Kentucky and Webbinelloidea Stewart and Lampe, 1947, emend. Conkin and Conkin, 1970 (as the species W. similis Stewart and Lampe, 1947, emend. Conkin and Conkin, 1970) from the Middle Ordovician Kimmswick Limestone of Missouri and the Late Ordovician Maysville Group of Kentucky.

Ordovician foraminifera have simple tests, the oldest known being the tubular (with proloculus) Hyperammina. By the Middle Ordovician, the hemitubular attached Tolypammina, the colonial (composed of spherical chambers) Sorosphaera, and the planoconvex Webbinelloidea (with a single "pin prick" inverted aperture) had evolved. A significant increase in the number of species occurred in the Late Ordovician along with the development of the irregularly-branched tubular Rhabdammina, the globular Blastammina, the spherical Thurammina with apertural projections, and the prolate biapertured Amphitremoida.

We allot several forms previously reported from the Ordovician as follows: Bathysiphon exiguus Moreman, 1930 is probably based on a sponge spicule fragment and Kerionammina fava Moreman, 1933 is referable to Incertae Sedis. Webbinella tholus Moreman, 1933 is a Hemisphaerammina Loeblich and Tappan, 1957, emend. Conkin and Conkin, 1979. Fairliella Summerson, 1958 and Metamorphina Browne (in Browne and Schott, 1963) are synonyms of Hemisphaerammina. Amphicervicis Mound, 1961 is a synonym of Tholosina Rhumbler, 1895. Raibosammina mica Moreman, 1930 is a synonym of R. aspera Moreman, 1930.

## Silurian

Descriptive studies of North American Silurian foraminifera began with the pioneer works of Moreman (1930 and 1933), Ireland (1939), Grubbs (1939), Stewart and Priddy (1941), and Dunn (1942). After two decades, five additional studies appeared: Mound (1961), Browne and Schott (1963), McClellan (1966), Mound (1968), and McClellan (1973). A few non-descriptive notices of Silurian foraminifera have been published as well: Dunn (1931), Miller (1956), and Conkin and Conkin (1960).

Figures 2 and 3 present a summary of genera and species and their stratigraphic distribution in the Silurian.

A massive systematic revision of Silurian foraminifera must be undertaken, based on a restudy of type materials utilized in previous systematic works. This study must be intercontinental in scope, including the important works of Eisenack (1932, 1937, and 1954), for the North American-European Silurian foraminiferal faunas are essentially cosmopolitan. The non-availability of Ireland's (1939) and Dunn's (1942) types has been the major block to study and elucidation of the affinities of the bulk of the species and some of the genera. It is, then, understandable that many of the generic and specific synonymies prepared by various workers (Stewart and Priddy, 1941; Mound, 1961 and 1968; Browne and Schott, 1963; and McClellan, 1966 and 1973) since 1942 have not noticeably clarified the affinities of most of the species, and many of the genera, of Silurian foraminifera; however, much of this material is now available for study. Additional assessment of the occurrences of Silurian foraminifera placed in their precise stratigraphic-paleontologic context in geographically widespread deposits must be accomplished concurrently with the systematic revision before Silurian foraminifera can be utilized for their ultimate purpose in geology, age determination and correlation. Indeed, progress in paleontologic understanding and stratigraphic utilization of Early Paleozoic foraminifera in particular, and Paleozoic in general, await such broad integrated paleontologic-stratigraphic studies.

Evolutionary advance by foraminifera in the Silurian is indicated by the innovation of the irregularly-coiled Glomospira and Lituotuba, the planispirally-coiled Ammodiscus, the high-spiralled Turritellella, and the labyrinthic Oryctoderma. It is important to note that the multiple-chambered, trochoidally-spiralled genus Trochammina did not evolve before Early Mississippian (Kinderhookian); Trochammina prima Stewart and Priddy, 1941 is a junior synonym of Hemisphaerammina osgoodensis (Stewart and Priddy), 1941. The only other Trochammina reported from the Silurian is T. bursaria Chapman, 1933, from the State of Victoria, Australia; Conkin and Conkin (1968) have suggested this form is conspecific with the radiolarian, Thuramminoides sphaeroidalis. Further, the form also described from the Silurian of Victoria, Hemigordius lilydalensis Parr, 1933 is an indeterminate arenaceous foraminiferan; it does not belong to the calcareous Hemigordius.

## Devonian

Figures 3 and 4 present the stratigraphic distribution of Devonian species. Helderbergian foraminifera have been described from Oklahoma by Moreman (1933) and Ireland (1939) and from Nevada by McClellan (1973). Conkin and Conkin (1977) considered Hemisphaerammina geometrica Eisenack, 1954 useful for separating the Early from Middle Devonian in New York and Tennessee. Ireland's (1939) foraminiferans from the Haragan Formation of Oklahoma may be Late Silurian Pridolian rather than Helderbergian, but are here placed in the Early Devonian for convenience. Inauris tubulata Conkin, Conkin, and Thurman, 1979, in any frequency of occurrence is definitive of late Early Devonian (middle Onesquethawan-Emsian) in Indiana, Kentucky, and Ohio (Conkin,

Conkin, and Thurman, 1979); it does range into the early part of the late Onesquethawan (Eifelian), but is extremely rare therein. A Semitextularia (complex, multi-chambered, multiple-apertured, calcareous genus)-earlandiinid (calcareous isomorph of the agglutinate genus Hyperammina) Zone marks the early late Onesquethawan; although these foraminiferans are not truly agglutinate, they are easily identified and are most useful for recognition of this part of the stratigraphic section. In older literature no absolute boundary was drawn between study of agglutinate and calcareous "smaller foraminiferans", those for which thin sectioning was not considered necessary for identification.

The first multi-chambered forms with chambers enlarging in size are found in the Middle Devonian. Oxinoxis ligula (Gutschick), 1962, emend. Conkin and Conkin, 1964, begins with the single-chambered form A in the Cazenovian (Skaneateles) and ranges into the early Osagean; the multi-chambered form B begins with poor development of a two-chambered form in the Tioughniogan (Ludlowville) while forms of O. ligula form B (with two or more chambers) are present in the Late Devonian and Early Mississippian (Kinderhookian). The multi-chambered rectilinear Reophax begins in the Hamiltonian.

Conkin (1961) recorded the non-diagnostic Hyperammina casteri Conkin, 1961, Ammodiscus semiconstrictus Waters, 1927, and Trepeilopsis recurvidens Gutschick and Treckman, 1959, emend. Conkin, Conkin, and Canis, 1968, from the Portwood Formation (either Middle Devonian Taghanican or Late Devonian Genessean) of the New Albany Group. T. recurvidens is well developed in the Late Devonian Saverton Shale in Illinois and Three Forks Formation of Montana. Summerson (1958) described Weikkoella as a new genus of arenaceous foraminifera (with two species) from the Middle Devonian of Ohio. Conkin, Conkin, Sawa, and Kern (1970) have shown Weikkoella to be the silicified cast of the oosporangial membrane of the charophyte alga (stonewort), Moellerina greenei Ulrich, 1886, emend. Conkin, Conkin, Sawa, and Kern, 1970. Conkin and Conkin (1970b) revised the genus Webbinelloidea Stewart and Lampe, 1947, recognizing it possesses a single "pin-prick" aperture and placed 14 species previously described by Stewart and Lampe (1947) and Summerson (1958) under the type species W. similis.

The most completely described Late Devonian foraminiferal fauna is that of the latest Famennian Louisiana Limestone of Missouri and Illinois (Conkin and Conkin, 1964a) in which five families, nine genera, and 18 species were described. Conkin (1961) reported Ammodiscus longexsertus (Gutschick and Treckman), 1969, A. semiconstrictus, Hyperammina casteri, H. rockfordensis? Gutschick and Treckman, 1959, and Trepeilopsis recurvidens from the Upper Olentangy Shale (Selmier Member of the New Albany Shale) of eastern Kentucky, and Conkin, Conkin, Walton, and Neff (1981) added Hippocrepinella sp., Reophax sp., and Thurammina congesta Gutschick, Weiner, and Young, 1961. A few specimens of non-diagnostic Hyperammina casteri, Reophax sp., and Tolypammina sp. occur in the Three Lick interval (Chagrin Member of the Cleveland Shale) of the Ohio Shale in Rowan County, Kentucky.

The spiralled, attached Trepeilopsis and compactly-winding Ammovertella arose in the latest Devonian from the more loosely arranged Tolypammina. Ammovertella and Trepeilopsis are most commonly encountered in Kinderhookian and younger strata.

"Textularia" is present in the Paleozoic, beginning in the Devonian; Textularia sensu stricto seems to begin in the Pennsylvanian and ranges to the Recent. "Palaeotextularia", without the inner "fibrous wall" of Paleotextularia, is present beginning in the Devonian and ranges at least into the Meramecian.

## Mississippian

The Mississippian agglutinate foraminiferal faunas (Figures 3, 5-8) are the best known of all in the Paleozoic as a result of the descriptive efforts of Conkin (1954 and 1961), Gutschick and Treckman (1959), Gutschick, Weiner, and Young (1961), Gutschick (1962), Conkin and Conkin (1964b, 1971, and 1981), Conkin and Ciesielski (1973), Conkin, Conkin, and Pike (1965), Conkin, Conkin, and MacDonald (1963), and Conkin, Conkin, Walton, and Neff (1981). Conkin (1961) made the initial attempt to use agglutinate foraminifera as a basis for age determination on a regional scale and set up a preliminary scheme of foraminiferal division of the Mississippian System in the east-central United States based on agglutinate foraminifera; these attempts have been extended into the Mississippian type area of Missouri and Illinois and westward to New Mexico (Conkin and Conkin, 1982). We have stressed the precise stratigraphic placement of described foraminiferans with close attention to stratigraphic detail, including the positions of disconformities of the magnitude of paracontinuities (Conkin and Conkin, 1973b). In fact, agglutinate foraminifera combined with paracontinuities allow precise placement of the Devonian-Mississippian boundary in the eastern United States, as well as the Kinderhookian-Osagean boundary in southern Indiana, Kentucky, and Tennessee, as grain per grain contacts (Conkin and Conkin, 1975).

Conkin (1961) and Conkin and Conkin (1967 and 1981) recognized the following foraminiferal phylogenetic lineages (Figure 3) useful in age determination and correlation within the Upper Devonian-Lower Mississippian: Hyperammina rockfordensis-H. kentuckyensis; Reophax lachrymosus-R. raymoorei; Ammobaculites leptos-A. gutschicki; and Oxinoxis ligula form A-form B.

A large number of agglutinate foraminiferal species are present in the Mississippian (Figure 5), and the Lower Mississippian can be separated into Kinderhookian and Osagean on the basis of agglutinates as outlined in Figure 6.

## Kinderhookian

Ammobaculites and Trochammina are first known in the Lower Mississippian (Kinderhookian). The Kinderhookian is especially characterized by Hyperammina rockfordensis and Ammobaculites leptos Gutschick and Treckman, 1959, emend. Conkin, Conkin, and Canis, 1968; however, *A. leptos* ranges into the lower Osagean. *H. rockfordensis*, reported (Conkin, 1961) from the Bedford Shale (Late Devonian or Early Mississippian) and from the lowest Osagean, is essentially a Kinderhookian species, and in any frequency of occurrence is definitive of the Kinderhookian. Singularly, *H. rockfordensis* is visible to the unaided eye in fresh samples of the Kinderhookian Jacobs Chapel Shale.

Osagean

*Hyperammina kentuckyensis* Conkin, 1954, restricted to the Osagean, appears in early Osagean, is well developed in the middle Osagean, and is best developed in late Osagean. This species is intercontinental in occurrence, for *H. kentuckyensis guangxiensis* was described (Conkin and Wang, 1980) from the late Yanguanian (Osagean) Shizexu and Kinling formations of South China; further, *H. kentuckyensis* is present in correlative positions in western Europe.

*Ammobaculites gutschicki* Conkin, 1961, perhaps an evolutionary derivitive from *A. leptos*, occurs rarely in the Kinderhookian, but characterizes the Osagean.

The first representative of the Palaeotextulariidae, *Cribrostomum*, is present outside North America in beds correlative with the Osagean. The primitive miliolacean-like (but agglutinate) forms previously referred to *Agathammina* might better be placed in the Ammodiscidae as suggested by Loeblich and Tappan (1964); this kind of *Agathammina* has been reported as *A. mississippiana* Conkin, 1961 from the Osagean and Chesterian of Kentucky and the genus continues into the Pennsylvanian and Permian (Cushman and Waters, 1930).

Middle Mississippian (Meramecian)

The Meramecian strata are dominated by carbonate lithologies with calcareous foraminifera which may be described as "larger foraminiferans" requiring thin section examination of wall structure for generic and specific determination. Figure 5 presents the Meramecian species and their stratigraphic distribution.

Outside North America, three palaeotextulariid genera, *Deckerella*, *Palaeobigenerina* and *Palaeotextularia* occur in Meramecian correlatives. The microgranular, calcareous *Earlandia consternatio* Conkin, 1961, occurs first in the middle Meramecian (Warsaw-Salem); it may be considered the calcareous isomorph of *Hyperammina kentuckyensis* Conkin, 1954. *E. consternatio* is present in the correlative stratigraphic position in the Lisbon Formation of Alaska; it does range into the Chesterian as well.

Late Mississippian (Chesterian)

The Chesterian strata bear the first *Climacammina* (*C. missippiana* Conkin, 1961) and *Palaeotextularia* (*P. asper* Cooper, 1947) as well as the distinctive *Hyperammina neoglabra* Conkin, 1954 which is best developed in the Pennsylvanian, particularly in the Desmoinesian Providence Limestone of western Kentucky. Figure 8 presents the Chesterian species and their stratigraphic distribution.

## Pennsylvanian-Permian

More papers have been published on Pennsylvanian foraminifera than those on any other Paleozoic system, but they have been stratigraphically scattered and no concerted effort has been expended to use them for age determination and correlation. Nearly one-third of these papers are authored by J. A. Cushman alone (1930) or Cushman and Waters (1927a, 1927b, 1928b, 1928d, and 1930) and are almost exclusively systematic. Other workers (Galloway and Harlton, 1928; Galloway and Ryniker, 1930; Harlton, 1927, 1928, and 1933; Marple, 1955, Roth and Skinner, 1930; St. Jean, 1957; Warthin, 1930, Waters, 1927 and 1928; and Toomey, 1972) essentially have followed the lead of Cushman and associates. A few papers, stressing paleoecology, have appeared (Croneis and Toomey, 1965; Henbest, 1958 and 1963; and Toomey, 1972). The closest approach to integrated paleontologic-stratigraphic studies in the Pennsylvanian are those of H. J. Plummer (1930 and 1945) and Ireland (1956).

It is obvious that the Pennsylvanian faunas need to be restudied, and additional faunas searched for and described, with individual species placed in their precise stratigraphic context, for only then can meaningful attempts be made to use these foraminiferans for age determination and correlation.

Very little has been written on North American Permian agglutinate foraminifera (Cushman and Waters, 1928a and 1928c; and Spandel, 1902); no stratigraphic treatment has appeared.

Most of the agglutinate genera present in the Mississippian continue on into the Pennsylvanian and Permian (Figures 7 and 9), and beyond into the Recent. A number of the species continue into Pennsylvanian and Permian. The dominant elements in the Pennsylvanian and Permian foraminiferal faunas are "larger" calcareous forms, and some of these bear a thin, scattered veneer of adventitious material on the outside of the basically calcareous tests; thus, these may be considered partially agglutinate. These tests usually have multiple-layered walls (particularly most of the textulariids of older terminology); for study of these "larger foraminifera" (with agglutinate veneers), the works of Cummings (1955 and 1956) are essential.

The globotextulariinid *Mooreinella* is restricted to the Pennsylvanian. *Textularia* (*sensu stricto*) arose in the Pennsylvanian. *Verneuilina* has its lowest and only Paleozoic occurrence in the Upper Pennsylvanian of Kansas (Ireland, 1956). The palaeotextulariid *Monogenerina* is recorded from the Middle and Upper Pennsylvanian of the United States, but is apparently restricted to the Permian elsewhere. In

North America, the spiroplectamminid genus, *Spiroplectammina*, begins in Middle Pennsylvanian and ranges into the Permian. The textulariinid *Bigenerina* is present in the Upper Pennsylvanian and the palaeotextulariid *Cribrogenerina* arose in Late Pennsylvanian and continued into the Permian to which it is apparently restricted outside of North America.

## Paleozoic Generic Summary

Exercising our best judgement, we preliminarily recognize in the Paleozoic 67 genera, distributed as follows (Figure 9): 54 agglutinate; four calcareous (*Earlandia*, an earlandiinid, probably *Earlandinita* Cummings, 1955, *Semitextularia*, and *Pseudopalmula*) and nine "larger foraminifera" with a veneer of agglutinate material, *Palaeotextularia*, *Climacammina*, *Mooreinella*, *Cribrostomum*, *Deckerella*, *Palaeobigenerina*, *Monogenerina*, *Bigenerina*, and *Cribrogenerina*.

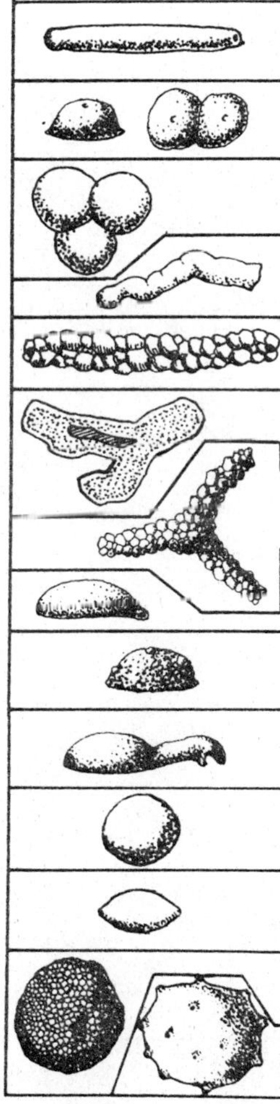

| SPECIES | ORDOVICIAN | | | | |
|---|---|---|---|---|---|
| | LOWER | MIDDLE | UPPER | | |
| | | | VIOLA | MAYSVILLE | RICHMOND |
| *HYPERAMMINA MINUTA* | X | | | | |
| *WEBBINELLOIDEA SIMILIS* | | X | | X | |
| *SOROSPHAERA TRICELLA* | | X | | X | X |
| *TOLYPAMMINA* SPP. | | X | | X | X |
| *MARSIPELLA AGGREGATA* | | | X | | |
| *RAIBOSAMMINA ASPERA* | | | X | | |
| *RHABDAMMINA TRIFURCATA* | | | X | | |
| *THOLOSINA ELONGATA* | | | X | | |
| *HEMISPHAERAMMINA THOLA* | | | X | X | X |
| *TOLYPAMMINA* N. SP. | | | | X | |
| *PSAMMOSPHAERA CAVA* | | | | X | |
| *AMPHITREMOIDA* SP. | | | | X | X |
| *BLASTAMMINA* SP. | | | | | X |
| *THURAMMINA* SP. | | | | | X |

Figure 1. Stratigraphic occurrence of species of North American Ordovician agglutinate foraminifera.

| Species | SILURIAN LOWER | SILURIAN MIDDLE | SILURIAN UPPER |
|---|---|---|---|
| COLONAMMINA BITUBA | X | | |
| ORDOVICINA TYPA | X | | |
| PSAMMOSAPHERA ARCUATA | X | | |
| SACCAMMINA JOLIETENSIS | X | | |
| SACCAMMINA? OVATA | X | | |
| SOROSPHAERA MULTICELLA | X | | |
| STEGNAMMINA MOREMANI | X | | |
| THURAMMINA? HEXACTINELLIDA | X | | |
| GLOMOSPIRA ARTICULOSA | X | X | |
| GLOMOSPIRA GORDIALIS | X | X | |
| LITUOTUBA EXSERTA | X | X | |
| LITUOTUBA INFLATA | X | X | |
| LAGENAMMINA SPHAERICA | X | X | |
| LAGENAMMINA STILLA | X | X | |
| PSEUDASTRORHIZA IRREGULARIS | X | X | |
| SOROSPHAERA OSGOODENSIS | X | X | |
| SOROSPHAERA TRICELLA | X | X | |
| THOLOSINA ELLIPTICA | X | X | |
| THOLOSINA HEMISPHAERICA | X | X | |
| THURAMMINA CORONATA | X | X | |
| TOLYPAMMINA TORTUOSA | X | X | |
| THURAMMINA ELLIPTICA | X | X | |
| THURAMMINA FOERSTEI | X | X | |
| THURAMMINA IRREGULARIS | X | X | |
| THURAMMINA LIMBATA DISCIFORMIS | X | X | |
| THURAMMINA PARVITUBA | X | X | |
| TURRITELLELLA FISHERI | X | X | |
| TURRITELLELLA OSGOODENSIS | X | X | |
| AMMODISCUS EXSERTUS | X | X | X |
| AMMODISCUS INCERTUS | X | X | X |
| STOMASPHAERA BRASSFIELDENSIS | X | X | X |
| BATHYSIPHON EXIGUUS | X | X | X |
| HYPERAMMINA CURVA | X | X | X |
| PSAMMOSPHAERA CAVA | X | X | X |
| SOROSPHAERA CONFUSA | X | X | X |
| STEGNAMMINA CYLINDRICA | X | X | X |
| AMMODISCUS CONSTRICTUS | | X | |
| AMMODISCUS BIFURCATA | | X | |
| AMMODISCUS? FURCA | | X | |
| ASCHEMONELLA BASTILLENSIS | | X | |
| CERATAMMINA CORNUCOPIA | | X | |
| COLONAMMINA VERRUCA | | X | |
| GLOMOSPIRA SILURIANA | | X | |
| HEMISPHAERAMMINA CORONATA | | X | |
| HEMISPHAERAMMINA? GIBBOSA | | X | |
| HEMISPHAERAMMINA GEOMETRICA | | X | |
| HYPERAMMINA? DEMINUTIONIS | | X | |
| HYPERAMMINA HARRISI | | X | |
| HYPERAMMINA HASTULA | | X | |
| LAGENAMMINA DISTORTA | | X | |
| LITUOTUBA ELONGATA | | X | |
| ORDOVICINA BICUSPIDATA | | X | |
| ORDOVICINA CYLINDRICA | | X | |
| ORYCTODERMA RARA | | X | |
| PSEUDASTRORHIZA DELICATA | | X | |
| PSAMMOSPHAERA ANGULARIS | | X | |
| PSAMMOSPHAERA SUBSPHAERICA | | X | |
| SACCAMMINA ACUTA | | X | |
| SACCAMMINA ASPERA | | X | |
| SACCAMMINA BIOSCULATA | | X | |
| SACCAMMINA? CORNUTA | | X | |
| SACCAMMINA MOREMANI | | X | |
| THOLOSINA SEDENTATA | | X | |
| THURAMMINA CRESCENTRICA | | X | |
| THURAMMINA ECHINATA | | X | |
| THURAMMINA FURCATA | | X | |
| THURAMMINA HEXAGONA | | X | |
| THURAMMINA INFLATA | | X | |
| THURAMMINA LIMBATA | | X | |
| THURAMMINA MAGNA | | X | |
| THURAMMINA PAPILLATA MONTICULIFERA | | X | |
| THURAMMINA PHASELA | | X | |
| THURAMMINA POLYGONA | | X | |
| THURAMMINA SLOCOMI | | X | |
| THURAMMINA SUBSPHAERICA | | X | |
| THURAMMINA TRANSVERSALIS | | X | |
| THURAMMINA TRIANGULARIS | | X | |
| THURAMMINA TUBULATA | | X | |
| THOLOSINA? CONVEXA | | X | |
| THOLOSINA ELONGATA | | X | |
| THURAMMINA QUADRITUBULATA | | X | |
| THURAMMINA ARCUATA | | X | |
| THURAMMINA JUBATA | | X | |
| THURAMMINA CF. T. PAPILLATA | | X | |
| RAIBOSAMMINA ASPERA | | X | |
| WEBBINELLOIDEA SIMILIS | | X | X |
| SOROSPHAERA? INFLATA | | X | X |
| RHABDAMMINA TRIFURCATA | | X | X |
| HEMISPHAERAMMINA THOLA | | X | X |
| STOMASPHAERA WALDRONENSIS | | X | X |
| ANICTOSPHAERA PROGRESSA | | | X |
| ATELIKAMARA INCOMPOSITA | | | X |
| SOROSPHAERA COMPACTA | | | X |
| THURAMMINA QUADRITUBULATA | | | X |

Figure 2. Stratigraphic occurrence of species of North American Silurian agglutinate foraminifera.

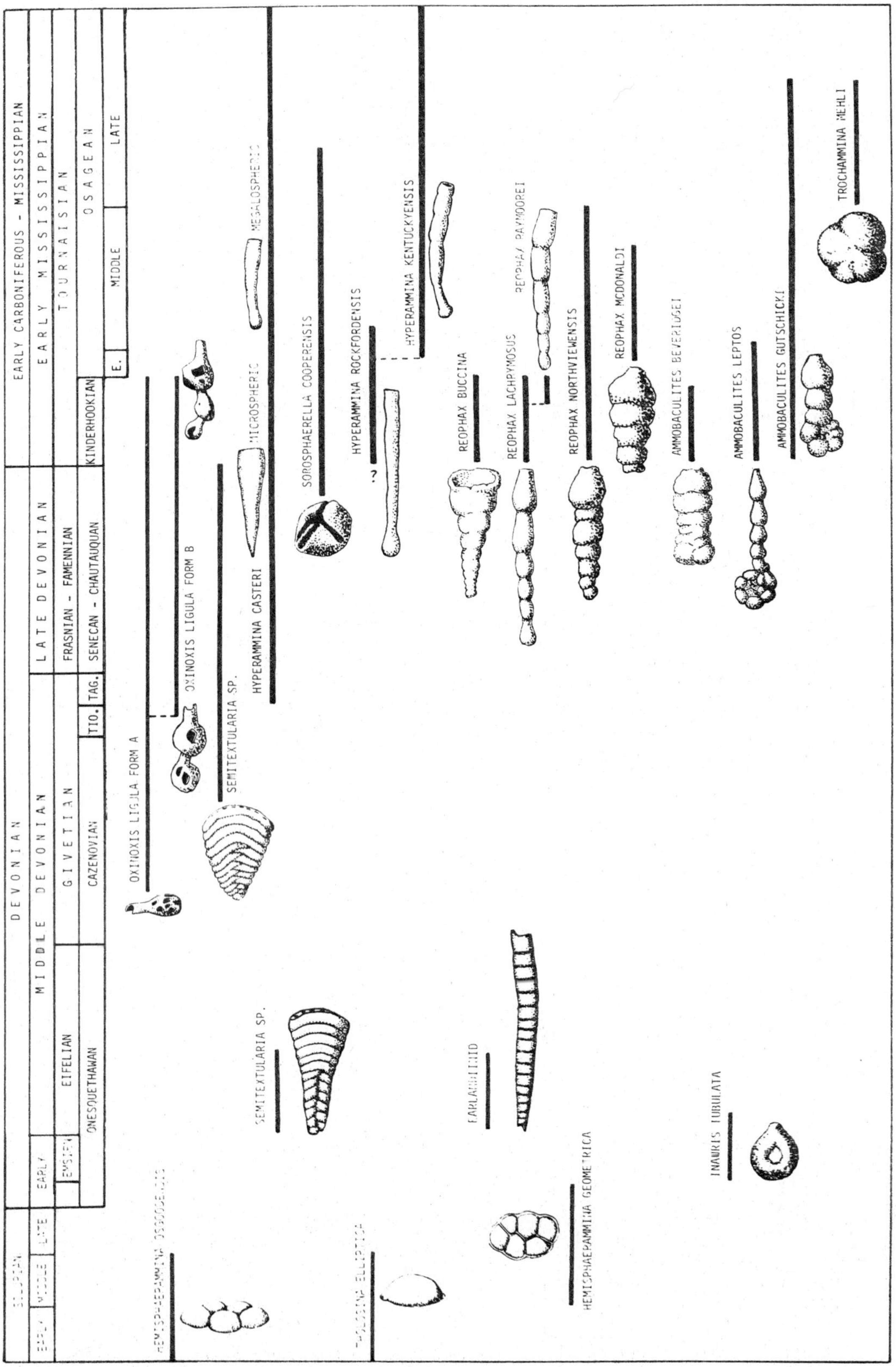


Figure 3. Ranges of selected stratigraphically useful species, including some evolutionary lineages, in the Silurian through Mississippian.

| | DEVONIAN | | | | |
|---|---|---|---|---|---|
| | HELDERBERG | SCHOHARIE | ONONDAGA | HAMILTON | UPPER |
| HEMISPHAERAMMINA QUADRIPARTITA | X | | | | |
| BATHYSIPHON SP. | X | | | | |
| CERATAMMINA CORNUCOPIA | X | | | | |
| STEGNAMMINA CYLINDRICA | X | | | | |
| PSAMMOSPHAERA CAVA | X | | | | |
| RAIBOSAMMINA ASPERA | X | | | | |
| SOROSPHAERA CONFUSA | X | | | | |
| ANICTOSPHAERA PROGRESSA | X | | | | |
| SOROSPHAERA COMPACTA | X | | | | |
| THURAMMINA ARCUATA | X | | | | |
| HEMISPHAERAMMINA THOLA | X | | | | |
| STORTHOSPHAERA MALLORYI | X | | | | |
| WEBBINELLOIDEA SIMILIS | X | | X | X | |
| SOROSPHAERA TRICELLA | X | | X | X | |
| RHABDAMMINA SP. | X | | | X | |
| BLASTAMMINA SP. | | X | | | |
| SACCAMMINA SP. | | X | | | |
| INAURIS TUBULATA | | X | X | | |
| HEMISPHAERAMMINA DICANTHA | | | X | | |
| HEMISPHAERAMMINA? BIPARTITA | | | X | | |
| HEMISPHAERAMMINA? MULTICARINATA | | | X | | |
| SEMITEXTULARIA THOMASI | | | | X | X |
| SACCORHIZA FLEXILIRAMOSA | | | | | X |
| GLOMOSPIRA BIPLANA | | | | | X |
| AMMODISCUS LONGEXSERTUS | | | | | X |
| ASCHEMONELLA LOUISIANA | | | | | X |
| AMPHITREMOIDA EISENACKI | | | | | X |
| AMPHITREMOIDA HUFFMANI | | | | | X |
| AMPHITREMOIDA SP. | | | | | X |
| ORYCTODERMA PSAMMOSPHAERIFORMIS | | | | | X |
| HYPERAMMINA SAPPINGTONENSIS | | | | | X |
| OXINOXIS LIGULA | | | | | X |
| PSAMMOSPHAERA? SP. A | | | | | X |
| PSAMMOSPHAERA? SP. B | | | | | X |
| PSAMMOSPHAERA? SP. C | | | | | X |
| THURAMMINA ADAMSI | | | | | X |
| THURAMMINA STRICKLERI | | | | | X |
| TOLYPAMMINA BULBOSA | | | | | X |
| TOLYPAMMINA CYCLOPS | | | | | X |
| TOLYPAMMINA GERSTERENSIS | | | | | X |
| TOLYPAMMINA JACOBSCHAPELENSIS | | | | | X |
| AMMODISCUS EXSERTUS | | | | | X |
| AMMODISCUS SEMICONSTRICTUS | | | | | X |
| BLASTAMMINA EISENACKI | | | | | X |
| HIPPOCREPINELLA? SP. | | | | | X |
| REOPHAX LACHRYMOSUS | | | | | X |
| REOPHAX BUCCINA | | | | | X |
| [?] SACCAMMINA HOWEI | | | | | X |
| SOROSPHAERELLA COOPERENSIS | | | | | X |
| TOLYPAMMINA FRIZZELLI | | | | | X |
| THURAMMINA CF. T. CONGESTA | | | | | X |
| TREPEILOPSIS RECURVIDENS | | | | | X |
| HYPERAMMINA CASTERI | | | | | ? |
| HYPERAMMINA ROCKFORDENSIS | | | | | ? |

Figure 4. Stratigraphic occurrence of species of North American Devonian agglutinate foraminifera.

| | MISSISSIPPIAN | | | | | | |
|---|---|---|---|---|---|---|---|
| | KIND. | | OSAGEAN | | | MIDDLE | UPPER |
| | L. | U. | L. | M. | U. | | |
| AMMOVERTELLA SP. | X | X | | | | | |
| BLASTAMMINA EISENACKI | X | X | | | | | |
| HYPERAMMINA COMPACTA | X | X | | | | | |
| HYPERAMMINA GRACILENTA | X | X | | | | | |
| HYPERAMMINA NITIDA | X | X | | | | | |
| HYPERAMMINA SAPPINGTONENSIS | X | X | | | | | |
| OXINOXIS LIGULA B | X | X | | | | | |
| PSAMMOSPHAERA SP. A | X | X | | | | | |
| REOPHAX BUCCINA | X | X | | | | | |
| REOPHAX SP. | X | X | | | | | |
| RHABDAMMINA SP. | X | X | | | | | |
| THURAMMINA FURCATA | X | X | | | | | |
| TOLYPAMMINA BOTONUNCUS | X | X | | | | | |
| AMMOBACULITES CHAPPELENSIS | X | X | X | | | | |
| AMMOBACULITES LEPTOS | X | X | X | | | | |
| AMMODISCELLA SP. | X | X | X | | | | |
| AMMOVERTELLA LISAE | X | X | X | | | | |
| AMMOVERTELLA CF. A. PRIMAPARVA | X | X | X | | | | |
| HYPERAMMINA CONSTRICTA | X | X | X | | | | |
| LITUOTUBA SEMIPLANA | X | X | ? | | | | |
| OXINOXIS LIGULA A | X | X | X | | | | |
| PSEUDASTRORHIZA CONICA | X | X | X | | | | |
| PSEUDASTRORHIZA DELICATA | X | X | X | | | | |
| REOPHAX CALATHUS | X | X | X | | | | |
| REOPHAX NORTHVIEWENSIS | X | X | X | | | | |
| THURAMMINA CONGESTA | X | X | X | | | | |
| REOPHAX LACHRYMOSUS | X | X | ? | | | | |
| THURAMMINA? TRIRADIATA | X | X | X | | | | |
| TOLYPAMMINA BRANSONI | X | X | X | | | | |
| TOLYPAMMINA BULBOSA | X | X | X | | | | |
| TOLYPAMMINA CYCLOPS | X | X | X | | | | |
| TOLYPAMMINA GERSTERENSIS | X | X | X | | | | |
| TREPEILOPSIS GLOMOSPIROIDES | X | X | X | | | | |
| AMMOBACULITES GUTSCHICKI | X | X | X | X | | | |
| AMMODISCUS LONGEXSERTUS | X | X | X | X | | | |
| HYPERAMMINA ROCKFORDENSIS | X | X | X | X | | | |
| SOROSPHAERELLA COOPERENSIS | X | X | | | X | | |
| TOLYPAMMINA FRIZZELLI | X | X | | | X | | |
| TROCAMMINA MEHLI | X | X | | | X | | |
| WEBBINELLOIDEA SIMILIS | X | X | | | X | | |
| AMMOBACULITES GUTSCHICKI | X | X | X | X | X | | |
| AMMOVERTELLA CF. A. INCLUSA | X | X | X | X | X | | |
| AMMOVERTELLA PIKEI | X | X | X | | X | | |
| GLOMOSPIRA ARTICULOSA | X | X | X | X | X | | |
| HYPERAMMINA CASTERI | X | X | X | X | X | X | X |
| TOLYPAMMINA JACOBSCHAPELENSIS | X | X | X | X | X | | |

| | MISSISSIPPIAN | | | | | | |
|---|---|---|---|---|---|---|---|
| | KIND. | | OSAGEAN | | | MIDDLE | UPPER |
| | L. | U. | L. | M. | U. | | |
| TREPEILOPSIS RECURVIDENS | X | X | X | X | X | | |
| PROTEONINA CUMBERLANDIAE | X | X | X | X | | X | |
| PROTEONINA WALLINGFORDENSIS | X | | | | | | |
| AMMODISCUS EXSERTUS | X | | | | | | |
| AMMODISCUS SEMICONSTRICTUS | X | | | | | | |
| AMMOBACULITES BEVERIDGEI | X | X | | | | | |
| OXINOXIS SWALLOWI | X | X | | | | | |
| REOPHAX RAYMOOREI | X | X | | | | | |
| SACCAMMINA HOWEI | X | X | | | | | |
| TREPEILOPSIS PRODIGALIS | X | X | | | | | |
| SOROSPHAERA PAPILLA | X | X | X | | | | |
| AMPHITREMOIDA HUFFMANI | | X | | | | | |
| PSEUDASTRORHIZA BACCULA | | X | | | | | |
| PSEUDASTRORHIZA DIGITATA | | X | | | | | |
| PSEUDASTRORHIZA LANCEOLATA | | X | | | | | |
| HYPERAMMINA CONICA | | X | | | | | |
| LAGENAMMINA SPHERICA | | X | | | | | |
| TOLYPAMMINA CONTINUUS | | X | | | | | |
| TOLYPAMMINA EXTENDA | | X | | | | | |
| TOLYPAMMINA LAOCOON | | X | | | | | |
| TOLYPAMMINA TORTUOSA | | X | | | | | |
| THURAMMINA DIFORAMENS | | X | | | | | |
| THURAMMINA ECHINADA | | X | | | | | |
| THURAMMINA FURCATA | | X | | | | | |
| THURAMMINA LIMBATA | | X | | | | | |
| THURAMMINA PUSTULOSA | | X | | | | | |
| THURAMMINA QUADRITUBULATA | | X | | | | | |
| THURAMMINA TUBULATA | | X | | | | | |
| THURAMMINA ARENACORNA | | ? | X | | | | |
| TOLYPAMMINA ROTULA | | X | ? | | | | |
| AMMOVERTELLA LABYRINTHA | | | X | | | | |
| CERATAMMINA CORNUCOPIA | | | X | | | | |
| ORYCTODERMA PALAEOZOICA | | | X | | | | |
| TROCHAMMINA OHIOENSIS | | | X | | | | |
| REOPHAX MACDONALDI | | | X | X | | | |
| REOPHAX MINUTISSIMUS | | | X | X | | | |
| HYPERAMMINA KENTUCKYENSIS | | | X | X | X | | |
| REOPHAX ASPER | | | | X | | | |
| AGATHAMMINA MISSISSIPPIANA | | | | X | X | | X |
| BLASTAMMINA SP. | | | | | X | | |
| REOPHAX CF. R. ARENATUS | | | | | X | | |
| EARLANDIA CONSTERNATIO | | | | | | X | X |
| CLIMACAMMINA MISSISSIPPIANA | | | | | | | X |
| PALAEOTEXTULARIA ASPER | | | | | | | X |
| REOPHAX KUNKLERENSIS | | | | | | | X |
| TREPEILOPSIS MISSISSIPPIANA | | | | | | | X |

Figure 5. Stratigraphic occurrence of species of North American Mississippian agglutinate foraminifera.

| | EARLY MISS. | | | | |
|---|---|---|---|---|---|
| | KIND. | | OSAGEAN | | |
| | E. | L. | E. | M. | L. |
| OXINOXIS SWALLOWI | X | | | | |
| AMMOBACULITES BEVERIDGEI | X | X | | | |
| AMPHITREMOIDA HUFFMANI | X | X | | | |
| BLASTAMMINA EISENACKI | X | X | | | |
| HYPERAMMINA COMPACTA | X | X | | | |
| HYPERAMMINA GRACILENTA | X | X | | | |
| HYPERAMMINA NITIDA | X | X | | | |
| HYPERAMMINA SAPPINGTONENSIS | X | X | | | |
| REOPHAX BUCCINA | X | X | | | |
| REOPHAX CALATHUS | X | X | | | |
| SACCAMMINA HOWEI | X | X | | | |
| THURAMMINA FURCATA | X | X | | | |
| TOLYPAMMINA BOTONUNCUS | X | X | | | |
| TREPEILOPSIS PRODIGALIS | X | X | | | |
| REOPHAX RAYMOOREI | | X | | | |
| HEMISPHAERAMMINA CECILLALICKERI | | X | | | |
| HYPERAMMINA CONICA | | X | | | |
| HYPERAMMINA DAVIDI | | X | | | |
| HYPERAMMINA DEBORAHAE | | X | | | |
| LAGENAMMINA SPHAERICA | | X | | | |
| PSEUDASTRORHIZA BACCULA | | X | | | |
| PSEUDASTRORHIZA DIGITATA | | X | | | |
| PSEUDASTRORHIZA DILLONI | | X | | | |
| PSEUDASTRORHIZA LANCEOLATA | | X | | | |
| THURAMMINA DIFORAMENS | | X | | | |
| THURAMMINA ECHINADA | | X | | | |
| THURAMMINA LIMBATA | | X | | | |
| THURAMMINA PUSTULOSA | | X | | | |
| THURAMMINA QUADRITUBULATA | | X | | | |
| TOLYPAMMINA CONTINUUS | | X | | | |
| TOLYPAMMINA EXTENDA | | X | | | |
| TOLYPAMMINA LAOCOON | | X | | | |
| AMMOVERTELLA LABYRINTHA | | | X | | |
| ORYCTODERMA PALAEOZOICA | | | X | | |
| TROCHAMMINA OHIOENSIS | | | X | | |
| PROTEONINA WALLINGFORDENSIS | | | X | X | |
| REOPHAX MCDONALDI | | | X | X | |
| REOPHAX MINUTISSIMUS | | | X | X | |
| "AGATHAMMINA" MISSISSIPPIANA | | | X | | X |
| AMMOBACULITES GUTSCHICKI | | | X | X | X |
| HYPERAMMINA KENTUCKYENSIS | | | X | X | X |

Figure 6. Differentiation of Kinderhookian from Osagean by agglutinate foraminifera.

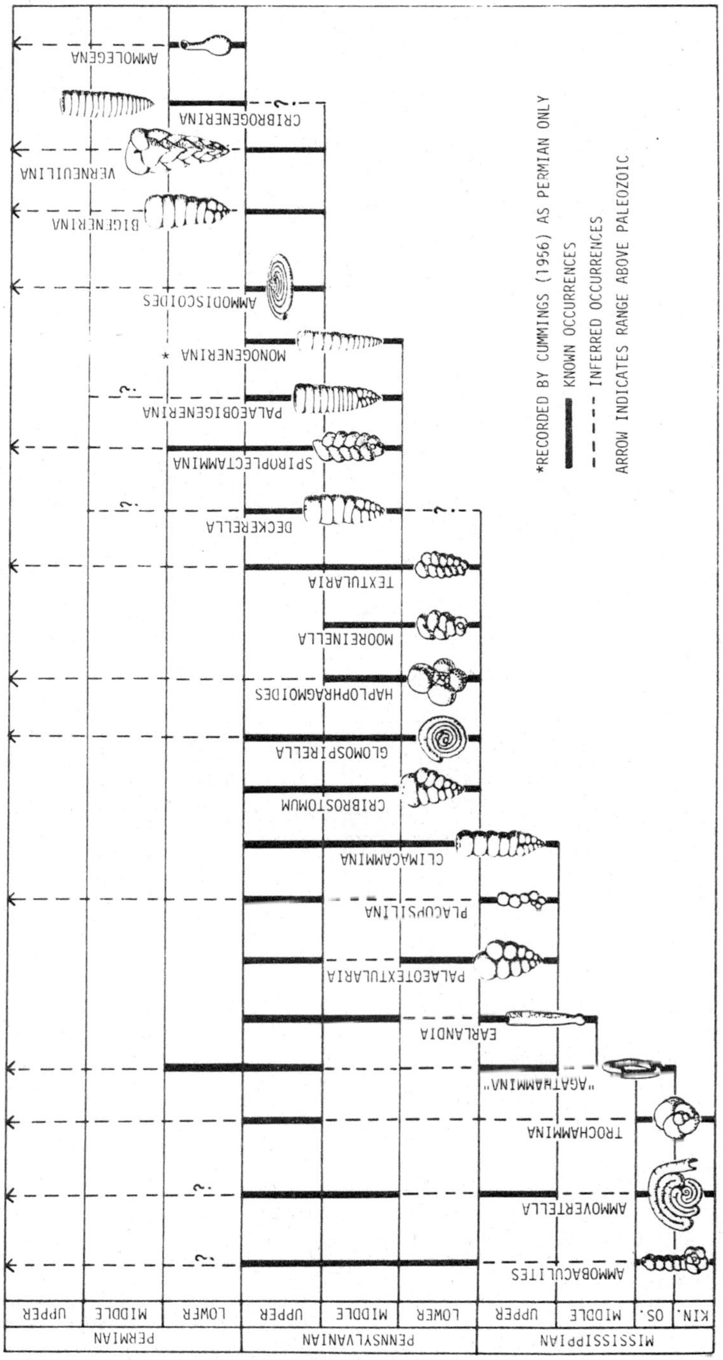

Figure 7. Range of selected genera in the Mississippian through Permian of North America.

| | MISS. | PENN-SYLVANIAN | | | PERMIAN | | |
|---|---|---|---|---|---|---|---|
| | U. | L. | M. | U. | L. | M. | U. |
| AMMODISCUS EXSERTUS | X | | | | | | |
| EARLANDIA CONSTERNATIO | X | | | | | | |
| HYPERAMMINA CASTERI | X | | | | | | |
| PALAEOTEXTULARIA ASPER | X | | | | | | |
| REOPHAX KUNKLERENSIS | X | | | | | | |
| CLIMACCAMINA MISSISSIPPIANA | X | | | | | | |
| HYPERAMMINA GLABRA | X | | X | X | | | |
| AMMODISCUS SEMICONSTRICTUS | X | X | X | X | | | |
| HYPERAMMINA GRACILIS RUGOSA | | X | | X | | | |
| REOPHAX BENDENSIS | | X | | | | | |
| PALEOTEXTULARIA SP. | | X | | | | | |
| CLIMACAMMINA MAGNA | | ? | ? | | | | |
| DECKERELLA GOESSI | | ? | ? | | | | |
| MOOREINELLA BISERIALIS | | X | X | | | | |
| HAPLOPHRAGMOIDES CONFRAGOSUS | | X | X | | | | |
| HYPERAMMINA EXPANSUS | | X | X | | | | |
| PROTEONINA CERVICIFERA | | X | X | | | | |
| REOPHAX MINUTISSIMUS | | X | X | | | | |
| HYPERAMMINA CLAVACOIDEA | | X | X | | | | |
| HYPERAMMINA SPINESCENS | | X | X | | | | |
| CLIMACAMMINA? MARBLENSE | | X | | ? | | | |
| GLOMOSPIRA SIMPLEX | | X | | X | | | |
| HYPERAMMINA BULBOSA | | X | X | X | | | |
| HYPERAMMINA GRACILIS | | X | X | X | | | |
| AMMOBACULITES MINUTUS | | X | X | X | | | |
| GLOMOSPIRA ARTICULOSA | | X | X | X | | | |
| GLOMOSPIRELLA UMBILICATA | | X | X | X | | | |
| REOPHAX EMACIATUS | | X | X | X | | | |
| REOPHAX FITTSI | | X | X | X | | | |
| REOPHAX GLENNENSIS | | X | X | X | | | |
| CRIBROSTOMUM? JEFFERSONENSIS | | ? | ? | ? | | | |
| SPIROPLECTAMMINA SUTTONENSIS | | | ? | ? | ? | | |
| AMMOBACULITES SUTTONENSIS | | | ? | ? | ? | | |
| AMMOBACULITES TEXTURATA | | | ? | ? | ? | | |
| AMMOBACULITES STORMI | | | X | | | | |
| AMMOVERTELLA INVERSA | | | X | | | | |
| CLIMACAMMINA? CUSHMANI | | | X | | | | |
| DECKERELLA CLAVATA | | | X | | | | |
| EARLANDIA BULBOSA | | | X | | | | |
| EARLANDIA MINUTA | | | X | | | | |
| HYPERAMMINA PROTEA | | | X | | | | |
| REOPHAX EXPATIATUS | | | X | | | | |
| REOPHAX TUMIDULUS | | | X | | | | |
| GLOMOSPIRA COMPRESSA | | | X | | | | |
| GLOMOSPIRA DUPLEX | | | X | | | | |
| GLOMOSPIRA UMBILICATA | | | X | | | | |
| HYPERAMMINA ELEGANTISSIMA | | | X | | | | |
| TREPEILOPSIS GRANDIS | | | X | | | | |
| TREPEILOPSIS SPIRANS | | | X | | | | |
| TEXTULARIA EXRAYENSIS | | | X | | | | |
| TURRITELLELLA GRANDIS | | | X | | | | |
| AMMOBACULITES SPIRANS | | | X | X | | | |
| AMMOBACULITES STENOMICA | | | X | X | | | |
| GLOMOSPIRA PUSILLA | | | X | X | | | |
| HYPERAMMINA NEOGLABRA | | | X | X | | | |
| MONOGENERINA GRANDIS | | | X | X | | | |
| PALAEOBIGENERINA PEREXIGUA | | | X | X | | | |

| | MISS. | PENN-SYLVANIAN | | | PERMIAN | | |
|---|---|---|---|---|---|---|---|
| | U. | L. | M. | U. | L. | M. | U. |
| REOPHAX ARENATUS | | | X | X | | | |
| REOPHAX ASPER | | | X | X | | | |
| AMMOBACULITES GRACILIS | | | | X | | | |
| AMMOBACULITES INCONSPICUA | | | | X | | | |
| AMMOBACULITES MAGNIGRANULATUS | | | | X | | | |
| AMMOBACULITES PARALLEUS | | | | X | | | |
| AMMOBACULITES POWERSI | | | | X | | | |
| AMMOBACULITES RECTA | | | | X | | | |
| AMMODISCUS ANNULARIS | | | | X | | | |
| AMMODISCELLA VIRGILENSIS | | | | X | | | |
| AMMODISCOIDES CONICA | | | | X | | | |
| AMMOVERTELLA ELEVATA | | | | X | | | |
| AMMOVERTELLA INCLUSA | | | | X | | | |
| AMMOVERTELLA PRIMAPARVA | | | | X | | | |
| AMMOVERTELLA TORNELLA | | | | X | | | |
| ASTRORHIZA? SP. | | | | X | | | |
| BIGENERINA[?] VIRGILENSIS | | | | X | | | |
| BIGENERINA CISCOENSIS | | | | X | | | |
| CLIMACAMMINA CYLINDRICA | | | | X | | | |
| CLIMACAMMINA[?] LUCILLEAE | | | | X | | | |
| CLIMACAMMINA MOELLERI | | | | X | | | |
| DECKERELLA LAHEEI | | | | X | | | |
| EARLANDIA PERPARVA | | | | X | | | |
| GLOMOSPIRA DIVERSA | | | | X | | | |
| GLOMOSPIRA MONOGRANULATA | | | | X | | | |
| GLOMOSPIRA REVERSA | | | | X | | | |
| GLOMOSPIRA SIMPLEX | | | | X | | | |
| GLOMOSPIRA SPINOSA | | | | X | | | |
| CRIBROSTOMUM? ATTENUATA | | | | X | | | |
| HYPERAMMINA ELEGANS | | | | X | | | |
| PALAEOTEXTULARIA GRAHAMENSIS | | | | X | | | |
| PLACOPSILINA CISCOENSIS | | | | X | | | |
| PSAMMOSPHAERA CAVA? | | | | X | | | |
| SACCAMMINIS MULTICELLA | | | | X | | | |
| TEXTULARIA CORNUTA | | | | X | | | |
| TEXTULARIA FUSCALIGNENSIS | | | | X | | | |
| SPIROPLECTAMMINA CASTENSIS | | | | X | | | |
| SPIROPLECTAMMINA CLAVATA | | | | X | | | |
| TEXTULARIA EXEMIA | | | | X | | | |
| TEXTULARIA[?] SP. | | | | X | | | |
| THURAMMINA DIFORAMENS | | | | X | | | |
| THURAMMINA ELLIPTICA | | | | X | | | |
| THURAMMINA LAWRENCENSIS | | | | X | | | |
| THURAMMINA RECTANGULARIS | | | | X | | | |
| THURAMMINA VERRUCOSA | | | | X | | | |
| TOLYPAMMINA DELICATULA | | | | X | | | |
| TOLYPAMMINA EXTENDA | | | | X | | | |
| TOLYPAMMINA NODOSA | | | | X | | | |
| TOLYPAMMINA SERPENS | | | | X | | | |
| TROCHAMMINA ARENOSA | | | | X | | | |
| TROCHAMMINA GRAHAMENSIS | | | | X | | | |
| VERNEUILIANA VIRGILENSIS | | | | X | | | |
| AMMOVERTELLA LABYRINTHA | | | | X | ? | | |
| AGATHAMMINA PROTEA | | | | | X | | |
| CRIBROGENERINA KRIZI | | | | | X | | |

Figure 8. Stratigraphic occurrence of species of North American agglutinate foraminifera in the Upper Mississippian through Permian.

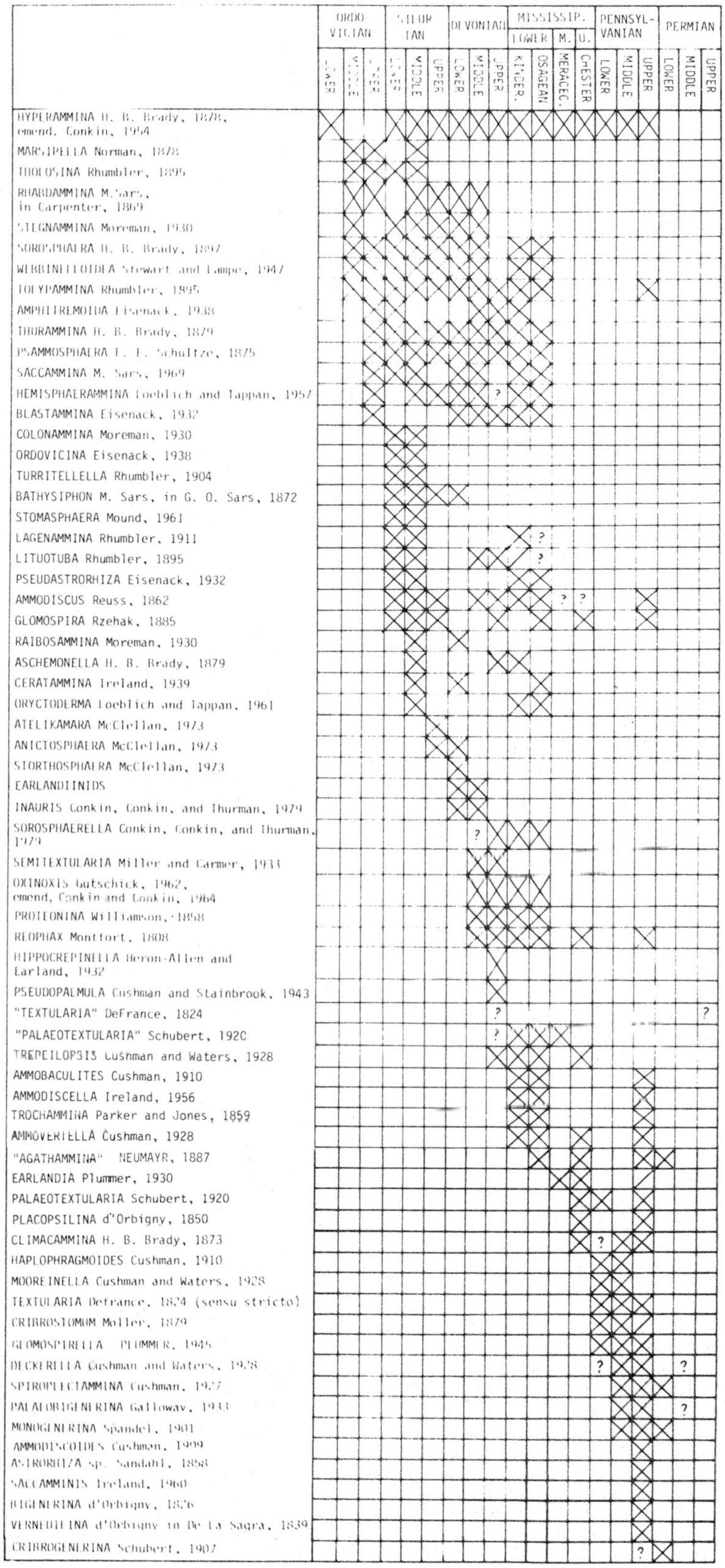


Figure 9. Stratigraphic occurrence of genera of North American Paleozoic agglutinate foraminifera.

# REFERENCES

Adams, C. G. 1967. Tertiary foraminifera in the Tethyan, American and Indo-Pacific provinces, In: C. G. Adams and D. B. Ager (eds.), Aspects of Tethyan biogeography. System. Assoc., Publ. 7: 195-217.

Albani, A. D. and K. R. Johnson. 1975. Resolution of foraminiferal biotopes in Broken Bay, New South Wales. Geol. Soc. Australia J. 22: 435-446.

Albers, C. C. and others. 1966. Foraminiferal ecological zones of the Gulf Coast: Progress report of the New Orleans Paleoecologic Committee Gulf Coast Section, Soc. Econ. Paleontol. Mineral. Gulf Coast Assoc. Geol. Soc. Trans., 16: 345-348.

Alexandrowicz, S. W. 1969. Thuramminoides sphaeroidalis Plummer (foraminifera) from Cambrian beds of the vicinity of Sandomierz. Annls. Soc. Geol. Pologne, 39: 27-34.

Alldredge, A. L. and B. M. Jones. 1973. Hastigerina pelagica: Foraminiferal habitat for planktonic dinoflagellates. Mar. Biol., 22: 131-135.

Alvarez, L. W., W. Alvarez, F. Asaro and H. V. Michel. 1980. Extraterrestrial cause for Cretaceous-Tertiary extinction. Science, 208: 1095-1108.

Alvarez, W., M. A. Arthur, A. G. Fischer, W. Lowrie, G. Napoleone, I. Premoli Silva and W. M. Roggenthen. 1977. Upper Cretaceous - Paleocene magnetic stratigraphy at Gubbio, Italy. V. Type section for the late Cretaceous - Paleocene geomagnetic reversal time scale. Geol. Soc. Amer. Bull., 88: 383-389.

Alvarez, W. A. et al. 1982. Iridium anomaly approximately synchronous with terminal Eocene extinctions. Science, 216: 886-888.

Anderson, J. B. 1968. Ecology of foraminifera from Mississippi Sound and surrounding waters. Alabama Acad. Sci. J., 39: 261-269.

Anderson, O. R. and A. W. H. Bé. 1976a. A cytochemical fine structure study of phagotrophy in a planktonic foraminifer, Hastigerina pelagica (d'Orbigny). Biol. Bull., 151: 437-449.

Anderson, O. R. and A. W. H. Bé. 1976b. The ultrastructure of a planktonic foraminifer, Globigerinoides sacculifer (Brady) and its symbiotic dinoflagellates. J. Foram. Res., 6(1): 1-21.

Anderson, O. R. and A. W. H. Bé. 1978. Recent advances in foraminiferan fine structure research, In: Hedley, R. H. and C. G. Adams (eds.), "Foraminifera", 3: 121-202. Acad. Press, London.

Anderson, O. R., M. Spindler, A. W. H. Bé and C. Hemleben. 1979. Trophic activity of planktonic foraminifera. J. Mar. Biol. Assoc. U.K., 59: 791-799.

Anglada, R. and A. J. O. Randrianasolo. 1971. Contribution à l'étude systématique du sous-ordre des Miliolina (Foraminiferida). Ann. univ. Provence, Sci., 46: 161-176.

Arthur, M. A. and S. O. Schlanger. 1979. Cretaceous "Oceanic Anoxic Events" as casual factors in development of reef-reservoired giant oil fields. Amer. Assoc. Petrol. Geol. Bull., 63: 870-885.

Balakhmatova, V. T. 1972. K sistematike semeystva Ataxophragmiidae Schwager, 1877. [On systematics of the family Ataxophragmiidae Schwager, 1877.] Akad. Nauk SSSR, Vopr. Mikropaleont., 5: 70-74.

Bandy, O. L. 1964. Foraminiferal biofacies in sediments of Gulf of Batabano, Cuba and their geologic significance. Amer. Assoc. Petrol. Geol., Bull. 48: 1666-1679.

Bandy, O. L. 1967. Cretaceous planktonic foraminiferal zonation. Micropaleontology, 13: 1-31.

Bandy, O. L. 1972. Origin and development of Globorotalia (Turborotalia) pachyderma. Micropaleontology, 18: 294-318.

Bandy, O. L. 1975. Messinian evaporite deposition and the Miocene/ Pliocene boundary, Paquasia-Capodarso sections, Sicily, In: T. Saito, and L.H. Burckle (eds.) Late Neogene epoch boundaries. Micropaleontology Spec. Pub. No. 1: 49-63.

Bandy, O. L., J. C. Ingle, Jr. and J. M. Resig. 1964. Foraminifera, Los Angeles County outfall area, California. Limnol. Oceanogr., 9: 124-137.

Bandy, O. L., J. C. Ingle, Jr. and J. M. Resig. 1965a. Modification of foraminiferal distribution by the Orange County outfall, California. Ocean Sci. Ocean Engineer., Trans. Marine Technol. Soc., 1965: 54-76.

Bandy, O. L., J. C. Ingle, Jr. and J. M. Resig. 1965b. Foraminiferal trends, Hyperion outfall, California. Limnol. Oceanogr., 10: 314-332.

Banner, F. T. 1966. Morfologiya, klassifikatsiya i stratigraficheskoe znachenie spirotsiklinid. [Morphology, classification and stratigraphic significance of the Spirocyclinidae.] Adad. Nauk. SSSR, Vopr. Mikropaleontol., 10: 201-224.

Banner, F. T. 1970. A synopsis of the Spirocyclinidae. Revta. Esp. Micropaleontol., 2: 243-290.

Banner, F. T. 1971. A new genus of the Planorbulinidae, an endoparasite of another foraminifer: Rev. Esp. Micropaleontol., 3: 113-128.

Banner, F. T. and W. H. Blow. 1960. Some primary types of species belonging to the superfamily Globigerinacea. Contr. Cushman Found. Foram. Res., 11: 1-41.

Banner, F. T. and C. P. G. Pereira. 1981. Some biserial and triserial agglutinated smaller foraminifera: their wall structure and its significance. J. Foram. Res., 11: 85-117.

Barker, P. F. and J. Burrell. 1977. The opening of Drake Passage. Mar. Geol., 24: 15-34.

Barnard, T. and F. T. Banner. 1980. The Ataxophragmiidae of England: Part I. Albian-Cenomanian Arenobulimina and Crenaverneuilina. Revta. Esp. Micropaleontol., 12: 383-430.

Barr, F. T. and W. A. Berggren. 1982. Lower Tertiary biostratigraphy and tectonics of northeastern Libya. 2nd Symposium on the Geology of Libya, Tripoli: 163-192.

Bates, J. M. and R. S. Spencer. 1979. Modification of foraminiferal trends by the Chesapeake-Elizabeth sewage outfall, Virginia Beach, Virginia. J. Foram. Res., 9: 125-140.

Batjes, D. A. J. 1958. Foraminifera of the Oligocene of Belgium. Inst. Roy. Sci. Natur. Belg. Mem. 143: 1-187.

Baum, G., P. R. Vail and J. Hardenbol. 1982. Unconformities and depositional sequences in relationship to eustatic sea-level changes: Gulf and Atlantic coastal plains. IGCP Proj. 174 Field Conference, Baton Rouge, Jan. 19-29, 1981.

Bé, A. W. H. 1968. Shell porosity of Recent planktonic foraminifera as a climatic index. Science, 161 (3844): 881-884.

Bé, A. W. H. 1980. Gametogenic calcification in a spinose planktonic foraminifer, Globigerinoides sacculifer (Brady). Mar. Micropaleontol. 5(3): 283-310.

Bé, A. W. H. and O. R. Anderson. 1976. Gametogenesis in planktonic foraminifera. Science, 192 (4242): 890-892.

Bé, A. W. H., D. A. Caron and O. R. Anderson. 1981. Effects of feeding frequency on life processes of the planktonic foraminifer Globigerinoides sacculifer (Brady) in laboratory culture. J. Mar. Biol. Assoc. U.K., 61(1): 257-277.

Bé, A. W. H. and D. B. Ericson. 1963. Aspects of calcification in planktonic foraminifera, In: Comparative Biology of Calcified Tissues. New York Acad. Sci. Trans., 109 (1): 65-81.

Bé, A. W. H., W. E. Frerichs, R. Harrison and M. E. Heiman. 1976. Variability in test porosity of Orbulina universa d'Orbigny at two Indian Ocean localities: Prog. Micropaleontol. Asano Mem. Vol.: 1-11.

Bé, A. W. H., C. Hemleben, O. R. Anderson, M. Spindler, J. Hacunda and S. Tuntivate-Choy. 1977. Laboratory and field observations of living planktonic foraminifera. Micropaleontology, 23(2): 155-179.

Bé, A. W. H., C. Hemleben, O. R. Anderson and M. Spindler. 1979. Chamber formation in planktonic foraminifera. Micropaleontology, 25(3): 294-307.

Bé, A. W. H., C. Hemleben, O. R. Anderson and M. Spindler. 1980. Pore structures in planktonic foraminifera. J. Foram. Res., 10(2): 117-128.

Bé, A. W. H. and L. Lott. 1964. Shell growth and structure of planktonic foraminifera. Science, 145(3634): 823-824.

Bé, A. W. H. and H. Spero. 1981. Shell regeneration and biological recovery of planktonic foraminifera after physical injury induced in laboratory culture. Micropaleontology, 27(3): 305-316.

Bé, A. W. H., H. Spero and O. R. Anderson. 1982. Effects of symbiont elimination and reinfection on the life processes of the planktonic foraminifer Globigerinoides sacculifer. Mar. Biol.

Bé, A. W. H. and D. S. Tolderlund. 1971. Distribution and ecology of living planktonic foraminifera in surface waters of the Atlantic and Indian Oceans, In: B. M. Funnell and W. R. Riedel, (eds.), The Micropalaeontology of Oceans: 105-149, Cambridge University Press.

Beckmann, J. P. 1960. Distribution of benthonic foraminifera at the Cretaceous-Tertiary boundary of Trinidad (West Indies). Rept. Intern. Geol. Congr., 21st sess., Norden, Copenhagen, Part 5: 57-69.

Belyaeva, N. V. 1964. Distribution of planktonic foraminifera in the water and on the floor of the Indian Ocean. Trudy Inst. Okeanol. Akad. Nauk SSSR., 69: 12-83.

Benson, R. H. 1975. The origin of the psychrosphere as recorded in changes of deep-sea ostracode assemblages. Lethaia, 8: 69-83.

Berger, W. H., and E. L. Winterer. 1974. Plate stratigraphy and the fluctuating carbonate line, In: K. J. Hsu and H. C. Jenkyns (eds.), Pelagic sediments on land and under the sea. Intl. Assoc. Sedimentol. Spec. Publ., 1: 11-48.

Berggren, W. A. 1962. Some planktonic foraminifera from the Maestrichtian and type Danian Stages of southern Scandinavia. Stockholm Contrib. Geol., 9: 1-106.

Berggren, W. A. 1969. Cenozoic chronostratigraphy, planktonic foraminiferal zonation and the radiometric time scale. Nature, 224: 1072-1075.

Berggren, W. A. 1971. Tertiary boundaries and correlations, In: B. M. Gunnell, and W. R. Riedel (eds.), Micropaleontology of the Oceans: 693-809. Cambridge Univ. Press.

Berggren, W. A. 1972a. Cenozoic biostratigraphy and paleobiogeography of the North Atlantic, Init. Rept. Deep Sea Drill. Proj., 12: 965-1001.

Berggren, W. A. 1972b. A Cenozoic time-scale - some implications for

regional geology and paleobiology. Lethaia, 5: 195-215.
Berggren, W. A. 1974a. Late Paleocene-Early Eocene benthonic foraminiferal biostratigraphy and paleoecology of Rockall Bank. Micropaleontology, 20: 426-448.
Berggren, W. A. 1974b. Paleocene benthonic foraminiferal biostratigraphy, biogeography and paleoecology of Libya and Mali. Micropaleontology, 20: 449-465.
Berggren, W. A. 1977. North Atlantic Cenozoic foraminifera, In: F. M. Swain (ed.), Stratigraphic Micropaleontology of Atlantic Basins and Borderlands: 389-410, Elsevier, Amsterdam.
Berggren, W. A. and J. Aubert. 1975. Paleocene benthonic foraminiferal biostratigraphy and paleoecology of Atlantic-Tethyan regions: Midway-type fauna. Palaeogeogr. Palaeoclimatol., Palaeoecol., 18: 73-192.
Berggren, W. A. and J. Aubert. 1976. Late Paleogene (Late Eocene and Oligocene) benthonic foraminiferal biostratigraphy and paleobathymetry of Rockall Bank and Hatton-Rockdall basin. Micropaleontology 22: 307-326.
Berggren, W. A. and J. Aubert. in press. Paleogene benthonic foraminiferal biostratigraphy and bathymetry of the central Coast Ranges of California. U.S. Geol. Surv. Prof. Paper.
Berggren, W. A. and B. U. Haq. 1976. The Andalusian Stage (Late Miocene): biostratigraphy, biochronology and paleoecology. Palaeogeogr., Palaeoclimatol., Palaeoecol., 20: 67-129.
Berggren, W. A. and C. D. Hollister. 1977. Plate tectonics and paleo-circulation - commotion in the ocean. Tectonophysics, 38: 11-48.
Berggren, W. A. and J. D. Phillips. 1971. The influence of continental drift on the distribution of Cenozoic benthonic foraminifera in the Caribbean and Mediterranean regions, In: C. Gray (ed.), The Geology of Libya, Catholic Press, Beirut: 265-299.
Berggren, W. A. and D. Schnitker. in press. Cenozoic marine environments in the North Atlantic and Norwegian-Greenland Sea, In: S. Saxov (ed.), Structure and development of the Greenland-Scotland Ridge, NATO Advanced Research Institute, Bressanone, Italy, May 11-16, 1981. Plenum Press, New York.
Berggren, W. A., et al. 1976. The El Cuervo section (Andalusia, Spain): micropaleontological anatomy of an Early Miocene lower bathyal deposit. Mar. Micropaleontol., 1: 195-247.
Berggren, W. A. and J. A. Van Couvering. 1974. The Late Neogene: biostratigraphy, geochronology and paleoclimatology of the last 15 million years in marine and continental sequences. Palaeogeogr., Palaeoclimatol., Palaeoecol., 16: 1-126.
Berggren, W. A. and J. A. Van Couvering. 1978. Biochronology, In: G. V. Cohee, M. F. Glaessner, and H. D. Hedberg (eds.), The geologic time scale, studies in geology. Amer. Assoc. Petrol. Geol. Mem. 6: 39-55.
Bergstresser, T. J. 1982. Planktonic foraminifera from the Upper Cretaceous Pierre Shale at Red Bird, Wyoming. J. Foram. Res., in press.
Bermudez, P. J. and F. C. de Rivero. 1963. Estudio systematico de los foraminiferos quitinosos, microgranulares y arenaceos. Univ. Cent. Venezuela, Ed. Bibl. 14, Colec. Cien. Biol. 1, Caracas, 398 p.
Betjeman, K. J. 1969. Recent foraminifera from the western continental shelf of Western Australia. Contr. Cushman Found. Foram. Res., 20: 119-138.
Bilyard, G. R. 1974. The feeding habits and ecology of Dentalium entale simpsoni Henderson. Veliger, 17: 126-138.

Blainville, H. M. D. de. 1825. Manuel de malacologie et de conchyliologie. (1837), F. G. Levrault, Paris, 664 p.
Blanc, P.-L., et al. 1980. North Atlantic deep water formed by the later middle Miocene. Nature, 283: 553-555.
Blow, W. H. 1969. Late Middle Eocene to Recent planktonic foraminiferal biostratigraphy, In: P. Brönnimann, and H. H. Renz (eds.), First Int. Conf. Plankt. Microfossils, Geneva: 199-421.
Blow, W. H. 1979. The Cainozoic Globigerinida. A study of the morphology taxonomy, evolutionary relationships and the stratigraphical distribution of some Globigerinida (mainly Globigerinacea). Text, pt. II, sec. 2, Danian to Oligocene planktonic foraminiferal biostratigraphy. ix + 753-1413 p., E. J. Brill, Leiden.
Bolli, H. M. 1957a. The genera Praeglobotruncana, Rotalipora, Globotruncana, and Abathomphalus in the Upper Cretaceous of Trinidad, B. W. I. U.S. Natl. Mus. Bull., 215: 51-60.
Bolli, H. M. 1957b. The genera Globigerina and Globorotalia in the Paleocene-lower Eocene Lizard Springs Formation of Trinidad, B. W. I. U.S. Natl. Mus. Bull., 215: 61-68.
Bolli, H. M. 1957c. Planktonic foraminifera from the Oligocene-Miocene Cipero and Lengua Formations of Trinidad, B. W. I. U.S. Natl. Mus. Bull., 215: 97-213.
Bolli, H. M. 1966. Zonation of Cretaceous to Pliocene marine sediments based on planktonic foraminifera. Bol. Inf. Asoz. Venez. Geol. Min. e Petr., 9: 3-32.
Boltovskoy, E. 1965a. Twilight of foraminiferology. J. Paleontol., 39: 383-390.
Boltovskoy, E. 1965b. Los foraminiferos Recientes; biologia, metodos de estudio, aplicacion oceanografica. EUDEBA, Buenos Aires, 510 p.
Boltovskoy, E. 1966. Depth at which foraminifera can survive in sediments. Contr. Cushman Found. Foram. Res., 17: 43-45.
Boltovskoy, E. 1980. On the benthonic bathyal-zone foraminifera as stratigraphic guide fossils. J. Foram. Res., 10: 163-172.
Boltovskoy, E., G. Giussani, S. Watanabe and R. Wright. 1980. Atlas of Benthic Shelf Foraminifera of the Southwest Atlantic. 147 p., Dr. W. Junk, The Hague.
Boltovskoy, E. and H. Lena. 1971. The foraminifera (except family Allogromiidae) which dwell in fresh water. J. Foram. Res. 1: 71-76.
Boltovskoy, E. and R. Wright. 1976. Recent Foraminifera. 515 p., Dr. W. Junk, The Hague.
Bradshaw, J. S. 1955. Preliminary laboratory experiments on ecology of foraminiferal populations. Micropaleontology, 1: 351-358.
Bradshaw, J. S. 1959. Ecology of living planktonic foraminifera in the North and Equatorial Pacific Ocean. Contr. Cushmanman Found. Foram. Res., 10(2): 25-64.
Bradshaw, J. S. 1961. Laboratory experiments on the ecology of foraminifera. Contr. Cushman Found. Foram. Res. 12: 87-106.
Brady, H. B. 1876. A monograph of Carboniferous and Permian foraminifera (the genus Fusulina excepted). Paleontogr. Soc. Mon. 30: 1-166.
Brady, H. B. 1884. Report on the foraminifera dredged by H.M.S. Challenger, during the years 1873-1876. Rept. Sci. Res. Voy. H.M.S. Challenger, Zool. 9: 1-814.
Brand, T. E. and J. H. Lipps. 1982. Foraminifera in the trophic structure of shallow-water Antarctic marine communities. J. Foram. Res. 12: 96-104.
Brasier, M. D. 1975. The ecology and distribution of Recent foraminifera from the reefs and shoals around Barbuda, West Indies. J. Foram. Res.,

5: 193-210.
Bremer, M. L. and G. P. Lohnmann. in press. Evidence for primary control of the distribution of certain Atlantic Ocean benthonic foraminifera by degree of carbonate saturation. Deep Sea Research.
Briggs, J. C. 1974. Marine Zoogeography. McGraw-Hill, New York, 475 p.
Brönnimann, P., L. Zaninetti, F. Bozorgnia, G. R. Dashti and A. Moshtaghian. 1971. Lithostratigraphy and foraminifera of the Upper Triassic Naiband Formation, Iran. Revue Micropaleontol., 14: 7-16.
Brouwer, J. 1965. Agglutinated foraminiferal faunas from some turbidite sequences. I. II. K. Nederl. Akad. Wetensch. Proc., ser. B, 68: 309-334.
Browne, R. G. and V. J. Schott. 1963. Arenaceous foraminifera from the Osgood Formation Osgood, Indiana. Bull. Amer. Paleontol. 46: 191-234.
Buchanan, J. B. and R. H. Hedley. 1960. A contribution to the biology of Astrorhiza limicola. J. Mar. Biol. Assoc. U.K., 39: 549-560.
Buchardt, B. and H. J. Hansen. 1977. Oxygen isotope fractionation and algal symbiosis in benthic foraminifera. Bull. Geol. Soc. Denmark, 26: 185-194.
Buzas, M. A. 1967. An application of canonical analysis as a method for comparing faunal areas. Anim. Ecol., 36: 563-577.
Buzas, M. A. 1970. On the quantification of biofacies. Proc. N. Amer. Paleontol. Conv. 1969. B: 101-116.
Buzas, M. A. 1972. Biofacies analysis of presence or absence data through canonical variate analysis. J. Paleontol., 46: 55-57.
Buzas, M. A. 1974. Vertical distribution of Ammobaculites in the Rhode River, Maryland, J. Foram. Res., 4: 144-147.
Buzas, M. A. 1978. Foraminifera as prey for benthic deposit feeders: results of predator exclusion experiments. J. Mar. Res., 36: 617-625.
Buzas, M. A. 1979. Quantitative biofacies analysis, In: J. H. Lips, W. H. Berger, M. A. Buzas, R. G. Douglas and C. A. Ross. Foraminiferal Ecology and Paleoecology. Soc. Econ. Paleontol. Mineral. Short Course No. 6: 11-20, Houston, 1979.
Buzas, M. A. 1982. Regulation of foraminiferal densities by predation in the Indian River, Florida. J. Foram. Res., 12: 66-71.
Buzas, M. A. and S. J. Culver. 1979. Foraminifera: distribution of provinces in the western North Atlantic. Science 209: 687-689.

Caron, D. A., A. W. H. Bé and O. R. Anderson. 1982. Effects of variations in light intensity on life processes of the planktonic foraminifer Globigerinoides sacculifer in laboratory culture. J. Mar. Biol. Assoc. U.K., 62(2): 435-451.
Carpenter, W. B., W. K. Parker and T. R. Jones. 1862. Introduction to the study of foraminifera. Roy. Soc. London, 319 p.
Cebulski, D. E. 1969. Foraminiferal populations and faunas in barrier-reef tract and lagoon, British Honduras. Amer. Assoc. Petrol. Geol. Mem., 11: 311-328.
Chapman, E. and W. J. Parr. 1936. A classification of the foraminifera. Proc. Roy. Soc. Vict. n. ser., 14: 139-151.
Charletta, A. C. Eocene benthic foraminiferal paleoecology and paleo-bathymetry of the New Jersey continental margin. Unpubl. Ph.D. dissert., Rutgers University: 73 p.
Christiansen, B. 1958. The foraminifer fauna in the Drobak Sound in the Oslo Fjord (Norway). Nutt Mag. Zool., 6: 5-91.
Christiansen, B. 1964. Spiculosiphon radiata, a new foraminifera from

northern Norway. Astarte, 25: 1-8.
Cifelli, R. 1982. Textural observations on some living species of planktonic foraminifera. Smithsonian Contr. Paleobiol. 45: 1-44.
Closs, D. 1963. Foraminiferos e Tecamebas da Lagoa dos Patos (R. G. S.) Bol. escol. Geol. Univ. Rio Grande do Sul, Porto Alegre, 11: 1-130.
Closs, D. and V. M. F. Medeiros. 1965. New observations on the ecological subdivision of the Patos Lagoon in southern Brazil. Bol. Inst. Cienc. Nat. Univ. Rio Grande do Sul, Porto Alegre, 24: 7-33.
Collinson, P. 1980. Vertical distribution of foraminifera off the coast of Northumberland, England. J. Foram. Res., 10: 75-78.
Conkin, J. E. 1954. Hyperammina kentuckyensis n. sp. from the Mississippian of Kentucky, and discussion of Hyperammina and Hyperamminoides. Contrib. Cushman Found. Foram. Res., 5: 165-169.
Conkin, J. E. 1961. Mississippian smaller foraminifera of Kentucky, southern Indiana, northern Tennessee, and south-central Ohio. Bull. Amer. Paleontol., 43: 131-368.
Conkin, J. E. and P. Ciesielski. 1973. Lower Mississippian (Kinderhookian) arenaceous foraminifera from the Maury Formation at Gipsy, Limestone County, Alabama. Alabama Geol. Surv. Bull. 103: 1-54.
Conkin, J. E. and B. M. Conkin. 1960. Silurian and Devonian foraminifera of Kentucky. Geol. Soc. Amer. Abst. S.E. Sec., p. 89.
Conkin, J. E. and B. M. Conkin. 1964a. Devonian foraminifera: Part I, the Louisiana Limestone of Missouri and Illinois. Bull. Amer. Paleontol., 43: 53-105.
Conkin, J. E. and B. M. Conkin. 1964b. Mississippian foraminifera of the United States: Part 1 - the Northview Formation of Missouri. Micropaleontology, 10: 19-46.
Conkin, J. E. and B. M. Conkin. 1965. Ordovician (Richmondian) foraminifera from Oklahoma, Missouri, Illinois, and Kentucky. Oklahoma Geol. Notes, 25: 207-221.
Conkin, J. E. and B. M. Conkin. 1967. Arenaceous foraminifera as a key to Upper Devonian-Lower Mississippian relationships in the type Mississippian area, In: Teichert, C. and E. L. Yochelson, (eds.), Essays in Paleontology and Stratigraphy. Kansas Univ. Dept. Geol., Spec. Publ., No. 2: 85-101.
Conkin, J. E. and B. M. Conkin. 1968. A revision of some Upper Devonian foraminifera from Western Australia. Palaeontology, 11: 601-609.
Conkin, J. E. and B. M. Conkin. 1970a. North American Kinderhookian (Lower Mississippian) arenaceous foraminifera. 6th Internat. Congr. Strat. Geol. Carbonif., Sheffield. 2: 575-584.
Conkin, J. E. and B. M. Conkin. 1970b. Middle Devonian arenaceous foraminifera of central Ohio: Part 1 - Revision of the genus Webbinelloidea Stewart and Lampe, 1947. Micropaleontology, 16: 1-14.
Conkin, J. E. and B. M. Conkin. 1971. Arenaceous foraminifera from the Lower Mississippian Banff Formation of the Crowsnest area, British Columbia-Alberta, Canada. 7th Internat. Cong. Strat. and Geol. Carbonif., Krefeld, p. 39.
Conkin, J. E. and B. M. Conkin. 1973a. Pre-Pennsylvanian foraminifera of North America. 12th Internat. Geol. Congr., India, p. 319-335.
Conkin, J. E. and B. M. Conkin. 1973b. The paracontinuity and the determination of the Devonian-Mississippian boundary in the type Lower Mississippian area of North America. Univ. Louisville Stud. Paleontol. Stratig., 1: 1-34.
Conkin, J. E. and B. M. Conkin. 1977. Paleozoic smaller foraminifera of

the North Atlantic Borderlands, In: Swain, F. M., Stratigraphic Micropaleontology of Atlantic Basin and Borderlands, p. 49-59, Elsevier, Amsterdam.

Conkin, J. E. and B. M. Conkin. 1979. North American Ordovician agglutinate foraminifera. Univ. Louisville Stud. Paleontol. Stratig., 8: 1-24.

Conkin, J. E. and B. M. Conkin. 1981. Early Mississippian (Kinderhookian) smaller foraminifera from the McCraney Limestone of Missouri and Illinois. Univ. Louisville Stud. Paleontol. Stratig., 15: 1-39.

Conkin, J. E. and B. M. Conkin. 1982. Agglutinate foraminifera from the Early Mississippian (Kinderhookian) Caballero Formation of New Mexico. 9th Internat. Cong. Strat. Geol. Carbonif. C. R., in press.

Conkin, J. E., B. M. Conkin and W. F. Canis. 1968. Mississippian foraminifera of the United States: Part 3: The limestones of the Chouteau Group in Missouri and Illinois. Micropaleontology, 14: 133-178.

Conkin, J. E., B. M. Conkin, and D. E. McDonand. 1963. Mississippian smaller foraminifera from the Southern Peninsula of Michigan. Micropaleontology, 9: 215-227.

Conkin, J. E., B. M. Conkin and J. W. Pike. 1965. Mississippian foraminifera of the United States: Part 2 - The Hannibal Formation of northeastern Missouri and western Illinois. Micropaleontology, 11: 335-359.

Conkin, J. E., B. M. Conkin, T. Sawa and M. Kern. 1970. Middle Devonian Moellerina greenei Zone and suppression of the genus Weikkoella Summerson, 1958. Micropaleontology, 16: 399-406.

Conkin, J. E., B. M. Conkin and E. D. Thurman. 1979. Inauris and Sorosphaerella, new genera of Paleozoic agglutinate foraminifera, and their stratigraphic significance. Univ. Louisville Stud. Paleontol. Stratig., 10: 1-12.

Conkin, J. E., B. M. Conkin, M. M. Walton and E. D. Neff. 1981. Devonian and Early Mississippian smaller foraminiferans of southern Indiana and northwestern Kentucky. Geol. Soc. Amer. Guide. Field Trip, No. 2: 87-112.

Conkin, J. E. and W. MacFarquhar. 1971. Foraminiferal genus Pelosina in the Caribbean and its relationships to the Paleozoic problematic genus Thuramminoides. Trans. Fifth Carib. Geol. Conf., Geol. Bull., 5: 137-139.

Conkin, J. E. and Ke-liang Wang. 1980. Chinese Paleozoic agglutinate foraminifera and their stratigraphic significance: Part 1 - Hyperammina kentuckyensis guangxiensis, new subspecies, from South China. Univ. Louisville Stud. Paleontol. Stratig., 14: 1-14.

Coomans, H. E. 1962. The marine mollusk fauna of the Virginian area as a basis for defining zoogeographical provinces. Beaufortia, 9: 83-104.

Cooper, C. L. 1947. Upper Kinkaid (Mississippian) microfauna from Johnson County, Illinois. J. Paleontol., 21: 81-94.

Corliss, B. H. 1979a. Response of deep-sea benthonic foraminifera to development of the psychrosphere near the Eocene/Oligocene boundary. Nature, 282: 63-65.

Corliss, B. H. 1979b. Quaternary Antarctic bottom water history; benthonic foraminiferal evidence from the southeast Indian Ocean. Quat. Res., 12: 271-289.

Corliss, B. H. 1981. Deep-sea benthonic foraminiferal faunal turnover near the Eocene/Oligocene boundary. Mar. Micropaleontol., 6: 367-384.

Corliss, B. H. in press. Quaternary circulation of the Antarctic Circumpolar Current.
Croneis, C. and D. F. Toomey. 1965. Gunsight (Virgilian) wewokellid sponges and their depositional environment. J. Paleontol. 39: 1-16.
Culver, S. J. 1980. Bibliography of North American Recent benthic foraminifera. J. Foram. Res., 10: 286-302.
Culver, S. J. and M. A. Buzas. 1980. Ditribution of recent benthic foraminifera off the North American Atlantic Coast. Smithsonian Contr. Mar. Sci. 6: 1-512.
Culver, S. J. and M. A. Buzas. 1981a. Recent benthic foraminiferal provinces on the Atlantic continental margin of North America. J. Foram. Res., 11: 217-240.
Culver, S. J. and M. A. Buzas. 1981b. Distribution of recent benthic foraminifera in the Gulf of Mexico. Smithsonian Contr. Mar. Sci., 8: 1-898.
Culver, S. J. and M. A. Buzas. 1981c. Foraminifera: distribution of provinces in the Gulf of Mexico. Nature, 290: 328-329.
Culver, S. J. and M. A. Buzas. 1982a. Distribution of recent benthic foraminifera in the Caribbean region. Smithsonian Contr. Mar. Sci. (13): (in press).
Culver, S. J. and M. A. Buzas. 1982b. Recent benthic foraminiferal provinces between Newfoundland and Yucatan. Geol. Soc. Amer. Bull., 93: 269-277.
Culver, S. J. and M. A. Buzas. in press. Recent benthic foraminiferal provinces in the Gulf of Mexico. J. Foram. Res.
Cummings, R. H. 1955. Nodosinella Brady, 1876 and associated Upper Paleozoic genera. Micropaleontology, 1: 221-238.
Cummings, R. H. 1956. Revision of the Upper Paleozoic textularid foraminifera. Micropaleontology, 2: 201-242.
Curry, W. B. and G. P. Lohmann. in press. Change in the distribution of $^{13}C$ in deep water of the North Atlantic during the last glaciation.
Cushman, J. A. 1910-1916. A monograph of the foraminifera of the North Pacific Ocean. U.S. Natl. Mus. Bull., 71(1-6): 1-59.
Cushman, J. A. 1918-1931. The foraminifera of the Atlantic Ocean. U.S. Natl. Mus. Bull. 104(108): 1-1064.
Cushman, J. A. 1918. The foraminifera of the Atlantic Ocean, Part 1, Astrorhizidae. U.S. Natl. Mus. Bull. 104(1): 1-111.
Cushman, J. A. 1926. The foraminifera of the Velasco Shale of the Tampico Embayment. Amer. Assoc. Petrol. Geol. Bull., 10: 583.
Cushman, J. A. 1927. An outline of a re-classification of the foraminifera. Contr. Cushman Lab. Foram. Res., 3: 1-105.
Cushman, J. A. 1928a. Foraminifera du Stampien du Basin de Paris. Bull. Soc. Sci. Seine-et-oise, 11 p.
Cushman, J. A. 1928b. Foraminifera, their classification and economic use. Cushman Lab Foram. Res. Spec. Publ. 1: 1-401.
Cushman, J. A. 1930. Notes on early Paleozoic foraminifera. Contrib. Cushman Lab. Foram. Res., 6: 43-44.
Cushman, J. A. 1940. Foraminifera Their Classification and Economic Use. 3rd ed. Harvard Univ. Press, Cambridge, 535 p.
Cushman, J. A. 1948. Foraminifera: Their Classification and Economic Use. 4th Ed. Harvard Univ. Press. Cambridge, 588 p.
Cushman, J. A. 1950. Cushman Card Catalog of Foraminifera. U.S. Geol. Surv., Washington, 33 16 mm microfilm reels.
Cushman, J. A. 1951. Paleocene foraminifera of the Gulf Coast region of

the United States and adjacent areas. U.S. Geol. Surv. Prof. Paper 232: 1-75.

Cushman, J. A. and E. D. Cahill. 1933. Miocene foraminifera of the coastal plain of the United States. U.S. Geol. Surv. Prof. Paper 175-A: 1-50.

Cushman, J. A. and H. H. Renz. 1946. The foraminiferal fauna of the Lizard Springs Formation of Trinidad, British West Indies. Cushman Lab. Foram. Res. Spec. Publ., 18: 1-48.

Cushman, J. A. and J. A. Waters. 1927a. Pennsylvanian foraminifera from Michigan. Contrib. Cushman Lab. Foram. Res., 3: 107-110.

Cushman, J. A. and J. A. Waters. 1927b. Arenaceous Paleozoic foraminifera from Texas. Contrib. Cushman Lab. Foram. Res., 3: 146-153.

Cushman, J. A. and J. A. Waters. 1928a. Upper Paleozoic foraminifera from Sutton County, Texas. J. Paleontol., 2: 358-371.

Cushman, J. A. and J. A. Waters. 1928b. The development of _Climacammina_ and its allies in the Pennsylvanian of Texas. J. Paleontol., 2: 119-130.

Cushman, J. A. and J. A. Waters. 1928c. Some foraminifera from the Pennsylvanian and Permian of Texas. Contrib. Cushman Lab. Foram. Res., 4: 31-55.

Cushman, J. A. and J. A. Waters. 1928d. Additional Cisco foraminifera from Texas. Contrib. Cushman. Lab. Foram. Res., 4: 6-67.

Cushman, J. A. and J. A. Waters. 1930. Foraminifera of the Cisco Group of Texas. Texas Univ. Bull., 3019: 22-80.

Dana, J. D. 1853a. Crustacea. U.S. Exploring Expedition, Report 1838-1842. 14: 690-1618.

Dana, J. D. 1853b. On an isothermal oceanic chart illustrating the geographical distribution of marine animals. Amer. J. Sci., 16: 153-167, 314-327.

Delaca, T. E. in press. Use of dissolved amino acids by the foraminifer _Notodendrodes antarctikos_. Amer. Zool.

Delaca, T. E., D. M. Karl and J. H. Lipps. 1981. Direct use of dissolved organic carbon by agglutinated benthic foraminifera. Nature 289: 287-289.

Delaca, T. E. and J. H. Lipps. 1972. The mechanism and adaptive significcance of attachment and substrate pitting in the foraminiferan _Rosalina globularis_ (d'Orbigny). J. Foram. Res., 2: 68-72.

Delaca, T. E., J. H. Lipps and R. R. Hessler. 1980. The morphology and ecology of a new large agglutinated Antarctic foraminifer (Textulariina: Notodendrodidae nov.). J. Linnaean Soc. London, Zool., 69: 205-224.

Delage, Y. and E. Hérouard. 1896. Traité de zoologie concrète. Tome 1. La cellule et les Protozoaires. 868 p., Schleicher Feres, Paris.

Dietrich, G., et al. 1980. General Oceanography, 2nd ed. 626 p., John Wiley, New York.

Douglas, R. G. 1973. Benthonic foraminiferal biostratigraphy in the central North Pacific, Leg 17, Deep Sea Drilling Project. Init. Rept. Deep Drill. Proj., 17: 607-671.

Douglas, R. G. 1981. Paleoecology of continental margin basins: A modern case history from the borderland of Southern California, In: R. G. Douglas, I. P. Colburn, and D. S. Gorsline (eds.), Depositional systems of active continental margin basins: short course notes: Soc. Econ. Paleontol. Mineral., Pacific Section, Los Angeles, p. 121-156.

Douglas, R. G. and S. M. Savin. 1975. Oxygen and carbon isotope analyses of Tertiary from Shatsky Rise and other sites in the North Pacific Ocean, Init. Rept. Deep Sea Drill. Proj., 32: 509-520.

Douglas, R. G. and S. M. Savin. 1978. Oxygen isotopic evidence for the depth stratification of Tertiary and Cretaceous planktonic foraminifera. Mar. Micropaleontol., 3: 175-196.

Douglas, R. and F. Woodruff. 1981. Deep-sea benthic foraminifera, In: C. Emiliani (ed.), The Oceanic Lithosphere. The Sea, 7: 1233-1327. Interscience, New York.

Duguay, L. E. and D. L. Taylor. 1978. Primary production and calcification by the soritid foraminifer Archais angulatus (Fitchell and Moll). J. Protozool., 25: 356-361.

Dujardin, F. 1835. Observations nouvelles sur less prêtendus céphalopodes microscopiques. Ann. Sci. Nat. ser. 2, 3: 312-314.

Dunn, P. H. 1931. The foraminifera of the Bainbridge. Ohio J. Sci., 31; 279.

Dunn, P. H. 1942. Silurian foraminifera of the Mississippi Basin. J. Paleontol., 16: 317-342.

Duplessy, J. -C., P. -L. Blanc and A. W. H. Bé. 1981. Oxygen-18 enrichment of planktonic foraminifera due to gametogenic calcification below the euphotic zone. Science, 213: 1247-1250.

Duplessy, J. -C., M. Moyes and C. Pujol. 1980. Deep water formation in the North Atlantic Ocean during the last ice age. Nature, 286: 479-482.

Eicher, D. L. 1967. Cenomanian and Turonian foraminifera from the western interior of the United States. Proc. 1 Int. Conf. plankt. foram: 163-176, Leiden, E. J. Brill.

Eisenack, A. 1932. Neue Mikrofossilen des baltischen Silurs. II. Palaeontol. Z. 14: 257-277.

Eisenack, A. 1937. Neue Microfossilen des baltischen Silurs. IV. Palaeontol. Z. 19: 217-243.

Eisenack, A. 1954. Foraminiferen aus dem baltischen Silurs. Senk. Leth. 35: 51-72.

Eldredge, N. and S. J. Gould. 1972. Punctuated equilibria: an alternative to phyletic gradualism, In: T. J. M. Schopt (ed.), Models in Paleobiology: 82-115, Freeman, Cooper and Co., San Francisco, Calif.

Ellis, B. F. and A. R. Messina. 1940-1982. Catalogue of the foraminifera. Amer. Mus. Nat. Hist., New York; Supplements (yearly).

El-Naggar, Z. R. 1971. On the classification, evolution and stratigraphical distribution of the Globigerinacea, In: A. Farinacci, Proc. II Planktonic Conf. 1: 421-476, Edizioni Tecnoscienze, Rome.

Ekman, S. 1953. Zoogeography of the Sea. 417 p. Sidgwick and Jackson, London.

Fleisher, R. L. 1974. Cenozoic planktonic foraminifera and biostratigraphy, Arabian Sea Deep Sea Drilling Project, Leg 23A. Init. Rept. Deep Sea Drill Proj., 23A: 1001-1072.

Frankel, L. 1972. Subsurface reproduction in foraminifera. J. Paleontol., 46: 62-65.

Frankel, L. 1975. Subsurface feeding in foraminifera. J. Paleontol., 49: 563-565.

Franz, D. R. and A. S. Merrill. 1980. Molluscan distribution patterns on the continental shelf of the Middle Atlantic Bight (Northwest Atlantic). Malacologia, 19: 209-225.

Frerichs, W. E. 1979. Planktonic foraminifera from the Sage Breaks Shale, Centennial Valley, Wyoming. J. Foram. Res., 9: 159-184.
Frerichs, W. E. 1980a. Age of the western interior Clioscaphites chouteauensis Zone. J. Paleontol., 54: 366-370.
Frerichs, W. E. 1980b. Paleoceanographic control of planktonic foraminiferal distributions in the Upper Cretaceous Niobrara Formation of the western interior. Cushman Found. Foram. Res. Spec. Pub., 19: 196-201.
Frerichs, W. E. and P. R. Adams. 1973. Correlation of the Hilliard Formation with the Niobrara Formation: Wyoming Geol. Assoc., Guidebook, 25th Field Conf: 187-192.
Frerichs, W. E., C. C. Atherton and P. N. Shive. 1975. Planktonic foraminifera from the chalky members of the Niobrara Formation, Centennial Valley, Wyoming. J. Foram. Res., 5: 294-312.
Frerichs, W. E. and J. J. Bergstresser. 1982. Planktonic foraminifera from the Late Cretaceous Fort Hays Limestone of the Western Interior (in preparation).
Frerichs, W. E. and N. B. Dring. 1981. Planktonic foraminifera from the Smoky Hill Shale of West Central Kansas. J. Foram. Res., 11: 47-69.
Frerichs, W. E. and R. Ely. 1978. Test porosity as a paleoenvironmental tool in the Late Cretaceous of the Western Interior. Contrib. to Geol. 16: 89-94.
Frerichs, W. E. and C. H. Gaskill. 1978. Textilaria americana Ehrenberg, type series of Heterohelix. J. Foram. Res., 8: 143-146.
Frerichs, W. E., M. E. Heiman, L. E. Borgman and A. W. H. Bé. 1971. Latitudinal variations in planktonic foraminiferal test porosity: Part 1 optical studies. J. Foram. Res., 2: 6-13.
Frerichs, W. E., E. M. Pokras and M. J. Evetts. 1977. The genus Hastigerinoides and its significance in the biostratigraphy of the western interior. J. Foram. Res., 7: 149-156.

Galloway, J. J. 1933. A manual of foraminifera. James Furman Kemp Memorial Ser., Pub. 1. 483 p., Principia Press, Bloomington, Ind.
Galloway, J. J. and B. H. Harlton. 1928. Some Pennsylvanian foraminifera of Oklahoma with special reference to the genus Orobias. J. Paleontol., 2: 338-357.
Galloway, J. J. and C. Ryniker. 1930. Foraminifera from the Atoka Formation of Oklahoma. Oklahoma Geol. Surv., Circ. 21: 1-37.
Gerlach, S. A. 1972. Substratum: General introduction, In: O. Kinne (ed.), Marine Ecology, 1(3): 1245-1250, John Wiley, New York.
Gevirtz, J. L., R. A. Park and G. M. Friedman. 1971. Paraecology of benthic foraminifera and associated micro-organisms of the continental shelf off Long Island, New York. J. Paleontol., 45: 153-177.
Gibson, T. G. 1967. Stratigraphy and paleoenvironment of phosphatic Miocene strata of North Carolina. Geol. Soc. Amer. Bull., 78: 631-650.
Gibson, T. G. 1968. Stratigraphy and paleoenvironment of the phosphatic Miocene strata of North Carolina: Reply. Geol. Soc. Amer. Bull., 79: 1437-1448.
Gibson, T. G. and M. A. Buzas. 1973. Species diversity: patterns in modern and Miocene foraminifera of the eastern margin of North America. Geol. Soc. Amer. Bull., 84: 217-238.
Glaessner, M. F. 1945. Principles of micropalaeontology. Melbourne Univ. Press, Melbourne, 296 p.
Golik, A. and F. B. Phleger. 1977. Benthonic foraminifera from the Gulf of Panama. J. Foram. Res., 7: 83-99.

Gould, S. J. and N. Eldredge. 1977. Punctuated equilibria: the tempo and mode of evolution reconsidered. Paleobiology, 3: 115-151.

Gradstein, F. M. and W. A. Berggren. 1981. Flysch-type agglutinated foraminiferal assemblages and the Maestrichtian through Paleogene history of the Labrador and North Seas. Mar. Micropaleontol., 6: 211-268.

Gradstein, F. M., K. G. Miller and W. A. Berggren. 1981. Depth independence of flysch-type agglutinated foraminifera. First Workshop on Arenaceous Foraminifera, September 7-9, Amsterdam: 16-17.

Gradstein, F. M. and S. P. Srivastava. 1980. Aspects of Cenozoic stratigraphy and paleoceanography of the Labrador Sea and Baffin Bay. Palaeogeogr., Palaeoclimatol., Palaeoecol., 30: 261-295.

Grell, K. G. 1973. Protozoology. Springer-Verlag, New York, 554 p.

Grell, K. G. 1979. Cytogenetic systems and evolution in foraminifera. J. Foram. Res., 9: 1-13.

Grice, G. D. and A. C. Hart. 1962. The abundance, seasonal occurence and distribution of the epizooplankton between New York and Bermuda. Ecol. Monogr., 32: 287-309.

Grigyalis, A. A. 1978. O vysshikh taksonakh foraminifer. [Concerning higher taxa of the foraminifera.] Paleont. Zh., 1978 (1): 3-12.

Grubbs, D. M. 1939. Fauna of the Niagaran nodules of the Chicago area. J. Paleontol., 16: 317-342.

Gutschick, R. C. 1960. Early Mississippian (Lower Carboniferous-Tournaisian) micropaleontology in the United States. 21st Internat. Geol. Congr., Pre-Quaternary Micropaleontol., 114-134.

Gutschick, R. C. 1962. Arenaceous foraminifera from oncolites in the Mississippian Sappington Formation of Montana. J. Paleontol., 36: 1291-1304.

Gutschick, R. C. and J. E. Treckman. 1959. Arenaceous foraminifera from the Rockford Limestone of northern Indiana. J. Paleontol., 33: 229-250.

Gutschick, R. C., J. L. Weiner and L. Young. 1961. Lower Mississippian arenaceous foraminifera from Oklahoma, Texas and Montana. J. Paleontol., 36: 1193-1221.

Haake, F. -W. 1977. Living benthic foraminifera in the Adriatic Sea: Influence of water depth and sediment. J. Foram Res., 7: 62-75.

Hallock, P. 1979. Trends in test shape with depth in large, symbiont-bearing foraminifera. J. Foram. Res., 9: 61-69.

Hallock, P. 1981a. Light dependence in Amphistegina. J. Foram. Res., 11: 40-46.

Hallock, P. 1981b. Algal symbiosis: a mathematical analysis. Mar. Biol., 62: 249-255.

Hallock, P. 1981c. Production of carbonate sediments by selected large benthic foraminifera on two Pacific coral reefs. J. Sed. Petrol., 51: 467-474.

Hallock, P. 1982. Evolution and extinction in larger foraminifera, Proc. 3d North Amer. Paleontol. Conv.

Hallock, P. and H. J. Hansen. 1979. Depth adaptation in Amphistegina: change in lamellar thickness. Bull. Geol. Soc. Denmark, 27: 99-104.

Hansen, H. J. and B. Buchardt. 1977. depth distribution of Amphistegina in the Gulf of Elat, Israel. Utrecht Micropaleontol. bull., 15: 205-239.

Hansen, H. J. and P. Dalberg. 1979. Symbiotic algae in milioline foraminifera: $CO_2$ uptake and shell adaptations. Bull. Geol. Soc.

Denmark, 28: 47-55.
Hansen, H. J. and Z. Reiss. 1972. Scanning electron microscopy of some asterigerinid foraminifera. J. Foram. Res., 2: 191-199.
Hansen, H. J., Z. Reiss and N. Schneidermann. 1969. Ultramicrostructure of bilamellar walls in Foraminiferida. Rev. Esp. Micropaleontol., 1(3): 293-316.
Haq, B. U. 1981. Paleogene paleoceanography: Early Cenozoic oceans revisited. Oceanol. Acta, Colloque, C4, Geologie des Ocean: 71-82.
Haq, B. U., I. Premoli-Silva and G. P. Lohmann. 1977. Calcareous plankton paleobiogeographic evidence for major climatic fluctuations in the Early Cenozoic Atlantic Ocean. J. Geophys. Res., 82: 3861-3876.
Hardenbol, J., R. R. Vail and J. Ferrer. 1981. Interpreting paleoenvironments, subsidence history, and sea-level changes of passive margins from seismic and biostratigraphy. Oceanol. Acta, Colloque C3, Geology of continental margins: 33-44.
Harlton, B. H. 1927. Some Pennsylvanian foraminifera of the Glen Formation of southern Oklahoma. J. Paleontol., 1: 15-27.
Harlton, B. H. 1928. Pennsylvanian foraminifera of Oklahoma and Texas. J. Paleontol., 1: 305-310.
Harlton, B. H. 1933. Micropaleontology of the Pennsylvanian Johns Valley Shale of the uachita Mountains, Oklahoma and its relationship to the Mississippian Caney Shale. J. Paleontol., 7: 3-29.
Haynes, J. R. 1965. Symbiosis, wall structure and habitat in foraminifera. Contrib. Cushman Found. Foram. Res., 16: 40-43.
Haynes, J. R. 1981. Foraminifera. John Wiley, New York, 433 p.
Hayward, B. W. and M. A. Buzas. 1979. Taxonomy and paleoecology of Early Miocene benthic foraminifera of North New Zealand and the North Tasman Sea. Smithsonian Contrib. Paleobiol. 36: 1-154.
Hazel, J. E. 1971. Paleoclimatology of the Yorktown Formation (Upper Miocene and Lower Pliocene) of Virginia and North Carolina, In: H. J. Oertli (ed.) Paleoecologie Ostracodes. Bulletin Centre Recherche Pau-SNPA, 5 Suppl: 361-375.
Hedgpeth, J. W. 1957. Marine biogeography, In: J. W. Hedgpeth (ed.), Treatise on marine ecology and paleoecology. Geol. Soc. Amer. 67 (1): 359-382.
Hemleben, C. 1969. Zur Morphogenese planktonischer Foraminiferen. Zitteliana, 1: 91-133.
Hemleben, C., A. W. H. Bé, O. R. Anderson and S. Tuntivate. 1977. Test morphology, organic layers and chamber formation of the planktonic foraminifer Globorotalia memardii (d'Orbigny). J. Foram. Res., 7(1): 1-25.
Henbest, L. G. 1958. Ecology and life association of fossil algae and foraminifera in a Pennsylvanian limestone, McAlester, Oklahoma. Contrib. Cushman Lab. Foram. Res., 9: 104-111.
Henbest, L. G. 1963. Biology, mineralogy, and diagenesis of some late Paleozoic sedentary foraminifera and algal-foraminiferal colonies. Cushman Found. Foram. Res. Spec. Publ., 6: 1-44.
Hessler, R. R. 1974. The structure of deep benthic communities from central oceanic waters, In: The biology of the oceanic Pacific, Oregon State Univ. Press, p. 79-93.
Hickman, C. S. and J. H. Lipps. in press. Foraminiferivory: selective ingestion of foraminifera and test alterations produced by the neogastropod Olivella. J. Foram. Res.
Hofker, J. 1951. The foraminifera of the Siboga Expedition. Pt. 3.

Siboga Exped., E. J. Brill, Leiden, 513 p.
Hoglund, H. 1947. Foraminifera in the Gullmar Fjord and the Skagerak. Zool. Bidg. Uppsala, 26: 1-328.
Hohenegger, J. and W. Piller. 1977. Die Stellung der Involutinidae Bütschli and Spirillinidae Reuss im System der Foraminiferen. N. Jb. Geol. Palaeontol. Mh., 7: 407-418.
Hornibrook, N. de B. 1966. New Zealand Tertiary microfossil zonation, correlation and climate, In: K. Hatai (ed.), Tertiary correlations and climatic changes in the Pacific. 11th Pacific Sci. Congress-Tokyo, Symposium No. 25: 29-39.
Hottinger, L. 1977. Distribution of larger Peneroplidae, Borelis and Nummulitidae in the Gulf of Elat, Red Sea. Utrecht Micropaleontol. Bull., 15: 35-109.
Hsu, K. J., M. B. Cita and W. B. F. Ryan. 1972. The origin of the Mediterranean evaporites. Init. Rept. Deep Sea Drill. Proj. 13: 1203-1231.
Hsu, K. J., W. B. F. Ryan and M. B. Cita. 1973. Late Miocene desiccation of the Mediterranean. Nature, 242: 240-244.

Ingle, J. C., G. Keller and R. L. Kolpack. 1980. Benthic foraminiferal biofacies, sediments and water masses of the southern Peru-Chile Trench area, southeastern Pacific Ocean. Micropaleontology, 26: 113-150.
Ireland, H. A. 1939. Devonian and Silurian foraminifera from Oklahoma. J. Paleontol., 13: 190-202.
Ireland, H. A. 1956. Upper Pennsylvanian arenaceous foraminifera from Kansas. J. Paleontol., 30: 831-864.

Jahn, T. L. and R. a. Rinaldi. 1959. Protoplasmic movement in the foraminiferan, Allogromia laticollaris; and a theory of its mechanism. Biol. Bull., 117: 100-118.
Jenkins, D. G. 1966. Planktonic foraminiferal zones and new taxa from the Danian to Lower Miocene of New Zealand. New Zealand J. Geol. Geophys., 8: 1088-1126.
Jenkins, D. G. 1966. Planktonic foraminiferal datum planes in the Pacific and Trinidad Tertiary. New Zealand J. Geol. Geophys., 9: 424-427.
Jenkins, D. G. 1967. Planktonic foraminiferal zones and new taxa from the lower Miocene to the Pleistocene of New Zealand. New Zealand J. Geol. Geophys., 10: 1064-1078.
Jenkins, D. G. 1971. New Zealand Cenozoic planktonic foraminifera. New Zealand Geol. Surv. Paleontol. Bull. 42: 1-278.
Jenkins, D. G. 1975. Cenozoic planktonic foraminiferal biostratigraphy of the southwestern pacific and Tasman Sea - DSDP Leg 29. Init. Rept. Deep Sea Drill Proj., 29: 449-467.
Jenkins, D. G. and W. N. Orr. 1972. Planktonic foraminiferal biostratigraphy of the eastern equatorial Pacific, Leg 9. Init. Rept. Deep Sea Drill Proj., 9: 1059-1196.

Kahler, F. and G. Kahler. 1966-1967. Fusulinida (Foraminiferida) Teil 1-4. Fossil. Catal. I. Anim., Pars 111-114. 1-974.
Kanmera, K., K. Ishii and R. Toriyama. 1976. The evolution and extinction patterns of Permian Fusulinaceans. Contrib. Geology and Paleontology of Southeast Asia, T. Kobayashi and W. Hashimoto (eds.), Univ. Tokyo Press, 17: 129-154.
Kauffman, E. G. and R. W. Scott. 1976. Basic concepts of community

ecology and paleoecology, In: R. W. Scott and R. R. West (eds.), Structure and Classification of Paleocommunities: 1-28, Dowden Hutchinson and Ross, Stroudsburg, Pennsylvania.

Keigwin, L. D. 1980. Palaeoceanographic change in the Pacific at the Eocene-Oligocene boundary. Nature, 287- 722-725.

Kellogg, T. B. 1975. Late Quaternary climatic changes in the Norwegian and Greenland Sea, In: G. Weller and S. A. Bowling (eds.), Climate of the Arctic, Univ. Alaska, Fairbanks: 3-36.

Kellough, G. R. 1959. Biostratigraphy and paleoecology of Midway Formation along Tehvacan Greek, Limestone County, Texas. Trans. Gulf Coast Assoc. Geol. Soc., 9: 146-160.

Kellough, G. R. 1965. Paleoecology of the Foramniferida of the Wills Point Formation (Midway Group) in northeast Texas. Trans. Gulf Coast Assoc. Geol. Soc., 15: 73-153.

Kennett, J. P. 1973. Middle and Late Cenozoic planktonic foraminiferal biostratigraphy of the southwestern Pacific - DSDP Leg 21. Init. Rept. Deep Sea Drill. Proj. 21: 575-640.

Kennett, J. P. 1977. Cenozoic evolution of Antarctic glaciation, the circum-Antarctic Current and their impact on global paleoceanography. J. Geophys. Res., 82: 3843-3860.

Kennett, J. P. 1978. The development of planktonic biogeography in the southern ocean during the Cenozoic. Mar. Micropaleontol., 3: 310-346.

Kennett, J. P. 1980. Paleoceanographic and biogeographic evolution of the Southern Ocean during the Cenozoic, and Cenozoic microfossil datums. Palaeogeogr. Palaeoclimatol. Palaeoecol., 31: 123-152.

Kennett, J. P., R. E. Houtz, P. B. Andrews, A. R. Edwards, V. A. Gostin, M. Hajos, M. A. Hampton, D. G. Jenkins, S. V. Margolis, A. T. Ovenshine, K. Perch-Nielsen. 1975. Cenozoic paleoceanography in the southwest Pacific Ocean, Antarctic glaciation and the development of the circum-Antarctic current. Init. Rept. Deep Sea Drill. Proj., 29: 1155-1166.

Kennett, J. P. and N. J. Shackleton. 1976. Oxygen isotope evidence for the development of the psychrosphere 38 Myr ago. Nature, 260: 513-515.

Kent, D. V. 1977. An estimate of the duration of the faunal change at the Cretaceous - Tertiary boundary. Geology, 5: 769-771.

Kinne, O. 1963. The effects of temperature and salinity on marine and brackish water animals. I Temperature. Oceanogr. Mar. Biol. Ann. Rev., 1: 301-340.

Kinne, O. 1970. Temperature: General introduction, In: O. Kinne (ed.), Marine Ecology, v. 1(1): 321-346, John Wiley, New York.

Kinne, O. 1971. Salinity: Animals - Invertebrates. In: O. Kinne (ed.), Marine Ecology, v. 1 (2), John Wiley, New York.

Kjellesvig, E. N. 1934. Trenton foraminifera from New York. Geol. Soc. Amer. Proc., 1933: 340.

Krashenninikov, V. A. 1973. Cretaceous benthonic foraminifera, Leg 20, Deep Sea Drilling Project. Init. Rept. Deep Sea Drill. Proj., 20: 205-221.

Krashenninikov, V. A. 1974. Upper Cretaceous benthonic agglutinated foraminifera, Leg 27, Deep Sea Drilling Project. Init. Rept. Deep Sea Drill. Proj., 27: 631-661.

Kroopnick, P. 1974. Correlation between $^{13}C$ and $CO_2$ in surface waters and atmospheric $CO_2$. Earth Planet. Sci. Lett., 22: 397-403.

Kroopnick, P. 1980. The distribution of $^{13}C$ in the Atlantic Ocean.

Earth Planet. Sci. Lett., 49: 469-484.
Kroopnick, P., R. F. Weiss and H. Craig. 1972. Total $CO_2$, $^{13}C$, and dissolved oxygen-$^{18}O$ at Geosecs II in the North Atlantic. Earth Planet. Sci. Lett., 16: 103-110.

Lagoe, M. B. 1979. Recent benthic foraminiferal biofacies in the Arctic Ocean. Micropaleontology, 25: 214-224.
Lamarck, J. B. 1804. Suite des mémoires sur les fossiles des environs de Paris. Ann. Mus. Nat. Hist. Nat., Paris, 5: 179-188.
Lamb. G. M. 1972. Distribution of Holocene Foraminiferida in Mobile Bay and the effect of salinity changes. Geol. Surv. Alabama Circ. 2: 1-12.
Lankford, R. R. and F. B. Phleger. 1973. Foraminifera from the nearshore turbulent zone, western North America. J. Foram. Res., 3: 101-132.
LeCalvez, J. 1938. Recherches sur les Foraminifères. I. Development et reproduction. Arch. Zool. Exp. Gen., 80: 163-333.
LeCalvez, J. 1947. Entosolenia marginata, foraminifère apogamique ectoparasite d'un autre foraminifère Dissorbis vilardeboanus. C. R. Hebd. Seanc. Acad. Sci., Paris, 224: 1448-1450.
Lee, J. J. 1974. Towards understanding the niche of the foraminifera, In: R. H. Hedley and C. G. Adams (eds.), Foraminifera, 1: 207-260, Academic Press, London.
Lee, J. J. 1980. Nutrition and physiology of the foraminifera. Biochem. Physiol. Protozoa, 2nd. Ed., 3: 43-66.
Lee, J. J., et al. 1961. Growth and physiology of foraminifera in the laboratory: Part 1- Collection and maintenance. Micropaleontology 7: 461-466.
Lehmann, U. 1971. Jaws, radula, and crop of Arnioceras (Ammonoidea). Palaeontology, 14: 338-341.
Levine, N. D., J. O. Corliss, F. E. G. Cox, G. Deroux, J. Grain, B. M. Honigberg, G. F. Leedale, A. R. Loeblich, III, J. Lom, D. Lynn, E. G. Merinfeld, F. C. Page, G. Poljansky, V. Sprague, J. Vavra and F. G. Wallace. 1980. A newly revised classification of the Protozoa. J. Protozool. 27: 37-58.
Lipps, J. H. 1964. Miocene planktonic foraminifera from Newport Bay, California. Tulane Stud. Geol. 2: 109-133.
Lipps, J. H. 1966. Wall structure, systematics, and phylogeny studies of Cenozoic planktonic foraminifera. J. Paleontol. 40: 1257-1274.
Lipps, J. H. 1975. Feeding strategies and test function in foraminifera. Benthonics 75 Abstracts, Dalhousie Univ., Halifax, p. 26.
Lipps, J. H. 1981. What, if anything, is micropaleontology? Paleobiology, 7: 167-199.
Lipps, J. H. in press. Biotic interactions in benthonic foraminifera, In: M. J. Tevesz and P. McCall (eds.), Biotic interactions in Recent and fossil communities, Pergamon Press.
Lipps, J. H. and M. G. Erskian. 1969. Plastogamy in foraminifera: Glabratella ornatissima (Cushman). J. Protozool., 16: 422-425.
Lipps, J. H. and J. W. Valentine. 1970. The role of foraminifera in the trophic structure of marine communities. Lethaia, 3: 279-286.
Loeblich, A. R., Jr. and H. Tappan. 1961. Cretaceous planktonic foraminifera. Part 1- Cenomanian. Micropaleontology 7: 257-304.
Loeblich, A. R., Jr. and H. Tappan. 1964. Sarcodina, chiefly "Thecamoebians" and Foraminiferida, In: R. C. Moore (ed.), Treatise on Invertebrate Paleontology. Geol. Soc. Amer. and Univ. Kans. Press, pt.

C, Protista 2: 1-900.
Lohmann, G. P. 1978b. Abyssal benthonic foraminifera as hydrographic indicators in the western South Atlantic Ocean. J. Foram. Res., 8: 6-34.
Longoria, J. F. 1974. Stratigraphic, morphologic, and taxonomic studies of Aptian planktonic foraminifera. Rev. Esp. Micropaleontol., No. Extr. Diciembre 1974: 1-134.
Longoria, J. F. and M. A. Gamper. 1975. The classification and evolution of Cretaceous planktonic foraminifera. Part 1: The superfamily Hedbergelloidea. Rev. Esp. Micropaleontol., No. Esp. Enero 1975: 61-96.
Lopez, E. 1979. Algal chloroplasts in the protoplasm of three species of benthic foraminifera: Taxonomic affinity, viability and persistence. Mar. Biol., 53: 201-211.
Luterbacher, H. P. and I. Premoli-Silva. 1964. Biostratigrafia del limite Cretaceo-Terziario nell Appennino Centrale. Riv. Ital. Paleont. Strat., 70: 67-128.
Lutze, G. F. 1965. Zur Foraminiferen-fauna der Ostsee. Meyniana, 15: 75-142.
Lutze, G. F. and G. Wefer. 1980. Habitat and asexual reproduction of Cyclorbicula compressa (d'Orbigny), Soritidae. J. Foram. Res., 10: 251-260.

Malmgren, B. A. and J. P. Kennett. 1981. Phyletic gradualism in a Late Cenozoic planktonic foraminiferal lineage: DSDP Site 284, southwest Pacific. Paleobiology, 7: 230-240.
Marple, M. F. 1955. Small foraminifera of the Pottsville Formation in Ohio. Ohio. J. Sci., 55: 81-89.
Marszalak, D. S. 1969. Observations on Iridia diaphana, a marine foraminifer. J. Protozool., 16: 599-611.
Matera, N. J. and J. J. Lee. 1972. Environmental factors affecting the standing crop of foraminifera in sublittoral and psammolittoral communities of a Long Island salt marsh. Mar. Biol., 14: 89-103.
Matthews, R. K. and R. Z. Poore. 1980. Tertiary $^{18}O$ record and glacio-eustastic sea-level fluctuations. Geology, 8: 501-504.
McClellan, W. A. 1966. Arenaceous foraminifera from the Waldron Shale (Niagaran) of southeast Indiana. Bull. Amer. Paleontol., 50: 447-518.
McClellan, W. A. 1973. Siluro-Devonian microfaunal biostratigraphy in Nevada. Bull. Amer. Paleontol., 62: 236-374.
McCulloch, Irene. 1977. Qualitative observations on Recent foraminiferal tests with emphasis on the Eastern Pacific. Pt. I-III, Univ. Southern California, Los Angeles, 1079 p.
McEnery, M. E. and J. J. Lee. 1981. Cytological and fine structural studies of three species of symbiont-bearing larger foraminifera from the Red Sea. Micropaleontology, 27: 71-83.
McGlasson, R. H. 1959. Foraminiferal biofacies around Santa Catalina Island, California. Micropaleontology, 5: 217-240.
McGowran, B. 1968. Reclassification of early Globorotalia. Micropaleontology, 14: 179-198.
Mello, J. R. and M. A. Buzas. 1968. An application of cluster analysis as a method of determining biofacies. J. Paleontol., 42: 747-758.
Menzies, R. J., R. Y. George and G. T. Rowe. 1973. Abyssal Environment and Ecology of the World Oceans. John Wiley and Sons, New York, 488 p.
Miklukho-Maklay, A. D. 1963. Verkniy Paleozoy sredney Azii. [Upper Paleozoic of central Asia.] Leningrad Univ., Leningrad, 328 p.

Miller, H. W., Jr. 1956. The index value of Silurian foraminifera, and some new forms from wells in Kansas. J. Paleontol., 30: 1350-1359.
Miller, K. G. in press. Late Paleogene paleoceanography of the Bay of Biscay: benthic foraminiferal evidence. Mar. Micropaleontol.
Miller, K. G. and W. B. Curry. 1982. Eocene to Oligocene benthic foraminiferal isotopic record in the Bay of Biscay. Nature, 296: 347-350.
Miller, K. G. and C. P. Lohmann. 1982. Environmental distribution of Recent benthic foraminifera on the northeast United States continental slope. Geol. Soc. Amer. Bull., 93: 200-206.
Miller, K. G. and B. E. Tucholke. in press. Development of Cenozoic abyssal circulation south of the Greenland-Scotland Ridge, In: S. Saxov (ed.), Structure and development of the Greenland-Scotland Ridge, NATO Advanced Research Institute, Bressanone, Italy, May 11-16, 1981. Plenum Press, New York.
Miller, K. G., W. B. Curry and D. R. Ostermann. in press. Late Paleogene benthic foraminiferal paleoceanography of the Goban Spur, DSDP Leg 80. Init. Rept. Deep Sea Drill. Proj., 80.
Miller, K. G., F. M. Gradstein and W. A. Berggren. 1982. Late Cretaceous to Early Tertiary agglutinated benthic foraminifera in the Labrador Sea. Micropaleontology, 28: 1-30.
Millne-Edwards, H. 1838. Memoire sur la distribution geographique des Crustaces. Ann. Sci. Natur. (Zool.), 10: 139-174.
Monger, J. W. H. and C. A. Ross. 1971. Distribution of fusulinaceans in the western Canadian Cordillera. Canadian J. Earth Sci., 8: 259-278.
Montfort, Denys de. 1808. Conchyliologie systêmatique et classification methodique des coquilles. 1: 1-409.
Moore, T. C., et al., 1978. Cenozoic hiatuses in pelagic sediments. Micropaleontology, 24: 113-138.
Moore, W. E. 1952. Preliminary report on the occurrence of Ordovician foraminifera near Catawba, Virginia. Virginia J. Sci., 3: 334.
Moreman, W. L. 1930. Arenaceous foraminifera from Ordovician and Silurian limestones of Oklahoma. J. Paleontol., 4: 42-59.
Moreman, W. L. 1933. Arenaceous foraminifera from the Lower Paleozoic rocks of Oklahoma. J. Paleontol., 7: 393-397.
Moullade, M. 1964. Pour une simplification de la taxinomie des foraminifères appartenant à la superfamille des Globigerinacea. C. R. Somm. Soc. Géol. Fr., 2: 58-60.
Moullade, M. 1965. Contribution au problème de la classification des Orbitolinidae (Foraminiferida, Lituolacea). C. R. Hebd. Séanc. Acad. Sci. Paris, 260: 4031-4034.
Mound, M. C. 1961. Arenaceous foraminifera from the Brassfield Limestone (Albion) of southeastern Indiana. Ind. Geol. Surv. Bull., 23: 1-38.
Mound, M. C. 1968. Arenaceous Foraminiferida and zonation of the Silurian rocks of northern Indiana. Ind. Geol. Surv. Bull. 38: 1-126.
Muller, W. A. 1975. Competition for food and other niche-related studies of three species of salt-marsh foraminifera. Mar. Biol., 31: 339-351.
Muller, W. A. and J. J. Lee. 1969. Apparent indispensability of bacteria in foraminiferal nutrition. J. Protozool. 16: 471-478.
Murray, J. and A. F. Renard. 1891. Deep-sea deposits based on the specimens collected during the voyage of H.M.S. Challenger in the years 1872-1876. Rept. Voy. Challenger, Longmans, London, 525 p.
Murray, J. 1897. On the distribution of the pelagic foraminifera at the surface and on the floor of the ocean. Nat. Sci. (Ecol.), 11: 17-27.

Murray, J. W. 1963. Ecological experiments on foraminifera. J. Mar. Biol. Assoc. U.K. 43: 621-642.

Murray, J. W. 1971a. Living foraminiferids of tidal marshes: a review. J. Foram. Res. 1: 153-161.

Murray, J. W. 1971b. An Atlas of British Recent Foraminiferids. Elsevier, New York, 244 p.

Murray, J. W. 1973. Distribution and ecology of living benthic foraminiferids. Crane, Russak and Company, New York, 274 p.

Murray, J. W. 1979. Cenozoic biostratigraphy and paleoecology of sites 403 to 406 based on the foraminifers. Init. Rept. Deep Sea Drill. Proj., 48: 415-430.

Myers, E. H. 1935. The life history of Patellina corrugata Williamson, a foraminifer. Bull. Scripps Inst. Oceanogr., Univ. California, Tech. Ser., 3(15): 355-392.

Myers, E. H. 1940. Observations on the origin and fate of flagellated gametes in multiple tests of Discorbis (foraminifera). J. Mar. Biol. Assoc. U.K., 24: 201-226.

Myers, E. H. 1943. Biology, ecology, and morphogenesis of a pelagic foraminifer. Stanford Univ. Pubs., Biol. Sci., 9(1): 1-40.

Myers, E. H. and W. S. Cole. 1957. Foraminifera, In: J. W. Hedgpeth (ed.), Treatise on Marine Ecology and Paleoecology. Geol. Soc. Amer. Mem. 67(1): 1075-1082.

Natland, M. L. 1933. The temperature- and depth-distribution of some Recent and fossil foraminifera in the southern California region. Scripps Inst. Oceanogr. Bull., Tech. Ser. 3: 225-230.

Norton, R. D. 1930. Ecologic relations of some foraminifera. Scripps Inst. Oceanogr. Bull., Tech. Ser. 2: 331-388.

Nyholm, K.-G. 1962. A study of the foraminifer Gypsina. Zool. Bidrag fran Uppsala, 33: 201-207.

Ode, H. 1974. Distribution and records of the marine Mollusca in the northwest Gulf of Mexico, Part 1: Crassatellacea-Poromyacea: Texas Conchol., 11: 3-8.

Olsson, I. 1976. Distribution and ecology of the foraminiferan Ammotium cassis (Parker) in some Swedish estuaries. Zoon, 4: 137-147.

Olsson, R. K. 1960. Latest Cretaceous and earliest Tertiary stratigraphy of New Jersey coastal plain. J. Paleontol., 47: 643-665.

Olsson, R. K. 1970. Planktonic foraminifera from base of Tertiary, Millers Ferry, Alabama. J. Paleontol., 44: 598-604.

Olsson, R. K., K. G. Miller and T. E. Ungrady. 1980. Late Oligocene transgression of middle Atlantic Coastal Plain. Geology 8: 549-554.

Orbigny, A. d'. 1826. Tableau méthodique de la classe des céphalopodes. Ann. Sci. Nat. 7: 245-314.

Orbigny, A. d'. 1839. Foraminifères, In: R. de la Sagra, Histoire physique, politique, et naturelle de l'ile de Cuba. 224 p.

Orr, W. N. 1967. Secondary calcification in the foraminiferal genus Globorotalia. Science, 157: 1554-1555.

Otvos, E. G. 1978. Calcareous benthic foraminiferal fauna in a very low salinity setting, Lake Pontchartrain, Louisiana. J. Foram., Res., 8: 262-269.

Owen, S. R. J. 1867. On the surface-fauna of mid-ocean. J. Linnean Soc. Lond. (Zoology), 9: 147.

Ozawa, T. 1970. Notes on the phylogeny and classification of the super-

family Verbeekinoidea. Mem. Fac. Sci. Kyushu Univ., ser. D., Geol., 20: 17-58.
Parker, F. L. 1948. Foraminifera of the continental shelf from the Gulf of Maine to Maryland. Harvard Coll. Mus. Comp. Zool. Bull., 100: 213-241.
Parker, F. L. 1952. Foraminiferal distribution in the Long Island Sound-Buzzards Bay area. Harvard Coll. Mus. Comp. Zool. Bull. 106: 427-473.
Parsons, B. and J. G. Sclater. 1977. An analysis of the variation of ocean floor bathymetry and heat flow with age. J. Geophys. Res., 82: 803-827.
Pessagno, E. A., Jr. 1967. Upper Cretaceous planktonic foraminifera from the western Gulf Coastal Plain. Palaeontograph. Amer. 5: 245-445.
Pessagno, E. A., Jr. and K. Miyano. 1968. Notes on the wall structure of the Globigerinacea. Micropaleontology, 14(1): 38-50.
Peterson, L. C. and G. P. Lohmann. 1982. Major change in Atlantic deep and bottom waters 700,000 yr ago: benthonic foraminiferal evidence from the South Atlantic. Quat. Res., 17: 26-38.
Pflum, C. E. and W. E. Frerichs. 1976. Gulf of Mexico deep-water foraminifers. Cushman Found. Foram. Res. Spec. Pub. no. 14: 1-125.
Phleger, F. B. 1952. Foraminiferal ecology off Portsmouth, New Hamsphire. Harvard Coll. Mus. Comp. Zool. Bull., 106: 315-390.
Phleger, F. B. 1954. Ecology of foraminifera and associated micro-organisms from Mississippi Sound and environs. Bull. Amer. Assoc. Petrol. Geol. 38: 584-647.
Phleger, F. B. 1960. Ecology and Distribution of Recent Foraminifera. Johns Hopkins Press, Baltimore, 297 p.
Phleger, F. B. 1964. Foraminiferal ecology and marine geology. Mar. Geol., 1: 16-43.
Phleger, F. B. 1970. Foraminiferal populations and marine marsh processes. Limnol. Oceanogr., 15: 522-534.
Pirlet, H. and R. Conil. 1973. L'evolution des Archaediscidae viseen. Bull. Soc. Belge Géol. Paleontol. Hydrol., 82: 241-300.
Plummer, H. J. 1927. Foraminifera of the Midway Formation in Texas. Univ. Texas Bull., 2644: 3-206.
Plummer, H. J. 1930. Calcareous foraminifera in the Brownwood Shale near Bridgeport, Texas. Univ. Texas Bull. 3019: 5-21.
Plummer, H. J. 1945. Smaller foraminifera in the Marble Falls, Smithwick, and lower Strawn strata around the Llano Uplift in Texas. Texas Univ. Bull. 4401: 209-271.
Poag, C. W. 1966. Paynes Hammock (Lower Miocene?) foraminifera of Alabama and Mississippi. Micropaleontology, 12: 393-440.
Poag, C. W. 1981. Ecologic Atlas of Benthic Foraminifera of the Gulf of Mexico. Mar. Sci. Inst. Woods Hole, Massachusetts. 174 p.
Poag, C. W. 1982. Environmental implications of test-to-substrate attachment among some modern sublittoral foraminifera. Geol. Soc. Amer. Bull. 93: 252-268.
Poag, C. W. in press. Neogene stratigraphy of the submerged U.S. Atlantic margin.
Pokorný, V. 1958. Grundzüge der zoologischen Mikropaläontologie, Bd. 1: 1-582, VEB Deutsch. Ver. Wissen., Berlin.
Poore, R. Z. 1976. Microfossil correlation of California Lower Tertiary sections: a comparison. U.S. Geol. Surv. Prof. Paper 743-F: F1-8.
Postuma, J. A. 1971. Manual of Planktonic Foraminifera. Elsevier Publ. Co. Amsterdam, 412 p.

Putrya, F. S. 1970. K poznaniyu Yurskikh Lentikulinid zapadnoy Sibiri. [On information of Jurassic Lenticulinidae of western Siberia.] Paleont. Zn. 1970 (4): 29-45.

Rauzer-Chernousova, D. M. and A. V. Fursenko. 1959. Osnovy Paleontologii, Obshchaya Chast. Prosteyshie. [Principles of Paleontology. Part 1, Protozoa.] Akad. Nauk SSSR, 368 p.

Reid, J. L. and R. J. Lynn. 1971. On the influence of the Norwegian-Greenland Sea and Weddell seas upon the bottom waters of the Indian and Pacific Oceans. Deep-Sea Res., 18: 1063-1088.

Reiss, Z. 1957. The Bilamellidea, nov. superfam., and remarks on Cretaceous globorotaliids. Contr. Cushman Found., Foram. Res., 8(4): 127-145.

Reiss, Z. 1963. Reclassification of perforate foraminifera. Israel Geol. Surv. Bull., 35: 1-111.

Reiss, Z. 1977. Foraminiferal research in the Gulf of Elat-Aqaba - a review. Utrecht Micropaleontol. Bull., 15: 7-25.

Resig, J. M. 1974. Recent foraminifera from a landlocked Hawaiian lake. J. Foram. Res., 4: 69-76.

Reuss, A. E. 1862. Entwurf einer systematischen Zusammenstellung der Foraminiferen. K. Akad. Wiss. Wien, Math.-Naturwiss. Cl., Sitz. 44 (Jahrg. 1861): 355-396.

Rhumbler, L. 1911. Die Foraminiferen (Thalamophoren) der Plankton-Expedition; Teil 1. Die allgemeinen Organisationsverhältnisse der Foraminiferen. Plankton-Exped. Humboldt-Stiftung, Ergebn. 3, L, C, 1: 1-331.

Roberts, D. G., L. Montadert and R. C. Searle. 1979. The western Rockall Plateau: stratigraphy and structural evolution. Init. Rept. Deep Sea Drill. Proj., 48: 1061-1088.

Ross, C. A. 1967. Development of fusulinid (Foraminiferida) faunal realms. J. Paleontol., 41: 1341-1354.

Ross, C. A. 1972. Paleobiological analysis of fusulinacean (Foraminifera) shell morphology. J. Paleontol., 46: 719-728.

Ross, C. A. and J. R. P. Ross. 1978. Adaptive evolution in the soritids Marginopora and Amphisorus (Foraminiferida). Scan. Elect. Micros. 2: 53-60.

Ross, C. A. and J. R. P. Ross. 1981. Biogeographic influences on Late Palaeozoic faunal distributions, In: C. P. Larwood and C. Nielsen (eds.), Recent and Fossil Bryozoa, Intl. Bryozool, Assoc., Olsen and Olsen, Freidensborg, Denmark, 199-212.

Roth, R. and J. Skinner. 1930. The fauna of the McCoy Formation, Pennsylvanian of Colorado. J. Paleontol., 4: 332-352.

Rottger, R. 1973. Die Ektoplasmahulle von Heterostegina depressa (Foraminifera: Nummulitidae). Mar. Biol., 21: 127-138.

Rottger, R. 1976. Ecological observations of Heterostegina depressa (Foraminifera, Nummulitidae) in the laboratory and in its natural habitat. Maritime Sed. Spec. Pub., 1: 75-80.

Rozovskaya, S. E. 1969. K revizii otryada Fusulinida. [On a revision of the Order Fusulinida.] Paleont. Zh. 1969 (3): 34-44.

Rozovskaya, S. E. 1975. Sostav, sistema i filogeniya otryada Fuzulinida. [Composition, phylogeny and systematics of the Order Fusulinida.] Akad. Nauk SSSR, Trudy Paleont. Inst. 149: 1-267.

Rubenstein, D. I. and M. A. R. Koehl. 1977. The mechanisms of filter feeding: some theoretical considerations. Amer. Natur., 111: 981-994.

Saidova, Kh. M. 1975. Bentosyne foraminifery Tikhogo Okeana, Chast' I-III. Akad. Nauk SSSR, Inst. Okeanol. P. P. Shirshova, Moscow, 290 p.
Saidova, Kh. M. 1981. O sovremennom sostoyanii sistemynadvidovykh taksonov Kaynozoyskikh bentosnykh foraminfer. [On the current state of the system of supraspecific taxonomy of Cenozoic benthonic foraminifera.] Akad. Nauk SSSR. Inst. Okeanol. P. P. Shirshova, Moscow, 73 p.
Saito, T. 1977. Late Cenozoic planktonic foraminiferal datum levels: the present state of knowledge toward accomplishing Pan-Pacific correlation, Proc. First Int. Congr. Pacific Neogene Stratig. Tokyo, 61-80.
Saito, T., P. R. Thompson and D. Breger. 1976. Skeletal ultramicro-structure of some elongate-chambered planktonic foraminifera and related species, In: Y. Takayanagi and T. Saito (eds.), Progress in Micropaleontology, Amer. Mus. Nat. Hist., New York, 278-304.
Saito, T., P. R. Thompson and D. Breger. 1981. Systematic index of Recent and Pleistocene planktonic foraminfera. Univ. Tokyo Press, Tokyo, 190 p.
Sanders, H. L. and R. R. Hessler. 1969. Ecology of the deep-sea benthos. Science, 163: 1419-1424.
Sandon, H. 1932. The food of Protozoa. Egyptian Univ., Publ. Fac. Sci., Misr-Sokkar Press. Cairo, 187 p.
Savin, S. M. 1977. The history of the Earth's surface temperature during the past 100 million years. Ann. Rev. Earth Planet. Sci., 5: 319-355.
Savin, S. M., R. G. Douglas and F. G. Stehli. 1975. Tertiary marine paleotemperatures. Geol. Soc. Amer. Bull., 86: 1499-1510.
Schafer, C. T. 1971. Sampling and spatial distribution of benthonic foraminifera. Limnol. Oceanogr., 16: 944-951.
Schafer, C. T. and F. E. Cole. 1982. Living benthic foraminifera distributions on the continental slope and rise east of Newfoundland, Canada. Geol. Soc. Amer. Bull., 93: 207-217.
Schnitker, D. 1970. Upper Miocene foraminifera from near Grimesland, Pitt County, North Carolina. N. Carolina Dept. Conserv. Devel., Raleigh, Spec. Publ. 5: 1-129.
Schnitker, D. 1971. Distribution of foraminifera on the North Carolina continental shelf. Tulane Stud. Geol. Paleontol., 8: 169-215.
Schnitker, D. 1974. Western Atlantic abyssal circulation during the past 120,000 years. Nature, 248: 385-387.
Schnitker, D. 1979a. Cenozoic deep water benthic foraminifera, Bay of Biscay. Init. Rept. Deep Sea Drill. Proj., 48: 377-413.
Schnitker, D. 1979b. The deep waters of the western North Atlantic during the past 24,000 years, and the re-initiation of the Western Boundary Undercurrent. Mar. Micropaleontol., 4: 265-280.
Schnitker, D. 1980a. North Atlantic oceanography as a possible cause of Antarctic glaciation and eutrophication. Nature, 284: 615-616.
Schnitker, D. 1980b. Global paleoceanography and its deep water linkage to the Antarctic glaciation. Earth-Sci. Rev., 16: 1-20.
Schnitker, D. 1980c. Quaternary deep-sea benthic foraminifers and bottom water masses. Ann. Rev. Earth Planet. Sci., 8: 343-370.
Schultze, M. S. 1854. Über den Organismus de Polythalamien (Foraminiferen), nebst Bemerkungen über die Rhizopoden im allgemeinen. Wilhelm Engelmann, Leipzig, 68 p.
Sclater, J. G., D. Abbott and J. Thiede. 1977. Paleobathymetry and sediments of the Indian Ocean. In: J. R. Heirtzler, H. M. Bolli, and T. A. Davis (eds.), Indian Ocean geology and biostratigraphy, Amer. Geophys. Union, Washington, D.C., 25-59.

Sclater, J. G., R. N. Anderson and N. L. Bell. 1971. Elevation of ridges and evolution of the central eastern Pacific. J. Geophys. Res., 76: 7888-7915.
Sclater, J. G., S. Hellinger, and C. Tapscott. 1977. The paleobathymetry of the Atlantic Ocean from the Jurassic to present. J. Geol., 85: 509-552.
Scott, D. B. and F. S. Medioli. 1978. Vertical zonations of marsh foraminifera as accurate indicators of former sea-levels. Nature, 272: 529-531.
Scott, G. H. 1982. Tempo and stratigraphic record of speciation in Globorotalia puncticulata. J. Foram. Res., 12: 1-12.
Sen Gupta, B. K. 1971. The benthonic foraminifera of the Tail of the Grand Banks. Micropaleontology, 17: 69-98.
Sen Gupta, B. K. 1972. Distribution of Holocene benthonic foraminifera on the Atlantic continental shelf of North America. 24th Intl. Geol. Cong. Montreal, section 8: 125-134.
Sen Gupta, B. K. 1977. Depth-distribution of modern benthic foraminifera on continental shelves of the World Ocean. Indian J. Earth Sci., 4: 60-83.
Sen Gupta, B. K. and W. B. Hayes. 1979. Recognition of Holocene benthic foraminiferal facies by recurrent group analysis. J. Foram. Res., 9: 233-245.
Sen Gupta, B. K. and R. T. Kilbourne. 1974. Diversity of benthic foraminifera on the Georgia continental shelf. Geol. Soc. Amer. Bull., 85: 969-972.
Sen Gupta, B. K. and R. T. Kilbourne. 1976. Depth distribution of benthic foraminifera on the Georgia continental shelf. Maritime Sed. Spec. Pub., 1: 25-38.
Sen Gupta, B. K., R. F. Lee and M. S. May. 1981. Upwelling and an unusual assemblage of benthic foraminifera on the northern Florida continental slope. J. Paleontol., 55: 853-857.
Sen Gupta, B. K. and C. T. Schafer. 1973. Holocene benthonic foraminifera in leeward bays of St. Lucia, West Indies. Micropaleontology, 19: 341-365.
Sen Gupta, B. K. and D. P. Strickert. 1982. Living benthic foraminifera of the Florida-Hatteras slope: Distribution trends and anomalies. Geol. Soc. Amer. Bull., 93: 218-224.
Shackleton, N. J. and J. P. Kennett. 1975. Paleotemperature history of Cenozoic and the initiation of Antarctic glaciation: oxygen and carbon isotope analyses in DSDP sites 277, 279, and 281. Init. Rept. Deep Sea Drill. Proj., 29: 743-755.
Sheehan, R. and F. T. Banner. 1972. The pseudopodia of Elphidium incertum. Rev. Esp. Micropaleontol., 4: 31-63.
Sherborn, C. D. 1888. A bibliography of the foraminifera, recent and fossil, from 1765-1888; with notes explanatory of some of the rare and little known publications. Dulau and Company, London. 152 p.
Sherborn, C. D. 1893-96. An idex to the genera and species of the foraminifera. Smithsonian Misc. Coll., (a) 1893, no. 856: 1-240; (b) 1896, no. 1031: 341-485.
Shive, P. N. and W. E. Frerichs. 1974. Paleomagnetism of the chalky members of the Niobrara Formation in Colorado and Kansas. J. Geophys. Res., 79; 3001-3008.
Shonman, D. and J. W. Nybakken. 1978. Food preferences, food availability and food resource partitioning in two sympatric species of

cephalaspidean opisthobranchs. Veliger, 21: 120-126.
Sigal, J. 1952. Ordre des Foraminifera, In: J. Piveteau (ed.), Traité de Paléontologie, 1: 133-301, Masson & cie., Paris.
Sissingh, W. 1978. Microfossil biostratigraphy and stage stratotypes of the Cretaceous. Geol. en Mijnbouw, 57: 433-440.
Sliter, W. V. 1965. Laboratory experiments on the life cycle and ecologic controls of Rosalina globularis d'Orbigny. J. Protozool., 12: 210-215.
Sliter, W. V. 1971. Predation on benthic foraminifers. J. Foram. Res., 1: 20-29.
Sliter, W. V. 1975. Foraminiferal life and residue assemblages from Cretaceous slope deposits. Geol. Soc. Amer. Bull., 86: 897-906.
Smit, J. 1977. Discovery of a planktonic foraminiferal association between the Abathomphalus mayaroensis Zone and the "Globigerina" eugubina Zone at the Cretaceous Tertiary boundary in the Barranco Del Gredero (Caravaca, SE Spain): A preliminary report. I & II. Proc. Kon. Ned. Akad. Weten., ser. B. 80: 280-301.
Smith, P. B. 1964. Ecology of benthonic species. U.S. Geol. Surv. Prof. Paper, 429-A: A1-A39.
Spandel, E. 1902. Die Foraminiferen des Permo-Carbon von Hooser, Kansas, Nord Amerika, In: Festschr. Nat. ges Nurenberg, 1901: 175-194.
Spindler, M. 1980. The pelagic gulfweed Sargassum natans as a habitat for benthic foraminifera Planorbulina acervalis and Rosalina globularis. N. Jb. Geol. Palaontol. Mh., 9: 569-580.
Spindler, M., C. Hemleben, A. W. H. Bé and O. R. Anderson. 1979. Lunar periodicity of reproduction in the planktonic foraminifer Hastigerina pelagica. Mar. Ecol. Prog. Ser., 1: 61-64.
Srinivasan, M. S. and J. P. Kennett. 1976. Evolution and phenotypic variation in the Late Cenozoic Neogloboquadrina dutertrei plexus. In: Y. Takayanagi, and T. Saito (eds.), Progress in Micropaleontology, Micropaleontol. Press Spec. Pub. 2: 329-355.
Srinivasan, M. S. and J. P. Kennett. 1981a. Neogene planktonic foraminiferal biostratigraphy and evolution: equatorial to subantarctic, south Pacific. Mar. Micropaleontol., 6: 499-533.
Srinivasan, M. S. and J. P. Kennett. 1981b. A review of Neogene planktonic foraminiferal biostratigraphy: applications in the equatorial and south Pacific. Soc. Econ. Paleontol. Mineral. Spec. Publ. No. 32: 395-432.
St. Jean, J. 1957. A Middle Pennsylvanian foraminiferal fauna from Dubois County, Indiana. Indiana Dept. Conserv. Geol. Surv. Bull., 10: 1-66.
Stainforth, R. M., J. L. Lamb, H. Luterbacher, J. H. Beard and R. M. Jeffords. 1975. Cenozoic planktonic foraminiferal zonation and characteristics of index forms. Univ. Kansas Paleontol. Contr., Art. 62: 1-425.
Stewart, G. A. and R. R. Priddy. 1941. Arenaceous foraminifera from the Niagaran rocks of Ohio and Indiana. J. Paleontol., 15: 366-375.
Stewart, G. A. and L. Lampe. 1947. Foraminifera from the Middle Devonian bone beds in Ohio. J. Paleontol., 21: 529-536.
Streeter, S. S. 1973. Bottom water and benthonic foraminifera in the Atlantic - glacial interglacial contrasts. Quat. Res. 3: 131-141.
Streeter, S. S. and S. A. Lavery. 1982. Holocene and latest glacial benthic foraminifera from the slope and rise off eastern North America. Geol. Soc. Amer. Bull., 93: 190-199.
Streeter, S. S. and N. J. Shackleton. 1979. Paleocirculation of the deep North Atlantic: 150,000-year record of benthic foraminifera and oxygen-

18. Science, 203: 168-171.
Summerson, C. H. 1958. Arenaceous foraminifera from the Middle Devonian limestones of Ohio. J. Paleontol., 32: 544-588.

Talwani, M. and O. Eldholm. 1977. Evolution of the Norwegian-Greenland Sea. Geol. Soc. Amer. Bull., 88: 969-999.
Tappan, H. 1971. Foraminiferida. McGraw-Hill Encyc. Sci. Tech. 5: 467-475.
Tappan, H. and A. R. Loeblich, Jr. 1966. Mikrostruktura stenki rakovin i sistematika nadsemeystva Discorbacea (Foraminiferida) [Microstructure of the shell wall and systematics of the superfamily Discorbacea (Foraminiferida).] Akad. Nauk SSSR, Vopr. Mikropaleont. 10: 375-392.
Tappan, H. and A. R. Loeblich, Jr. 1982. Granuloreticulosa, In: S. P. Parker (ed.), Synoposis and clasification of living organisms, 1: 527-552, McGraw-Hill, New York.
Tendal, O. S. and R. R. Hessler. 1977. An introduction to the biology and systematics of Komokiacea (Textulariina, Foraminiferida). Galathea Rept. 14: 165-194.
Thompson, M. L. 1964. Fusulinacea, In: A. R. Leblich, Jr. and H. Tappan, Sarcodina, chiefly "Thecamoebians" and Foraminiferida. Treatise on Invertebrate Paleontology, Pt. C., Protista 2 (R. C. Moore, ed.), Geol. Soc. Amer. and Univ. Kansas Press, Lawrence, p. 358-436.
Thorson, G. 1966. Some factors influencing the recruitment and establishment of marine benthic communities. Neth. J. Sea Res. 3: 267-293.
Tjalsma, R. C. 1982. What happens to deep-sea benthic foraminifera near the Eocene/Oligocene boundary? IGCP Proj. 174 Field Conference, Baton Rouge, Jan. 19-29, 1981.
Tjalsma, R. C. and G. P. Lohmann. 1982. Paleocene-Eocene bathyal and abyssal benthic foraminifera from the Atlantic Ocean. Micropaleontology Spec. Publ., 4.
Todd, R. 1965. A new Rosalina (foraminifera) parasitic on a bivalve. Deep-Sea Res., 12: 831-837.
Tolderlund, D. S. and A. W. H. Bé. 1971. Seasonal distribution of planktonic foraminifera in the western North Atlantic. Micropaleontology, 17(3): 297-329.
Toomey, D. F. 1972. The biota of the Pennsylvanian (Virgilian) Leavenworth Limestone, midcontinental region. Part 3: Distribution of calcareous foraminifera. J. Paleontol., 46: 276-278.
Toriyama, R. 1960. Tentative classification of fusulinid foraminifera. Fossils, 1: 34-38.
Towe, K. M. 1971. Lamellar wall construction in planktonic foraminifera, In: A. Farinacci (ed.), Proc. 2nd Planktonic Conf. Rome: Edizioni Tecnoscienza, 2: 1213-1224.
Twain, M. 1897. Cable from London to the Associated Press in response to his premature obituary.

Uchio, T. 1960. Ecology of living benthonic foraminifera from the San Diego area. Cushman Found. Foram. Res. Spec. Pap. 5: 1-72.
Udvardy, M. D. F. 1975. A classification of the biogeographical provinces of the world. IUCN Occasional Paper No. 18: 1-48.
Ulleberg, K. 1974. Foraminifera and stratigraphy of the Viborg Formation in Sofienlund, Denmark. Bull. Geol. Soc. Denmark, 23: 269-291.

Vail, P. R. and J. Hardenbol. 1979. Sea-level changes during the Tertiary. Oceanus, 22: 71-79.
Vail, P. R. and J. Hardenbol. 1981. Effect of sea-level changes on the shelf-slope boundary. Bull. Amer. Assoc. Petrol. Geol., 65: 1003.
Vail, P. R. and R. M. Mitchum. 1980. Global cycles of sea-level change and their role in exploration. Proc. Tenth World Petrol. Congr., 2, Expl. Supply and Demand: 95-104.
Vail, P. R., et al., 1977. Seismic stratigraphy and global changes of sea level, In: Payton, C. E. (ed), Seismic strtaigraphy--applications to hydrocarbon exploration. Amer. Assoc. Petrol. Geol. Mem. 26: 49-205.
Valentine, J. W. 1968. The evolution of ecological units above the population level. J. Paleontol. 42: 253-267.
Van Couvering, J. A., et al. 1976. The terminal Miocene event. Mar. Micropaleontol., 1: 263-286.
Van Couvering, J. A., et al. 1981. The terminal Eocene event and the Polish connection. Palaeogeogr., Palaeoclimatol., Palaeoecol., 36: 321-362.
VandenBerg, J. and A. A. H. Wonders. 1979. Paleomagnetism of Late Mesozoic pelagic limestones from the Southern Alps. Geol. Ultraiectina 20: 125-136.
Van Hinte, J. E. 1965. Some foraminifera and correlation of the type Campanian. Proc. 2nd W. Afr. Micropal. Coll. Ibadan 1965: 86-92.
Van Hinte, J. E. 1976. A Cretaceous time scale. Bull. Amer. Assoc. Petrol. Geol., 60: 498-516.
Van Hinte, J. E. 1978. Geohistory analysis--application of micropaleontology in exploration geology. Bull. Amer. Assoc. Petrol. Geol., 62: 201-222.
Van Hinte, J. E. 1979. The Coniacian, Santonian and Campanian stratotypes. Lethaia, 12: 183-187.
Veerbeck, J. W. 1977. Calcareous nannoplankton biostratigraphy of Middle and Upper Cretaceous deposits in Tunisia, southern Spain and France. Utrecht Micropal. Bull. 16: 1-157.
Veerbeck, J. W. 1978. Phylogeny, classification and biostratigraphic distribution of keeled Rotaliporinae. Proc. Kon. Ned. Akad. Wet., ser. B, 81: 113-143.

Walton, W. R. 1964. Recent foraminiferal ecology and paleoecology, In: J. Imbrie and N. D. Newell (eds.), Approaches to Paleoecology, John Wiley, New York, 151-237.
Warthin, A. S., Jr. 1930. Micropaleontology of the Wetumka, Wewoka, and Holdenville Formations. Oklahoma Geol. Surv. Bull., 53: 1-94.
Waters, J. A. 1927. A group of foraminifera from the Dornick Hills Formation of the Ardmore Basin (Oklahoma). J. Paleontol., 1: 129-133.
Waters, J. A. 1928. A group of foraminifera from the Canyon division of the Pennsylvanian formations in Texas. J. Paleontol., 1: 271-275.
Watts, A. B. and M. S. Steckler. 1979. Subsidence and eustacy at the continental margin of eastern North America, In: M. Talwani, W. F. Hay, and W. B. F. Ryan (eds.), Deep drilling results in the Atlantic Ocean: continental margins and paleoenvironment. Amer. Geophys.Union, M. Ewing Ser., 3: 218-234.
Webb, P. N. 1973. Upper Cretaceous-Paleocene foraminifera from Site 208 (Lord Howe Rise, Tasman Sea), DSDP Leg 21. Init. Rept. Deep Sea Drill. Proj., 21: 541-573.
Wefer, G. 1976. Environmental effects on growth rates of benthic fora-

minifera (shallow water, Baltic Sea). Maritime Sed. Spec. Pub., 1: 39-50.

Wefer, G., J. S. Killingley and G. F. Lutze. 1981. Stable isotopes in Recent larger foraminifera. Palaeogeogr., Palaeoclimatol., Palaeoecol., 33: 253-270.

Williams, D. F., R. Rottger, R. Schmaljohann and L. Keigwin. 1981. Oxygen and carbon isotopic fractionation and algal symbiosis in the benthic foraminiferan Heterostegina depressa. Palaeogeogr., Palaeoclimatol., Palaeoecol., 33: 231-251.

Wonders, A. A. H. 1980. Middle and Late Cretaceous planktonic foraminifera of the western Mediterranean area. Utrecht Micropaleontol., Bull., 15: 7-157.

Woodruff, F., and R. G. Douglas. 1981. Response of deep-sea benthic foraminifera to Miocene paleoclimatic events. DSDP Site 289. Mar. Micropaleontol., 6: 617-632.

Woodruff, F., S. M. Savin and R. G. Douglas. 1981. Miocene stable isotope record: a detailed Pacific deep ocean study and its paleoclimatic implications. Science, 212: 665-668.

Woodward, S. P. 1856. A Manual of the Mollusca, or, a Rudimentary Treatise of Recent and Fossil Shells. John Weale, London, 524 p.

Worthington, L. V. 1976. On the North Atlantic circulation. J. Hopkins Oceanographic Studies, no. 6: 1-110.

Wright, R. 1979. Benthic foraminiferal repopulation of the Mediterranean after the Messinian (Late Miocene) event. Palaeogeogr. Palaeoclimatol., Palaeoecol., 29: 189-214.

Zaninetti, L. 1976. Les foraminifères du Trias. Essai de synthèse et corrélation entre les domaines Mésogéens Européen et Asiatique. Riv. Ital. Paleontol. Stratig., 82: 1-258.

Zhang Linxin, Wang Yujin and Wang Jianhua. 1981. Classification of Fusulinida. Selected Papers 1st Conv. Micropalaeontol. Soc. China: 30-36.